T0348845

Earthquake Hazard, Risk, and Disasters

Hazards and Disasters Series

Earthquake Hazard, Risk, and Disasters

Series Editor

John F. Shroder
Emeritus Professor of Geography and Geology
Department of Geography and Geology
University of Nebraska at Omaha
Omaha, NE 68182

Volume Editor

Max Wyss
Professor Emeritus of Geophysics
University of Alaska; Expert
International Centre for Earth Simulation
Geneva, Switzerland

AMSTERDAM • BOSTON • HEIDELBERG • LONDON
NEW YORK • OXFORD • PARIS • SAN DIEGO
SAN FRANCISCO • SINGAPORE • SYDNEY • TOKYO

ELSEVIER

Elsevier
225 Wyman Street, Waltham, MA 02451, USA
Radarweg 29, PO Box 211, 1000 AE Amsterdam, Netherlands
The Boulevard, Langford Lane, Kidlington, Oxford OX5 1GB, UK

Library of Congress Cataloging-in-Publication Data
Application submitted

British Library Cataloguing in Publication Data
A catalogue record for this book is available from the British Library

Library of Congress Control Number: 2014944359

ISBN: 978-0-12-394848-9

For information on all Elsevier publications
visit our web site at store.elsevier.com

Working together
to grow libraries in
developing countries

www.elsevier.com • www.bookaid.org

Contents

4.   The Most Useful Countermeasure Against Giant
      Earthquakes and Tsunamis—What We Learned
      From Interviews of 164 Tsunami Survivors
      Mizuho Ishida and Masataka Ando

5.   Aggravated Earthquake Risk in South Asia:
      Engineering versus Human Nature
      Roger Bilham

6.   Ten Years of Real-time Earthquake Loss Alerts
      Max Wyss

7. Forecasting Seismic Risk as an Earthquake Sequence Happens

*J. Douglas Zechar, Marcus Herrmann, Thomas van Stiphout
and Stefan Wiemer*

8. How to Render Schools Safe in Developing Countries?

*Amod M. Dixit, Surya P. Acharya, Surya N. Shrestha
and Ranjan Dhungel*

9. The Socioeconomic Impact of Earthquake Disasters

James E. Daniell

Contributors

Surya P. Acharya, National Society for Earthquake Technology — Nepal (NSET)

John G. Anderson, Nevada Seismological Laboratory & Department of Geological Sciences and Engineering, University of Nevada, NV, USA

Masataka Ando, Institute of Earth Sciences, Academia Sinica, Taiwan

Kuvvet Atakan, Department of Earth Science, University of Bergen, Bergen, Norway

John Bevington, ImageCat Ltd., Centrepoint House, Guildford, Surrey, UK

Glenn P. Biasi, Nevada Seismological Laboratory & Department of Geological Sciences and Engineering, University of Nevada, NV, USA

Roger Bilham, CIRES and Geological Sciences, University of Colorado, Boulder, CO, USA

James N. Brune, Seismological Laboratory, University of Nevada Reno, Reno, NV, USA

Victor Chebrov, Kamchatka Branch of Geophysical Survey, RAS, Petropavlovsk-Kamchatsky, Russia

James E. Daniell, Center for Disaster Management and Risk Reduction Technology; Geophysical Institute, Karlsruhe Institute of Technology, Hertzstrasse, Karlsruhe, Germany; General Sir John Monash Scholar, The General Sir John Monash Foundation, Melbourne, Victoria, Australia; SOS Earthquakes, Earthquake-Report.com web service, Cederstraat, Mechelen, Belgium

Ranjan Dhungel, National Society for Earthquake Technology — Nepal (NSET)

Amod M. Dixit, National Society for Earthquake Technology — Nepal (NSET)

M. Eineder, German Aerospace Center (DLR), Earth Observation Center (EOC), Oberpfaffenhofen, Germany

C. Geiß, German Aerospace Center (DLR), Earth Observation Center (EOC), Oberpfaffenhofen, Germany

Matthew C. Gerstenberger, GNS Science, Lower Hutt, New Zealand

Marcus Herrmann, Swiss Seismological Service, ETH Zurich, Zurich, Switzerland

Mitsuyuki Hoshiba, Meteorological Research Institute, The Japan Meteorological Agency, Tsukuba, Japan

Charles Huyck, ImageCat Inc., Oceangate, CA, USA

Mizuho Ishida, Earthquake and Tsunami Research Project for Disaster Prevention, JAMSTEC, Japan

Vladimir G. Kossobokov, The Abdus Salam International Centre for Theoretical Physics — SAND Group, Trieste, Italy; Institute of Earthquake Prediction Theory and Mathematical Geophysics, Russian Academy of Sciences, Moscow, Russian Federation; Institut de Physique du Globe de Paris, France; International Seismic Safety Organization, ISSO

Mustapha Meghraoui, Institut de Physique du Globe, UMR 7516, University of Strasbourg, France

Gero W. Michel, CRO & Head of Risk Analytics, Montpelier Re, Hamilton, HM HX Bermuda

Anastasia Nekrasova, The Abdus Salam International Centre for Theoretical Physics — SAND Group, Trieste, Italy; Institute of Earthquake Prediction Theory and Mathematical Geophysics, Russian Academy of Sciences, Moscow, Russian Federation

Giuliano Panza, Department of Geosciences, University of Trieste, Trieste, Italy; The Abdus Salam International Centre for Theoretical Physics — SAND Group, Trieste, Italy; China Earthquake Administration, Institute of Geophysics, Beijing, China; International Seismic Safety Organization, ISSO

Imtiyaz A. Parvez, CSIR Centre for Mathematical Modelling and Computer Simulation, Bangalore, India

Antonella Peresan, Department of Geosciences, University of Trieste, Trieste, Italy; The Abdus Salam International Centre for Theoretical Physics — SAND Group, Trieste, Italy; International Seismic Safety Organization, ISSO

M. Pittore, GFZ German Research Centre for Geosciences, Potsdam, Germany

Philippe Rosset, WAPMERR, Geneva, Switzerland

K. Saito, GFDRR Global Facility for Disaster Reduction and Recovery, World Bank, Washington, DC, USA

Tom Schacher, Swiss Agency for Development and Cooperation (SDC)'s Corps for Humanitarian Aid, Switzerland

Danijel Schorlemmer, GFZ German Research Centre for Geosciences, Potsdam, Germany; University of Southern California, Los Angeles, CA, USA

Surya N. Shrestha, National Society for Earthquake Technology — Nepal (NSET)

E. So, University of Cambridge, Cambridge, UK

Gennady Sobolev, Institute of Physics of the Earth, RAS, Moscow

Mark W. Stirling, GNS Science, Lower Hutt, New Zealand

H. Taubenböck, German Aerospace Center (DLR), Earth Observation Center (EOC), Oberpfaffenhofen, Germany

Stavros V. Tolis, Geoseismic G.P., Athens, Greece

Thomas van Stiphout, Independent Researcher, Zurich, Switzerland

Enrica Verrucci, ImageCat Ltd., Centrepoint House, Guildford, Surrey, UK

M. Wieland, GFZ German Research Centre for Geosciences, Potsdam, Germany

Stefan Wiemer, Swiss Seismological Service, ETH Zurich, Zurich, Switzerland

Zhongliang Wu, Institute of Geophysics, China Earthquake Administration, Beijing, People's Republic of China

Max Wyss, University of Alaska; International Centre for Earth Simulation, Geneva, Switzerland

J. Douglas Zechar, Swiss Seismological Service, ETH Zurich, Zurich, Switzerland

Hazards are processes that produce danger to human life and infrastructure. Risks are the potential or possibilities that something bad will happen because of the hazards. Disasters are that quite unpleasant result of the hazard occurrence that caused destruction of lives and infrastructure. Hazards, risks, and disasters have been coming under increasing strong scientific scrutiny in recent decades as a result of a combination of numerous unfortunate factors, many of which are quite out of control as a result of human actions. At the top of the list of exacerbating factors to any hazard, of course, is the tragic exponential population growth that is clearly not possible to maintain indefinitely on a finite Earth. As our planet is covered ever more with humans, any natural or human-caused (un-natural?) hazardous process is increasingly likely to adversely impact life and construction systems. The volumes on hazards, risks, and disasters that we present here are thus an attempt to increase understandings about how to best deal with these problems, even while we all recognize the inherent difficulties of even slowing down the rates of such processes as other compounding situations spiral on out of control, such as exploding population growth and rampant environmental degradation.

Some natural hazardous processes, such as volcanos and earthquakes that emanate from deep within the Earth's interior, are in no way affected by human actions, but a number of others are closely related to factors affected or controlled by humanity, even if however unwitting. Chief among these, of course, are climate-controlling factors, and no small measure of these can be exacerbated by the now obvious ongoing climate change at hand (Hay, 2013). Pervasive range and forest fires caused by human-enhanced or induced droughts and fuel loadings, mega-flooding into sprawling urban complexes on floodplains and coastal cities, biological threats from locust plagues, and other ecological disasters gone awry; all of these and many others are but a small part of the potentials for catastrophic risk that loom at many different scales, from the local to planet girdling.

In fact, the denial of possible planet-wide catastrophic risk (Rees, 2013) as exaggerated jeremiads in media landscapes saturated with sensational science stories and end-of-the-world Hollywood productions is perhaps quite understandable, even if simplistically short-sighted. The "end-of-days" tropes promoted by the shaggy-minded prophets of doom have been with us for centuries, mainly because of Biblical verse written in the early Iron Age during remarkably pacific times of only limited environmental change. Nowadays however, the

Armageddon enthusiasts appear to want the worst for the rest of us in order to validate their death desires and justify their holy books. Unfortunately we are all entering times when just a few individuals could actually trigger societal breakdown by error or terror, if Mother Nature does not do it for us first. Thus we enter contemporaneous times of considerable peril that present needs for close attention.

These volumes we address here about hazards, risks, and disasters are not exhaustive dissertations about all the dangerous possibilities faced by the ever-burgeoning human populations, but they do address the more common natural perils that people face, even while we leave aside (for now) the thinking about higher-level existential threats from such things as bio- or cyber-technologies, artificial intelligence, ecological collapse, or runaway climate catastrophes.

In contemplating existential risk (Rossbacher, 2013) we have lately come to realize that the new existentialist philosophy is no longer the old sense of disorientation or confusion at the apparently meaninglessness or hopelessly absurd worlds of the past. Instead it is an increasing realization that serious changes by humans appear to be afoot that even threatens all life on the planet (Kolbert, 2014; Newitz, 2013). In the geological times of the Late Cretaceous an asteroid collision with Earth wiped out the dinosaurs and much other life; at the present time by contrast, humanity itself appears to be the asteroid.

Misanthropic viewpoints aside, however, an increased understanding of all levels and types of the more common natural hazards would seem a useful endeavor to enhance knowledge accessibility, even while we attempt to figure out how to extract ourselves and other life from the perils produced by the strong climate change so obviously underway. Our intent in these volumes is to show the latest good thinking about the more common endogenetic and exogenetic processes and their roles as threats to everyday human existence. In this fashion, the chapter authors and volume editors have undertaken to show overviews and more focused assessments of many of the chief obvious threats at hand that have been repeatedly shown on screen and print media in recent years. As this century develops, we may come to wish that these examples of hazards, risks, and disasters are not somehow eclipsed by truly existential threats of a more pervasive nature. The future always hangs in the balance of opposing forces; the ever-lurking, but mindless threats from an implacable nature, or heedless bureaucracies countered only sometimes in small ways by the clumsy and often febrile attempts by individual humans to improve our little lots in life. Only through improved education and understanding will any of us have a chance against such strong odds; perhaps these volumes will add some small measure of assistance in this regard.

Specifically in this volume, earthquakes are an occurrence fairly well understood by many educated people in the world, even if they live in seismically quiescent regions. Nonetheless in seismically active areas, because of the common long temporal gaps between many events, it is common for people to overlook the necessity for strong building codes and proper behavior in the event of major seismic events. This is especially true where the twin scourges of poverty and corruption combine to produce shoddy construction and multiple levels of bribery to cut construction costs. The result can be atrociously inflated casualty rates and the expunging of whole regions of buildings in major seismic events. Certainly the most tragic result in these cases is the pancake flattening of so many schools with children in them, as happened in Pakistan in 2005 and China in 2008. It would seem that any society that cannot at least protect its most vulnerable children from these terrible hazards cannot presume to have much of a secure future.

The chapters presented in this volume represent the best current information that Editor Max Wyss has been able to gather together from his many colleagues as they collectively attempt to bring greater enlightenment about the so-deadly processes of seismicity. I was especially impressed by the ongoing attempts to predict at least some aspects of the potentials for further seismic effects in the future in especially vulnerable areas. This reflects well upon this community of seismic experts, whom many might expect to be intimidated into silence by the travesty of the Italian court's decision to convict a group of scientists of manslaughter for 6-year prison sentences in relation to the 2009 earthquake in L'Aquila for failing to adequately warn of the impending event. The communication of risk is what this volume is most concerned with, and in that sense, this volume adds to the effort by scientists to continue to try to improve communications about hazards, risks, and disasters. I am pleased with the result, even while we all recognize the great difficulties that science has to provide objective information about natural hazards that the public can actually use to modify human behavior—a most difficult task indeed. Perhaps this volume will succeed in at least a small measure.

John (Jack) Shroder
Editor-in-Chief

REFERENCES

W.W. Hay, Experimenting on a Small Planet: A Scholarly Entertainment, Springer-Verlag, Berlin, 2013, 983 p.

E. Kolbert, The Sixth Extinction: An Unnatural History, Henry Holt & Company, NY, 2014, 319 p.

A. Newitz, Scatter, Adapt, and Remember, Doubleday, NY, 2013, 305 p.

M. Rees, Denial of catastrophic risks, Science 339 (6124) (2013) 1123.

L.A. Rossbacher, Contemplating existential risk, Earth, Geologic Column 58 (10) (October, 2013) 64.

Acknowledgments

I thank the anonymous reviewers for their comments and Louisa Hutchins for managing the assembly of the book.

Introduction to Earthquake Hazard, Risk, and Disasters: Why a Book on Earthquake Problems Now?

Two megaearthquakes that generated great tsunamis, killing hundreds of thousands during the last decade, also shocked millions of people, who watched on television the horrific devastation filmed by owners of smart phones, who miraculously survived. These extraordinary events changed the thinking of experts as well as that of the public. The inadequate protection of the Dai Ichi nuclear power plant prompted the Swiss population to outlaw nuclear generation of energy by a nationwide vote.

Seismologists had to revise their thinking. When C. F. Richter, in his classic book "Elementary Seismology" defined the term "great earthquake" as one with magnitude M8+, there were no M9+ earthquakes known. Now that two of them have ruptured 1,000 and 650 km long plate boundary segments, respectively, we need to coin the new expression "megaearthquake" for referring to them, an expression that fits the current age of superlatives.

More important is the new understanding of ruptures along faults that these two events have taught us. The maximum credible earthquake (MCE), which a fault is capable of, forms a key input for estimating the seismic hazard near a fault. Because faults are segmented and plate boundaries, like the Pacific coast of Japan, mostly rupture in limited segments, generating M8 class earthquakes, seismologists have not been bold enough to consider the possibility that M(MCE) may be larger than 9. Previously, it was thought that the greatest earthquakes, like the ones in Chile (1960) and Alaska (1964), were only possible along straight segments of subduction zones. Now, one has to consider the possibility that plate boundaries like the Himalayas and the Pacific coast of Mexico may surprise us with megaearthquakes rupturing through the segments usually generating "only" great earthquakes.

Chapters in this book describe the state of the art in new and important tools and methods to understand the earthquake hazard and to reduce the risk. Paleoseismology (*Meghraoui and Atakan*) is the primary tool for hunting for evidence concerning megaearthquakes of the past approximately 10,000 years. Space techniques allow the mapping of deformations of the Earth's surface with centimeters, even millimeters, accuracy (*Taubenboeck et al.*), which allows the construction of detailed models for past earthquakes and provides maps of strain accumulation for future earthquakes. In addition, satellite images greatly facilitate mapping the damage in the wakes of disasters, enabling an effective response on an informed basis (*Hyuck et al.*). In addition to these technological advances, simple well-designed approaches to reconstruction in devastated areas are much needed (*Schacher*).

Advances in early warning (*Hoshiba*) make it possible to shut down dangerous processes, while the high amplitude seismic waves are approaching. Once these waves have hit population centers, real-time earthquake loss assessments can now estimate reasonably reliably the numbers of casualties that probably resulted within about an hour of the earthquake (*Wyss*). This enables first responders to mount rescue efforts commensurate with the extent of the disaster.

The dream of predicting earthquakes reliably has not been realized yet, but attempts to make progress in this field are described in three chapters (*Wu; Sobolev and Chebrov; Kossobokov*). The shift away from predicting to forecasting is presented by *Schorlemmer and Gerstenberger and by Zechar et al.*

The current controversy concerning the method and results of estimating seismic hazard and risk is addressed in detail. *Stirling* argues the case of the standard method of estimating seismic hazard, whereas *Panza et al.* present the objections to what they consider an inadequate, even incorrect method. Although deterministic estimates of the hazard have some advantages, *Michel* explains the need of insurers to calculate the hazard and risk probabilistically. *Anderson et al.* summarize the ingenious method of using the presence of precariously balanced rocks to estimate the upper limit of ground accelerations that could have occurred locally during the last approximately 10,000 years.

Earthquake engineering is most useful in reducing the risk the population is exposed to by designing new structures so they will resist strong ground shaking (*Tolis*). However, special techniques have to be developed and taught in regions where the construction materials and skills are limited (*Dixit et al.*). Unfortunately, the best efforts of earthquake engineers are nixed, if greedy developers and companies find ways of ignoring building codes (*Bilham*). A related problem influencing damage patterns is that of often unknown soil conditions beneath the built environment (*Parvez and Rosset*).

The impact of earthquake disasters on the population and the socioeconomic consequences is examined by *Daniell* and the responses of the Japanese coastal population in the face of the approaching megatsunami are analyzed by *Ishida and Ando*. With these topics, this book covers the most significant advances in the struggle to slow the increasing earthquake casualties we experience because of population growth.

Max Wyss
University of Alaska; International Centre for Earth Simulation,
Geneva, Switzerland

Chapter 1

Remote Sensing for Disaster Response: A Rapid, Image-Based Perspective

Charles Huyck [1], Enrica Verrucci [2] and John Bevington [2]

[1] *ImageCat Inc., Oceangate, CA, USA,* [2] *ImageCat Ltd., Centrepoint House, Guildford, Surrey, UK*

ABSTRACT

In the midst of responding to a disaster, emergency managers typically lack actionable information. Remote sensing has the potential to help emergency managers streamline response and recovery by providing: (1) a backdrop of situational awareness—which can be invaluable for assessing likely impacts—and (2) a means to assess the distribution and magnitude of damage. Effective use of remote sensing requires careful selection, acquisition, analysis, and distribution of data and results. Although best practices are often gleaned through trial and error as an event unfolds, the integration of remote sensing techniques in the standard response protocols is far more effective when undertaken outside of response and recovery. This chapter provides practical examples and guidelines for emergency managers and other stakeholders exploring the capabilities of remote sensing in emergency response activities.

1.1 INTRODUCTION

In the immediate aftermath of a disaster, accurate and timely information is essential for coordinating emergency response activities and supporting early-recovery operations. In these uncertain and pressured times, decision makers are focused on understanding the severity of the event in order to most effectively coordinate response activities. However, information available to decision makers in a disaster's aftermath comes to them as pieces of the puzzle, providing only a partial portrait of the event. Considering the numerous detrimental consequences of uninformed or uncertain decisions on the overall efficiency of disaster response, this lack of information constitutes a great risk for decision makers. With the ability to provide periodic, synoptic observations, images acquired from satellite or aerial remote sensors have the potential to fill the gap in information in the early hours and days of a disaster. Information derived from these images can verify the magnitude and spatial extent of damage in a timely manner and with limited costs.

The vast array of techniques that remote sensing technology offers, when used alongside geographical information system (GIS) software, can reduce uncertainty and serve as a catalyzing agent for information acquisition and distribution. In order to convey the full potential of these advanced technologies to date, this chapter provides a discussion of the practical use of remote sensing in support of decision making following natural disasters. This discussion is enhanced by a set of real-world examples of remote sensing and GIS techniques supporting response and early recovery in most of the major disaster events from the past decade. The chapter also describes the main advances in the field, including best practices and acknowledges the limitations of the use of remotely sensed data.

1.2 REMOTE SENSING AND DISASTER RESPONSE

Following a disaster, a need exists to quickly understand the scope of the event and to initiate and coordinate early disaster response. When suitable preparation is in place, information derived from remotely sensed imagery can establish a common operating picture and allow communication between responders to proceed smoothly and effectively. Remote sensing technologies can be applied extensively to disaster response—starting with early response (e.g., preevent monitoring, early situational assessment) and moving to long-term recovery. Following a simplified timeline (Figure 1.1), this section examines the benefits of using remotely sensed data in the different phases of disaster response. The section also explains which methods and data types are the most suitable for the end objective of each phase.

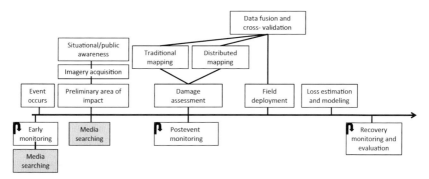

FIGURE 1.1 Simplified timeline listing possible uses of remote sensing technologies and data to support disaster response. Gray fields are not performed with remote sensing. Fields showing the ↱ symbol require a multitemporal acquisition of remotely sensed data.

1.2.1 Early Monitoring and Determination of Preliminary Area of Impact

Determination of the Area of Impact (AOI) consists of an iterative process that starts when an event is approaching (in the case of predictive events, such as hurricanes), when the event occurs (e.g., the main shock of an earthquake) or begins to unfold, and continues until the recovery phase. The preliminary AOI is usually constructed using a fusion of several sources. Media reports (with information posted on social media websites increasingly utilized) and modeled scenario events are the most used sources in the disaster-response field. For earthquakes, the United States Geological Survey (USGS) provides Prompt Assessment of Global Earthquakes for Response (PAGER) and World Agency of Planetary Monitoring and Earthquake Risk Reduction (WAPMERR) provides Quake Loss Assessment for Response and Mitigation (QLARM) alerts. These usually constitute the first source of information available to the public; both provide maps estimating the affected area and are distributed by email, short messaging service, and Twitter as well as posted, respectively, on the USGS and WAPMERR websites.

A preliminary AOI can be used to plan and instigate satellite or aerial data acquisition missions. It is common practice to include as wide a region as possible within the preliminary AOI in order to obtain the full-scale variability of the event within the selected boundary/boundaries.

For unfolding events, such as fires or floods, multitemporal remotely sensed imagery (imagery captured over the same location at several points in time, usually days apart) can be used to monitor how the event is spreading and aid decision makers in developing and implementing mitigation strategies. For more sudden events, such as earthquakes, remote sensing imagery can provide detailed information before any ground survey can take place, especially in mountainous terrain or in locations with security restrictions, reducing misunderstandings and miscommunications that may arise from a lack of information.

Fighting wildfires in the western United States is a compelling example of how response effectiveness increases with remote sensing data monitoring and proper planning. As extensive areas of the region are largely isolated and inaccessible, remote sensing data (either satellite or aerial or obtained by unmanned aerial vehicles) can provide a basis for establishing the AOI and assess burn area, the urban proximity, and the potential human and financial exposure.

The final goal of the early assessment phase is essentially to define a boundary that delimits the affected area, so that preliminary information on the event can be generated. Low-resolution data[1] (data with a spatial resolution of >30 m) and

---

1. For the purposes of this chapter, spatial resolution is classed into four categories: low (>30 m), medium (10–30 m), high (1–10 m), and very high (<1 m).

Flooding in Central Thailand

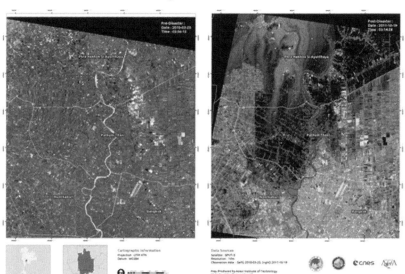

FIGURE 1.2   Preevent (left) and postevent (right) imageries from the SPOT-5 satellite sensor (Disaster Charter, 2011a). In these false-color images, red areas depict vegetation and urban areas are seen as gray/white areas. The black/green in the postevent image is floodwater spreading from the north. *Imagery copyright 2011 Centre National d'Etudes Spatiales. Map produced by the Asian Institute of Technology.*

techniques of automatic extraction are best suited for this purpose. At such an early stage, moderate resolution data (10−30-m spatial resolution), much of which is available free of charge and without restriction, are generally preferred over more costly high (<10 m) or very-high resolution (VHR—sub-1-m) data, the use of which could well be considered a waste of resources. Automatic methods of extraction are also more useful than a detailed analysis as rapidity is the most important factor to take into account at this point of the event response. Disasters that provide warning can be systematically monitored as they progress using multitemporal imagery. Moderate resolution data are invaluable for such events. An example from the 2011 Thailand floods is provided in Figure 1.2.

1.2.2 Situational Awareness/Public Awareness

As postdisaster response begins, search and rescue (S&R) activities, logistics planning, and monitoring require careful coordination. At this stage, the need for a detailed representation of the magnitude and the spatial distribution of damage becomes even greater. Remotely sensed data are therefore of great value to the response community at this juncture, adding great detail and accuracy to media-based and modeled estimates of preliminary damage. When acquired and processed in a timely manner, remotely sensed imagery is an ideal resource for conveying the severity of an event to decision makers, local communities, and the wider public. During the period in which data are being prepared for analysis and before damage assessment can start, image data can be used to provide situational-awareness maps and to identify potential cascading effects.

In general, responders have found postevent imagery indispensable for situational assessment and public awareness, especially when combined with preevent imagery. Details of an event's impact on specific regions and communities, as well as information about disabled infrastructure and services, provide enough solid information for incident commanders and emergency managers to begin deploying resources and making decisions with confidence. Imagery can also be used to express the magnitude and scope of the disaster to the public through mass media. The emergency management community can use these data to notify the public of vital information such as evacuation zones, shelter locations, and transportation impediments. Increasingly, television and newspaper reports use Google Earth preevent imagery to provide an overview of the unfolding event and context to stories they are reporting.

A method increasingly being used in the response community to attain a greater intelligibility of the information gained from remote sensing is to overlay a common referencing system, such as the national grid. Data so formatted are optimal resources to aid communication of the extent and magnitude of the event to the affected community. Responders adopted this strategy after 9/11, for example. Reference grids were used to track fires, remove debris, and coordinate virtually all activities in the area (Huyck and Adams, 2002). The value associated with these maps largely depends on how quickly they can be acquired, processed, interpreted, and disseminated. For smaller events, the process is usually completed within days; however, major events often require several weeks.

On January 12, 2010, a 7.0 magnitude earthquake struck Haiti. The Haitian Government estimated a death toll of >220,000 people (USGS, 2014), and some of Haiti's most populous areas suffered mass destruction. The international community responded immediately to launch extensive S&R missions and provide emergency assistance. The disaster also encouraged numerous awareness building and training activities (Schacher, 2014—Chapter 3 of this book). Due to the damage to local infrastructure and professional capacity, the traditional disaster-response systems employed by relief actors in Haiti lacked the capacity to enable information sharing among teams of responders from the international community and to distribute results in a timely manner. In this context, the implementation of crowd-sourced, distributed damage mapping provided great benefits, in terms of both rapidity and operational efficiency, making it easier for the relief organizations to be better informed about the extent and intensity of the damage (Ghosh et al., 2011).

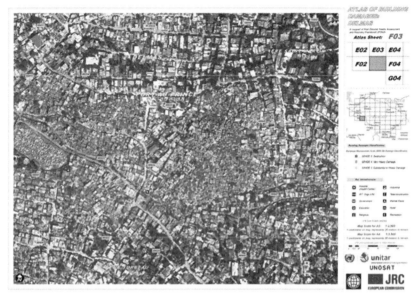

FIGURE 1.3   Remote sensing imagery can be used as base maps for situational-awareness products for coordinating response and relief efforts. The Haiti Building Damage Atlas was one such product generated after the 2010 earthquake. Developed in support of the Postdisaster Needs Assessment (PDNA), the atlas represents a joint analysis by the United Nations Institute for Training and Research (UNITAR) Operational Satellite Applications Program (UNOSAT), the European Commission (EC) Joint Research Center (JRC), and the World Bank.

Figure 1.3 provides an example of grid-based mapping for the 2010 Haiti earthquake. An overview map with postevent imagery as the background map was divided into subsections, which were mapped in detail. The effort provided the basis for a coordinated response organized between international and multilateral organizations.

Imagery can also be used to alert nongovernmental organizations (NGOs) and international organizations working in disaster response, in addition to the global community itself. When distributed through the media, images such as those shown in Figures 1.4 and 1.5 are particularly striking and can stimulate outpourings of donations as well as the deployment of international aid or volunteer work (Laituri and Kodrich, 2008). This aspect might seem trivial at first; however, the availability of material resources is the main driving force behind timely and appropriate postdisaster response and recovery, and ultimately resilience. A major success story since the dawn of high and VHR resolution, commercial satellites has been the institutionalization of the International Charter for Space and Major Disasters (the Charter) in 2000. The Charter aims to provide a unified system of space-data acquisition for all satellite image providers and delivers these images to governments affected by

FIGURE 1.4   Catastrophic damage to individual structures is apparent in very-high resolution imagery. Digitalglobe's Quickbird-02 satellite sensor showing Banda Aceh, Indonesia, before and after the great 2004 Indian Ocean earthquake and tsunami.

natural or man-made disasters. Data are used for response and relief activities globally. In its first decade, member Agencies provided data for 292 separate disasters (Figure 1.6).

Situational-awareness maps are most effective when combined with imagery of very-high spatial resolution. VHR satellite acquisition as well as aerial data can be used to develop these products. The detail of the analysis in creating the maps, however, can greatly vary according to how quickly the processed data must be distributed and the level of accuracy required by the receiving party. The information that situational-awareness maps contain (e.g., boundaries of inaccessible areas and blocked roads) should be sufficiently simple to be correctly interpreted automatically or with limited visual interpretation from the analyst.

1.2.3 Damage Assessment

Damage assessment is the phase of disaster response for which the applications of remote sensing techniques are best known. Many remote sensing technologies can be applied to assess the damage after natural and man-made events. The choices of data type and the most suitable techniques depend primarily on the level of detail required, but also on specific characteristics of the investigated hazard (e.g., fire is better detected with thermal bands; however, high-resolution data are required for assessing damage at a per-building level). Table 1.1 provides examples of remotely sensed damage detection for several hazard types.

A key advantage of damage detection using remote sensing data is the immediate context provided by the imagery. In the early aftermath of a disaster, when detailed in situ surveys are not an option, remote sensing is the only way to obtain a comprehensive view of the damage and prioritize areas to be inspected. Remote sensing analysis can also detect pockets of severe damage or extensive areas of light damage, which may go otherwise

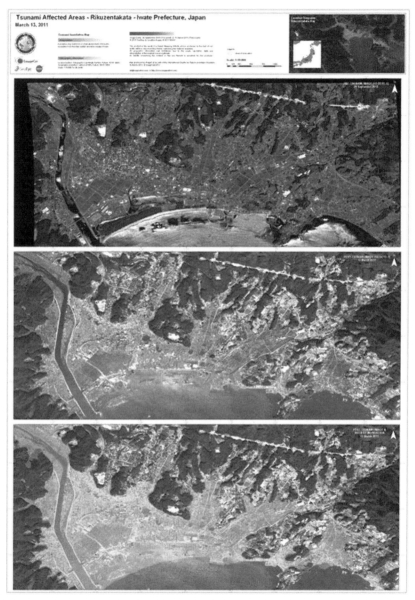

FIGURE 1.5 Delineation of the extent of tsunami damage in Japan following the 2011 Tōhoku Japan earthquake and tsunami. For insurance purposes, damage for the entire coast of Japan was delineated within days. *Disaster Charter (2011b).*

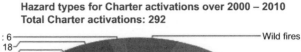

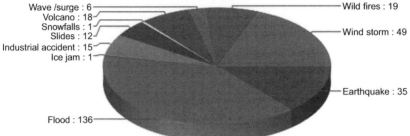

FIGURE 1.6 Disaster mapping from remote sensing in action—statistics from the first decade of International Charter activations. Flood was the predominant hazard type and accounted for >75 percent of all activations when combined with windstorm and earthquake. *Disaster Charter (2013).*

unnoticed. While several remote sensing techniques can be applied to damage detection, automated methods are usually more suited to the early postdisaster phase when a quick overview of the worst affected areas is required. These rapid determinations have high inherent uncertainties. Therefore, these methods are best suited for determining the affected area's perimeter while initially assessing the magnitude of damage. They can then be used by

TABLE 1.1 Examples of Damage Detection Using Remotely Sensed Data

Hazard	Detected Damage
Fires	Burn area, building-level damage assessment, blocked roadways, damaged critical infrastructure
Floods	Inundation extent, impacted buildings, blocked roads, submerged bridges, overtopped levees and dams, key inundated facilities such as power plants or factories, crop damage
Wind	Wind speed estimation from observed damage, levee failure, debris estimation, damage to ports, power outage area, forestry impact, damage to offshore oil facilities/spill extent
Tsunami or surge	Debris, wrack line, impacted buildings, severity of building damage (i.e., catastrophic or inundation), damaged roadways, damage to the Fukushima nuclear power plant
Earthquake	Earthquake fault rupture and displacement, liquefaction, inundation from subsidence, landslides, destroyed buildings, building damage by severity, damage to critical infrastructure—such as bridges, power plants, wharfs, damaged roadways, transportation impediments

decision makers to quickly decide whether a disaster declaration is needed and to assess appropriate resource requirements, including tasking satellites or aerial missions to capture finer-resolution imagery. Detailed damage assessment, especially when conducted at the per-building level, requires more accurate methods. Expert visual interpretation of optical data, both satellite or aerial, is often the most appropriate method. When using this form of visual intelligence, it is important to bear in mind that the detail at which damage can be detected is largely dependent upon the spatial resolution of the imagery itself. Therefore, data resolution must be chosen according to the size of the smallest object for which the damage assessment is required.

Even if costlier, VHR satellite products (with a spatial resolution finer than 1 m) are to be prioritized over moderate resolution products when the assessment of damage to individual structures is required. This choice is fully justifiable at this stage, as any lessons to be learned from a disaster starts with a detailed portrait of the spatial location and magnitude of the damage. The resolution of VHR imagery is also appropriate for detecting disruption to transportation networks and to identify open spaces to be used for locating shelters. Moving from response and relief stages into recovery, the same data can be used for a preliminary assessment of resource needs for reconstruction and for supporting planning strategies. These data also allow for the creation of on-the-spot realistic contingency plans given the status of the surrounding environment.

Despite ever higher spatial resolution offered by improved satellite and aerial sensors, some damage will always go undetected. Imagery captured from directly overhead (where the observational zenith angle is close to zero, or nadir) does not provide adequate visual perspective for the detection of minor structural damage, such as cracks in walls (Booth et al., 2011). When viewed from the nadir, moderate and major structural damage, especially to masonry buildings, is more obvious due to the presence of rubble or debris. However, when the damage is major or there have been catastrophic failures of building structures, visual interpretation of remotely sensed imagery leaves little chance of overestimation. In the early phases of damage assessment, the motivation is to save lives, prevent further casualties and failures, and allocate appropriate resources. Therefore, identifying heavily damaged buildings is more critical than assessing the full range of damage grades in the study area. Once this first assessment has been conducted, additional imagery, such as oblique-view aerial, or ground-based in situ data can help refine the final map product of detailed damage.

1.2.4 Distributed Damage Assessment

As a result of the rapid diffusion of virtual globes and web-GIS systems in the past decade, distributed interpretation of imagery for damage assessment is becoming common practice. Distributed damage assessment—or "crowd sourcing"—consists of taking a large mapping task and breaking it into pieces that are manageable by a single analyst and assessed separately by a distributed group, most often through an online environment. Distributed mapping, along with cloud computing, makes it possible to perform complex and otherwise time-consuming interpretive tasks rapidly. Distributed assessment involves the use of multiple analysts performing damage interpretations collaboratively, with a web-GIS platform serving the preevent and postevent image data and the users being allocated a small subset of the study area. Once each subset area is completed, the results of the interpretation are fed into a central database where quality assurance and user analysis are performed.

A prominent early example of this paradigm from 2007 was the work of untrained, volunteer analysts on the Amazon Mechanical Turk, aimed at searching for the wreckage of aviator Steve Fossett's plane. The project was ultimately unsuccessful in its goal, but managed to employ >50,000 analysts. The technique was used again in response to the 2010 Haiti Earthquake. Commissioned by the World Bank, the Global Earth Observation Catastrophe Assessment Network (GEO-CAN), established by ImageCat Inc., amassed and trained >600 image processing professionals to identify collapsed and heavily damaged structures for over 1,000 square kilometers of affected area (Corbane et al., 2011). The resulting maps documenting >30,000 damaged buildings were used to support a joint Postdisaster Needs Assessment (PDNA) by the Haitian Government, the World Bank, United Nations and EC (Ghosh et al., 2011; Figure 1.7). Distributed assessment following disasters has continued in

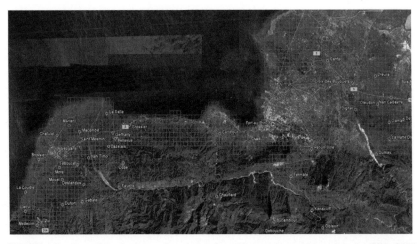

FIGURE 1.7   Over 600 scientists and engineers from 23 countries participated in the GEO-CAN crowd-sourced mapping of building damage following the 2010 Haiti Earthquake. Over 30,000 buildings—marked in yellow—were identified as needing to be replaced over an area >1000 km^2. This area was divided into equal size grid cells (shown here in red) that were individually allocated to GEO-CAN volunteers.

more recent events with professional and nonprofessional (public) networks now being used. However, distributed mapping is not always possible. Imagery data are commonly owned with licensing restrictions and therefore may not be disseminated widely and openly, yet the recent acquisition of crowd sourcing experts Tomnod by satellite imaging company DigitalGlobe demonstrates the future potential of this model of damage assessment.

Crowd sourcing or distributed damage assessment greatly favors rapidity in production of data over detail and accuracy. However, this allows emergency responders to get a preliminary damage assessment much more rapidly than do field surveys and/or traditional anecdotal damage assessment methods. As processed data are automatically stored and distributed online through web services directly to the image analysts, the technique has become more efficient and organized than traditional methods and does not require the use of complex software or the capacity to store big data sets on servers.

1.2.5 Field Deployment

The concept of "remote" sensing might convey the idea of perception and analysis from a distance. Although generally the case, data derived from remotely sensed imagery are, in reality, used to support field surveys much more than an average user may be aware. As the general public is getting more accustomed to the use of portable devices for navigation and geolocation services, the number of users drawing on remote sensing data for in situ field operations is constantly growing.

Deploying remote sensing data and derived information for disaster response in the field is a powerful support tool for field teams. Remote sensing data can be either loaded onto mobile devices or laptops or served through GPRS/3G/4G communications, if available, into GIS or web-mapping applications. When integrated with GPS, field tools can be coopted as navigation and notation devices. Analysts can use imagery to navigate to areas of interest or systematically record damage states using standardized digital or paper collection tools. When used in the early postdisaster phase, remote sensing data can serve a large number of purposes, including navigation to the areas of interest based on interpretation of the imagery itself, or through linked GPS location feeds and redirection based on the presence of obstacles, such as damaged bridges or landslides. Scientists from the global earthquake model (GEM), a global collaborative foundation for the advancement of seismic risk assessment, have developed a series of field data capture tools that use remote sensing. The GEM Mobile Tools incorporate both images as background maps (for visual assessment and for navigation purposes) as well as data derived from remote sensing prior to the field deployment, including damage data sets from imagery or locations of specific sampled locations for ground-based assessment (Bevington et al., 2012).

With these modern systems for "on-the-go" damage assessment, as notes are taken either by directly marking on the map or filling in text notations, observations are immediately tied to precise coordinate locations and can be exported directly into common software for their visualization (e.g., Google Earth) or interpretation, and further manipulation in GIS software. When working on cloud computing, another technical revolution of the last decade, data can be directly stored on a server so that all teams are always aware of the position of the other teams and of the areas for which damage assessment is still necessary, thus avoiding any risk of duplication. As exporting the data is so easy, maps can easily be produced and distributed to the local officials and to other teams or served online for a more widespread distribution.

Many of the modern tools and applications for collecting data make use of the recent tablet computing and smart phone boom. This hardware has built-in GPS receivers and camera technology. As such, when GPS-enabled digital photographs are taken in the field and uploaded to a website with postevent imagery, end users are able to see a powerful combination of a ground and bird's eye view of damage. In the satellite imagery, users see both the regional extent of damage and have also the opportunity to rapidly grasp the degree of damage severity and the geographic constraints of each particular region, such as key impassable transportation networks, bodies of water, mountains, and landslides. When simultaneously examined with mapped samples of ground damage, the microview and macroview merge to powerfully illustrate the extent and distribution of damage. When presented within a GIS platform, users can overlay GIS data and query information in a given area quickly and effectively.

Just as data from remote sensing inform field workers, information gathered from field deployments can contribute to the creation of protocols for future remote damage detection. With the modern resources and optimizing systems of crowd sourcing, web-GIS platforms can be customized to allow citizens to act as sensors and upload images via social networks to help populate rapid assessment damage maps. Even if unverified and not usually taken with high-resolution cameras, crowd-sourcing data are a very important resource, as it is not possible for field teams to acquire digital photographs of damage as rapidly as the affected population. When standardized and formalized into protocols, these data could greatly enhance prioritization of S&R activities. These data can allow responders to assess where the damage is the greatest, what type of buildings they are looking at, what infrastructure is remaining, what hazards may exist, where they might stage resources, and so forth. While S&R teams are waiting to be deployed, information technology groups could collect and organize data for them to review en route so that they can anticipate the conditions in which they will be working.

Deploying remote sensing data with field tools allows for the validation of preliminary damage estimates, situational awareness, and widespread dissemination of a common operating picture. In-field data collection helps consistently reduce the uncertainty of the damage assessment conducted using only nadir imagery by adding a contextual understanding of the situation as seen from the ground. Many challenges, however, exist to building a robust system for field deployment. Longevity of the operating system, reliance on internet connectivity, ease of use in the field, battery life of hardware, and data storage space are just a few commonly experienced issues with field data collection. Although a detailed evaluation of these considerations is outside the scope of this chapter, it is important to recognize that when systems are built, the ability to view both preevent and postevent imageries combined with derived GIS overlays is necessary in large-scale events.

1.2.6 Postdisaster and Recovery Monitoring

As described previously, image data, when acquired periodically, can be used to check the status of evolving hazards or damage from cascading events (e.g., landslides or fire fire-following earthquakes), as well as to monitor and evaluate the speed and efficacy of the long-term recovery process. Floods, fires, infrastructure damage, migrating populations, temporary settlement in tent cities, debris clearance, staging and removal, power restoration, levee status, and other important evolving conditions can be checked in fine detail when images are acquired regularly. Examples of such use include the monitoring of levees following Hurricane Katrina and the Fukushima Daiichi Nuclear power plant following the Tōhoku, Japan, earthquake and tsunami, illustrated in Figure 1.8. Of course, multitemporal monitoring can commence before the disaster and inform the preevent vulnerability and resilience assessment of facilities or neighborhoods. As continuous assessment of locations can be costly, this type of application is usually limited to high-value facilities for which cascading failures may have catastrophic consequences.

In the postdisaster recovery phase, multitemporal image sets are frequently used to perform monitoring and evaluation of recovery. Imagery and the resulting derived data can support decision making in planning resilient long-term recovery, enabling communities to improve upon previous planning decisions. Both aerial and satellite imageries can be applied to this purpose. Indicators of recovery can be measured and monitored from imagery. Major changes in the landscape (e.g., new urbanization, extensive demolition and reconstruction, and deforestation/vegetation regrowth) can be easily captured by using moderate resolution data. VHR imagery may be used for applications requiring a greater level of detail (e.g., debris removal, per-building analysis of demolition/reconstruction, clearance and state of roads, and construction and removal of emergency shelters). Satellites are the perfect platform for such applications as revisit rates can be programed into task orders (i.e., to capture an image every three months), meaning

FIGURE 1.8 DigitalGlobe's Quickbird-02 satellite monitored the Fukushima Daiichi nuclear power plant following damage caused by the 2011 Tōhoku, Japan, earthquake and tsunami. This is a prime example of the ability of remote sensing to inform decision-making processes where it would not be possible to monitor damage directly with in situ field teams. *Photo credit: DigitalGlobe FirstLook.*

increasing cost efficiencies over custom aerial acquisitions. End users of such applications have included the British Red Cross, UN-HABITAT and the World Bank to track investment and support ongoing recovery programs (ReBuilDD, 2011). Much potential also exists for local-level decision makers to adopt remote sensing to support regional or community rebuilding activities. Projects such as the EC Framework 7 Project *SENSUM* (SENSUM, 2013) are focusing on expanding open-source tool kits for systematic and independent recovery assessment from satellite sensors. Explaining the differential patterns of recovery provides a powerful tool for decision makers to verify the efficiency and efficacy of their investments and understand the link between differential recovery patterns and the social vulnerability and resilience of the affected population (Figure 1.9).

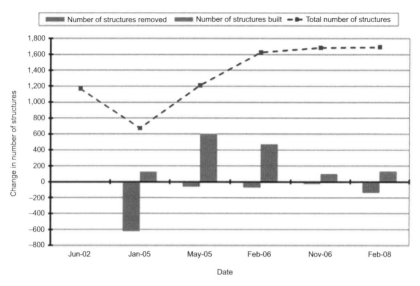

FIGURE 1.9 Long-term recovery monitoring from satellite imagery—statistics on building changes derived from multitemporal imagery following the 2004 Indian Ocean Tsunami for the village of Baan Nam Khem, Thailand. Remote sensing provides an independent and systematic method for measuring, monitoring, and evaluating recovery (ReBuilDD, 2011).

1.2.7 Loss Estimation and Modeling

Modeling platforms such as HAZUS-MH®, the GEM OpenQuake model, and others can greatly benefit from integration with data derived from remote sensing (Huyck et al., 2006). Imagery can be exploited to extract building inventories, or to allocate existing building inventories spatially. For example, if data are collected by census tract or block, but the hazard is highly localized such as with coastal hazards or riverine flooding, remote sensing can be used to distribute building stock to areas where buildings are actually present. Where key building parameters such as building size or height are important, they can be sampled locally. Default mapping schemes (statistical assignments of building type by occupancy) can be updated based on remote visual inspection (Figure 1.10).

In addition to exposure, remote sensing data can be used to calibrate loss estimations directly after an event. Loss models are based on statistical assumptions of what happens in a typical scenario—either based on empirical evidence or testing. Wind, water, fire, and ground motion can vary significantly from predicted estimates. Analysts can examine remotely sensed imagery and estimate the severity of the hazard that caused the damage. For example, if the rate that wind speed dissipates is called into question for a given event, analysts could quantify the wind speed from buildings with well-known and highly predictable responses. These samples can be used to interpolate a new hazard surface.

FIGURE 1.10 Exposure data can be gleaned from remote sensing data and fed into loss estimation models to determine parameters such as the number of buildings and square footage. This figure shows estimated numbers of steel-framed moment-frame buildings over seven stories in (a) Rio de Janeiro, Brazil, (b) Sao Paulo, Brazil, and (c) Buenos Aires, Argentina. These data are derived from remote-sensing building classification. *Photo credit: ImageCat. Background map courtesy of Bing.*

1.3 LIMITATIONS, UNCERTAINTIES, AND BEST PRACTICE

1.3.1 Technical Considerations

Several factors limit the type and quality of information extracted from satellite and aerial images. Some limitations are linked to the technical characteristics of the sensor, while others depend on the acquisition mode used for each scene. The spatial and temporal resolutions of the sensor are key parameters that the user must understand to make an informed decision over which sensor is the most appropriate. The spatial resolution influences the extent of the area covered by a single acquisition as well as the level of detail in the image. Images covering larger areas will provide less detail, as every object of the captured scene will be represented at a smaller scale. Conversely, high-resolution imagery will have finer image detail, but each scene will cover a smaller area of the globe.

Temporal resolution relates to the length of the time between successive acquisitions. Images with low temporal resolutions have long revisit periods. Hence, when conditions of acquisitions are not optimal (e.g., excessive cloud cover for optical imagery), the time windows between useable acquisitions can increase substantially. Depending on the extent of the area to cover, useful collections might happen on a daily basis. Large regions, however, are generally more difficult to image. The user must therefore consider the trade-off between: (1) imaging the entire affected area and (2) covering only the areas with the highest impact with more frequent acquisition. The first strategy can be used to monitor big changes; the second is surely more useful for frequent monitoring. Characteristically, a fusion of sensors will improve the utility of the image set, with moderate spatial resolution used in the first instance to identify significant areas of change. Sensors with a finer spatial resolution can then be targeted at only the significant areas.

Acquisition parameters, which may vary greatly from scene to scene, are also important to be considered. Cloud cover is a typical limitation of optical collection of satellite imagery. Clouds are symptomatically associated with hurricanes and storm events and may be present for several days over areas affected by intense rainfall and flooding. Similarly, imaging through smoke, after wildfires or fire-following earthquakes is also quite difficult. Synthetic aperture radar (SAR) satellite data can be used to acquire images in the case of intense cloud cover or smoke. However, in comparison to optical data, radar images are more difficult to interpret by inexperienced users and generally require sophisticated image processing software to be analyzed. Because of the easily identifiable response of water to the radar signal, SAR data are commonly used for delineating flooded areas. However, other damage interpretation applications (e.g., earthquake damage assessment) may be limited to skilled radar users.

The angle of acquisition is a factor that cannot be downplayed in importance when selecting multiple scenes for change detection analysis. Near-nadir images, which have an angle of acquisition close to zero, are preferable for applications requiring object-by-object comparisons, especially where high-rise buildings are present. Oblique-view images captured by airborne sensors are valuable for distinguishing moderate levels of damage because the walls of buildings can be analyzed. Following the 2010 Haiti earthquake, oblique imagery collected by Pictometry International Corporation was used to validate damage assessment from nadir imagery and validate the presence of soft-story ("pancaking") collapses and moderate damage to walls and roofs (Booth et al., 2011). However, this type of imagery requires trained specialists and intensive image collection from the aircraft and so is rarely used in immediate postdisaster damage assessment.

1.3.2 User Interpretation

Uncertainty in the estimation of damage is also a recurring issue when performing damage assessment. Uncertainty in the interpretation of imagery is dependent on the spatial resolution of the base image, a parameter that should be carefully chosen to meet the ultimate needs of the end users. If the task is to identify burn areas following a wildfire, then a moderate resolution may be appropriate; however, a finer-resolution imagery would be required for site-specific observations.

The scale of the image is an aspect that the end user has some control over, as additional aerial or VHR images can be collected at a later date. However, there are other sources of uncertainty that cannot be managed or predicted. Each disaster produces a unique pattern of damage, so each damage assessment will have peculiar sources of uncertainty that will only become apparent during the validation phase of the final product. For instance, some building structural types or specific characteristics of local soil conditions can produce damage patterns that are too subtle to be detected solely by remotely sensed data, especially if acquired at the nadir. The 2010 Canterbury, New Zealand, earthquake was a remarkable example of subtle, building-foundation damage due to liquefaction, of which identification was particularly difficult from remote sensing imagery, even when captured from an oblique angle.

When using near-nadir or oblique imagery as a sole data source, the exterior of buildings may still appear undamaged while, in reality, the structure is not structurally sound. Significant damage to the interior walls and fixtures is rarely apparent externally. Moderate external damage can also be difficult to identify, where exterior cracks are hidden by vegetation, external cladding, or large objects. In spite of these limitations, the amount of overestimation, or "false positives", is very low where damage is moderate to extreme (e.g., partial to total collapse).

1.3.3 Best Practice

Previous sections have explained how remote sensing technologies can efficiently support emergency response operations, from the very early phases of a disaster, right through to long-term recovery. It was also highlighted how most images can be interpreted fairly accurately by nonspecialists who can perform basic visual intelligence tasks. However, in order to exploit the full potential of satellite data, some areas of best practice have been highlighted.

1.3.4 Timing of Data Acquisition and Data Type

Data type and acquisition time are crucial choices for emergency managers. End users must be cognizant of the final purpose of the data acquisition and decide which sensor (e.g., satellite—optical or radar, airborne) and which spatial resolution (e.g., low, moderate, high, VHR) are the most appropriate. Timing is also critical. For rapid-onset disasters, imagery must be acquired before the start of clean-up operations or some damage may be missed. Authorization for acquisition should be immediate once a disaster is declared and satellites and aerial resources should be tasked by the lead response agency. For events with warning (e.g., floods, fires, and hurricanes), imagery should be acquired at the peak of the event, in order to capture the full-extent of the damage at a particular location with a single scene.

1.3.5 Protocol Adjustment

Emergency management agencies interested in investigating the potential of remote sensing technologies for their operations should consider introducing remote sensing training, or hiring remote sensing specialists in their general operating procedures. The standard protocols for response should, therefore, be amended to include a remote sensing preparedness checklist to assure readiness in data acquisition, distribution, processing, and dissemination. Remote sensing protocols need to be explicit, succinct, and clear in order to allow rapid and efficient staff training. Yet these should be flexible enough to allow for changes due to differences in disasters. A list of potential mapping products should be provided. For each product, requirements in terms of data needs, technical skills, and procedures should be clearly detailed and tracked in the resulting metadata of the product. The optimal application (e.g., rapid contingency plan, per-building damage assessment) of each final product, as well as the potential limitations, should also be specified.

1.3.6 Licenses and Data Ownership Issues

Licenses and data ownership issues can represent an impediment to the widespread distribution and analysis of remotely sensed data. Access to base imagery represents a crucial requirement in order to achieve rapidity of image processing and distribution of the final products. Emergency managers and damage-assessment coordinators should, therefore, identify the best methods to share base images among the recruited analysts. This aspect is not trivial, as commercial resellers might impose limitations on the distribution of base images.

1.3.7 Open Access to Data and Information

Open access to data and information leads to a widespread distribution and reduces the time required for the analysis. The key benefits of openness are

1. Transparency—Analysts and responders do not trust information provided on a need to know basis. Background imagery is required to make an assessment.
2. Crowd sourcing—Open data allow analysts to harness crowd-sourced interpretation efforts.
3. Synergistic byproducts—If data are distributed openly and effectively, others will build byproducts that promote effective use.
4. Redundancy—Open data have led to multiple distribution points, which alleviates the pressure on any single agency.
5. Simplified data access to all the interested parties—The more widely data are distributed, the more likely they will be discovered by those who need them. This includes the media and general public.

1.3.8 Crowd-Sourced Interpretation

In large distributed events such as Hurricanes Katrina, Sandy or the 2010 Haiti Earthquake, imagery can yield much more critical information than time permits a single analyst or small group of analysts to uncover. Emerging methods such as cloud computing and crowd sourcing can streamline the dissemination of both imagery and interpreted products considerably. With a cloud-supported web map service architecture, a very large number of analysts can simultaneously access preprocessed full resolution data and perform analyses that are automatically stored on the server for the successive phase of validation. Through collaborations with existing institutions such as Crisis Mappers, OpenStreetMap, or GEO-CAN, systems can be deployed that allow a "crowd" to provide a preliminary assessment of damage that can be combined with preliminary loss estimates or ground photographs in an online environment. This was pioneered in the previously mentioned response to the Haiti earthquake through the GEO-CAN damage assessment for the joint PDNA for building restoration. This approach has evolved yet further from GEO-CAN's use of a vetted, scientific group of analysts "a private crowd" to "public crowds" of general mapping and imagery enthusiasts (not limited to professional image analysts/engineers). The latter approach has been widely adopted by Tomnod, who analyze postevent imagery for global issues such as disaster response and diverse subjects such as searching for missing climbers or ships. The debate continues as to the merits and limitations of each approach; however, there is a fundamental need for analysis of the performance of each individual's interpretations and clear and succinct protocols are required to ensure that a standardized product is produced.

1.3.9 Supporting Field Missions

Google Android, Apple iPad/iPhone, and Windows-based field tools can be built to allow S&R and field teams to take data into the field. There are several open-source tools that are currently available and in use, such as those produced by GEM. Tiling imagery in such a way that it can be brought into the field with preliminary estimates of damage will increase the utility of the data and help build acceptance.

Internet connectivity is often sporadic following disasters. Therefore, analysts must assume that people needing data will not have connectivity and plans must be made accordingly. Distributing data on external hard drives by mail is an option when mail is functioning, but other contingencies are needed, and redundancy in distribution methods should also be sought to ensure the data reach the right people as quickly as possible. Data delivered to the field should be associated with robust metadata and preprocessed to use both within a GIS, image processing software, and online mapping technologies. The purpose of the data sets should be well defined as analysts have little time to figure out the relevance, accuracy, and completeness of such data sets, and thus, there is little trust in the final product. These products are a distraction at best and lead to poor emergency response decisions at worst.

1.3.10 Training

Several steps of data manipulation and interpretation are required to produce a truly informative map. To ensure that each operation is performed to the best possible standard, the trained staff must have different competencies. Image analysts are needed to preprocess the base data into usable data sets or GIS-ready products. Orthorectification, color balancing, or resampling may be required for raw data. Remote sensing experts are also required to produce rapid assessment maps, which are obtained by automatically extracting information from moderate resolution imagery. When damage interpretation is performed in a distributed environment, GIS software specialists must be tasked to create Web-GIS applications and portals for data distribution. GIS or statistical analysts will be required both for the data analysis and for the compilation of the statistical results. Managers should be well aware of the technical competency of their staff, and assign tasks accordingly (Huyck et al., 2006). Supervisors should also be in place to troubleshoot and provide guidance for visual assessment tasks.

1.3.11 Documentation of Mapping Experiences

During an emergency, mapping products are distributed to end users with diverse requirements. Asking for feedback to each of the groups involved can aid the identification of specific users' needs.

Analysts rarely receive feedback on the practical usefulness of the products provided to the end users. So, potential improvements are not addressed and misrepresentations often go unresolved. Documenting the successes and the lessons learned of each postdisaster mapping exercise can provide valuable information to emergency managers and to the whole community. Ideally, this effort of diligent documentation could set a new standard on postdisaster reporting, not only for communities in the United States but also internationally.

A comprehensive collection of maps and images used for emergency response could be established to allow responders and emergency managers, who are still not familiar with remote sensing, to be better informed of their data needs. This process could also lead to the identification of standard products in emergency management and to a better interoperability between emergency responders by establishing a common operating picture. Map samples could be widely distributed to emergency managers for training purposes in order to optimize data delivery procedures in times of disasters. The sample must be presented as the results of simple and reproducible techniques that should not require licensed software.

1.4 CONCLUSIONS

Remotely sensed images provide a unique ability to rapidly communicate the magnitude and spatial extent of damage from a disaster. Large amounts of information can be automatically or manually extracted from imagery as data that support decision making following disasters. These types of data are increasingly commissioned and used by governmental agencies (e.g., Federal Emergency Management Agency), multilateral organizations (e.g., World Bank), the insurance industry, and NGOs. This chapter has described several successes in the use of remote sensing for disaster management in recent years, with an acknowledgment of its limitations and recommendations for best practice.

Remote sensing has the unique ability to establish a common operating picture that can be updated with additional image acquisitions over the same area, and be used for a wide variety of users. However, it is one of a number of tools at the disposal of a disaster manager. It provides increased efficiencies and can reduce uncertainties following major events, especially in remote or inaccessible areas, and it is at its strongest when allied with other tools, such as catastrophe risk models or in situ observations. With foresight and proper staffing, these technologies and products can be formally integrated into the disaster-response process to streamline evacuation, disaster declarations, S&R, logistics, damage assessment, distribution of relief, and long-term recovery.

REFERENCES

Bevington, J., Crowley, H., Dell'Acqua, F., Eguchi, R., Huyck, C., Iannelli, G., Jordan, C., Morley, J., Parolai, S., Pittore, M., Porter, K., Saito, K., Sarabandi, P., Wieland, M., Wright, A., Wyss, M., 2012. Exposure data development for the global earthquake model: inventory data capture tools. In: Proceedings of the 15th World Conference on Earthquake Engineering, Lisbon, Portugal, Paper n. 5057.

Booth, E., Saito, K., Spence, R., Madabhushi, G., Eguchi, R.T., October 2011. Validating assessments of seismic damage made from remote sensing. Earthquake Spectra, Haiti Earthquake Special Issue 27 (S1), S157.

Corbane, C., Saito, K., Dell'Oro, L., Gill, S., Piard, B., Huyck, C., Kemper, T., Lemoine, G., Spence, R., Krishnan, R., Bjorgo, E., Senegas, O., Ghesquiere, F., Lallemant, D., Evans, G., Gartley, R., Toro, J., Ghosh, S., Svekla, W., Adams, B., Eguchi, R., 2011. A comprehensive analysis of building damage in the January 12, 2010 Mw7 Haiti earthquake using high-resolution satellite and aerial imagery. J. Am. Soc. Photogramm. Remote Sens. 77, 977−1009.

Disaster Charter, 2013. Disasters Statistics between 2000 and 2010. Available from: http://www.disasterscharter.org/web/charter/emdat.

Disaster Charter, 2011a. Flooding in Central Thailand. Map Product. Available from: http://www.disasterscharter.org/image/journal/article.jpg?img_id=109194&t=1319554561083.

Disaster Charter, 2011b. Tsunami Affected Areas—Rikuzentakata—Iwate Prefecture, Japan. Map Product. Available from: http://www.disasterscharter.org/image/journal/article.jpg?img_id=97049&t=1301047281630.

Ghosh, S., Huyck, C.K., Greene, M., Gill, S., Bevington, J., Svekla, W., DesRoches, R., Eguchi, R., 2011. Crowd-sourcing for rapid damage assessment: the global earth observation catastrophe assessment network (GEO-CAN). Earthquake Spectra 27, S179–S198.

Huyck, C.K., Adams, B.J., 2002. Emergency Response in the Wake of the World Trade Centre Attack: The Remote Sensing Perspective. In: MCEER Special Report Series: Engineering and Organizational Issues Related to the World Trade Center Attack, vol. 3.

Huyck, C.K., Adams, B.J., Ghosh, S., Eguchi, R.T., 2006. Suggestions for effective use of remote sensing data in emergency response. In: Proceedings of the 8NCEE, San Francisco.

Laituri, M., Kodrich, K., May 6, 2008. Online disaster response community: people as sensors of high magnitude disasters using internet GIS. Sensors 8, 3037–3055.

ReBuilDD, 2011. Remote Sensing for Built Environment Disasters and Development. Available from: http://rebuildd.org/.

Schacher, T., 2014. Disaster risk reduction through the training of masons and public information campaigns: experience of SDC's competence centre for reconstruction in Haiti. In: Wyss, M. (Ed.), Earthquake Hazard, Risk, and Disasters. Elsevier.

SENSUM, 2013. Framework to Integrate Space-Based and In-situ Sensing for Dynamic Vulnerability and Recovery Monitoring. Available from: http://www.sensum-project.eu/home.

USGS, 2014. Earthquakes with 50,000 or More Deaths. Available from: http://earthquake.usgs.gov/earthquakes/world/most_destructive.php.

Chapter 2

The Capabilities of Earth Observation to Contribute along the Risk Cycle

H. Taubenböck[1], C. Geiß[1], M. Wieland[2], M. Pittore[2], K. Saito[3], E. So[4] and M. Eineder[1]

[1] *German Aerospace Center (DLR), Earth Observation Center (EOC), Oberpfaffenhofen, Germany,* [2] *GFZ German Research Centre for Geosciences, Potsdam, Germany,* [3] *GFDRR Global Facility for Disaster Reduction and Recovery, World Bank, Washington, DC, USA,* [4] *University of Cambridge, Cambridge, UK*

ABSTRACT

The complexity of earthquake events and their manifold effects uncovers a strong niche for interdisciplinary and multidisciplinary analyses. Remote sensing data and methods are nowadays widely deployed to contribute information along the risk cycle. In this chapter, we document these contributions and discuss limitations simultaneously by means of an in-depth literature survey and presentation of selected examples. These include hazard-centered analysis such as site characterization and quantification of surface deformations in preevent and postevent applications. Furthermore, preevent seismic vulnerability-centered assessments of the built and natural environment are presented, which build upon the capability of remote sensing to map elements at risk, area wide. Lastly, damage assessment for postevent applications is discussed and completed by demonstrating recovery-monitoring capabilities.

2.1 INTRODUCTION

In 2011, 90 percent of the recorded natural catastrophes were weather-related; however, nearly two-thirds of economic losses and about half of insured losses were caused by geophysical events—principally from large earthquakes (Munich Re, 2012). The most destructive earthquakes of the year 2011 occurred on February 22 in Christchurch, New Zealand and on March 11 in Tohoku, Japan. These events underlined that the reliable deterministic prediction of earthquakes with sufficient lead time to handle and reduce impacts is

Earthquake Hazard, Risk, and Disasters. http://dx.doi.org/10.1016/B978-0-12-394848-9.00002-X

not yet feasible. Thus, the scientific community needs to seek new, innovative, and multidisciplinary ideas to reduce the impact of earthquakes hitting vulnerable areas across the globe.

This book stresses the need and the capability of today's earthquake-related researchers to team up for a step forward. It provides an overview on the broad advancements and also on the limitations of multidisciplinary research in the earthquake domain, for example, in seismology, geology, remote sensing, social sciences, economy, and governance. The overall goal of this chapter is to provide an overview of the capabilities and limitations of remote sensing to contribute to the risk cycle. More specifically, we examine opportunities to support earthquake exposure mapping, risk and vulnerability assessment, damage mapping, and recovery monitoring. We aim to do this by a short review of Earth observation (EO)-related scientific literature as well as by exemplifying EO products for the predisaster and postdisaster phases within the risk cycle.

As the scientific literature lacks common terminology for terms such as "risk", "hazard", "exposure", and "vulnerability", we shall briefly define our understanding for this chapter: A definition that is exhaustively used within the earthquake disaster community terms (seismic) "risk" as a function of (seismic) hazard, exposure, and (seismic) vulnerability (e.g., UN/ISDR, 2004; Birkmann, 2007; Taubenböck et al., 2008; Müller et al., 2011). Generally, "hazards" can be characterized by several constituent factors such as magnitude (only events are considered as extreme when some regular level is exceeded), duration (persistence of an event), speed of onset (time between the first occurrence of an event and its peak), temporal spacing (sequencing of events ranging from random to periodic), spatial extent (space covered by an event), and spatial dispersion pattern of distribution over the space the impact can occur (Gravely, 2001; quoted in Schneiderbauer and Ehrlich, 2004). In short, seismic hazard is understood as the probability of the levels of ground shaking resulting from earthquakes within a given time span.

The term "exposure" comprises elements such as buildings, critical infrastructure, and people, which can be adversely affected by an imposed hazard. Seismic "vulnerability" refers to the probability of loss given a level of ground shaking (Crowley and Pinho, 2012). Thus, vulnerability characterizes the degree of susceptibility of the elements exposed to a particular hazardous source (e.g., Geiß and Taubenböck, 2013 or Schneiderbauer and Ehrlich, 2004 for an extensive overview of definitions). These definitions are used in this chapter.

In the following section, we provide an overview on the current literature regarding the use of remote sensing in the phases of the risk cycle and document these capabilities with selected examples. The chapter is structured according to the timeline of a risk cycle. This incorporates preevent hazard analyses, preevent exposure mapping, and vulnerability assessment. Subsequently, the chapter discusses postevent hazard analysis as well as damage assessment and recovery monitoring.

2.2 CAPABILITIES OF REMOTE SENSING FOR ASSESSING AND MAPPING EARTHQUAKE RISK AND DAMAGE

2.2.1 Preevent Capabilities for Earthquake Hazard Assessment

Earthquakes are mainly caused by plate tectonics leading to strain or deformation along the plate boundaries. Depending on the local material properties of the fault surfaces, this strain may discharge in a harmless sliding motion, or it may accumulate over longer time periods leading to increased stress in the crust that will be discharged in a sudden fracture.

Unfortunately, remote sensing techniques cannot directly measure the most critical parameter, that is, the amount of accumulated stress in the crust. Indications are that some parameters can be directly measured with remote sensing techniques, which may provide useful information for earthquake prediction (e.g., Tronin, 2010), although, currently we are far from being able to predict earthquakes using satellite data, especially from an operational point of view. However, Deichmann et al. (2011) noted that remote sensing can contribute valuable information for microscale zonation by deriving information for producing geological, and soil maps. Digital surface models (DSMs) and multispectral imagery proved to be a valuable data source for detailed, spatially consistent, and thematically suitable site characterization (e.g., Shafique et al., 2012; Yong et al., 2008).

One example is a special EO technique called Interferometric Synthetic Aperture Radar (InSAR), which allows the measurement of centimeter displacements of the surface of the Earth, which permits the calculation of strain accumulation and inference of stress changes. The InSAR technique exploits the carrier phase signal contained in images taken from satellite-borne synthetic aperture microwave radars (SAR), which is a fine measure for the distance between each image pixel and the radar antenna (Bürgmann et al., 2000). By using multitemporal SAR data, preseismic land surface deformations in the order of centimeters and even millimeters can be measured based on the concepts and techniques of differential interferometric SAR (Stramondo et al., 2007).

The difficulty in observing strain accumulation arises from the horizontal plate motion, which is slow, small, and may be spatially distributed over several kilometers. For example, a relative plate motion of 1 centimeter per year distributed over a plate boundary area of 20 km results in a strain rate of 5.0×10^{-7} per year. Measuring this small signal with InSAR requires a very high accuracy over wide areas and an observation and integration of the motion over decadal intervals. Current SAR satellites are not designed to perform this multiyear task with the required accuracy. The main limitations are related to atmospheric wave propagation errors and phase decorrelation caused by natural land surface changes. As an alternative, dense global positioning system (GPS) networks such as GEONET in Japan (Yamagiwa et al., 2006; Miyazaki and Heki, 2001) and SCIGN in California/USA (Hudnut et al., 2002) are installed to perform deformation measurements in real time.

In order to enable global observation of strain accumulation, future space-borne InSAR missions are designed (Krieger et al., 2010), to provide a weekly global monitoring of surface displacements. This would be a great step forward in science and would allow—together with a knowledge of the crustal material properties—improved calculation of strain accumulation or better monitoring of crustal elevation or lowering. This yields the possibility of a more systematic understanding of the physical behavior during the buildup phase that leads to possible future earthquakes.

2.2.2 Preevent Capabilities for Earthquake Exposure Mapping and Vulnerability Assessment

Although the measurement of the critical parameters for earthquake hazard assessment or even prediction using EO data in an operational way is not yet solved, remote sensing provides a unique capability to map land cover and land use that can represent elements at risk. The intrinsic advantages of EO data are timeliness, applicability anywhere in the world, and large-area coverage. This includes the detection and characterization of exposed components of what primarily comprises the built environment. Multiscale data allow for mapping of the built environment at a global (e.g., Potere and Schneider, 2009; Esch et al., 2012; Pesaresi et al., 2013) to regional scale (e.g., Angel et al., 2005; Taubenböck et al., 2014). EO data allow the characterization of the built environment (e.g., Ehrlich et al., 2013; Poli and Caravaggi, 2013) with thematic details such as buildings, infrastructures, and lifelines at a local scale (e.g., Tralli et al., 2005; Taubenböck et al., 2009a). Studies prove that the indirect assessment of the vulnerability of the exposed elements is possible, especially by multidisciplinary approaches. Geiß et al. (in press), Taubenböck et al. (2009b), or Borzi et al. (2011) assess seismic building vulnerability by combining knowledge from remote sensing and civil engineering.

1. EO data are increasingly being used in the context of mapping exposure at various geometric levels, from the global to the local scale. During the past decades, EO sensors have been developed to a stage where global maps of urban areas—featuring a concentration of elements at risk—are made possible with a spatial resolution from 300 m to 2 km (Potere and Schneider, 2009). Current available global data sets including an "urban" class are, for example, MODIS500 and GRUMP (Goldewijk, 2001). The latter is based on a combination of optical remote sensing and ground-based data, allowing for a coarse localization of "exposed" elements (Figure 2.1).

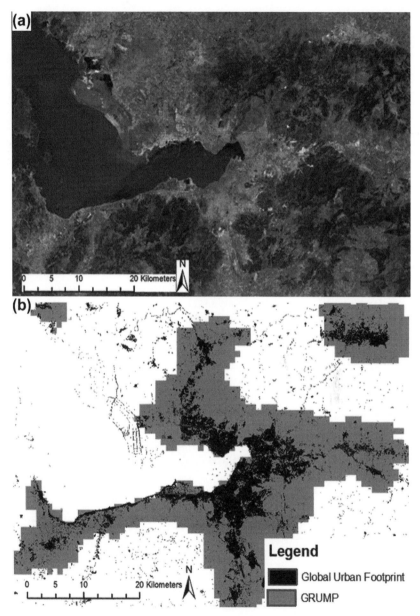

FIGURE 2.1   (a) Landsat data visualizing the urban area of Izmir, Turkey, and its periurban and rural surroundings; (b) urban footprint classification derived from TSX strip-map data (black) and the GRUMP "urban" class (red).

However, it is evident that the spatial resolution does not fully represent the high spatial variation of existing settlement patterns, especially in low-density rural areas. New EO initiatives aim to improve the spatial resolution of global maps of urban areas based on optical (Pesaresi et al., 2011; Pesaresi et al., 2012) and radar data (Esch et al., 2012, 2013; Gamba and Lisini, 2013; Miyazaki et al., 2012). Existing products are extended by a multitemporal component for monitoring urban growth (e.g., Angel et al., 2005; Taubenböck et al., 2012).

However, an adequate characterization of elements at risk requires a thematic and geometric refinement of geospatial information, but this is difficult to achieve. The latest generation of satellite sensors with spatial resolutions of 1 m and better enable researchers to compile (earthquake risk related) building inventories (e.g., Ehrlich et al., 2013; Wieland et al., 2012a). Characterizing exposed elements at risk by mapping physically homogeneous areas using the concept of urban structure types (e.g., Bochow, 2010; Hermosilla et al., 2014; Wurm et al., 2009) is also common. Parameters such as building densities, floor space indices, or vegetation fraction are commonly applied to discriminate different structural characteristics within the complex urban environment with the aim of relating these structural types to general classifications of physical vulnerability.

On the level of individual buildings, physical vulnerability parameters of buildings that can be determined from remote sensing consists of the building footprint, height, shape characteristics, roof materials, location, and structure type (e.g., Mueller et al., 2006; Geiß and Taubenböck, 2013). The period of construction can be estimated by using multitemporal data sets, for example, from the Landsat mission (with data available since 1972) at block level (e.g., Taubenböck et al., 2012). However, studies reveal the limitation for covering large areas due to data availability, costs, processing requirements, or limited automation.

2. Beyond mapping exposure, remote sensing also has the capability of allowing the assessment of the vulnerability of the elements at risk. The assessment of seismic vulnerability of buildings using remote sensing can be distinguished according to two different approaches: (1) by defining a direct relation, using vulnerability curves based on physical features that can be gained from remote sensing data (e.g., Taubenböck et al., 2009b; Borzi et al., 2011) or (2) by using remotely sensed data primarily for spatial interpolation and extrapolation of in situ surveys, by supervised regression and classification techniques (e.g., Borfecchia et al., 2010; Geiß et al., in press).

Remote sensing studies also show capabilities beyond physical vulnerability, such as mapping population exposed to natural hazards. The community standard for global population distribution at an approximate spatial resolution of 1 km is based on spatial data (among other EO data) and respective image-analysis technologies and a multivariable dasymmetric modeling approach to disaggregate census counts within an administrative boundary (e.g., Bhaduri et al., 2007). Spatial disaggregation approaches of population census data are a common research field in remote sensing to assess demographic exposure components (e.g., Dobson et al., 2000; Chen, 2002). In addition, approaches for spatial extrapolation of punctual population data have been examined (e.g., Taubenböck et al., 2007). These are useful when information on total population is based on outdated census data. For instance, Aubrecht et al. (2013) provided an overview on available multilevel geospatial information and modeling approaches from global to local scales that could serve as a spatially refined inventory for people living in disaster-prone areas. Furthermore, physical parameters of the built environment have also been used as proxies to assess the spatial distribution of social vulnerability (e.g., Ebert et al., 2009; Zeng et al., 2012). These approaches correlate urban typologies such as slum areas, for example, to income levels of people living there for an assessment of their vulnerability.

2.2.2.1 Mapping Exposure—Selected Examples

Where are the exposed areas? The question is simple; however, from a global or large-area perspective, the answer is complex. Many regions across the world are hampered by a lack of available and up-to-date geodata or the available data have insufficient geometric and thematic detail. This is further complicated by the fact that exposed elements are highly varying over space and time.

As mentioned in the section above, new initiatives are currently on-going, aiming at products that feature a better spatial resolution regarding the distribution of global settlements—thus, exposed areas. As one example, the global urban footprint initiative at the German Aerospace Center (DLR) uses globally available very-high-resolution radar images from the German TanDEM-X mission. The data are collected in the strip-map mode with a single polarization and a spatial resolution of approximately 3 m. The proposed method to identify settlements includes a specific preprocessing of the original intensity SAR data followed by an automated, threshold-based, image-analysis procedure (Esch et al., 2012). The preprocessing aims at providing additional texture information for classification. Highly textured regions are highlighted. They typically represent highly structured, heterogeneously built-up areas. The technique takes advantage of specific characteristics of urban SAR data, which show strong scattering due to double-bounce effects in these areas. In particular, the preprocessing focuses on the analysis of differences of local speckle characteristics ("speckle divergence"), which is used along with the original intensity information to automatically define urbanized areas.

The classification process continues with the identification of potential urban scatterers—the so-called urban seeds. Urban seeds are characterized by strong reflectance due to double-bounce effects. In combination with a heterogeneous neighborhood—a high speckle divergence related to the true texture compared to homogeneous areas. Details of this classification procedure have been presented in Esch et al. (2010, 2012). The result is an urban footprint classification. Accuracy assessments for the urban footprint classification have been carried out for various test sites across the globe (e.g., at urban areas, periurban and even rural surroundings) proving that settlements are detected with an accuracy of >80 percent (e.g., Taubenböck et al., 2011; Esch et al., 2012).

Figure 2.1 displays the differences of two selected data sets—GRUMP and Global Urban Footprint—to visualize the differences in spatial resolution for the example of the earthquake-prone city of Izmir in Turkey and its rural surroundings. The capability for mapping elements at risk becomes especially obvious by the high spatial resolution of the TanDEM-X data, which allows the detection of scattered settlements in rural areas with a low density. Thus, the capability to localize elements at risk for large areas is significantly refined.

Globally, a classification of the built environment would allow a localization and assessment of exposure, or elements at risk. A characterization of these elements at risk needs an increasing geometric and thematic detail. Remotely sensed data in the range of 1-m spatial resolutions enable one to approach this goal. Very-high-resolution optical satellites—such as Ikonos, Quickbird, WorldView I and II, or GeoEye—provide geometric detail of up to 41 cm, which makes it possible to identify and characterize even small urban structures such as single buildings or streets. Nonetheless, monoscopic spaceborne satellite imagery is mostly limited to two-dimensional analysis. However, in combination with DSMs obtained by airborne laser scanning or stereoscopic techniques, detailed information about the third dimension can be integrated.

The data sets mentioned above allow the mapping and characterizing of elements at risk. The exposed built environment can be described by parameters such as building size, height, volume, or roof type or allow for the derivation of structural parameters such as building density, floor space density, or building alignment (e.g., Crowley and Pinho, 2012; Poli and Caravaggi, 2013). Figure 2.2 shows a 3-D model of Cologne highlighting the downtown area (brown color) and a decreasing building density to the peripheral urban areas. The data set is derived using an object-based classification approach using multisensoral EO data (Wurm et al., 2011). This is an example of characterizing exposure.

2.2.2.2 Assessing Vulnerability—A Selected Example

As briefly reviewed above, EO data are increasingly being used in the context of inventory classification. Characteristics of exposed assets are derived, which can be linked to their vulnerability to seismic hazard. In the context of assessment of structural vulnerability, it is obvious that stability of buildings

FIGURE 2.2   Visualization of a three-dimensional city model of Cologne, Germany, allowing the quantification and physical characterization of the elements at risk—in this case buildings; the downtown area is highlighted in brown.

cannot directly be measured by EO data, but indirect relationships are used to correlate types of buildings with their affiliated behavior in the case of ground shaking. In this manner, any additionally sensed data, such as photographs of facades (e.g., construction material, lateral load resisting system) allow significant value adding to assess the likely seismic performance of a building.

Commonly used ground-based methods to analyze an exposed building stock with respect to its seismic resistance usually entail a detailed screening of single buildings by structural engineers (FEMA, 2002) and are highly labor intensive. They are, therefore, not suitable for accommodating to the rapidly changing spatiotemporal conditions in many present-day cities, and moreover do often not scale well with limited time and budget resources.

Novel ground-based sensing techniques, such as omnidirectional imaging, show a large potential for overcoming the limitations of commonly used screening methods. Coupling EO data and ground-based surveys therefore seem a valuable and optimal solution, especially when coupling with a sampling framework. This can create synergy effects, where each imaging source can be used to infer specific, scale-dependent information about an exposed building stock and its vulnerability. This would help reduce costs while retaining the representation of the area of interest. Sampling has the advantage that only subsets of a larger area of interest need to be analyzed in detail to estimate summary statistics for the whole area, or for well-defined strata if a stratified sampling is used.

Figure 2.3 shows a possible implementation of such a "sampling approach" (Wieland and Pittore, 2012). In a top-down analysis, the processing scheme moves from an aggregated neighborhood scale to a detailed per-building scale

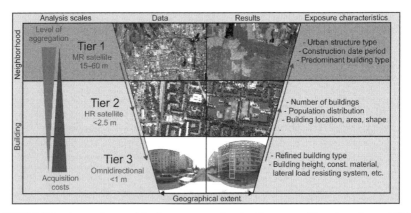

FIGURE 2.3　Approach to multiscale exposure estimation from ground-based sensing and EO data with considered analysis scales, input image data, and relevant exposure characteristics.

involving three analysis tiers that are related to each other and that are based on the analysis of different image types. Across this pyramidal searching, only the necessary data are acquired and processed, and the focused geographical extent is narrowed. The aim is to minimize acquisition costs and processing time by using EO data analysis to guide more detailed per-building surveys with ground-based sensing techniques.

In a "tier 1 analysis", a regional assessment at an aggregated neighborhood scale on the basis of medium spatial (30−60-m ground sampling distance (GSD)), but high temporal (revisit period of 16 days) and spectral (0.45−12.5 μm in four to seven spectral bands) resolution satellite images is carried out to stratify the urban environment into areas of homogeneous urban structures (Wieland et al., 2012a). A useful stratification is defined, based on the hypothesis that the building inventory composition, and therefore, the vulnerability model, follows Tobler's law of spatial autocorrelation. This hypothesis proposes that commonly neighboring buildings share several characteristics related to vulnerability, such as construction date, typology, material of construction, and occupancy, and can therefore be clustered together. Low-level image segmentation, high-level supervised image classification, and change-detection analysis are used to extract the strata of relatively homogeneous urban structure types. The resulting tier 1 exposure information layer provides aggregated information about the spatial distribution of predominant building types with associated structural characteristics and their approximate construction date.

A "tier 2 analysis" based of high-resolution optical EO data with a GSD <2.5 m at a per-building scale provides the location, area, and shape of individual buildings (Wieland et al., 2012b). It can further enrich the tier 1 exposure layer with information about the number and density of buildings aggregated for each area of homogeneous urban structure. The tier 1 or tier 2 exposure information layers, moreover, amass to define the strata composing a stratified sampling scheme. For each stratum, a sample is drawn, whose size is chosen following a proportional allocation scheme.

Inside the identified sample areas, a more detailed "tier 3 analysis" on a building scale is carried out by ground-based data captured with commonly used screening techniques or state-of-the-art omnidirectional imaging techniques. Omnidirectional imaging is particularly suited for rapid exposure data capturing since it acquires all visuals surrounding the camera (360° horizontally and almost 180° vertically). This makes data acquisition easy and fast, and above all, independent of an operator's point of view. A camera is mounted on the roof top of a car and navigated along predetermined sample routes. The analysis of ground-based omnidirectional images provides a street view of the objects of interest and therefore allows an automated or manual extraction of building characteristics such as height, lateral load resisting system, and construction material leading to a more differentiated building-type classification. An example of an automated feature extraction is given in Figure 2.4. From contiguous omnidirectional images, a 3D reconstruction of building façades can be obtained using optical flow measures (Wieland et al., 2012b). Based on automated georeferencing of the images, the information extracted from the omnidirectional images can be linked with building footprints extracted from EO data.

Given the assumption that each stratum is composed of a relatively homogeneous urban structure, the sampled information can be backpropagated in a subsequent of bottom-up approach from the building scale to the neighborhood scale using the corresponding stratum as a basis. The resulting tier 3 exposure information layer can therefore provide a refined picture of the exposed building stock for each stratum. This includes information about the actual composition of building types and their characteristics (Figure 2.5(a)).

A method that was recently proposed by Pittore and Wieland (2013) integrates vulnerability proxies derived from a combination of ground-based sensing and EO data analysis within a Bayesian network to derive a probabilistic classification of building vulnerability. A Bayesian approach is particularly

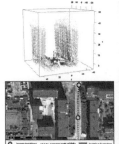

FIGURE 2.4   **Example of 3D reconstruction based on automated processing of omnidirectional images.**

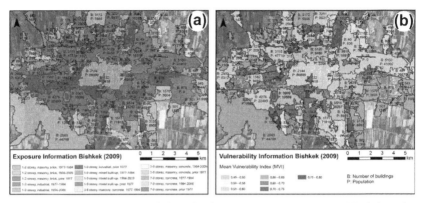

FIGURE 2.5   (a) Tier 3 exposure information layer; (b) probabilistic vulnerability information layer.

suitable for the integration of information from multiple imaging sources at varying scales with different accuracies, since it allows for the combining of available information about the buildings regardless of the source including considering uncertainties. For each building, successfully characterized by both omnidirectional and EO data analysis, a posterior probability distribution of its vulnerability classes is derived according to the European Macroseismic Scale 1998 (EMS-98) (Grünthal et al., 1998). The conditional probability table of the node that connects the building characteristics with the vulnerability classification is based on expert knowledge and can be extracted for sample typologies from the World Housing Encyclopedia reports (WHE) (World Housing Encyclopedia, 2011). The WHE reports are generated by local experts on building construction and describe in detail the most relevant building types, including their physical characteristics, and give an evaluation of their structural vulnerability in EMS-98. Figure 2.5(b) shows an aggregated vulnerability distribution for the city of Bishkek in Kyrgyzstan. The vulnerability assessment is based on a Bayesian information integration of sampled buildings. The posterior probability distribution of the occurrence of vulnerability classes has been averaged over all observed buildings per stratum and is described in terms of a scalar Mean Vulnerability Index (MVI) (Figure 2.5(b)) (see Pittore and Wieland, 2013 for details). The MVI ranges from 0 (lowest degree of vulnerability) to 1 (highest degree of vulnerability).

2.2.3 Postevent Capabilities of Remote Sensing for Earthquake Hazard Analysis

Large earthquakes cause surface displacements in the order of decimeters to meters, which can clearly be identified in SAR interferograms. The Landers earthquake in 1992 was the first major event captured by the European Remote Sensing Satellite ERS-1 and analyzed using the InSAR technique (Massonnet et al., 1993). The generation of a so-called coseismic interferogram requires one image before and one after the event. Although postevent images are easy to schedule—provided that the satellite is still in operation—preevent images are globally only available if a satellite mission regularly monitors all georisk areas. The probability of decorrelation errors due to natural land surface changes can be reduced significantly if the time differences between preevent and postevent data are in the order of weeks to months (Bürgmann et al., 2000). Therefore, only missions with a frequent and global observation plan can guarantee a suitable preseismic reference image in the case of a disaster. Nevertheless, several successful examples mapping earthquake surface displacements have been demonstrated in the past with available systems, for example, Tohoku-Oki 2011 (Simons et al., 2011) or Wenchuan 2008 (Shen et al., 2009).

Current high-resolution SAR missions such as TerraSAR-X or Radarsat-2 with limited capabilities in area mapping fulfill these requirements only partly. Upcoming missions such as the European Sentinel-1 satellite will have suitable observation plans. However, the phase quality of longer wavelength L-band (20 cm) systems is significantly better than the quality of C-band systems used by the European missions ERS-1/2, ENVISAT/ASAR or Sentinel-1. X-band systems experience an even faster decorrelation and are therefore less suitable for global long-term observation.

Sometimes, the displacement caused by an earthquake is so large—in the order of meters—that it reaches the resolution of the particular SAR imaging system. In this case, conventional image disparity measures can be used as an alternative to the highly sensitive interferometric technique, and both optical and radar images can be exploited to derive the displacement pattern around the epicenter. Also, in this case, radar images are favorable because they provide much a better pixel localization accuracy than do optical images (Eineder et al., 2011).

2.2.3.1 InSAR Analyses of the January 12, 2010, Haiti Earthquake—A Selected Example

Techniques to measure deformation of the surface of our planet have come a long way since the Middleton workshop in 1968, when NASA engineers asked earth scientists, if it would be useful if one could determine positions on the Earth within about 10—1 m from space. After more than a decade of efforts to reduce the uncertainties of position estimates to about 10 cm by the competing techniques of the Long Baseline Interferometry and the LAGEOS laser satellite, the GPS surpassed the more expensive and cumbersome earlier systems. The GPS soon provided a wealth of information about surface deformations of the Earth at the centimeter level. Two drawbacks of this otherwise outstanding tool are Hardware in the form of permanent and temporary stations is required, and deformation vectors are obtained only for points where an instrument is operated.

InSAR (Interferometry of Synthetic Aperture Radar) is the newest technique to map deformation fields of the Earth's surface. Its great advantage lies in the ability of mapping a continuous vector field of deformation without the deployment of instruments on the ground, provided the topography and nature of the surface allow coherent reflections. Drawbacks of the technique include the necessity of having more than one SAR image of the area of interest. Images before and after the deformation are required and exposures taken at different angles, usually during the ascending and descending paths, need to be available. The pitfalls and their solutions are described in a concise way by Wright et al. (2004).

For modeling earthquake sources, InSAR maps of surface deformations, brought about by rupture, provide the best possible data. Because the surface deformations are mapped in great detail, the source parameters, which cause the surface deformations observed, can be determined within small uncertainties. The parameters of the earthquake source are Position of the surface rupture (or rupture plane at depth), strike and dip of the fault plane, rake of the slip vector in the fault plane, depth extent of the rupture, and distribution of the slip within the rupture plane, including the location of largest energy release.

The superiority of earthquake source models based on InSAR data over models based on recorded seismic waves are the following: The location of the largest energy release and the extent of the rupture can be determined within an uncertainty of about 0.5 km by InSAR, whereas the uncertainty of tele-seismically located epicenters are approximately 25 km during approximately the first day after the event (Wyss, 2011), improving later to about 9 km. In addition, the estimate of rupture geometry by seismological techniques furnishes two perpendicular solutions because of the symmetry of wave radiation, whereas InSAR delivers the definition of the one and only correct orientation of the rupture plane.

Other tectonic parameters of the Earth's crust that would be desirable to know for understanding tectonics, such as estimating the time-dependent earthquake hazard, and possibly predicting earthquakes cannot be derived by any technique, not even InSAR. The most elusive among these parameters are the state of stress and the elastic energy available in the seismogenic crust (top 15−30 km of the Earth). An approximate estimate of the orientation of the stress tensor can be derived from the geometry of the source model. For more accurate estimates of the stress directions in a crustal volume, many earthquakes with varying source geometries and a careful analysis are required. The stress drop caused by the rupture is the only aspect of the state of stress that can be calculated approximately from the strain drop that is part of the source model constructed to fit the surface deformation measured by InSAR. The ambient stress level that remains after the stress dropped is unknown.

The after creep that follows many earthquakes can also be modeled based on the displacement fields derived from InSAR images. The partitioning of tectonic deformation along faults and especially plate boundaries into sudden slip in earthquake ruptures and aseismic creep is an important question. Since the discovery that faults are creeping (Smith and Wyss, 1968) the importance of this phenomenon has been increasingly appreciated by seismologists. Documenting aseismic tectonic deformation is an important task for which InSAR is one of the primary tools.

The big questions are How much energy was released during the earthquake and how much energy is still stored in the fault region and may cause dangerous aftershocks? SAR missions with frequent observations of georisk areas over long time spans could contribute to answer these questions. The proposed Tandem-L mission (Krieger et al., 2010), in conjunction with worldwide dense GPS networks, could provide such data in the future.

Figure 2.6 shows the interferometric fringes of the deformation caused by the destructive Haiti M_w 7 earthquake on January 12, 2010 (from Suresh et al., 2011). While these fringes visually reveal significant deformation in the line-of-sight direction to the L-band ALOS satellite, they do not yet provide a 3D surface deformation map. Fortunately, even if the ALOS sensor has a long repeat cycle of 46 days, for this event, first images became available on January 16 and 25, 2010. For Haiti, even faster, images from the German TerraSAR-X satellite could be tasked on January 14, just two days after the earthquake. The earlier described InSAR methods were not useful with TerraSAR-X images as the preevent images were too old for the short X-band wavelength (3.1 cm). However, because of the high image resolution, correlation measurements were successfully performed and revealed large displacements of up to 1 m. Surprisingly, the results did not initially meet the expectations from the model that predicted an east–west oriented motion along the Enriquillio fault. Instead, they revealed an unexpected upward

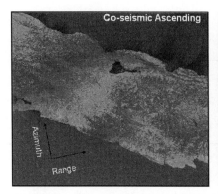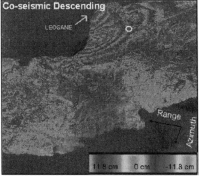

FIGURE 2.6   Fringes in L-band interferograms from ALOS Palsar data clearly show the line-of-sight deformation caused by the earthquake, but need to be processed further for a quantitative 3D interpretation (Left 28-02-2009/16-01-2010), (Right 09-03-2009/25-01-2010). *Based on Suresh et al. (2011).*

motion at the Leogange delta of about 80 cm. In this disaster, much of the infrastructure was destroyed, and satellite images could provide first indications of the damage (Huyck et al., 2014). More analysis examples on Haiti can be found on the GEO supersites webpage http://supersites. earthobservations.org (Last accessed 29.04.13.) (Figure 2.7).

2.2.4 Postevent Capabilities for Damage Assessment and Recovery Monitoring

Currently, remote sensing is one of the very few ways in which information on large affected areas can be captured immediately after an earthquake (Vu and Ban, 2010; Huyck et al., 2014; Wegscheider et al., 2013). It has become a valuable tool during emergency response (Dell'Acqua et al., 2009) by supporting decision makers with up-to-date spatial information (Joyce et al., 2009). Since its inception in 2000, the International Charter "Space and Major Disasters" (www.disasterscharter.org) has played a major role in facilitating the distribution of remotely sensed data for emergency response, with an ever-growing number of imagery being acquired and damage maps generated. The latter is carried out traditionally by professional analysts at processing/mapping/analyzing facilities such as UNOSAT (UNOSAT, 2011),

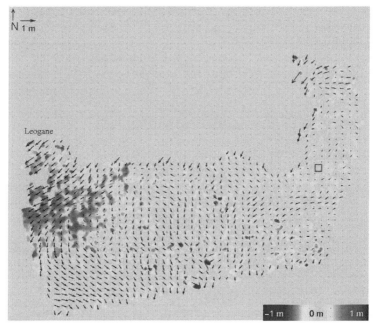

FIGURE 2.7   The 3D displacements of the Haiti M7 earthquake on January 12, 2010, measured with image correlation techniques from TerraSAR-X data (72.45°W−72.28°W, 18.40°N−18.65°N) *(Based on Suresh et al. (2011).).* Colors indicate vertical motion.

DLR-ZKI (Voigt et al., 2007), SERVIR (SERVIR, 2012), SERTIT (SERTIT, 2012), or e-GEOS (e-GEOS, 2012) and recently also by volunteers experienced in remote sensing by applying collaborative mapping (Kerle and Hoffman, 2013).

To date, a key use of remote sensing has been in postearthquake damage assessments. Dell'Acqua and Gamba (2012) discriminate "rapid damage assessment" and "detailed damage assessment". Rapid damage assessment reports the extent of the damaged area with different damage levels, typically aggregated according to a certain spatial unit (e.g., grid cells or urban blocks). These initial maps are "produced" very quickly after the event and "immediately disseminated" to decision makers. The detailed damage assessment reports, for example, information on damage grades at individual building level based on data with the best available resolution. These are produced "within the crisis time frame" of a couple of days or weeks (see Voigt et al., 2011 for lessons learned and a critical discussion regarding the numerous mapping products dealing with the January 12, 2010, Port-au-Prince Haiti earthquake).

Generally, rapid damage mapping is still primarily based on the manual interpretation of remotely sensed data in order to avoid long (pre) processing times (Trianni and Gamba, 2009) and to provide the needed accuracy and reliability of the results (Voigt et al., 2011). Most limitations arise due to data availability and misinterpretations (Kerle, 2010). Building damage is normally more accurately recognized for severely damaged structures and underestimated for slightly or moderately damaged structures. Furthermore, patterns of damage are more likely to be accurately mapped, whereas the differentiated delineation of the damage state for individual structures appears less feasible (Corbane et al., 2011a). It has been recognized that damaged buildings commonly still show roof characteristics in good order from the top-view perspective, thus limiting the capability of remote sensing for classifications without errors of omission.

In general, different sectors/stakeholders require different types of information; a lot of suspicion is still observed among decision makers about the usefulness of remote sensing for their work, and how it would complement other conventional methods on the ground. We are at a pivotal point where remote sensing either becomes mainstream in ex post activities, or not. To help within this process, data standards are needed, quality assurance needs to be standardized, and clear definitions of the products need to be communicated; currently, most of the rapid response activities are still done in a rather ad hoc manner.

2.2.4.1 Mapping Damage: The January 12, 2010, Earthquake in Haiti

As a very significant example, a damage map of the city of Port-au-Prince (Haiti) is shown in Figure 2.8 following the January 12, 2010, earthquake event. The pansharped GeoEye-1 image data were interpreted manually by

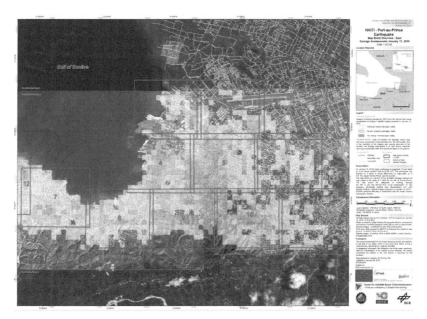

FIGURE 2.8   Overview damage map of the city of Port-au-Prince following the 2010 earthquake (Voigt et al., 2011) prepared by the Center for Satellite Based Crisis Information of the German Aerospace Center (accessible via DLR-ZKI, 2010). *Reproduced with permission from the American Society of Photogrammetry and Remote Sensing.*

professional analysts at the Center for Satellite Based Crisis Information of the German Aerospace Center (DLR-ZKI). The map indicates zones of damage and reveals the damage severity. Due to the wide disaster extent and heterogeneity of local urban structures, rapid damage assessment was carried out for quadratic grid cells of 250×250 m. For each of these, an average damage level according to three categories ($<10\%$, $10-40\%$, $>40\%$) was estimated (Voigt et al., 2011).

A building-by-building, manual interpretation was also carried out in Port-au-Prince jointly by UNOSAT, EU-JRC and the World Bank and numerous other collaborators, where $600 +$ volunteers were mobilized to carry out a crowd-sourced damage assessment. The results demonstrated that even with the heavily damaged buildings the manual interpretation results were underestimated by a significant amount (Corbane et al., 2011). A similar exercise was carried out following the Christchurch earthquake (Foulser-Piggott et al., 2013).

Besides manual image interpretation for rapid and detailed damage assessment, (semi)automated damage detection approaches are in the focus of the scientific community (see e.g., Geiß and Taubenböck, 2013 for an overview and Dell'Acqua and Gamba, 2012 for an in-depth review). Damage mapping approaches can be most likely categorized according to the remotely sensed data they are based on. For instance, Kaya et al. (2005) used solely optical, postevent imagery for damage estimation, whereas Balz and Liao (2010) utilized only postevent SAR data. However, most approaches compare pre-earthquake and postearthquake images, because it is believed that change-detection approaches deliver higher accuracy results (e.g., Li et al., 2008; Ehrlich et al., 2009). This is due to the fact that damage assessment is principally a change-detection problem, where the mapping classes are correlated with the level of damage undergone by the buildings (Trianni and Gamba, 2009). Based on optical preevent and postevent imagery, approaches to detect change and subsequently derive earthquake damage of buildings in an auto-mated way were developed (e.g., Kosugi et al., 2004; Turker and Sumer, 2008; Vu and Ban, 2010). Also preevent and postevent LiDAR data are incorporated in order to reach a high level of morphologic detail and accuracy (Vu et al., 2004; Li et al., 2008). Data acquisition of optical imagery shortly after an earthquake may be limited due to cloud coverage and weather conditions, and flight campaigns in remote areas can commonly not be carried out quickly. This makes the use of SAR-based approaches particularly useful. Thus, Hoffmann (2007), Gamba et al. (2007), Matsuoka and Yamazaki (2010), and Schmitt et al. (2010) assessed earthquake damage of built structures by using preevent and postevent SAR data, and Stramondo et al. (2006), Chini et al. (2009), and Brunner et al. (2010) combined preevent optical and postevent SAR imageries.

2.2.4.2 Recovery Monitoring Using EO Data

Generally, the recovery phases can be categorized into the immediate emergency-response phase where the primary concern is to save people's lives, followed by the short-term, medium-term, and long-term recovery and recon-struction planning, implementation, and evaluation phases. However, recovery is increasingly seen as a dynamic process, and the use of the concept "phases of recovery" is considered to oversimplify the process as in reality, the various roles—decision makers, rescue teams—and phases interact and overlap (Brown et al., 2008). In this section though, the phase concept is be adopted as a convenient way to describe the general progress in the application of EO data and processing techniques to the recovery processes.

During the short-term recovery phase, the priority is to provide temporary shelters and basic amenities for the survivors. It has been argued that providing temporary shelters is not an optimal solution, as survivors should be relocated but housed near, or on their original plot of land. However, in many emergency situations, it is still the case that temporary shelters are provided by the local and national governments as well as international agencies on a site earmarked for temporary shelters. In many cases, these clusters of temporary shelters spring up without much coordination. EO techniques developed for identifying refugee/IDP camps can be applied to estimate the number of temporary shelters (e.g., Kranz et al., 2010; Tiede et al., 2013). One example is shown in Figure 2.9, where the number of dwellings in an IDP camp in Darfur was successfully extracted using high-resolution optical images (Kemper et al., 2011).

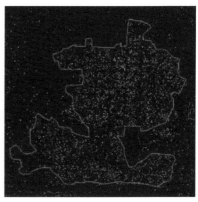

FIGURE 2.9 The number of dwellings in an IDP camp in Darfur was successfully extracted automatically using the Geoeye-1 image seen on the left. The image on the right shows the camp boundary, and the individual tents are shown as white dots. *Taken from Kemper et al. (2011)*.

For planning reconstruction, it is generally necessary to obtain baseline data of the area that is about to be reconstructed. Particularly, in developing countries where spatially referenced baseline data are lacking, EO data can be useful to establish this baseline for certain sectors, such as housing and transportation. The British Red Cross carried out a detailed survey of an area within Haiti's Port-au-Prince suburb called Delma 19, where prior to the earthquake in 2010, the majority of the buildings were informal settlements. Satellite images, together with ground surveys, were used to establish baseline data containing the following information: building footprints, names and other details of the occupants, construction type, number of stories, use type, together with the level of damage (British Red Cross, 2012). Ideally, baseline data would be collected prior to a disaster, so that they are already available and up to date when reconstruction planning commences. In Haiti and in subsequent disasters, Open Street Map (OSM) has been collecting baseline data by combining the tracing of imagery to obtain building footprints with ground surveys of individual buildings to collect attribute data. The World Bank's Open Data for Resilience Initiative (OpenDRI) is collaborating with OSM in many countries, particularly in south Asian cities, to create these spatial baseline data and make them available so that stakeholders can have access to this critical information for disaster risk management as well as a plethora of other issues including urban/rural development (OpenDRI, 2013).

Once the baseline is established and recovery plans are developed, generally speaking, the gear shifts toward the planning and implementation of the midterm to long-term recovery plans. Until recently, despite the large amount of international donors' funding that is provided for the implementation of the recovery plans, there has not been any monitoring and evaluation framework. The ReBuilDD group carried out studies on the feasibility of using EO data for recovery monitoring and evaluation, by first

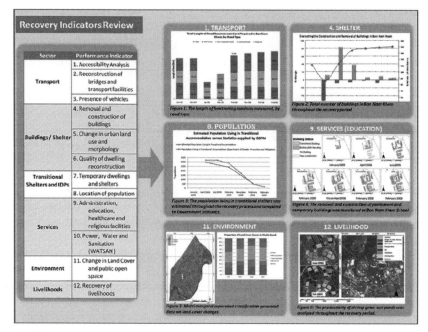

FIGURE 2.10 The recovery indicators developed by the ReBuilDD group to measure the progress of recovery in the sectors seen in the table. The outputs from the analysis could be provided as charts or maps as seen above.

developing a list of indicators to evaluate the progress of recovery against, based on international frameworks such as TRIAMS and MDG (Brown et al., forthcoming). This was followed by identifying and developing measures to monitor the implementation of recovery using EO data. Indicators for a wide-ranging number of sectors seen in Figure 2.10 such as the transportation, housing, environment, and livelihood sectors were developed. For each sector, a subindicator called performance indicator was defined, and ways to measure these indicators using EO data were developed. Case studies were demonstrated from the 2005 Kashmir (Pakistan) earthquake and the 2004 Indian Ocean tsunami-affected countries (Brown et al., 2010). The monitoring and validation of recovery are still a relatively new area of research, and the challenges it provides, including the complexity of the recovery process itself, is providing stimulus for future research.

2.3 CONCLUSION AND INFERRING SUGGESTIONS FOR EO ON EARTHQUAKE ANALYSIS

Remote sensing has the capability for substantial geospatial contributions to all stages within the risk cycle. These comprise "hazard analyses", "exposure mapping", and "vulnerability assessment for preevent applications". For the postevent case, remote sensing enables "hazard analysis" as well as "damage assessment" and "recovery monitoring". An in-depth overview on the remote sensing capabilities and limitations along the risk cycle is provided by the special issue "Remote sensing contributing to assess earthquake vulnerability and effects" (2013) published in *Natural Hazards*.

For "preevent hazard analyses", remote sensing contributes primarily to microscale zonation, and measurements of small displacements or deformations of the Earth's surface.

The capabilities of EO for "preevent exposure" and "vulnerability analysis" are manifold. These arise from the unique capability to map land cover/land use, that is, mapping elements at risk at various scales and related thematic detail. Furthermore, the vulnerability of the exposed elements, for example, building stability, or vulnerability of people can be characterized and assessed with remote sensing using indirect correlation methods. Although many products on exposure and vulnerability, based on EO data, have been proven useful, systematic studies combining all theoretically possible geospatial data for risk analysis as well as the provision of all these products for systematic evaluation of the maximum capability of remote sensing (e.g., EO, ground-based sensing) are still missing. With the upcoming future missions, this priority seems feasible due to explorative knowhow and better data availability; however, this goal will need coordinated progress in the science community.

In the "postevent phase" rapid mapping, based on remote sensing data, holds the capability to capture information on the immediate situation over large affected areas straight after an earthquake. During the last decade, it has thus become a valuable and commonly applied tool for emergency response. However, products are still not standardized, and quality control needs to be carried out in a systematic way. Clear definitions of products must be communicated, demonstrated, and harmonized with the various users from different science disciplines and stakeholders. With this scientific goal in mind and the advent of future EO missions, such as Sentinel, Radarsat constellation, Cartosat-3, and WorldView 3, the prospects should ensure a high temporal and spatial coverage for timelier and more accurate damage assessment. Thus, the scientific community needs to overcome the lack of understanding observed among decision makers about the usefulness of remote sensing for their work on the ground. We are at a pivotal point where remote sensing either becomes mainstream in ex post activities, or it does not.

The capability of remote sensing for systematic monitoring of "medium- and long-term recovery and reconstruction planning", "implementation", and "evaluation" phases is highly viable, as has been shown by the list of potential physical recovery indicators. Especially with large-area events EO data allow capturing of the progress of measures and support decision making with these overviews. In any case, however, the scientific research is still at an early stage.

Overall, it is evident that remote sensing has the capability of providing a large share of geospatial information for earthquake analysis and information for policy and decision makers involved in earthquake disaster management. However, as some examples highlighted in this chapter have shown, multidisciplinary and transdisciplinary approaches are required to add value to the progress of science and to overcome limitations of individual disciplines. This book is a perfect example of combining the different perspectives, data, methods, results of different disciplines to allow an overview and critique of all their capabilities and shortcomings, with an aim of engaging in a multidisciplinary approach to addressing all phases of the risk cycle in the future.

ACKNOWLEDGMENTS

This work has been cofunded by the European Commission under FP7 (Seventh Framework Programme) THEME [SPA.2012.1.1-04] Support to emergency response management—Grant agreement no: 312972; Project title: SENSUM. The research was partly funded by the Helmholtz-EOS (Earth Observation System) http://www.gdmc.nl/zlatanova/Gi4DM2010/gi4dm/Pdf/p187.pdf.

REFERENCES

Angel, S., Sheppard, S.C., Civco, D.L., 2005. The Dynamics of Global Urban Expansion. Transport and Urban Development Department, the World Bank, Washington, DC p. 102.

Aubrecht, C., Özceylan, D., Steinnocher, K., Freire, S., 2013. Multi-level geospatial modeling of human exposure patterns and vulnerability indicators. Nat. Hazards 68 (1), 147−164.

Balz, T., Liao, M., 2010. Building-damage detection using post-seismic high-resolution SAR satellite data. Int. J. Remote Sens. 31 (13), 3369−3391.

Bhaduri, B., Bright, E., Coleman, P., Urban, M., 2007. LandScan USA: a high resolution geospatial and temporal modeling approach for population distribution and dynamics. GeoJournal 69, 103−117.

Birkmann, J., 2007. Risk and vulnerability indicators at different scales: applicability, usefulness and policy implications. Environ. Hazards 7 (1), 20−31.

Bochow, M., 2010. Automatisierungspotenzial von Stadtbiotopkartierungen durch Methoden der Fernerkundung (Ph.D. thesis). Universität Osnabrück.

Borfecchia, F., Pollino, M., De Cecco, L., Lugari, A., Martini, S., La Porta, L., Ristoratore, E., Pascale, C., 2010. Active and passive remote sensing for supporting the evaluation of the urban seismic vulnerability. Ital. J. Remote Sens. 42 (3), 129−141.

Borzi, B., Dell'Acqua, F., Faravelli, M., Gamba, P., Lisini, G., Onida, M., Polli, D., 2011. Vulnerability study on a large industrial area using satellite remotely sensed images. Bull. Earthquake Eng. 9, 675−690.

British Red Cross, 2012. Regeneration of a Community. http://www.redcross.org.uk/What-we-do/Emergency-response/Recovering-from-disasters/Haiti-earthquake-2010/Recovery/Shelter (Last accessed 23.06.13.).

Brown, D., Bevington, J., Platt, S., Saito, K., Adams, B.J., Chenvidyakarn, T., Spence, R.J., Chuenpagdee, R., Khan, A., So, E. K. M. (to be published). Monitoring and Evaluating Post-disaster Recovery Using High-Resolution Satellite Imagery − Towards Standardised Indicators for Post-disaster Recovery, Disasters, forthcoming.

Brown, D., Saito, K., Platt, S., Chenvidyakarn, T., Spence, R., Adams, B., September 2008. Indicators for monitoring, measuring and evaluating post-disaster recovery. In: 6th International Workshop on the Use of Remote Sensing for Disaster Response, Pavia, Italy.

Brown, D., Platt, S., Bevington, J., Spence, R.J., Chendivyakarn, T., Saito, K., Adams, B., So, E., Khan, A., Chuenpagdee, R., 2010. Disaster Recovery Indicators: Guidelines for Monitoring and Evaluation. Available from: http://www.carltd.com/sites/carwebsite/files/Disaster%20Recovery%20Indicators.pdf (Last accessed 23.02.14.).

Brunner, D., Lemoine, G., Bruzzone, L., 2010. Earthquake damage assessment of buildings using VHR optical and SAR imagery. IEEE Trans. Geosci. Remote Sens. 48, 2403–2420.

Bürgmann, R., Rosen, P., Fielding, E., 2000. Synthetic aperture radar interferometry to measure Earth's surface topography and its deformation. Annu. Rev. Earth Planet. Sci. 28, 169–209.

Chen, K., 2002. An approach to linking remotely sensed data and areal census data. Int. J. Remote Sens. 23, 37–48.

Chini, M., Pierdicca, N., Emery, W.J., 2009. Exploiting SAR and VHR optical images to quantify damage caused by the 2003 Bam Earthquake. IEEE Trans. Geosci. Remote Sens. 47 (1), 145–152.

Crowley, H., Pinho, R., 2012. Global Earthquake Model: Community-Based Seismic Risk Assessment. Protection of Built Environments against Earthquakes, pp. 3–19.

Corbane, C., Saito, K., Dell'Oro, L., Eguchi, R., Adams, B., Bjorgo, E., Evans, G., Ghosh, S., Gartley, R., Ghesquiere, F., Gill, S., Kemper, T., Krishnan, R.S.G., Lemoine, G., Piard, B., Senegas, O., Spence, R., Svekla, W., Toro, J., October 2011a. A comprehensive analysis of building damage in the January 12, 2010 M7 Haiti earthquake using high-resolution satellite and aerial imagery, special issue on the 2010 Haiti earthquake. Photogramm. Eng. Remote Sens. 77 (10).

Corbane, C., Carrion, D., Lemoine, G., Broglia, M., 2011b. Comparison of damage assessment maps derived from very high spatial resolution satellite and aerial imagery produced for the Haiti 2010 earthquake. Earthquake Spectra 27, 199–218.

Deichmann, U., Ehrlich, D., Small, C., Zeug, G., 2011. Using High Resolution Satellite Data for the Identification of Urban Natural Disaster Risk. Global Facility for Disaster Reduction and Recovery, Washington, DC.

Dell'Acqua, F., Gamba, P., 2012. Remote sensing and earthquake damage assessment: experiences, limits, and perspectives. Proc. IEEE 100 (10), 2876–2890.

Dell'Acqua, F., Lisini, G., Gamba, P., 2009. Experiences in optical and SAR imagery analysis for damage assessment in the Wuhan, May 2008 Earthquake. IEEE Int. Geosci. Remote Sens. Symp. 1–5, 2417–2420.

DLR-ZKI, 2010. Damage Assessment Overview Map Port-au-Prince East-Earthquake, Haiti. URL: http://www.zki.dlr.de/map/1194 (accessed 09.04.13.).

Dobson, J.E., Bright, E.A., Coleman, P.R., Durfee, R.C., Worley, B.A., 2000. LandScan: a global population database for estimating populations at risk. Photogramm. Eng. Remote Sens. 66, 849–857.

Ebert, A., Kerle, N., Stein, A., 2009. Urban social vulnerability assessment with physical proxies and spatial metrics derived from air- and spaceborne imagery and GIS data. Nat. Hazards 48, 275–294.

e-GEOS, 2012. e-GEOS: Emergency Response Mapping. Available at: http://www.eurimage.com/applications/emergency.html (accessed 11.07.12.).

Ehrlich, D., Guo, H., Molch, K., Ma, J.W., Pesaresi, M., 2009. Identifying damage caused by the 2008 Wenchuan earthquake from VHR remote sensing data. Int. J. Digital Earth 2 (4), 309–326.

Ehrlich, D., Kemper, T., Blaes, X., Soille, P., 2013. Extracting building stock information from optical satellite imagery for mapping earthquake exposure and its vulnerability. Nat. Hazards 68 (1), 79–96.

Eineder, M., Minet, C., Steigenberger, P., Cong, X., Fritz, T., 2011. Imaging Geodesy—toward centimeter-level ranging accuracy with TerraSAR-X. IEEE Trans. Geosci. Remote Sens. ISSN: 0196-2892 49 (2), 661–671. http://dx.doi.org/10.1109/TGRS.2010.2060264. IEEE.

Esch, T., Marconcini, M., Felbier, A., Roth, A., Heldens, W., Huber, M., Schwinger, M., Taubenböck, H., Müller, A., Dech, S., 2013. Urban footprint processor – Fully automated processing chain generating settlement masks from global data of the TanDEM-X mission. In: IEEE Geoscience and Remote Sensing Letters 10 (6), 1617–1621.

Esch, T., Thiel, M., Schenk, A., Roth, A., Mehl, H., Dech, S., February 2010. Delineation of urban footprints from TerraSAR-X data by analyzing speckle characteristics and intensity information. IEEE Trans. Geosci. Remote Sens. 48 (2), 905–916.

Esch, T., Taubenböck, H., Roth, A., Heldens, W., Felbier, A., Thiel, M., Schmidt, M., Müller, A., Dech, S., 2012. TanDEM-X mission: new perspectives for the inventory and monitoring of global settlement patterns. J. Sel. Top. Appl. Earth Obs. Remote Sens. 6, 22.

FEMA 154, 2002. Rapid Visual Screening of Buildings for Potential Seismic Hazards: A Handbook. ATC, Washington, DC.

Foulser-Piggott, R., Spence, R.J., Brown, D., 2013. The Use of Remote Sensing for Building Damage Assessment Following 22nd February 2011 Christchurch Earthquake: The GEOCAN Study and its Validation. Internal report. Available from. Cambridge Architectural Research Ltd., UK.

Gamba, P., Dell'Acqua, F., Trianni, G., 2007. Rapid damage detection in the Bam area using multitemporal SAR and exploiting ancillary data. IEEE Trans. Geosci. Remote Sens. 45 (6), 1582–1589.

Gamba, P., Lisini, G., 2013. Fast and efficient urban extent extraction using ASAR wide swath mode data. IEEE Journal of Selected Topics in Applied Earth Observations and Remote Sensing 6 (5), 2184–2195.

Geiß, C., Taubenböck, H., 2013. Remote sensing contributing to assess earthquake risk: from a literature review towards a roadmap. Nat. Hazards 68 (1), 7–48.

Geiß, C., Taubenböck, H., Tyagunov, S., Tisch, A., Post, J., Lakes, T. Assessment of seismic vulnerability from space. Earthquake Spectra, http://dx.doi.org/10.1193/121812EQS350M.

Grünthal, G., Musson, R., Schwarz, J., Stucchi, M., 1998. European Macroseismic Scale 1998. European Seismological Commission.

Goldewijk, K., 2001. Estimating global land use change over the past 300 years: the HYDE database. Global Biogeochem. Cycles 15 (2), 417–434. http://dx.doi.org/10.1029/1999GB001232.

Gravely, D., 2001. Risk, Hazard and Disaster. University of Canterbury, New Zealand.

Hermosilla, T., Palomar-Vazquez, J., Balaguer-Beser, A., Balsa-Barreiro, J., Ruiz, L.A., 2014. Using street based metrics to characterize urban typologies. Comput. Environ. Urban Syst. 44, 68–79.

Hoffmann, J., 2007. Mapping damage during the Bam (Iran) earthquake using interferometric coherence. Int. J. Remote Sens. 28 (6), 1199–1216.

Hudnut, K.W., Bock, Y., Galetzka, J.E., Webb, F.H., Young, W.H., 2002. The Southern California integrated GPS network (SCIGN). In: Fujinawa, Y., Yoshida, A. (Eds.), Seismotectonics in Convergent Plate Boundary. Scientific Publishing Company (TERRAPUB), Tokyo, pp. 167–189.

Huyck, C., Verrucci, E., Bevington, J., 2014. Remote Sensing for Disaster Response: A Rapid, Image-Based Perspective. In: Shroder, J., Wyss, M. (Eds.), Earthquake Hazard, Risk, and Disasters. Elsevier, London, pp. 1−24.

Joyce, K., Wright, K., Samsonov, S., 2009. Remote sensing and the disaster management cycle. In: Jedlovec, G. (Ed.), Advances in geoscience and remote sensing, vol. 48(7). INTECH, pp. 317−346.

Kaya, S., Curran, P.J., Llewellyn, G., 2005. Post-earthquake building collapse: a comparison of government statistics and estimates derived from SPOT HRVIR data. Int. J. Remote Sens. 26 (13), 2731−2740.

Kemper, T., Jenerowicz, M., Pesaresi, M., Soille, P., March 2011. Enumeration of dwellings in Darfur camps from GeoEye-1 satellite image using mathematical morphology. IEEE J. Sel. Top. Appl. Earth Obs. Remote Sens. 4 (1), 8−15. Article number 5546897.

Kerle, N., 2010. Satellite-based damage mapping following the 2006 Indonesia earthquake—how accurate was it? Int. J. Appl. Earth Obs. Geoinf. 12 (6), 466−476.

Kerle, N., Hoffman, R.R., 2013. Collaborative damage mapping for emergency response: the role of cognitive systems engineering. Nat. Hazards Earth Syst. Sci. 13, 97−113.

Kosugi, Y., Sakamoto, M., Fukunishi, M., Lu, W., Doihara, T., Kakumoto, S., 2004. Urban change detection related to earthquakes using an adaptive nonlinear mapping of high-resolution images. IEEE Geosci. Remote Sens. Lett. 1 (3), 152−156.

Kranz, O., Gstaiger, V., Lang, S., Tiede, D., Zeug, G., Kemper, T., Vega Ezquieta, P., Clandillon, S., February 2−4, 2010. Different approaches for IDP camp analyses in West Darfur (Sudan) − a status report. In: 6th International Symposium on Geo-information for Disaster Management (Gi4DM), Torino, Italy.

Krieger, G., Hajnsek, I., Papathanassiou, K., Eineder, M., Younis, M., DeZan, F., Lopez-Dekker, P., Huber, S., Werner, M., Prats, P., Fiedler, H., Werninghaus, R., Freeman, A., Rosen, P., Hensley, S., Grafmüller, B., Bamler, R., Moreira, A., 2010. Tandem-L: a Mission for monitoring earth system dynamics with high resolution SAR interferometry. In: Proceedings of European Conference on Synthetic Aperture Radar (EUSAR). VDE Verlag GmbH, ISBN 978-3-8007-3272-2, pp. 506−509.

Li, M., Cheng, L., Gong, J., Liu, Y., Chen, Z., Li, F., Chen, G., Chen, D., Song, X., 2008. Post-earthquake assessment of building damage degree using LiDAR data and imagery. Sci. China Ser. E Technol. Sci. 51, 133−143.

Massonnet, D., et al., 1993. The displacement field of the Landers earthquake mapped by radar interferometry. Nature 364, 138−142.

Matsuoka, M., Yamazaki, F., November 18, 2010. Comparative analysis for detecting areas with building damage from several destructive earthquakes using satellite synthetic aperture radar images. J. Appl. Remote Sens. 4, 041867. http://dx.doi.org/10.1117/1.3525581.

Mueller, M., Segl, K., Heiden, U., Kaufmann, H., 2006. Potential of high-resolution satellite data in the context of vulnerability of buildings. Nat. Hazards 38, 247−258.

Müller, A., Reiter, J., Weiland, U., 2011. Assessment of urban vulnerability towards floods using an indicator based approach—a case study for Santiago de Chile. Nat. Hazards Earth Syst. Sci. 11, 2107−2123.

Munich Re, 2012. Review of Natural Catastrophes in 2011: Earthquake Result in Record Loss Year. Press release, 4th of January 2012. http://www.munichre.com/en/media_relations/press_releases/2012/2012_01_04_press_release.aspx.

Miyazaki, S., Heki, K., 2001. Crustal velocity field of southwest Japan: subduction and arc-arc collision. J. Geophys. Res. 106, 4305−4326.

Miyazaki, H., Shao, X., Iwao, K., Shibasaki, R., 2012. An automated method for global urban area mapping by integrating ASTER satellite images and GIS data. IEEE Journal of Selected Topics in Applied Earth Observations and Remote Sensing. http://dx.doi.org/10.1109/JSTARS.2012.2226563.

Open Data for Resilience Initiative (OpenDRI), 2013. Open Data for Resilience Initiative − Overview, Global Facility for Disaster Reduction and Recovery, the World Bank Group. Downloadable from. https://www.gfdrr.org/node/3116 (Last accessed 23.06.13).

Pesaresi, M., Blaes, X., Ehrlich, D., Ferri, S., Gueguen, L., Haag, F., Halkia, M., et al., 2012. *A Global Human Settlement Layer from Optical High Resolution Imagery*, European Union. p. 121. Publications Office of the European Union, 2012, Luxembourg.

Pesaresi, M., Ehrlich, D., Caravaggi, I., Kauffmann, M., Louvrier, C., 2011. Toward global automatic built-up area recognition using optical VHR imagery. *IEEE Journal of Selected Topics in Applied Earth Observations and Remote Sensing* 4 (4), 923−934.

Pesaresi, M., Guo, H., Blaes, X., Ehrlich, D., Ferri, S., Gueguen, L., et al., 2013. A global human settlement layer from optical HR/VHR RS data: concept and first results. IEEE J. Sel. Top. Appl. Earth Obs. Remote Sens. 6, 2102−2131.

Pittore, M., Wieland, M., 2013. Towards a rapid probabilistic seismic vulnerability assessment using satellite and ground-based remote sensing. Nat. Hazards 68 (1), 115−146. http://dx.doi.org/10.1007/s11069-012-0475-z.

Poli, P., Caravaggi, I., 2013. 3D modelling of large urban areas with stereo VHR satellite imagery − lessons learned. Nat. Hazards 68 (1), 53−78.

Potere, D., Schneider, A., 2009. Comparison of global urban maps. In: Gamba, P., Herold, M. (Eds.), Global Mapping of Human Settlements: Experiences, Data Sets, and Prospects. Taylor & Francis Group, pp. 269−308.

Schmitt, A., Wessel, B., Roth, A., 2010. Curvelet-based change detection on SAR images for natural disaster mapping. Photogramm. Fernerkundung Geoinform. 6, 463−474.

Schneiderbauer, S., Ehrlich, D., 2004. Risk, Hazard and People's Vulnerability to Natural Hazards. A Review of Definitions, Concepts and Data. Joint Research Centre, European Commission, EUR 21410.

SERTIT, 2012. SErvice Régional de Traitement d'Image et de Télédétection. Available at: http://sertit.u-strasbg.fr/english/en_welcome.htm (accessed 11.07.12.).

SERVIR, 2012. NASA Missions: SERVIR. Available at: http://www.nasa.gov/mission_pages/servir/index.html (accessed 11.07.12.).

Shafique, M., van der Meijde, M., van der Werff, H.M.A., 2012. Evaluation of remote sensing-based seismic site characterization using earthquake damage data. Terra Nova 24 (2), 123−129.

Shen, Z.-K., Sun, J., Zhang, P., Wan, Y., Wang, M., Bürgmann, R., Zeng, Y., Gan, W., Liao, H., Wang, Q., 2009. Slip maxima at fault junctions and rupturing of barriers during the 2008 Wenchuan earthquake. Nat. Geosci. 2, 718−724.

Simons, M., Minson, S.E., Sladen, A., Ortega, F., Jiang, J., Owen, S.E., Meng, L., Ampuero, J.-P., Wei, S., Chu, R., Helmberger, D.V., Kanamori, H., Hetland, E., Moore, A.W., Webb, F.H., 2011. The 2011 magnitude 9.0 Tohoku-Oki earthquake: mosaicking the megathrust from seconds to centuries. Science 332, 1421. http://dx.doi.org/10.1126/science.1206731.

Smith, S.W., Wyss, M., 1968. Displacement of the San Andreas fault subsequent to the 1968 Parkfield earthquake. In: Seismological Society of America Bulletin 58 (6), 1955−1973.

Stramondo, S., Bignami, C., Chini, M., Pierdicca, N., Tertulliani, A., 2006. Satellite radar and optical remote sensing for earthquake damage detection: results from different case studies. Int. J. Remote Sens. 27 (20), 4433−4447.

Stramondo, S., Saroli, M., Tolomei, C., Moro, M., Doumaz, F., Pesci, A., Loddo, F., Baldi, P., Boschi, E., 2007. Surface movements in Bologna (Po Plain-Italy) detected by multitemporal DInSAR. Remote Sens. Environ. 110, 304−316.

Suresh, G., Minet, C., Eineder, M., Parizzi, A., Yague-Martinez, N., 2011. Haiti 2010 earthquake: a 3D deformation analysis. In: Proceedings of the ESA Fringe Workshop 2011, pp. 1−6, Frascati, Italy.

Taubenböck, H., Esch, T., Felbier, A., Wiesner, M., Roth, A., Dech, S., 2012. Monitoring of mega cities from space. Remote Sensing of Environment 117, 162−176.

Taubenböck, H., Goseberg, N., Setiadi, N., Lämmel, G., Moder, F., Oczipka, M., Klüpfel, H., Wahl, R., Schlurmann, T., Strunz, G., Birkmann, J., Nagel, K., Siegert, F., Lehmann, F., Dech, S., Gress, A., Klein, R., 2009a. "Last-Mile" preparation for a potential disaster—interdisciplinary approach towards tsunami early warning and an evacuation information system for the coastal city of Padang, Indonesia. Nat. Hazards Earth Syst. Sci. 9, 1509−1528.

Taubenböck, H., Roth, A., Dech, S., Mehl, H., Münich, J.C., Stempniewski, L., Zschau, J., 2009b. Assessing building vulnerability using synergistically remote sensing and civil engineering. In: Kreck, A., Rumor, M., Zlatanova, S., Fendel, E. (Eds.), Urban and Regional Data Management. Taylor & Francis Group, London, pp. 287−300.

Taubenböck, H., Post, J., Roth, A., Zosseder, K., Strunz, G., Dech, S., 2008. A conceptual vulnerability and risk framework as outline to identify capabilities of remote sensing. Nat. Hazards Earth Syst. Sci. 8, 409−420.

Taubenböck, H., Roth, A., Dech, S., 2007. Linking structural urban characteristics derived from high resolution satellite data to population distribution. In: Coors, V., Rumor, M., Fendel, E., Zlatanova, S. (Eds.), Urban and Regional Data Management. Taylor & Francis Group, London, pp. 35−45.

Taubenböck, H., Wiesner, M., Felbier, A., Marconcini, M., Esch, T., Dech, S., 2014. New dimensions of urban landscapes: the spatio-temporal evolution from a polynuclei area to a mega-region based on remote sensing data. Appl. Geogr. 47, 137−153.

Taubenböck, H., Esch, T., Felbier, A., Roth, A., Dech, S., 2011. Pattern-based accuracy assessment of an urban footprint classification using TerraSAR-X data. IEEE Geosci. Remote Sen. Lett. 8 (2), 278−282.

Tiede, D., Fuereder, P., Lang, S., Hoelbling, D., Zeil, P., 2013. Automated analysis of satellite imagery to provide information products for Humanitarian relief operations in refugee camps from scientific development towards operational services. Photogramm. Fernerkundung Geoinform. 2013 (3), 185−195.

Tralli, D.M., Blom, R.G., Zlotnicki, V., Donnellan, A., Evans, D.E., 2005. Satellite remote sensing of earthquake, volcano, flood, landslide and coastal inundation hazards. ISPRS J. Photogram. Remote Sens. 59, 185−198.

Trianni, G., Gamba, P., 2009. Fast damage mapping in case of earthquakes using multitemporal SAR data. J. Real-Time Image Proc. 4, 195−203.

Tronin, A.A., 2010. Satellite remote sensing in seismology. A review. Remote Sens. 2, 124−150.

Turker, M., Sumer, E., 2008. Building-based damage detection due to earthquake using the watershed segmentation of the post-event aerial images. Int. J. Remote Sens. 29 (11), 3073−3089.

UN/ISDR, 2004. Living with Risk: A Global Review of Disaster Reduction Initiatives. United Nations/International.

UNOSAT, 2011. Unitar's Operational Satellite Applications Programme. Available at: http://www.unitar.org/unosat/ (accessed 19.09.11.), Strategy for Disaster Reduction, Geneva, Switzerland, UN Publications.

Voigt, S., Kemper, T., Riedlinger, T., Kiefl, R., Scholte, K., Mehl, H., 2007. Satellite image analysis for disaster and crisis-management support. IEEE Trans. Geosci. Remote Sens. 45 (6), 1520−1528.

Voigt, S., Schneiderhan, T., Twele, A., Gähler, M., Stein, E., Mehl, H., 2011. Rapid damage assessment and situation mapping: learning from the 2010 Haiti earthquake. Photogramm. Eng. Remote Sens. ISSN: 0099-1112 77 (9), 923−931. American Society for Photogrammetry and Remote Sensing.

Vu, T.T., Ban, Y., 2010. Context-based mapping of damaged buildings from high-resolution optical satellite images. Int. J. Remote Sens. 31 (13), 3411−3425.

Vu, T.T., Matsuoka, M., Yamazaki, F., 2004. LiDAR based change detection of buildings in dense urban areas. In: IEEE International Geoscience and Remote Sensing Symposium, September 2004, Anchorage, AK, USA, pp. 3413−3416.

Wegscheider, S., Schneiderhan, T., Mager, A., Zwenzner, H., Post, J., Strunz, G., 2013. Rapid mapping in support of emergency response after earthquake events. Nat. Hazards 68 (1), 181−196.

Wieland, M., Pittore, M., Parolai, S., Zschau, J., Moldobekov, B., Begaliev, U., 2012a. Estimating building inventory for rapid seismic vulnerability assessment: towards an integrated approach based on multi-source imaging. Soil Dyn. Earthquake Eng. 36, 70−83.

Wieland, M., Pittore, M., Parolai, S., Zschau, J., 2012b. Exposure estimation from multi-resolution optical satellite imagery for seismic risk assessment. ISPRS Int. J. Geoinform. 1, 69−88.

Wieland, M., Pittore, M., 2012. Remote sensing and omnidirectional imaging for efficient building inventory data capturing: application within the Earthquake Model Central Asia. In: Proceedings of the IEEE International Geoscience and Remote Sensing Symposium, Munich, 22−27.07.2012.

Wright, T.J., Parsons, B.E., Lu, Z., 2004. Toward mapping surface deformation in three dimensions using InSAR. Geophys. Res. Lett. 31. http://dx.doi.org/10.1029/2003GL018827.

World Housing Encyclopedia (Online). http://www.world-housing.net (accessed 14.04.11.).

Wurm, M., Taubenböck, H., Roth, A., Dech, S., 2009. Urban structuring using multisensoral remote sensing data. In: Proc. 2009 Urban Remote Sensing Joint Event (2009), Shanghai, China.

Wurm, M., Taubenböck, H., Schardt, M., Esch, T., Dech, S, 2011. Object-based image information fusion using multisensor earth observation data over urban areas. In: International Journal of Image and Data Fusion 2 (2), 121−147.

Wyss, M., Elashvili, M., Jorjiashvili, N., Javakhishvili, Z., 2011. Uncertainties in teleseismic epicenter estimates: implications for real-time loss estimate. Bull. Seismol. Soc. Amer. 101 (3), 1152−1161.

Yamagiwa, A., Hatanaka, Y., Atanaka, Y., Yutsudo, T., Miyahara, B., 2006. Real-time capability of GEONET system and its application to crust monitoring. Geospatial Inf. Auth. Jpn. Bull. 53-4.

Yong, A., Hough, S.E., Abrams, M.J., Cox, H.M., Wills, C.J., Simila, G.W., 2008. Site characterization using integrated imaging analysis methods on satellite data of the Islamabad, Pakistan, Region. Bull. Seismol. Soc. Am. 98 (6), 2679−2693.

Zeng, J., Zhu, Z.Y., Zhang, J.L., Ouyang, T.P., Qiu, S.F., Zou, Y., Zeng, T., 2012. Social vulnerability assessment of natural hazards on county-scale using high spatial resolution satellite imagery: a case study in the Luogang district of Guangzhou. South China Environ. Earth Sci. 65, 173−182.

Chapter 3

Disaster-Risk Reduction through the Training of Masons and Public Information Campaigns: Experience of SDC's "Competence Centre for Reconstruction" in Haiti

Tom Schacher

Swiss Agency for Development and Cooperation (SDC)'s Corps for Humanitarian Aid, Switzerland

ABSTRACT

Although the training of civil engineers in earthquake-resistant construction techniques has made much progress over the past decades, its impact on the overall safety of the rapidly growing urban areas in poor countries is often limited. Huge parts of these cities are built by small-scale contractors and private owners with no knowledge of seismic construction techniques and little to no involvement of engineers. In order to address the objective of disaster-risk reduction effectively, the training of masons and awareness raising among the population are of the utmost importance. Based on experience in Pakistan in the aftermath of the 2005 earthquake, in Haiti, the Swiss Agency for Development and Cooperation decided to create a Competence Centre for Reconstruction to achieve more disaster-resistant reconstruction through the training of masons and awareness building. This Haitian experience will be presented in detail in this chapter.

3.1 INTRODUCTION

Today, more than half of the world's population lives in urban agglomerations. A large proportion lives in the megacities of developing nations, many of which are located in areas at risk of natural disasters.

Many cities in developing countries possess a relatively small high-rise city center and a huge periphery of low-rise, one-to-three story buildings.

Earthquake Hazard, Risk, and Disasters. http://dx.doi.org/10.1016/B978-0-12-394848-9.00003-1

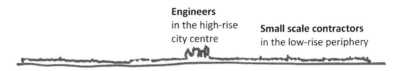

FIGURE 3.1 **Conceptual cross-section of cities in poor nations; qualified engineers prefer the lucrative high-rise city centers.**

Structural engineers are involved in the design of the (financially) more interesting high-rise buildings, whereas "normal" low-rise buildings are left to small-scale contractors without the involvement of costly engineering services (Figure 3.1). Even when engineers do become involved in the design of low-rise buildings, they usually consider the job done when construction drawings leave their offices. Proper implementation is the responsibility of the contractor.

Over the past decades, many praiseworthy efforts have gone into the training of engineers in earthquake-resistant construction techniques, in both rich and poor countries. Contractors and workers on the other hand, particularly in poor nations, have received little help. They are left on their own to seek training, mostly within their families or neighborhoods, very rarely in professional training schools.

If the low-rise buildings where most people live are left to the contractors alone, how can the seismic know-how of engineers have any influence on the end product? Equally, if engineers are not involved in the construction process, and contractors and workers do not know about earthquake-resistant building techniques, how can the vulnerability of the common urban building stock ever be reduced? (see also Dixit et al., 2014 in this book).

Postdisaster reconstruction phases provide a unique opportunity for reviewing and revising building habits and construction standards. Humanitarian aid operations can play a major role by ensuring the use of more disaster-resistant construction methods. For this to happen, the aid agencies would need funding that enables them to stay and work for a prolonged period of time rather than concentrating solely on victims' immediate needs and the provision of temporary shelters. Financial donors too frequently insist on quick results and limit their funding to a one-year postdisaster period, far too short for aid organizations to involve themselves in serious reconstruction work.

The Swiss Humanitarian Aid (SHA) as a government agency suffers much less from such constraints on its budget. Moreover since the SHA is physically present in the field with its own construction experts, it is able to promote better building methods through actual constructions and through training programs.

The following pages show the efforts of the Swiss Agency for Development and Cooperation (SDC) to improve construction standards in Haiti through training activities and public information campaigns.

3.2 CONTEXT

The earthquake of January 12, 2010, caused unprecedented devastation in Haiti. United Nation (UN) sources put the death toll at >230,000, with a further 300,000 injured. Nearly 1.5 million people lost their homes, and many are still living in temporary shelters.

The SHA, which is part of the Swiss Ministry of Foreign Affairs' Agency for Development and Cooperation (SDC), was active in Haiti for a number of years prior to the earthquake, supporting the school construction programs of nongovernmental organizations (NGOs). After a humanitarian response that ranks as the biggest in its history, the SHA decided on a two-tier follow-up program: the reconstruction of earthquake-resistant schools and the promotion of similar construction methods through the training of workers involved in the reconstruction process.

SDC being a government organization, any program would have to involve close cooperation with the Haitian authorities and the international community. To counterbalance the tendency of international aid organizations to focus on the most accessible areas, in this case, the capital Port-au-Prince, SDC opted for a decentralized approach. Both the school construction and the training program were to take place at the periphery of the earthquake-affected area. (see Huyck et al., 2014, Figure 7, in this book).

In July 2010, the SHA strengthened the Cooperation Office in Port-au-Prince with the addition of a Competence Centre for Reconstruction (CCR). The aim is to improve project quality and augment local capacities by providing technical and methodological support to the reconstruction efforts of SDC and its partners. Specifically, the CCR helps to improve the skills of construction workers by developing training modules in collaboration with the National Vocational Training Institute. The CCR also supports the Ministry of Public Works in developing and disseminating appropriate building techniques, improving general public awareness of earthquake and hurricane-resistant building methods.

3.3 IDENTIFICATION OF THE MOST APPROPRIATE CONSTRUCTION TECHNIQUE

Although there are many ways to improve earthquake resistance, most of the methods are beyond the financial and technical capacities of poor nations, particularly when it comes to ordinary low-rise housing. The choice of the most appropriate technique depends on a variety of factors: the availability and cost of construction materials, the know-how of local builders, and last but not least, government approval.

In Haiti where the earthquake struck an urban area built of modern materials such as concrete blocks, reinforced concrete frames, and full concrete slabs, the earthquake-resistant method would have to be based on

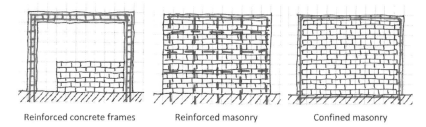

Reinforced concrete frames   Reinforced masonry   Confined masonry

FIGURE 3.2 **Three types of common earthquake-resistant structures (in red: steel reinforcement).**

precisely these materials. The use of timber for structural reinforcement at the beginning of the twentieth century in so-called "gingerbread houses", had to be excluded due to the extent of deforestation in Haiti and the prohibitive cost of importing timber of the insect- and fungus-resistant type used a century ago. In any case, construction workers no longer have the relevant know-how and would require disproportionate training efforts. Steel construction with lightweight insulation panels, as proposed by some companies, was also out of question, as it could not meet the need for simple, economical housing for the general population.

For cement-based techniques, the choices are limited: *unreinforced cement-block* masonry is not earthquake resistant and *reinforced-concrete frames* are too difficult and expensive to build properly (Charleson, 2003) with the additional problem of adding walls that would not interfere with structural behavior. As for two of the best-known earthquake-resistant building techniques—*reinforced masonry* and *confined masonry*—the former is difficult to implement correctly without close full-time supervision[1], whereas the second method is much more promising, being well known in Latin America as well as in India and in other parts of the world (Figure 3.2).

In our first contact with the Haitian Ministry of Public Works to discuss the introduction and promotion of *confined masonry* as the official housing reconstruction standard, the director in charge presented us with a draft guideline developed by a Canadian engineering company using an identical method. To our surprise, the illustrations used in the manual had been taken directly from our own manuals prepared some years before in Pakistan and published in India (Schacher, 2009) as well as from the reference manual on "confined masonry" prepared by our colleagues in Peru (Blondet, 2005). When the director in charge realized that much of his new manual was based on our own work, all doors opened and our collaboration was assured.

---

1. Personal communication from Marcial Blondet, Pontificia Universidad Católica del Perú (PUCP), Lima.

Even more important, *confined masonry* is not unknown in Haiti. Its execution however is typically poor and ignores the basic principles of earthquake-resistant construction. Introducing an official housing reconstruction standard would be of little value unless accompanied by widespread training and a public information campaign.

3.4 IDENTIFICATION OF PARTNERS FOR THE TRAINING OF MASONS

Many organizations involved in reconstruction offer "training", that is, they train workers to build shelters, or more rarely houses, exactly the way they are planned. In most cases, training of this kind is of no further use beyond this particular situation.

Training masons in earthquake-resistant construction at a national level requires a more thorough approach. In Haiti as in many other countries, masons are low on the social scale. "On-the-job" training tends to be provided by the family or neighbors. At best, professional training is offered by a very small number of private institutions on a local basis. Masonry training is not part of the curricula offered by national entities such as the Institut National de la Formation Professionnelle (INFP).

Since SDC's initiative focuses as much on the training of workers as on the promotion of a national seismic construction standard, it was imperative to work with a partner in the national education system who not only would ensure promotion of the same technical standard all over the country but would also ensure its lasting implementation and nationwide certification of the masons thus trained.

The National Institute for Professional Training (INFP) was selected as SDC's main partner and a training program developed in accordance with the administration's certification requirements. The Swiss Reconstruction Competence Centre prepared the training content and defined the setup, and INFP provided a school building in Petit Goâve, about 2 h west of the Port-au-Prince city center, and two additional trainers, also accepting responsibility for the trainees' final examinations.

The second institutional partner is the Ministry of Public Works (Ministère des Travaux Publiques, de Transport et de Communications; MTPTC), which guarantees the technical correctness of the training content.

Another partner, at some point, was the International Labor Organization (ILO) for whom further training material was developed to meet the needs of the workforce involved in the neighborhood reconstruction program of the capital.

The Red Cross societies from various countries, NGOs, and private national training entities were also important partners, providing trainees and materials for small houses to be built by the trainees, and which eventually asked the CCR to train their own trainers (Figure 3.3).

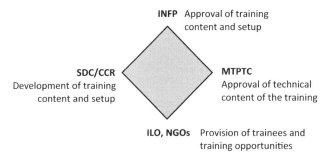

FIGURE 3.3 **Partners and their roles.**

3.5 DEVELOPMENT OF TRAINING CONTENT AND TRAINING SETUP

In the absence of preexisting masonry training, the INFP asked SDC to develop a permanent national multiyear training program. This was beyond the scope and abilities of the SDC/SHA as a humanitarian aid organization whose contribution is normally limited to a time frame of one to four years, and whose technical assistance tends to focus on reconstruction and a reduction of vulnerability to disasters. The development of full-scale, professional training programs, and the strengthening of local institutions over the long term, would be a job for the development branch of the SDC.

Given the limited time frame it was decided to focus efforts entirely on training in the confined-masonry technique and wherever necessary to include basic masonry training. In accordance with the INFP training style, a school-based program was chosen with theory lessons in classrooms, plus practical exercises.

Four trainers were recruited—an engineer and a technician by SDC and two technicians by the INFP—and familiarized with the training material. The classroom was equipped with chairs and a blackboard together with a generator, fans, a computer, and a digital projector; External space was hard surfaced for practical exercises, and the recruitment of trainees was begun. A small inscription fee was requested to cover the lunch provided. Within a few weeks, >100 candidates had been recruited, and these were divided into groups of 20–25 trainees.

The theory partook the form of 16 PowerPoint lessons (with annotations for the trainers) with 16 posters showing all the slides used previously, allowing participants to discuss specific issues after the lessons (Figure 3.4). These were much appreciated.

An additional manual with 100 practical exercises was provided to allow the trainers to go beyond conventional teaching methods. Three-dimensional demonstration items prepared for the lessons, being absolutely new to them, proved difficult to remember at the appropriate moment in the lesson (Figure 3.5).

FIGURE 3.4   **PowerPoint lessons with teacher manuals and posters.**

FIGURE 3.5   **3D demonstration items for lessons and practical outdoor exercises.**

The training course lasted 6 weeks, including a week for the final examinations. The practical part of the examination was conducted in teams of four to five persons, building on a test piece in the schoolyard. Theoretical know-how was tested orally and with the help of illustrations on which the students were to comment (Figure 3.6). Despite the initial reluctance of the INFP, written examinations were avoided that would have heavily disadvantaged masons with little or no formal education. This new approach, which was more suitable for adults with a variety of schooling backgrounds, worked very well.

The participation of trainees in classroom theory lessons was unexpectedly lively, due no doubt, to the ability of the head trainer. Some students even took notes during the classroom lessons. Participation in practical exercises was less consistent but acceptable (it is hard work to carry around concrete blocks!). The practical part of the final examinations showed that not all trainees in the first intake had been genuine masons. It was discovered

FIGURE 3.6   **Practical and theoretical examinations with illustrations to facilitate comments.**

that, as well as youngsters with no work experience, the trainees included a lawyer and a cook! Admission tests helped to screen the second intake of trainees so as to limit training to persons with a background in construction work.

3.6 ADAPTATION OF THE TRAINING TO VARYING SITUATIONS

With time, it was discovered that participants had difficulties in remembering even simple but important construction details, despite active participation in the classroom lessons. Trainees were not used to learning in a school environment through their eyes and ears, preferring hands-on on practical exercises. A Swiss NGO active in the construction of small confined-masonry houses for local earthquake victims agreed to provide houses to be rebuilt by trainees, hoping to be able to hire the workers so trained for the construction of additional houses at a later stage.

3.6.1 Masonry Training on Building Sites

The training program was therefore revised with Mondays dedicated to theory lessons in the classroom and the rest of the week dedicated to real work on building sites (Figure 3.7). The course duration was extended by one week, six days for the training and one extra for the examinations.

Additional training material was developed to bring theoretical knowledge to the building site: Posters with all relevant construction details were made in such a way that they could be attached to a string on the construction site and changed according to work progress (Figure 3.8). The same illustrations were collected in a small, pocked-size manual that the masons could carry with them and take home.

3.6.2 Training of Trainers with "the Monument"

Various professional training institutions were keen to include the confined-masonry technique in their curricula. They agreed to send teachers to a shortened version of the training course, being unable to accept the absence of

FIGURE 3.7   **Training on building sites.**

  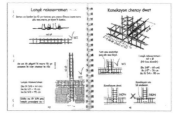

FIGURE 3.8   Training material: posters for the building site and manuals for the trainees.

FIGURE 3.9   "The monument": a full-scale example of the confined-masonry method.

teachers for seven weeks. Since the new trainees were professionals, a duration of only five weeks seemed feasible. However, experience has shown that it is not wise to forego the practical part of training.

It was decided to use the so-called "monument". The original idea was to create a full-scale example of confined-masonry construction as a demonstration tool, to be located in a public garden so that passers-by could see how earthquake-resistant construction should be done. Unable to find the time to present the idea to the authorities, it was decided just to build one next to the classroom for training purposes (Figure 3.9). When this first "monument" had to be demolished due to construction work at the school its loss seriously hampered training sessions. However, given the compelling need for practical training, it was rebuilt and has since been replicated in the training centers of partner institutions.

From April 2011 to May 2013, there have been 19 training sessions, and nearly 500 masons have been certified by the INFP, while 50 trainers are now on hand to provide masonry training in various locations.

3.7   A PUBLIC INFORMATION CAMPAIGN TO ACCOMPANY THE TRAINING

What is the use of training masons if no one will hire them? Having learned that doing the job properly requires materials of the right quality in sufficient quantity, properly trained masons might cause costs to rise. Why hire people who insist on more cement with quality aggregates when others are willing to work with half the amount and cheaper materials?

Only a public information campaign can provide the answers to these questions. Months after their training, it was found that masons often failed to use their know-how on actual building sites. When asked why, they replied that the contractor would not allow them to use suitable amounts of water, cement, or steel. So it was back to "normal". If even engineers, who often own the construction companies, do not see the advantage of working with trained masons, what hope is there for the general public to understand?

The first effort to reach the public was to produce calendars illustrating confined-masonry construction step by step. This was done over three consecutive years, varying illustrations to reach a wide spectrum of the public but always with the same message. The calendars (Figure 3.10) were a huge success, tens of thousands being distributed by various organizations including the UN Agencies United Nations Office for Project Services (UNOPS) and United Nations Stabilization Mission in Haiti (MINUSTAH).

The calendars were eventually transformed into pocket-sized booklets to be distributed as *aide-memoires* to the masons during training.

Together with the Ministry of Public Works and other partners such as UN Habitat it was decided to create a communication platform and launch a widespread information campaign using different media. A first action consisted of a series of billboard posters placed in key locations throughout the city over a period of several months (Figure 3.11). Each billboard showed a key aspect of the new construction method under a Ministry of Public Works heading. Haitians expect little from their government, so this provided the authorities with a welcome opportunity to show that they cared.

FIGURE 3.10 **Calendars showing construction details in a simple way, and an engineer explaining the illustrations to the public.**

FIGURE 3.11   **Billboard campaign in key city locations.**

A series of A3- and A2-sized posters in the same graphic style were placed on neighborhood walls (like election posters), or in NGO and administration offices (Figure 3.12). Each series focused on a particular aspect of construction that is often done in a way that is below standard.

The public was so enthusiastic about the billboards and posters that they were transformed into wall paintings, beating us to it and anticipating the next step in the campaign (Figure 3.13).

Three further ideas were developed:

- T-shirts with different messages to be distributed to the trainees and on the reconstruction sites of the 16/6 program by the ILO, a communication project that is yet to materialize.
- Two television (TV) spots to be broadcast by public and private TV stations. The original spots have been reworked to include suggestions from the MTPTC and members of the communication platform. However, as the

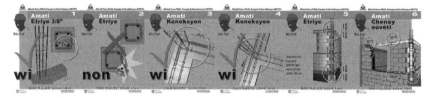

FIGURE 3.12   **Poster series showing good and bad construction detailing.**

FIGURE 3.13   **Technical posters transformed into wall paintings by the people.**

FIGURE 3.14 Various communication projects: T-shirts, TV spots, and instructions on cement bags.

project was developed without prior consultation with potential TV stations, the spots have not yet been aired.

- "Instructions for use" to be printed on every cement bag sold in the country. It seems odd that instructions are printed on every can of beans and not on cement bags of such vital use to the public. The MTPTC contacted the cement producers, who were happy to print the instructions on all new bags (Figure 3.14).

3.8 CONCLUSION AND LESSONS TO BE LEARNED

The training of construction workers and small-scale contractors, as well as awareness building for the general population, need to be viewed as genuine Disaster-Risk Reduction (DRR) activities on a par with other more traditional DRR activities.

As previously in Pakistan so in Haiti, experience has shown that such efforts are cost effective. A few foreign experts can have a major lasting impact, ensuring safer local construction. To be effective, the training and information materials need to be prepared well in advance. Experts must be chosen carefully to ensure that as well as technical proficiency (ranging from construction to graphics), they also have cultural sensitivity. Ideally, they should receive ongoing training and be incorporated in a permanent competence center.

3.8.1 Readymade Training Material

The decision to make the training of masons the backbone of the Reconstruction Competence Centre program in Haiti was inspired by experience in Pakistan 4 years earlier when SDC, together with the national Earthquake Reconstruction and Rehabilitation Authority and UN Habitat, was actively involved in the promotion and training of the population in earthquake-resistant reconstruction methods (Schacher, 2008). This made it possible to use existing materials and to build on the experience and good reputation acquired at the time in Pakistan.

The availability of proven training material, which only needs adapting to the new context rather than being newly created, is of the utmost importance. It is difficult to convince authorities of the need for training programs and information campaigns and their feasibility without the appropriate documentation from prior experience. Yet, it is difficult to find funds to prepare such documents before a new disaster occurs, since funds are only made available after disasters occur.

A permanent (re) construction competence center attached to a university would be the ideal place for the development of training and information material on construction techniques appropriate to different disaster prone countries. It would have to be staffed by people willing to leave for extended missions and could be financed through funds from permanent disaster-risk reduction initiatives.

3.8.2 Adaptation of the Material to Local Needs

Given the urban context in Haiti, the focus of the CCR activity in the first three years was on the promotion of a single construction method, the confined-masonry technique. This made it possible to fine-tune the training methods and spend more time in the all-important campaign to inform the general public.

Building materials, the competence of builders, and the structure of the building trade, can vary significantly from country to country. In Haiti, most small-scale construction companies consist of a single "boss" who knows where to find workers and tools, of which he possesses very few. He might not even know how to read drawings but is usually eager to learn. In other countries, even small companies might have a permanent core staff. But the workers might be very nationalistic and unwilling to accept training materials with illustrations representing people of a different culture (skin color, dress code, gender issues). Construction material also differs. In one country, the preferred material may be burnt bricks, while in another, concrete blocks might be the only material available (Carlevaro, 2013).

Illustrations therefore have to be adapted and documents translated into a local language. This takes not only time (including time to understand a minimum of the local context and culture) but also expertise, both in construction techniques and in the handling of graphical and layouting software.

The time available for aid missions should be used to adapt available materials rather than produce new ones. This is obvious in the medical domain, so why not in disaster-risk reduction activities? Experts should be "broad-spectrum", technical persons able to perceive and understand local cultural specificities and adapt the materials accordingly.

3.8.3 Identification of Partners

Training and information activities are only effective when done in collaboration with local partners who have a stake in the program and will ensure that the training and information reaches the target audience. Partners are needed at all levels, from the authorities who will ensure the official approval and mainstreaming of the technique, to local organizations to ensure implementation. These have the advantage of speaking the local language, knowing the priorities of their "clients", and most importantly, will always be available to monitor and guide the gradual cultural changes required for safer construction techniques.

The choice should fall on partners, public or private, who are already active in the field and have a strong institutional presence. They must be well known to and be trusted by the people. Having several partners for different aspects of the program can be an advantage in case one partner wishes to pull out of the program.

3.8.4 Training Methods

Masonry training should be as hands-on as possible. Theory is necessary but should be delivered in small doses, at a convenient time and in an appropriate manner. PowerPoint sessions should be limited to the training of trainers, except for short sessions showing masons the desired end result, or how things are done elsewhere. Local people often find it difficult to imagine such things, never having seen them before.

3.8.5 Allocation of a Correct Time Frame

The introduction of safer construction methods requires a cultural change, and a cultural change requires time. All concerned need time: for the trainees and the local population to understand and integrate new information, and for foreign experts to learn about local needs and local priorities.

An active presence in the country of several years has to be planned for. The more conventional short-term consulting missions are often not only a waste of time and money but they may also be perceived by local partners as based on arrogant "know-it-all" behavior. If well planned, the cost of long-term missions need not be disproportionate compared to that of short-term efforts that are personnel intensive.

REFERENCES

Blondet, M., 2005. Construction and Maintenance of Masonry Houses, for Masons and Craftsmen. Pontificia Universidad Católica del Perú, Lima.

Carlevaro, N., 2013. Rapport de passation: Volet Information et Volet Formation (End of mission handover report on training and information activities), Internal document, CCR, SDC-Haiti.

Charleson, A., 2003. Should reinforced concrete frames be banned? In: Earthquake Hazard Newsletter, vol. 7, No. 2. Victoria University of Wellington, New Zealand.

Dixit, A., et al., 2014. How to render schools safe in developing countries? In: Shroder, J., Wyss, M. (Eds.), Earthquake Hazard, Risk and Disasters. Elsevier, London, pp. 183–202.

Huyck, C., et al., 2014. Remote Sensing for disaster response: a rapid, image-based perspective. In: Shroder, J., Wyss, M. (Eds.), Earthquake Hazard, Risk and Disasters. Elsevier, London, pp. 1–24.

Schacher, T., 2008. Good Engineering Without Appropriate Communication Doesn't Lead to Seismic Risk Reduction: Some Thoughts about Appropriate Knowledge Transfer Tools, 14th World Conference on Earthquake Engineering. Beijing, China.

Schacher, T., 2009. Confined Masonry for One and Two Storey Buildings in Low-tech Environments: A Guidebook for Technicians and Artisans. NICEE, Indian Institute of Technology, Kanpur, India.

Chapter 4

The Most Useful Countermeasure Against Giant Earthquakes and Tsunamis— What We Learned From Interviews of 164 Tsunami Survivors

Mizuho Ishida [1] and Masataka Ando [2]

[1] *Earthquake and Tsunami Research Project for Disaster Prevention, JAMSTEC, Japan,* [2] *Institute of Earth Sciences, Academia Sinica, Taiwan*

ABSTRACT

In order to clarify people's behavior after the March 11, 2011, Tohoku-Oki earthquake (M_w 9.0), we interviewed 164 tsunami survivors in 10 cities mainly in Iwate and Miyagi prefectures and also city officials in charge of the disaster-prevention measures in Miyako and Kamaishi cities, Iwate prefecture, Tohoku region. During the 150 years prior to this event, three large tsunamis have struck the Tohoku region. Ever since these tsunamis, disaster mitigation efforts have been undertaken in the region, for example, constructions of the huge breakwaters, practicing of earthquake early warning schemes, holding periodical tsunami evacuation drills, and publicizing geohazard maps. Despite these efforts, 18,554 deaths and missing persons were reported by the National Police Headquarters on June 10, 2013. Our interviews revealed that only around 30 percent of interviewed people escaped immediately after the shaking of seismic ground motion subsided. Therefore, we searched through the interviews as to why so many people were slow or reluctant to evacuate.

We presupposed possible reasons for their slow reaction as follows: (1) the magnitude of the earthquakes expected in the offshore Tohoku region publicized by the governmental committee was much smaller than that of the 2011 Tohoku-oki earthquake; (2) the earthquake magnitude and tsunami height of the first warning issued by the Japan Meteorological Agency were much smaller than those of the actual event; (3) people living there made wrong conclusions or biased guesses of the magnitude of the tsunami height based on their experience of past tsunamis or on what they heard from their parents or grandparents; (4) many people had too much confidence in the breakwaters,

Earthquake Hazard, Risk, and Disasters. http://dx.doi.org/10.1016/B978-0-12-394848-9.00004-3

which were known to be the sturdiest constructions in the world. From our interview, we found that the reasons of (1) and (2) above are not real causes of their slow action. Reasons (3) and (4) together may have been the real causes of people's slow action. On the other hand, we found that very few schoolchildren (~ 1.8 percent of deaths and missing persons, although they constitute 8.5 percent of the population) lost their lives at schools in Iwate, Miyagi, and Fukushima prefectures. This indicates that the role of elementary schools should be reassessed in their role as emergency shelters, as well as the function of education and education about the earthquakes and tsunami hazards and natural hazards in general in the local communities.

4.1 INTRODUCTION

On March 11, 2011, a catastrophic earthquake, Tohoku-Oki earthquake (M_w 9.0), ruptured along almost the entire offshore region of the northeast coast of Japan. The magnitude of the main shock is the largest in Japan since 1868 (the Meiji era) when modern seismic observations started. The people living along the Pacific coast of the Tohoku district have often suffered from large earthquakes that have occurred near the northeastern coast of Japan. The general aspects of the damaging historical earthquakes in Tohoku for the past approximately 400 years have been documented by local knowledgeable people. Recent studies of the geological evidence of traces left by the earthquakes and tsunamis could be identified as an approximately 1,000-year-old tsunami event in the region, but the details of the damage could be quantified or estimated only for the events in the past 150 years or so (Usami, 2003).

During these 150 years, three large tsunamis struck the Tohoku region in 1896, 1933, and 1960. The death tolls from the 1896, 1933, and 1960 tsunamis amounted to 22,915, 3,064, and 110 along the Tohoku coast, respectively (Usami, 2003; Yamashita, 2005). By the tsunami generated in the 1896 earthquake, the broad area of the coastal region of the Tohoku district suffered serious damage. Prior to this earthquake, no general term existed to name this coastal region, and it had been referred to by its local town or county name. Afterward the word "Sanriku (meaning three counties)" came to be used before long, and the whole coastal region of the Tohoku district came to be called the "Sanriku Coast" (Yonechi and Imaizumi, 1994). The common names of the 1896 earthquake (M_w 8.3) and the 1933 earthquake (M_w 8.1) are "the Meiji Sanriku Earthquake Tsunami" and "the Sanriku Earthquake Tsunami", respectively, and the 1960 tsunami (M_w 9.5) was named "the Chile Earthquake Tsunami" by the Japan Meteorological Agency (JMA). For people who live in the northeastern Pacific coastal area, the three earthquakes became a story to be memorized as the earthquakes that caused severe tsunami damage to the Sanriku coast. Learned from the experience, mitigation efforts have been undertaken in the region by national and regional governments, for example, periodical tsunami evacuation drills, updates of regional geohazard maps, practicing of earthquake early warning (EEW) schemes, and constructions of sea embankments (Figure 4.1(e)), and breakwaters at the mouth of the bay.

Despite these long-term efforts, 18,554 deaths and missing persons due to the 2011 Tohoku-Oki earthquake were reported by the National Police Headquarters, Japan (as of June 10, 2013). We thought many lives could have been saved if people took appropriate action immediately after the strong ground shaking stopped. This could have been the case, because the tsunami arrived at the coast >25–30 min later after the shaking stopped and safe highlands could be reached within about 10–20 min on foot.

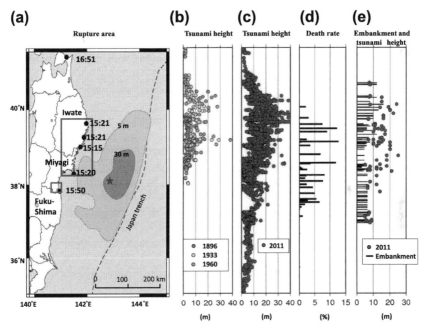

FIGURE 4.1   (a) The rupture area of the March 11, 2011, Tohoku-oki earthquake, including 5- and 30-m fault slip contours (Lee et al., 2011). The epicenter of the main shock (star) and the areas in which people were interviewed (rectangles) are shown. The arrival times of the large tsunami waves at the tide gauge stations (black circles) are shown as the local time (UT + 9 h). These waves arrived at inner-bay towns 3–5 min later. (b) Inundation heights (meters) along the Tohoku coast for the June 16, 1896, M_w 8.5, March 3, 1933, M_w 8.4 and May 23, 1960, M_w 9.5 (Chile) tsunamis (Usami, 2003). (c) The wave heights of the tsunami following the March 11, 2011, M_w 9.0 earthquake (The 2011 Tohoku Earthquake Tsunami Joint Survey Group, 2011). (d) The estimated death rate (in percent: the total number of deaths and missing persons divided by population in each location) in the inundated area of each municipality, following the 2011 Tohoku-Oki earthquake. The number of deaths and missing persons in each municipality and the inundation areas is provided by the Statistics Bureau and the Director General for Policy Planning of Japan (DGPPJ, 2012). (e) The heights of embankments and tsunami waves at the corresponding sites (Central Disaster Prevention Council, Cabinet Office, 2011). Only in the northernmost areas were the embankments higher than the incoming tsunami waves, but the tsunami still flooded through the gaps in the breakwaters (Iwate prefecture office).

In order to clarify people's behavior after the 2011 event, we interviewed 150 tsunami survivors in six cities in Iwate and Miyagi prefectures by visiting several times public evacuation shelters, houses, or on-sites in various affected areas (Ando et al., 2013). These interviews suggest that no small ratio of death toll could be ascribed to a failure in the functioning of modern mitigation technologies in the face of an unprecedented disaster. For example, the JMA underestimated the initial earthquake size and tsunami heights, so the EEW of the JMA might have worked the other way, and the sea embankments of a 4- to 15-m height (Figure 4.1(e)), and the breakwaters at the mouth of the bay built by the local government of Iwate prefecture under the special law of the "Countermeasures for earthquakes and tsunamis" (Central Disaster Management Council, 2010) may have given an excessive sense of security to the local people.

Although we are equipped with many modern technologies such as those mentioned above, we easily failed in the use of modern technologies to save the lives of people in the face of the unprecedented natural disaster. It is not sturdiness of technologies or constructions but robustness as a whole that works out in moments of crisis.

In the hope of finding robust means for disaster prevention from the citizens' point of view, we interviewed 164 tsunami survivors. Our previous paper (Ando et al., 2013) used 150 interviews, among which 30 interviews were carried out without attendance of the authors. In this paper, we analyzed reselected 164 tsunami survivors including a newly interviewed set of 44 tsunami survivors directly conducted by us, excluding the 30 interviewees without us. Eight of the 44 were local welfare officers, and four of the 44 were fire corps volunteers, in Iwate and Miyagi prefectures. We also interviewed four officials in charge of the disaster-prevention measures in Miyako and Kamaishi cities, Iwate Prefecture, Tohoku district in order to survey in detail the damage to houses and death rate in three cities and four towns, in particular, Taro town in Miyako city and Kamaishi city, both of which had been putting extensive resources for the tsunami mitigation since 1933 and 1978, respectively.

Based on the previous interviews and new surveys, we report here our additional findings on the people's reaction to this earthquake and make practical proposals for countermeasures to prevent disasters in the future.

4.2 LOCATIONS OF THE INTERVIEW CITIES AND CHARACTERISTICS OF THE INTERVIEWEES

In Iwate prefecture (Miyako, Yamada, Otsuchi, Kamaishi, Ofunato, and Rikuzentakata cities), we interviewed 127 persons, and 37 persons in Miyagi prefecture (Ishinomaki, Yamamoto, Minami-sanriku, and Kesen-numa cities) (Figure 4.1). The interview sites are shown in Figures 4.1(a) and 4.2, together with the rupture area of the Tohoku-Oki Earthquake with 5- and 30-m fault slip

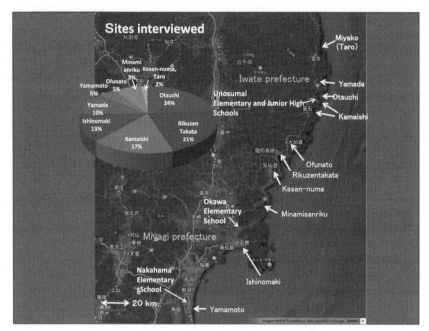

FIGURE 4.2   The 10 cities where we had carried out the interview survey and the ratio of person interviewed in the 10 cities. The triangles show the elementary schools.

contours (Lee et al., 2011). The tsunami >10—20 m in wave amplitude attacked many towns and cities on the Pacific coast where we conducted interviews. The arrival times of the largest tsunami waves recorded at tide gauge stations are shown in Figure 4.1. In the worst case, it was estimated that these waves had inundated the inner-bay town within 3—5 min of hitting the outer coast (personal communication).

We interviewed four officials in charge of the disaster-prevention measures in Miyako and Kamaishi cities, Iwate Prefecture, in March 2013, in order to know the details of the situation of suffering and tasks of the staffs of the city halls, such as information gathering and its dissemination to citizens and taking care of refugees. We did not count the responses of the four officials in the interview statistics because their actions were bound by duty. The result of the analysis of the present paper may not be very different from those of the previous paper, but we believe that the present analysis is based on more strictly controlled information or data than before.

The interviewees were 46 percent female and 54 percent male, and the age distribution is shown in Figure 4.3. The age distribution of the interviewees is biased toward senior ages, because from 10 am to 5 pm we interviewed people aged between 65 and 15 years who were out for work, clearing rubbles and searching for missing persons. The ratio of the seniors we could interview

Age distribution

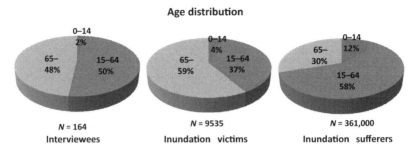

FIGURE 4.3 Demographics of the three age groups: 0—14, 15—64, and 65 years and older. The age distribution of the interviewees in this study, the death tools in the 10 studied cities, and the number of people in the 10 cities whose addresses were in the tsunami inundation areas are shown. The data used in the middle and right plots are after Tani (2012), "*N*" shows the number of samples. *N* in the middle is the total number of deaths in the inundated area and *N* on the right is the population in the inundation area in the 10 cities.

is >1.5 times larger than the actual ratio of the senior population in the devastated areas, where 30 percent of the population is 65 years or older. On the other hand, the death ratio of the people who were 65 years or older was 59 percent out of the total deaths, which is about more than three times higher than that of people below 65 years of age (Figure 4.3).

4.3 EVACUATION BEHAVIORS

As a typical feature of the coastal region of the Pacific side of the Tohoku region (especially in Miyagi and Iwate Prefectures), many small bays and headlands occur along the irregular coastline. In many locations, hills are present just behind dwellings located in narrow coastal lowlands. Therefore, in the bay and headland area, highlands or hills safe from the tsunami are within 5—20 min on foot at a normal pace from the tsunami-affected areas. In contrast, in the coastal plain areas, the tsunami inundated land up to 2—4 km from the coastline (Sawai et al., 2012).

However, the tsunami evacuation behaviors of the population were similar in both the areas. Most people did not think that a tsunami would come soon, even though they noticed that the shaking produced by the earthquake was extremely large.

Figure 4.4 shows the interviewees' first actions and the motivations for their subsequent actions after the earthquake ground motion has calmed down. Only 34 percent of interviewees evacuated soon after the earthquake and >50 percent of the interviewees were tidying rooms cluttered by the strong and minute-long shaking, or preparing for evacuation, and/or calling to check their family's safety, or chatting with neighbors (Figure 4.4(a)). Some of them went to harbors or rivers to watch the incoming tsunami (2 percent). Fire corps volunteers were carrying out their duties to close the water gates in charge, or

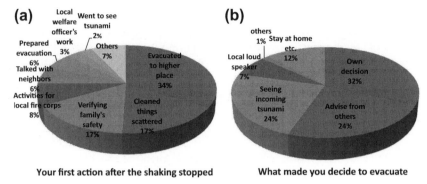

Your first action after the shaking stopped   **What made you decide to evacuate**

FIGURE 4.4 (a) The interviewees' first action. (b) The motivations of their action after the earthquake ground motion had calmed down.

alerting people to evacuate, through loudspeakers (8 percent). Local welfare officers visited each house assigned and checked the safety of the people living in the community (3 percent). Others were on their way home from shopping or their working place, or standing outside in a state of confusion (7 percent). Figure 4.4(b) shows what made people decide to evacuate. Only 32 percent of the interviewees decided to evacuate by themselves. Twenty-four percent of the interviewees evacuated because others urged them to do so, and 24 percent did so after they saw the tsunami coming (Figure 4.4(b)).

Only 33 percent of the interviewees were able to evacuate safely (Figure 4.5(a)), with 32 percent seeing the tsunami close behind them. The interviewees amounting to 21 percent felt very threatened, for instance, they

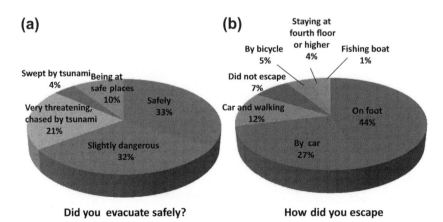

Did you evacuate safely?   **How did you escape**

FIGURE 4.5 (a) Interviewees evacuated safely or not. (b) Major transportation mode used in evacuation.

were chased by the tsunami, and their feet or their cars were partly submerged, and 4 percent were swept away by the tsunami but survived. More than 40 percent of the interviewees evacuated on foot (Figure 4.5(b)). Most of them are older than 65 years, and they said that it was not difficult to walk to their shelters, because they were just 10—15 min away. About 30 percent of the interviewees used cars, and two of them were swallowed up by the tsunami. They did not remember how they got out of the car, probably through the window. One of them could get out of the car, then climb up on one of the solid houses and stayed there overnight. Another person in a car drifted along with the water and was washed ashore accidentally where he was helped out by a passer-by. We do not go into detail about each interviewee's behavior here because this was detailed in Ando et al. (2013).

4.4 THE EFFECT OF THE CURRENT TECHNOLOGY FOR DISASTER PREVENTION

Our question is why 60 percent of the interviewees and more people did not evacuate immediately after the ground motion calmed down (Figure 4.6); about 30 percent of the interviewees were busy ascertaining the security of their family; about 30 percent of the interviewees did not consider the tsunami a possibility or expected the tsunami to be small; about 15 percent of the interviewees were engaged in their own work as local welfare officers who are helping elderly people living by themselves or people who cannot move without assistance, or as the member of the firefighting team, and about 7 percent of the interviewees believed that breakwaters that were built in the bay would protect them from the tsunami (Figure 4.6).

After the 1995 Kobe earthquake, nationwide efforts were enhanced for earthquake disaster prevention, as follows: (1) Publication of a probability map

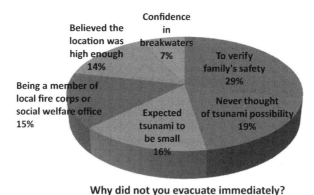

Why did not you evacuate immediately?

FIGURE 4.6 The reason why people did not evacuate immediately after the ground shaking subsided.

of ground motion intensity within 30 years; (2) Development of JMA's EEW system that has become operational since October 1, 2007; and (3) Estimation of the historical and future tsunami height and inundation area in Iwate prefecture for the 1896, 1933, and the anticipated Miyagi-Oki earthquakes. In addition to the countermeasures for the disaster prevention made after the 1995 Kobe earthquake, the lessons learned from the 1933 Sanriku Earthquake had prompted the Japanese government to take countermeasures against tsunami disasters, that is, (4) Constructions of sea embankments, breakwaters in the mouth of bays, water gates, and planting of trees along the coastal banks, etc., have been promoted after 1933.

In the following sections, we discuss the efficiency of four preventive measures stated above in relation to the people's evacuation behaviors.

4.4.1 The Earthquake Hazard Assessment

The first factor to be considered is that the earthquake hazard assessments issued by the Japanese Government underestimated the danger. The incorrect assessments by the Government may have been the cause of the residents' slow action.

Expected earthquake magnitude and the resultant hazard in northern Japan assessed and publicized by the Headquarters for Earthquake and Research Promotion, Cabinet Office of Japan (2010), were significantly smaller than that of the 2011 Tohoku-Oki earthquake. The probability map of ground motions equal to or larger than 6 lower on the JMA seismic intensity scale (around VII on the MM scale), for an event to be expected to occur with 99 percent probability within 30 years from 1st January 2010 (Headquarters for Earthquake Research Promotion, 2010) is shown in Figure 4.7(a). This map has been revised every year. The zone of an anticipated M7.5 Miyagiken-Oki earthquake is shown by a blue circle. The probability of the occurrence of the assumed event is 99 percent within 30 years following January 2010. The seismic intensity distribution resulting from the 2011 Tohoku-Oki earthquake (Hoshiba et al., 2011) is shown in Figure 4.7(b). These two intensity maps look completely different from each other in their patterns. However, only two among 164 interviewees had known about this map. One of them had seen it and another just heard about it. Both could not tell what it looks like. We asked the same question to officials in charge of the disaster-prevention measures in Miyako and Kamaishi cities, Iwate Prefecture. They also did not know anything about the probabilistic Seismic Hazard Maps for Japan.

Therefore, we conclude that the assessment issued by the Government did not affect the residents' action after the earthquake, even if the assessment was incorrect. Rather we must conclude that the information given by the Government or local governments had not been appropriately disseminated to the local residents.

(a)   **(b)**

Probabilistic seismic hazard maps for japan   Observed seismic intensity map
(2011)

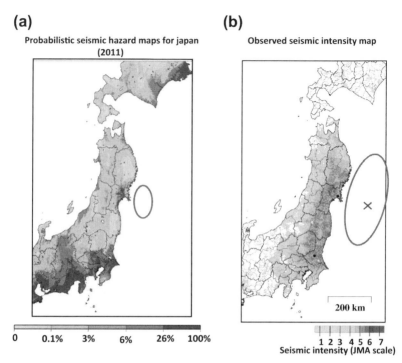

0   0.1%   3%   6%   26%   100%   1 2 3 4 5 6 7
                                                   Seismic intensity (JMA scale)

FIGURE 4.7   (a) Distribution map of the probability (scale given at the bottom) of ground motions equal to or larger than the seismic intensity of 6 lower on the Japan Meteorological Agency (JMA) seismic intensity scale (around VII on the MM scale), calculated to occur within 30 years from the present (Start date: January 1, 2010). The zone of large earthquakes anticipated, called the M7.5 Miyagiken-Oki earthquake, is shown by a blue circle. (b) Seismic intensity distribution of the 2011 Tohoku-Oki earthquake, the epicenter (X-mark) (Hoshiba et al., 2011) and the fault area (a blue oval) shown in Figure 4.1 (JMA seismic intensity is classified into 10 categories, namely, 0–4, and 5 lower, 5 upper, 6 lower, 6 upper, and 7).

4.4.2 JMA Earthquake Early Warning

The second factor to be considered in the analysis of people's behavior after the 2011 Tohoku-Oki earthquake is that the earthquake magnitude and tsunami height of the first warning issued by the JMA EEW were smaller than those of the actual event. This could be a reason that the majority of the local residents thought that breakwaters would protect them. The region of "warning" and "forecast" defined by the EEW and distribution of seismic intensities were estimated and issued in quasireal time. The JMA immediately transmitted the information to the Nippon Housou Kyokai (NHK, Japan Broadcasting Corporation) only once, and the NHK issued the EEW announcements through the TV and radio all over the country. The contents of the information are the origin time, the hypocenter location, the name of the place of the epicenter, and the name of the area where the intensity of 5 lower or greater and intensity of 4 are estimated. The EEW was issued 15 s earlier than the strong ground motion (intensity of 5 lower or greater on the JMA scale; this is almost the equivalent of VI of the MM scale that would arrive everywhere in the district) (Hoshiba et al., 2011).

On the other hand, the information of the expected tsunami height is transmitted to both the prefecture and the NHK. But the forecasted tsunami height in the first alert information was significantly smaller than that of the actual one. The information was updated by the JMA as follows: M7.9 (MJMA: Magnitude determined by JMA) was issued at 14:49, M8.4 (MJMA) at 16:00, M8.8 (M_w) at 17:30, March 11, and M9.0 (M_w) at 12:55, March 13. Therefore, the tsunami heights initially estimated by the JMA were 3 m at Iwate Prefecture and 6 m at Miyagi Prefecture, respectively; thus, local residents might think that breakwaters would protect them. We do not discuss here the reasons of the delay of the tsunami information, because it was already discussed in Ando et al. (2013). However, the first and also corrected information did not reach the local residents because of the blackout of electric power, which happened soon after the big shake. Consequently, most residents were unable to get any information through the TV or radio (Figure 4.8(a)).

Although >50 percent of the interviewees could not get any information about the tsunami, about 30 percent of the interviewees received the tsunami warning through the local community radio called Bosai-musen (the disaster-prevention radio if literally translated) (Figure 4.8(b)). The broadcast had been made on the battery-powered system that was operated in the local municipal offices during emergency times such as earthquakes, fires, or serious interruption of life lines. Bosai-musen receivers had been installed in each house by the local governments. The tsunami warning was also broadcast through the administration's loudspeakers, but it was rather difficult for people to hear it because of the echoes reflected from hills and house walls. However, only about 25 percent of the interviewees who could receive the tsunami warning

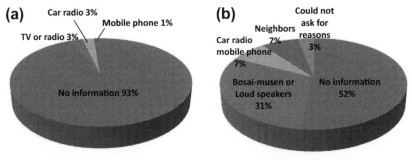

(a) Did you receive the Earthquake Early Warning

(b) How did you get the tsunami warning

FIGURE 4.8   **How did people get the earthquake and tsunami information issued by the Japan Meteorological Agency.** (a) Earthquake Early Warning. (b) Tsunami warning.

through the Bosai-musen or with administration loudspeakers decided to evacuate when they heard it. In other words, only 7 percent of the total interviewees followed the advice transmitted by the Bosai-musen or the administration's loudspeakers (Figure 4.4(b)). Only 7 percent of the interviewees received the tsunami information from a car radio or a mobile phone, and 7 percent of the interviewees were informed of it by neighbors (Figure 4.8(b)).

Our interview suggests that fortunately or unfortunately, the early warning issued by the JMA did not reach most of the people. The incorrect information did not affect the people's behavior and action very much, although the system itself must have been designed to save people from a disastrous earthquake like this one.

4.4.3 Role of Prior Knowledge and Disaster Assessment

The third question is whether most residents made a reasonable guess before the earthquake about the height and inundation area of the incoming tsunami based on their past experience or stories which they heard from their parents or grandparents. If prior knowledge about earthquakes and tsunamis did help them to survive, we want to know what kind of knowledge or experience they had before the earthquake. More than 40 percent of the interviewees experienced the 1960 Chile earthquake tsunami, and >20 percent of the interviewees heard about past tsunamis such as the 1960, 1933, and 1896 events, but 17 percent of the interviewees did not know about a tsunami at all (Figure 4.9(a)). About 70 percent of the interviewees knew that a tsunami could occur after the earthquake, but only 19 percent of the interviewees anticipated the large scale tsunami, 25 percent of interviewees thought of a smaller scale tsunami like the 1960 event (Figure 4.1(b)) or a tsunami alerted by the JMA in the past 4 years (Ando et al., 2013), and 20 percent of the interviewees did not know the fact

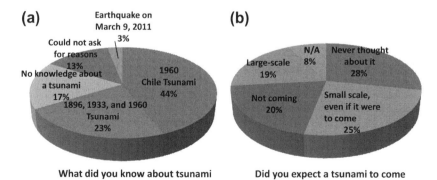

What did you know about tsunami   Did you expect a tsunami to come

FIGURE 4.9 **Knowledge and forecast of the tsunami height.** (a) Knowledge of the conventional tsunami. (b) Forecast of the tsunami height just after the 2011 Tohoku-Oki earthquake.

that tsunamis are closely linked with big earthquakes, nor imagined at all that the tsunami would occur after this event (Figure 4.9(b)).

Only about 30 percent of the interviewees who knew of the 1960 Chile earthquake tsunami (tsunami induced by the Chilean earthquake) evacuated to a higher place soon after the shaking stopped. This is about 12 percent of the total interviewees. Because the 1960 tsunami was not very high (Figure 4.1(b)) and disaster-prevention facilities such as the sea embankment, breakwater, dike and flood gate, were built after 1960, the experience of the 1960 tsunami did not prompt people to flee. They said that they did not feel any shaking during the 1960 tsunami, but felt the strong ground shaking of the 2011 earthquake. But they did not understand what the difference between the two ground motions meant.

The department of general affairs in Iwate Prefecture carried out in 2004 a rather detailed simulation study of tsunami heights and inundation areas for the cases of the 1896, 1933, and the anticipated Miyagi-Oki earthquake. The study covers 13 areas, including Taro, Yamada, and Otsuch towns and Kamaishi, Ofunato, and Rikuzentakata cities, Iwate Prefecture. Based on the simulations for the above three earthquakes, the assessments of damage were made for eight (2^3 cases) different cases. Each case has three different parameters: (1) occurrence time of the earthquake (nighttime in winter or daytime in summer); (2) existence of disaster-prevention facilities (yes or no); and (3) the required time for people to take refuge in the disaster-prevention facilities (35 or 40 min). The number of fatalities, seriously injured persons, and medium-grade injured persons were also obtained from the assessment. These results were published in 2006 (The Department of General Affairs, Iwate Prefecture, 2006) and available through the internet.

In the following, we briefly summarize the result. The worst case was observed for the situation in which the earthquake occurred in the daytime in midsummer before the tsunami disaster-prevention facilities were built, and the required time to reach a refuge is assumed to be 40 min. Because the beach in each town is crowded with bathers during the summer time, it is expected that the escape of the population from the beach to a safer place becomes difficult. According to the assessment, the number of deaths and seriously injured persons in Iwate Prefecture were 1,014 and 632, respectively.

The least damage was observed for the case in which the earthquake occurred in winter at night after the tsunami disaster-prevention facilities were built and the required time to reach a refuge is 35 min. The number of deaths and seriously injured persons were estimated as 100 and 49, respectively. However, these numbers in the assessment published in 2006 by the Iwate Prefecture were much smaller than the real numbers (5,823 for deaths) for the case of the 2011 tsunami associated with the Tohoku-Oki Earthquake.

It is not clear as to how these results of the assessment were put to practical use for the sake of the residents in the Tohoku district. We asked whether officials in charge of the disaster-prevention measures in Miyako city and

Kamaishi city (both in Iwate prefecture) knew about the report. They had known that the department of general affairs of Iwate Prefecture did the damage estimation for Iwate prefecture but not about details of the content. So they could not inform citizens widely about the report. But one of the interviewees told us that he had seen the inundation simulation maps in Kamaishi city with and without the effect of tsunami disaster-prevention facilities estimated by the Kamaishi City Office. Then, we asked some fire corps volunteers and local welfare officers if they knew about the damage assessment for the anticipated earthquake made by Iwate prefecture. They did not know anything about it. If they knew even a little about this assessment, their behavior and also the general public's behavior after the earthquake might have been different and many lives would have been saved.

The following is a story that a fire corps volunteers told us: When he felt the shaking of the earthquake, he jumped out of the office to close the gates of the embankment of which he was in charge. He did not have any earthquake information at all then. Regardless of the size of an earthquake, this was the necessary action described in their manual. After the gates were closed, he and his colleague went back to the volunteer fire station to report their activities and they then began to patrol the town to tell the people through a loud speaker to run away to a safe place. But some people ignored their warning and went to watch the sea. Some of them were swept away by the tsunami. On the way back from the embankment's gates with his colleague by car, he saw his mother watching him in front of his house. Both of them did not expect at that time that a large tsunami would come soon. Finally, his car was partially submerged, and the house was washed away by the tsunami. Fortunately, he and his mother were able to meet at the shelter the next day.

A local welfare officer, 70 years old, also told us that he was chased by the tsunami while visiting an area he was in charge of, but fortuitously, he was able to just barely escape the tsunami. All the people for whom he had responsibility had been already evacuated when he visited them, because the area that he is in charge of is located in the peripheral coastal plain where the people had enough time to evacuate, as the tsunami came about 1 h after the earthquake occurred. He himself never thought about the tsunami coming after the earthquake until he saw black smoke raised up by the tsunami. He did not know at all that a tsunami came after an earthquake, because he came to live in Tohoku only about 20 years ago.

The total number of the fatalities or missing persons of fire corps volunteers were 69 in our survey areas according to the reports by the regional government, which are Miyako, Kamaishi, Kesen, Ishinomaki, and Iwaki cities, Iwate prefecture, and 70 percent (50) of those sacrificed their lives while on duty. We believe that this kind of situation must be improved. At present, a committee of the Fire and Disaster Management Agency (Fire Disaster Management Agency, 2011) is examining how to keep the fire-fighting team safe. In order to achieve the safety of both the general public and the people whose duty is to guard the public, we need a quantitative and accurate damage assessment. In the case of the 2011 Tohoku-Oki Earthquake, we had a rather detailed assessment on a model earthquake before 2011. But the results of the assessment were not well circulated to the people such as firefighter volunteers. Although the assessment made by Iwate Prefecture itself predicted a substantially lower degree of the damage by the earthquake, it must have been very helpful for many people to survive, if it was appropriately utilized.

4.5 CONSTRUCTION OF SEA EMBANKMENTS OR BREAKWATERS IN THE BAY

During the last 150 years, large tsunamis have struck the Tohoku district. Therefore, disaster mitigation efforts have been undertaken in this region. These include the construction of extraordinarily big sea embankments, water gates, and breakwaters in the mouths of bays, etc.

Taro town (Figure 4.2), a small town in Miyako city, suffered from heavy damage by the 1896 Meiji Sanriku Earthquake Tsunami and the 1933 Sanriku Earthquake Tsunami; the inundation height in this district was 14.6 m during the 1896 event (M_t 8.2) and 10.1 m during the 1933 event (M_w 8.1). The casualties by the 1896 and 1933 tsunamis amounted to 1867 and 911, and the death rates were 83 percent and 32 percent, respectively (Yamashita, 2003). The plan of the sea embankment construction was made to put an end to the repetition of the history of such disasters. The construction started in 1934 and was completed in 1958. The embankment with an X-shaped cross-section is 1350 m in length, about 7.7 m in height, and its width at the bottom and top is 25 and 3 m, respectively. When the 1960 Chile Earthquake Tsunami occurred, there were no casualties in this district. Thus, the sea embankment was extended and completed in 1966. The total length of the X-shape sea embankment became 2,433 m. However, the 1960 tsunami was small, and the tsunami did not arrive at the sea embankment according to the survey of the inundation area of Taro town by the JMA (JMA, 1961). Thus, the effectiveness of this embankment is doubtful for the 1960 tsunami.

Yamada town (Figure 4.2) also suffered from heavy damage by the 1933 tsunami and constructed a tide barrier after the 1933 event. However, the 1960 tsunami flooded from the break of the barrier into Yamada town, and water did not readily disappear after the flood. After this event, sea embankments of a 4- to 9-m height were constructed along the shoreline.

Only a smaller scale sea embankment was constructed in Otsuchi town (Figure 4.2) after the 1933 tsunami. Then, after the 1960 tsunami, sea embankments about 6 m high and a large flood gate were constructed.

After Ofunato city had been damaged by both the 1896 and 1933 tsunamis, the city complex was moved to the highland area so that the districts would not be hit by a tsunami (Ishiwatari, 2011; Nakajima and Tanaka,

2011). However, the 1960 tsunami brought great damage to Ofunato city again. After this event, the construction of a big breakwater at the mouth of the bay started in 1962 and was completed in 1968. The maximum water depth was 38 m. The efficiency of the breakwater had been assessed by Horikawa and Nishimura (1969) based on the numerical simulation for the cases of 1896, 1933, and 1960 tsunamis in the bay area. They showed that the breakwater could have reduced the tsunami height to about a half for a teleseismic event such as the 1960 Chile tsunami, but they also pointed out that in the case of near earthquake events such as the 1896 and 1933 tsunamis, the efficiency of the breakwater would be much more variable than the teleseismic case, because the spatial changes of near-coast topography would very effectively interact with the incoming tsunami's directivities and wavelengths generated by near earthquakes such as the 1896 and 1933 tsunamis.

Kamaishi city also started to construct the breakwaters at the mouth of the Kamaishi bay in 1978 and completed this in 2008. This is the deepest breakwater in the world. The maximum depth of the water is 63 m, and the top of the breakwater is 4 m above the mean sea level. This breakwater was registered in the Guinness Book of World Records as the world's hugest dike.

Rikuzentakata city also suffered serious damage from the 1896, 1933, and 1960 tsunamis. After the 1933 tsunami, Rikuzentakata city raised the ground level of a part of low-lying land by approximately 3.5 m and surrounded the front and the side with banks of a 6.5-m height. But the city was damaged by the 1960 tsunami, so Rikuzentakata city prepared a double bank such as the front line bank that was 3 m high and the second line bank that was 4.5 m high to sandwich the famous big pine grove called Takada Matsubara as a coastal levee. In addition, they constructed a water gate at the estuary of the Kesen River, and the river banks on both the sides of the river.

Yamamoto town lies on a flat land, and Japanese black pines had been planted along the shore to protect against the tsunami after the 1960 event. This was maintained as a policy for disaster reduction in the country (Central Disaster Management Council, 2010).

4.6 DISCUSSIONS ON THE EFFECT OF THE BREAKWATERS

As described above, the cities, towns, and villages facing the Pacific coast of the Tohoku district have undertaken various efforts of long-term measures against tsunamis (Central Disaster Management Council, 2010). People had believed that the breakwaters constructed along the coast would protect them from tsunamis. After the 2011 Tohoku-Oki Earthquake, higher-than-expected tsunami waves hit the coastal zones, flooding over the embankments, causing devastating damage. In order to show the damage distribution in each city and town, we subdivided the cities and towns into smaller subareas (Figure 4.10). Death rates obtained by us are rather higher than those publicized. We used the number of fatalities and that of the people who are supposed to live in the damaged houses, because people will lose their life if they did not escape from the damaged house, not the house in the inundation area.

We surveyed in detail the damage of houses and the death rate in the four towns and three cities described above: Taro, Yamada, Otsuchi, and Yamamoto towns and Kamaishi, Ofuanto, and Rikuzentakata cities. Figure 4.10 shows the death rates in subareas in each town and city. Each percentage is estimated by using the total number of deaths and missing persons divided by the number of persons who lived in the houses destroyed largely or completely by the earthquake or tsunami in subareas. The death rates in Otsuchi town (Figure 4.10(c)) and Rikuzentakata city (Figure 4.10(f)) are about 13 percent on the average, rather higher than four other towns. In Taro town (Figure 4.10(a)), Yamada town (Figure 4.10(b)), Yamamoto town (Figure 4.10(f)), and Kamaishi city (Figure 4.10(e)), the death rates are around 8 percent. But the death rate in Ofunato city (Figure 4.10(d)) is about 5 percent, which is considerably lower than that of six other cities and towns even though the tsunami with a maximum height >10 m attacked Ofunato city. This tsunami height is almost the same as that observed in Kamaishi city where much severe damage and high death rates were recorded (Ministry of Land, Infrastructure, Transport and Tourism, 2011). Because the 1896 and 1933 tsunamis brought greater damage to Ofunato city, disaster-prevention city construction has been pushed forward in Ofunato city. Furthermore, the 1960 tsunami brought large damage again to the city; a fire-prevention building zone had been installed in the main business area to the seaside of the railroad and the construction of fire-proof buildings was promoted there, because there was little damage to the steel-reinforced concrete constructions by the 1960 tsunami. Six refuge buildings were established in the business area (Ishiwatari, 2011; Nakajima and Tanaka, 2011). Furthermore, the people ran away to the hill immediately after the tsunami because disaster-prevention education was accomplished thoroughly (Ishiwatari, 2011).

On the other hand, Otsuchi town and Rikuzentakata city suffered heavy damage by the tsunami that ran up along the river and overflowed its bank and also by the tsunami wave that arrived directly from the shore. Both Otsuchi town and Rikuzentakata city lie on a flat plain along a river. A large flood gate and many other flood gates of Otsuchi town were closed soon after the earthquake by the local fire corps, but the dike of the river was broken by the tsunami and it surged into the town. Almost the same situation as that observed in Otsuchi town occurred in Rikuzentakata city (Ando et al., 2013), and the damage became heavier because floods came from two directions, from the river and from the shore. Thus, the ratio of the fatalities to the population in the inundated area became very high in these two locations.

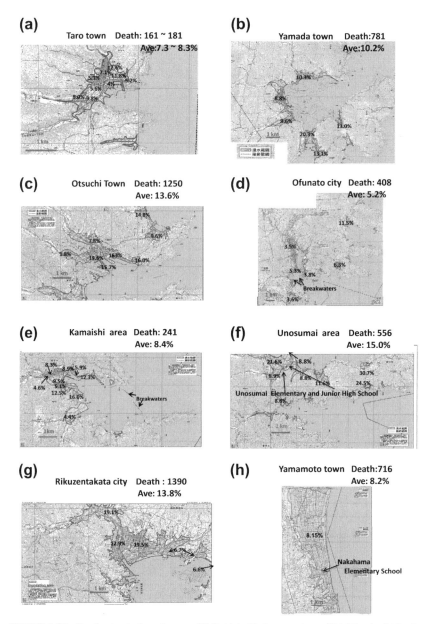

FIGURE 4.10 Death rates in the subareas of 7 districts. Each percentage of fatalities is obtained by using the total number of deaths and missing persons divided by the number of persons living in the houses destroyed completely or on a large scale by the earthquake or the tsunami (Tables 4.1–4.9). The number of persons who lived in the destroyed house is estimated by using the average number of family members in a subregion, which is obtained by the population and the number of the households. Inundation areas (shaded areas) are downloaded from a server of the Geographical Survey Institute of Japan. The total number of deaths and missing persons and

TABLE 4.1 Death Rate in the Subareas of Seven Districts

No. of deaths	Total number of deaths or missing persons in the main part of the cities and towns
E & T	Total number of houses destroyed completely or on a large scale by the earthquake or the tsunami for Taro, Yamada, Ofunato, Kamaishi, and Yamaoto districts and that of the houses destroyed completely or partially for Otsuchi and Rikuzentakata districts
No. of houses	The number of houses in a subregion
Population	Population in a subregion
Average number	Mean number of household members provided from the population divided by the number of houses
Estimated population	The total number of people who lived in the destroyed houses (E & T). The number is obtained by multiplying the above average number by the number of destroyed houses (E & T)
Inundation death rate	This is obtained from the ratio of the "death population" to the "estimated population". If the number of fatalities is less than five, we did not use the data in such a subregion because it is lacking in statistical reliability

The data in parentheses show the reference date. The number of "inundation death rate" in the individual table was plotted in Figure 10(a)–(h).

Before the 2011 Tohoku-Oki Earthquake, it had been widely believed that the tsunami measures of Kamaishi city and Taro town were perfect. The municipal officers in Kamaishi city told us that they had been very proud of the breakwaters in the Kamaishi bay when those were completed. Because they had heard from the public administration in Iwate prefecture that Kamaishi city would be guarded from the tsunami by the sturdy breakwaters that would last for 100 years. About 50 percent of the interviewees in Kamaishi city said that the breakwaters built in the mouth of the bay will protect the city. However, the breakwaters were broken by the first attack of the tsunami, and both Taro town and Kamaishi city suffered heavy damage in the neighborhood of the shore and the riverside. Nevertheless, a group of tsunami experts (Takahashi et al., 2011) still claim that the breakwaters functioned to delay the arrival time of the tsunami by 5 min from 35 min, which is the tsunami arrival time without the breakwaters and to reduce the tsunami height from 13.7 to 8 m. They also claim that many people were

death ratio in an inundation area in each city and town is shown on the top of the figure and the death ratios into smaller subareas are shown in figures. (a) Taro town, Miyako City, (b) Yamada Town, (c) Otsuchi Town, (d) Ofunato City, (e) Kamaishi area, Kamaishi City, (f) Unosumai area, Kamaishi City, (g) Rikuzentakata City, and (h) Yamamoto Town.

TABLE 4.2 Taro Town

Name of Town	kawamuko	Tategamori	Tanaka	Kobayashi-1	Mukouyama	Aratani	Aosari	Nohara	Otobe/2
No of deaths	23	17	10	6	12	36	7	30	20
No. of houses	211	119	64	48	46	190	38	91	56
Population	579	333	180	120	123	509	134	255	159
Death rate	4	5.1	5.6	5	9.8	7.1	5.2	11.8	12.6

The death rates in the subareas were obtained from the ratio of the "No. of deaths" to "Population", because we could not get E & T in subareas. Then, the average death rate in the town was obtained by using the E & T (756) and No. of deaths (161–181) in Toro Town which was provided by the town office. Data for the Subregions were provided personally from a member of the Japan Bosai-si Society in Iwate prefecture (July 16, 2011).
Source: Data obtained personally from a member of the Japan Bousaisi Society (March 13, 2013).

TABLE 4.3 Yamada Town

Name of Town	Osawa	Yamada	Orikasa	Funakoshi	Tanohama	Oura
No of deaths	131	314	113	86	103	34
E & T	467	1,404	508	151	327	108
No. of houses	770	2,571	1,045	770	514	355
Population	2,105	6,553	2,697	2,166	1,237	862
Average number	2.73	2.55	2.58	2.81	2.41	2.43
Estimated population	1,274.9	3,580.2	1,310.6	424.3	788.1	262.4
Inundation death rate	10.3	8.8	8.6	20.3	13.1	13

Source: Data were compiled at the Yamada-machi Government Office (May 31, 2013).

able to escape by virtue of a 5-min delay of the tsunami arrival. Although it is doubtful to believe that the 5-min delay from 35 min was very helpful for people to escape, it is ironical to know that the giant construction of the breakwater functioned for delaying the tsunami arrival by only 5 min. The inundation area in the Kamaishi area (Figure 4.10(e)) is <1 km to the hill and people had enough time to escape after the major shaking stopped, because the tsunami arrived 30 min after the earthquake. Municipal officers in Kamaishi city said that they ordered evacuation through a loud speaker, and continued calling for evacuation through the Bosai-musen radio. But people did not respond promptly.

In conclusion, we believe contrary to the opinion of Takahashi et al. (2011) that the breakwater in Kamaishi city was not very useful for the residents there. Rather than being useful, the erroneous belief of the sturdiness of the breakwater and overreliance on the breakwater may have led to a slower response to the order to evacuate and eventually led to a heavier death toll in Kamaishi city.

As described in the previous section, Taro town had a great wall of embankment. However, the embankment of the seaside collapsed during the first attack of the tsunami and the tsunami that went up the river overflowed from the upper reaches of the river dike. The official in Miyako city, who is also in charge of disaster prevention in Taro town, said that this was not unexpected for them, because they had known since its construction that the embankment of the seaside was not built as firmly as that of the inside. But the power of the tsunami was weakened by that of the seaside, and the water level inside became low and water flow became very slow when the tsunami drew back. Because most damage of the built environment

TABLE 4.4 Otsuchi Town

Name of Town	Machikata	Sakuragi-cho	Komakura	Sawayama	Ando	Akahama	Kirikiri	Namiita	Kozuchi
No of deaths	668	24	42	79	218	95	100	24	3
E & T	1,421	542	107	390	558	237	400	58	4
No of houses	1,853	579	110	1,195	824	371	954	143	200
Population	4,483	1,421	272	3,104	1,953	938	2,475	404	499
Average number	2.4	2.5	2.5	2.6	2.4	2.5	2.6	2.8	2.5
Estimated population	3,410.4	1,355	267.5	1,014	1,339.2	592.5	1,040	162.4	10
Inundation death rate	19.6	1.8	15.7	7.8	16.3	16	9.6	14.8	30

Source: Data were obtained through the internet from the homepage of Otsuchi Town (November 30, 2011).

TABLE 4.5 Ofunato City

Name of Town	Sakari-cho	Ofunato-cho	Akasaki-cho	Matsuzaki-cho	Ryori	Okirai
No of deaths	15	156	58	64	27	88
E & T	151	1,090	571	659	158	284
Population	3,554	10,047	5,069	4,718	2,754	3,213
Average number	2.7	2.7	2.7	2.7	2.7	2.7
Estimated population	434.7	2,943	1,541.7	1,779.3	425.6	766.8
Inundation death rate	3.5	5.3	3.8	3.6	6.3	11.5

The mean number of household members in each subregion was not obtained, because the number of houses in a subregion was not obtained. The mean number of the household members in Ofunato City was not used in individual subregions.
Source: Data were compiled by the Ofunato Municipal Office (February 28, 2013).

was caused when the tsunami drew back, it is considered that the embankment of the seaside functioned to reduce the damage. His opinion may be correct, but we still suspect that the overconfidence on sea embankments or breakwaters delayed the timing for the local residents to evacuate, although there might be other unexplained reasons that influenced their behaviors.

The sea embankments in Yamada town also collapsed in the first tsunami. A local welfare officer saw on the way to visiting a person requiring assistance that many people were watching the sea over the breakwater and did not escape. He heard somebody shouting that a tsunami went over the embankment and he could escape from the tsunami. He never even imagined that the tsunami would overflow the embankment.

In Yamamoto town, there was no embankment, but a forest of Japanese black pine and rice fields stretched along the coast. The residents called the forest a windbreak forest but not a tide-water control forest. A local welfare officer said that he did not think about the tsunami at all and was chased by the tsunami, and another officer said that he was swept over by the tsunami but managed to survive.

When we compare the fatality ratio of the above two towns, Yamamoto town, which has no embankment, and Yamada town, which has a big embankment, we do not see much difference between them. This may indicate that the embankments and breakwaters were not very useful for disaster

TABLE 4.6 Kamaishi Area in Kamaishi City

Name of Town	Shinhama-cho	higashimae-cho	Hama-cho	Minato-cho	Tadagoe-cho	Oo-machi	Owatari-cho	matsubara-cho	Ureishi-cho	Heita-cho
No of deaths	14	15	38	6	47	20	12	22	45	22
E & T	57	120	203	52	239	121	123	89	132	201
No. of houses	103	187	447	57	341	287	282	230	402	1,389
Population	205	394	940	129	704	574	594	454	856	3,410
Average number	1.99	2.11	2.1	2.26	2.06	2	2.11	1.97	2.13	2.46
Estimated population	113.4	253.2	426.3	117.5	492.3	242	259.5	175.3	281.2	494.5
Inundation death rate	12.3	5.9	8.9	5.1	9.5	8.3	4.6	12.5	16	4.4

Source: Data were compiled by Kamaishi Municipal Office (January 22, 2013).

TABLE 4.7 Unosumai Area in Kamaishi City

Name of Town	Unosumai 1-19	Unosumai 20-22	Unosumai 23-30	Ryouish-cho	Katagishi-cho 1-9	Hakozaki shirahama	Hakozaki 4	Hakozaki 5-12
No of deaths	320	14	34	45	33	42	7	61
E & T	601	120	496	218	156	47	10	195
No. of houses	990	67	469	261	275	133	28	273
Population	2,437	173	1,164	614	662	387	80	734
Average number	2.46	2.58	2.48	2.35	2.41	2.91	2.86	2.69
Estimated population	1,478.5	160	496	512.3	376	136.8	28.6	524.6
Inundation death rate	21.6	8.8	6.9	8.8	8.8	30.7	24.5	11.6

Source: Data were compiled by the Kamaishi Municipal Office (January 22, 2013).

TABLE 4.8 Rikuzentakata City

Name of Town	Takata-cho	Otomo-cho	Kesen-cho	Yonesaki-cho	Takekoma-cho	Hirota-cho
No of deaths	1,015	49	206	79	41	40
E & T	1,895	247	534	319	76	333
No. of houses	2,783	634	1,102	747	425	1,101
Population	7,641	1,911	3,287	2,754	1,203	3,532
Average number	2.75	3.01	2.98	3.69	2.83	3.21
Estimated population	5,211.3	743.5	1,591.3	2,277.1	215.1	1,068.9
Inundation death rate	19.5	6.6	12.9	6.7	19.1	3.7

Source: Data were obtained through the internet from the homepage of Rikuzentakata City (January 22, 2013).

TABLE 4.9 Yamamoto Town

Name of Town	Yamamoto-cho
No of deaths	716
E & T	2,751
No of houses	5,235
Population	16,704
Average number	3.19
Estimated population	8,778
Inundation death rate	8.15

Source: Data were obtained through the internet from the homepage of Yamamoto Town (March 13, 2013).

reduction in the case of the Tohoku-Oki Earthquake. Moreover, we found that many of the embankments, breakwaters, and river dikes were broken by the tsunami and did not function as planned. Therefore, overreliance on these structures is very risky in such a case as the Tohoku-Oki earthquake. We must be aware that any structure, even if it is constructed carefully, would be worn out due to repeated stress.

4.7 ROLE OF ELEMENTARY SCHOOLS IN DISASTER PREVENTION

It is surprising to know that no schoolchildren lost their lives at the schools in Iwate Prefecture (Kanno, 2011). The death rates of the schoolchildren in two other prefectures are also extremely low. It is more surprising when we consider that 131 of 1,035 schools in this area were inundated by the tsunami, and that 90 percent of the schoolchildren were in school at the time of the earthquake. But 639 children, students, and teachers were killed in three prefectures by the tsunami of this earthquake (Ministry of Education, Culture, Sports, Science and Technology (MEXT), 2012). Particularly, the death toll (445 persons) in Miyagi prefecture greatly outnumbers those in Iwate (107 persons) and Fukushima (87 persons) prefectures. The high death toll in Miyagi prefecture includes a tragic case of Ohkawa elementary School in Ishinomaki city in Miyagi prefecture, where 74 among 108 children and 10 among 13 teachers were swept over by the tsunami. This tragedy seems to have been caused by a teacher's inappropriate instruction after the earthquake, based on which schoolchildren just waited in the schoolyard until they were attacked by the tsunami, in spite of the fact that there was a safe hill near the school (Ikegami and Kato, 2012). Why they did not flee until 1 min before the tsunami arrival was investigated by Okawa Elementary School accident examination Committee and the report of the Committee was very recently submitted to the Board of Education, Ishinomaki City and its report has become available through Internet (The Board of Education, Ishinomaki City, 2014). This report lists 24 recommendations on countermeasures for disaster prevention; all of the recommendations are well considered and should be implemented to all schools in Japan.

On the other hand, there is a miracle escape story. The 570 students of the elementary and junior high school in Unosumai, Kamaishi city, Iwate Prefecture (see Figure 4.2 for the location), were well trained and promptly coped with the situation. All the children and teachers escaped to the high ground nearby as the danger of flooding was imminent, because these elementary and junior high schools were located in the lowland near the coast (Figure 4.10(f)). After they safely escaped, their school was flooded by the tsunami, but no lives were lost (Central Disaster Management Council, 2011).

These two examples show how important teachers' instructions are in such a case as in this earthquake. Although a few exceptions occurred, the teachers in almost all the schools guided schoolchildren and students to safer places and saved their lives. All the schools had their own risk management manuals, and 90 percent of them prescribed the refuge/evacuation plan at the time of an earthquake, and 50 percent of them had instructions for the tsunami case. Thus, disaster-prevention education had been made part of school education.

Nevertheless, a significant number of schoolchildren were lost in the Tsunami. According to the report by MEXT (2012), most of the students killed by the Tsunami were on the way back from the schools to their homes. Such a case cannot be avoided unless we had very versatile communication systems between schoolchildren and teachers.

In order to discover the role of elementary schools, we visited Nakahama Elementary School in Yamamoto town (Figures 4.2 and 4.10(f)) guided by a local welfare officer who had lived close to the school. She told us that all the houses around the school were swept away by the tsunami, and that only the elementary school was left standing. She also told us that all the children escaped into the storeroom on the third floor following the direction by the principal and they were saved. Considering this is just an example out of many other schools, we believe that elementary schools are really functioning shelters from earthquakes and tsunamis. This statement is substantiated by the fact that about 70 percent of the school buildings in this region are earthquake-resistant buildings (Fire Defense Management Agency, 2010). Moreover, elementary schools can play an important role because almost all of them are within walking distance from the residences and dwellers should be very familiar with the location of the school.

4.8 CONCLUSION

We examined the people's behavior during the March 11, 2011, Tohoku-Oki earthquake based on the interviews with 164 tsunami survivors in Iwate and Miyagi prefectures. We discovered to our surprise that most residents in this region were not informed of the Earthquake Probability Map by the Japanese Government, nor did they know about the rather detailed damage assessment prepared by the local government, nor did they hear about the EEW issued by the JMA immediately before the strong seismic ground motion arrived. Unfortunately, all information mentioned above consisted of underestimated values of the earthquake damage, the tsunami heights, and the degree of predicted damage by earthquakes compared with those of the real event. Therefore, the fact that the EEW was not properly transmitted had no influence on the people's reaction at the actual scenes of the earthquake and tsunami.

We also found that many people were very hesitant or seemingly reluctant to evacuate immediately after the shaking of the earthquake subsided, notwithstanding the disaster mitigation efforts that have been undertaken and practiced in the Tohoku region as a national policy, especially after the 1960 Chile Earthquake Tsunami happened. We presumed that most people in this region were very aware of the risk of big earthquakes and associated tsunamis. But our supposition was not correct. We concluded that most residents had estimated the height and inundation area of the incoming tsunami based on their past experience, and they believed that strong sea embankments or breakwaters would protect the city from the tsunami.

It became clear through our interviews that the people had too much confidence in the sea embankments and breakwaters. Unfortunately, these constructions did not perform their function as expected but might have led to the opposite: more damage to humans and buildings. However, the Reconstruction Agency of Japan (2011) is planning now to reconstruct much stronger sea embankments or breakwaters than there used to be, saying that these new constructions will save people. We have to recognize that any structure will get time worn and will need rebuilding or large-scale repairing every 5—60 years. This repeated repairing and maintenance of the large-scale constructions may require tremendous resources and does not seem to be very practical, considering that the tsunami such as the 2011 event is a quite rare event, such as occurring once in a millennium. The present study strongly casts doubts on the reconstruction plan, and we need to seek other more effective measures to protect people from earthquakes and tsunamis.

Although we are equipped with many modern technologies such as detailed numerical simulation or damage assessment and construction of huge embankment or breakwaters, the present study shows that the safety of people is not achieved by such modern technologies but with a simpler and more economical means: education on the mechanism of earthquakes and tsunami generation to the general public and utilization of elementary schools as refugees. We would like to point out here that education must include an uncertainty of the current knowledge of seismology and disaster-prevention technologies, because the seismology alone would not allow the prediction of the damaging earthquake or tsunami precisely, or in a timely manner in the foreseeable future.

ACKNOWLEDGMENTS

We thank many survivors who agreed to be interviewed, facing hardship at the evacuation shelters. I thank H. Mizutani and M. Wyss who kindly read the final manuscript, and provided us with many useful suggestions for improvement. We also thank the officials in charge of the disaster-prevention measures in Miyako and Kamaishi cities for providing data and participating in our interview. We appreciate the cooperation of the officials of Ofunato city and Yamada town for providing data and an anonymous member of the Japan Bosai-si Society in Iwate prefecture. We also appreciate S. Kitagawa who kindly introduced the officials in Miyako, Kamaishi, and Ofunato cities and S. Tsukada who explained to me about the JMA EEW. We thank S. Kitamura, Y. Aono, M. Takahashi, H. Anzai, and Y. Kunihiro for having cooperated in the interviews.

REFERENCES

Ando, M., Ishida, M., Hayashi, Y., Mizuki, C., Nishikawa, Y., Tu, Y., 2013. Interviewing insights regarding the fatalities inflicted by the 2011 Great East Japan Earthquake, Nat. Hazards Earth Syst. Sci. 13, 2173—2187. http://dx.doi.org/10.5194/nhess-13-2173-2013.

Central Disaster Management Council, 2010. Report of the Committee for Technical Investigation on Lessons Learned from Past Disasters — The 1960 Chile Earthquake Tsunami, 214 pp. (in Japanese).

Central Disaster Management Council, 2011. Report of the Committee for Technical Investigation on Countermeasures for Earthquake and Tsunamis Based on the Lessons Learned from the "2011 Off the Pacific Coast of Tohoku Earthquake", 50 pp. (in Japanese).

Central Disaster Prevention Council, September 30, 2011. Cabinet Office, Lessons Learned from the Great East Japan Earthquake.

Fire Disaster Management Agency, 2011. Study Meeting about the Ways of the Firefighting Team Activity at the Time of the Large-Scale Disaster on the Basis of the East Japan Great Earthquake Disaster. Reference Material, 94 pp. (in Japanese).

Headquarters for Earthquake Research Promotion, 2010. Distribution Map of Probability of Ground Motions Equal to or Larger than Seismic Intensity 6 Lower, Occurring within 30 Years from the Present (Start date: January 1, 2010), National Seismic Hazard Maps for Japan. Earthquake Research Committee.

Horikawa, K., Nishimura, H., 1969. The effect of the tsunami breakwater. In: The Proceeding of 16th Coastal Engineering Meeting, pp. 365−369 (in Japanese).

Hoshiba, M., Iwakiri, K., Hayashimoto, N., Shimoyama, T., Hirano, K., Yamada, Y., Ishigaki, Y., Kikuta, H., 2011. Outline of the 2011 off the Pacific coast of Tohoku earthquake (M_w 9.0) − Earthquake early warning and observed seismic intensity. Earth Planet. Space 63, 547−551.

Ikegami, M., Kato, Y., 2012. What Happened at that Time in Okawa Elementary School? Seishisha, 317 pp. (in Japanese). http://www.pref.iwate.jp/∼hp010801/tsunami/yosokuzu_index.htm.

Ishiwatari, A., 2011. The Human Damage Caused by the 2011 Tsunamis of Each East Japan Pacific Coast Cities, Towns and Villages. Report of Geological Society of Japan (in Japanese). http://www.geosociety.jp/hazard/content0065.html.

Japan Meteorological Agency, 1961. Report of the 1960 Chile Tsunami Earthquake. Technical Report of the Japan Meteorological Agency, No. 8, pp. 1−389, (in Japanese).

Kanno, H., 2011. Reconstruction of Education Activities at School. Report of Iwate Education Board, (in Japanese). http://www.mext.go.jp/b_menu/shingi/chukyo/chukyo9/shiryo/__icsFiles/afieldfile/2011/07/05/1308095_1.pdf.

Lee, S.-J., Huang, B.-S., Ando, M., Chiu, H.-C., Wang, J.-H., 2011. Evidence of large scale repeating slip during the 2011 Tohoku-Oki earthquake, Geophys. Res. Lett. 38, L19306. http://dx.doi.org/10.1029/2011GL049580, 2011.

Ministry of Education, Sports, Science and Technology—Japan, Research about the School Correspondence in the East Japan Great Earthquake Disaster, 51 pp., 2012 (in Japanese). http://www.mext.go.jp/a_menu/kenko/anzen/__icsFiles/afieldfile/2012/07/12/1323511_2.pdf.

Ministry of Land, Infrastructure, Transport and Tourism Report about the Tsunami Protection Effect of a Breakwater in the Entrance of Bay. 2011 (in Japanese). http://www.city.ofunato.iwate.jp/www/contents/1305074403730/files/2wankoubouhatei_0602.pdf.

Ministry of Land, Infrastructure, Transport and Tourism, 2011. About the Countermeasures of Tsunami in Harbors. Report of Port and Airport Department, Chubu Regional Bureau, pp. 1−19 (in Japanese). http://www.jgs-chubu.org/download/syn4/pdf/20/2-2.pdf.

Nakajima, N., Tanaka, A., June 2011. The past tsunami disaster and reconstruction planning in the Sanriku region. Urban Plann. 291, 45−48 (in Japanese).

Sawai, Y., Namegawa, Y., Okamura, Y., Satake, K., 2012. Challenges of anticipating the 2011 Tohoku earthquake and tsunami using coastal geology. Geophys. Res. Lett. 39, L21309. http://dx.doi.org/10.1029/2012GL053692, 2012.

Statics Bureau and Director General for Policy Planning of Japan (DGPPJ), March 29, 2012. Statics of Devastated Cities (in Japanese). http://www.stat.go.jp/info/shinsai/index.htm#kekka.

Takahashi, S., Kuriyama, Y., Tomita, T., Kawai, Y., Arikawa, T., Tatsumi, D., Negi, T., 2011. Urgent Survey for 2011 Great East Japan Earthquake and Tsunami Disaster in Ports and Coasts — Part I (Tsunami). An English abstract of the Technical Note of Port and Air Port Research Institute, No. 1231, 200 pp.

Tani, K., 2012. Distribution of the death tools and the death rate on the Great East Japan Earthquake, Paper of Dept. Geophys. Saitama Univ., 32, pp. 1—26 (in Japanese).

The Department of General Affairs, Iwate Prefecture, 2006. Report of Earthquake and Tsunami Simulation and the Assessment of Damage in Iwate Prefecture (Summary Version), 191 pp. (in Japanese). http://www.pref.iwate_jp/~hp01801/tsunami/yosokuzu_index.html.

The Board of Education, Ishinomaki city, 2014. Report of Okawa Elementary School Accident Inspection Committee, 233 pp. (in Japanese) (http://city.ishinomaki.lg.jp/cont/220101800/8425/201403164845.htm).

The 2011 Tohoku Earthquake Tsunami Joint Survey Group, 2011. Nation wide field survey of the 2011 off the Pacific coast of Tohoku earthquake tsunami. J. Jpn Soc. Civ. Eng. Ser. B 67, 63—66. http://www.coastal.jp/ttjt/.

Usami, T., 2003. Materials for Comprehensive List of Destructive Earthquakes in Japan, 416 - 2001, in Japanese. Univ. of Tokyo Press, Tokyo, 605 pp. (in Japanese).

Yamashita, F., 2003. Outline history of "town declaration of the tsunami disaster prevention" and large dike in Sanriku Coast, Taro town. Historical Earthquake 19, 165—171 (in Japanese).

Yamashita, F., 2005. Fare of Tsunami-Transmission of Sanriku Tsunami Tales. Tohoku Univ. Press (in Japanese).

Yonechi, F., Imaizumi, Y., 1994. Geographical and Sociological Problem about the Origin of the Place Name "Sanriku District", Department of Education annual report, Iwate University, 54, pp. 131—144 (in Japanese).

Chapter 5

Aggravated Earthquake Risk in South Asia: Engineering versus Human Nature

Roger Bilham

CIRES and Geological Sciences, University of Colorado, Boulder, CO, USA

ABSTRACT

One-third of the world's population lives in South Asia, an area that has lost 0.4 million people to earthquake deaths in the past 12 years, and more than 2 million in the past millennium when populations averaged one-tenth of their present levels. The large multicentury earthquake death toll results from a broad collisional zone where earthquakes with $M_w > 7.0$ are common inland. The purpose of this article is to examine some of the underlying reasons why populations are especially vulnerable to earthquakes in the region, and why a perceived disconnect exists between earthquake-resistant engineering and those populations most at risk from earthquakes. The article notes that urban growth and changes in building styles have rendered urban populations more vulnerable than in the past, but that there exist numerous hidden factors within the structure of societies that act to thwart the best intentions of seismologists and engineers to apply ubiquitous earthquake resistance. More than 80 percent of all earthquake deaths worldwide have occurred in nations where the mean per capita income is $<\$3,200$ per year. The Gross Domestic Product of a country influences its level of education, its financial ability to implement earthquake resistance, and the efficiency of its laws to regulate the welfare of its people. Although all these factors contribute to the resilience of building stock to shaking from earthquakes, in many cases, the absence of code enforcement in the building industry is directly responsible for the collapse of structures. Contractors who are anxious to maximize profits in societies where corrupt practices are endemic can avail themselves of numerous opportunities to circumvent not only earthquake-resistant regulations but also common-sense safe-assembly guidelines, and the resulting dwellings frequently become the homes of the urban poor. For this reason, many of the cities of Asia are disproportionately fragile, and can be expected to be the site of numerous future earthquake disasters.

Earthquake Hazard, Risk, and Disasters. http://dx.doi.org/10.1016/B978-0-12-394848-9.00005-5
103

5.1 INTRODUCTION: HAZARD, RISK, AND AGGRAVATED RISK

It is well to introduce this chapter with a brief definition of terms. "Earthquake hazards" are historical observations of the effect, or potential effects, of earthquakes on society in a specified region. Earthquake hazards are quantified by the severity and duration of the shaking produced by an earthquake (local accelerations, velocities, and displacements), and the indirect effect of these accelerations at the earth's surface (landslides, tsunami, and liquefaction). "Earthquake risks" are the estimates of the effects of future earthquakes in specified regions, based on insights from a catalog of historical seismic hazards. The historical and recent record of earthquake hazards can be used to develop quantitative probabilities of future shaking that guide engineers in the design of structures that will resist damage during future earthquakes.

In this chapter, I introduce the concept of "aggravated seismic risk", a term invoked to embrace those elements of seismic risk that are caused by the flawed mitigation of seismic risk by errors in seismic hazard assessment, or the neglect of an appropriate engineering response to known hazards.

Although earthquake risk can be reduced by human intervention, hazards cannot. But although hazards cannot be changed, the accuracy with which historical earthquakes can be quantified can often be improved by careful study. Sloppy studies of hazards can lead to exaggerated or underestimated estimates of risks. The world's written history is a fickle data logger subject to human memory and perception, and considerable effort has been expended in reassessing historical records of earthquakes.

Estimates of earthquake hazards provide the key shaking accelerations and durations used by engineers to construct buildings or structures (bridges, power stations, dams, and pipelines) to survive future earthquake shaking. The mandate of engineers is to construct buildings that do not collapse during earthquakes and therefore do not harm their occupants. A building can be constructed to withstand the strongest shaking, but the cost of a building increases with its resistance to collapse. Hence, engineers are required to minimize the cost of a structure as a function of its use. Hospitals, schools, and fire stations must be constructed to withstand the strongest shaking, and are expected to remain functional immediately after shaking has ceased. At the other extreme, it is possible to construct a building with minimal strength that will survive an earthquake permitting its occupants to escape injury, but which must be extensively repaired or torn down and rebuilt following the earthquake. The cost of a building immune to all earthquake damage may be 10 times the cost of a building with minimal resistance that must be repaired after an earthquake. The cost of minimal earthquake resistance (i.e., sufficient resistance to guarantee occupant survival) is often <10 percent of the cost of a similar structure that would otherwise collapse without earthquake-resistant features.

Earthquake risk is of major importance to insurance companies that must estimate their exposure to losses in regions where earthquakes are expected. Regions of a higher risk require higher insurance premiums. For this reason, insurance companies are eager to refine estimates of future risk.

In contrast, earthquake risk is rarely considered by home owners in cities or villages, or people renting dwellings and apartments. Many residents, even in hazardous seismic zones, have zero knowledge of how their house or apartment was constructed. Some owners may have added to, or modified, their homes without thinking whether those changes have made their homes more, or less, resistant to future earthquakes. This is an example of "aggravated seismic risk". A knowledge of future earthquake shaking may be known to parts of a society, but it is not reaching the user community.

More importantly, although construction guidelines exist in all parts of the world (and in many places these guidelines include codes related to earthquake risk), in most parts of the world, these guidelines were, and still are, not considered important by contractors or by authorities responsible for regulating building construction. The reason for this disinterest is that contractors, builders, and building inspectors have together, or in isolation, subconsciously applied their own estimate of seismic risk to constructions for which they are responsible.

A few hundred years ago, the nonimplementation of building codes was not as much a problem as it is now, because populations were sufficiently sparse for most people to live in single-story structures in rural communities. However, in the past 100 years, populations have increased by an order of magnitude, and populations are now concentrated in cities. The dense packing of people in cities has required a change in city construction from ground-level, single-family homes to multifamily housing in multistory buildings made of concrete and steel. This has led to a change in construction methods that are more vulnerable to earthquake shaking.

A measure of our awareness of the present vulnerability of the world building stock is that when an earthquake occurs near a major city, within half an hour of the main shock seismologists can calculate, using empirical methods, how many people are injured, how many are dead, and the approximate cost of reconstruction (Wyss et al., 2006; Jaiswal and Wald, 2010). Yet despite the rapidity of these empirical calculations, it often takes many days before survivors at the epicenter are able to count the dead and wounded, and many months or years before the true costs of reconstruction are known. This is not a success story—it is a measure of failure of two scientific disciplines (seismology and earthquake engineering), and in a broader context, a failure of a species to recognize that it inhabits a planet where some small fraction of its dwelling units will collapse every year due to earthquakes.

5.2 STATISTICS OF EARTHQUAKE FATALITIES

Many millions of people have died from building collapse in the past several thousands of years (Bilham, 2009; Holzer and Savage, 2013), but >80 percent of these deaths are recorded to have occurred in <12 percent of the world's land area. This region of high seismic risk is a broad zone that follows the Alpine/Himalayan collision zone along the southern edge of the EuroAsian plate (Figure 5.1). Earthquakes here resulted from the convergence of the African, Arabian, Indian, Australian, and Pacific plates toward the EuroAsian plate at velocities of 1−10 centimeters per year. The historically high death toll from earthquakes in the region is attributable to the high population density within the collision zone (England and Jackson, 2011) that includes several of the earliest centers of civilization.

Although populations in this region prior to 1800 remained relatively low, since 1800, they have increased by an order of magnitude, and many former villages are now major cities with population densities two or more orders of magnitude larger. Numerous megacities (population >8 million) now host more people than the populations of entire nations two centuries ago. This change in demographics has inevitably increased the potential for a future earthquake close to a large urban population to result in an unprecedented death toll and economic cost (Hough and Bilham, 2006; Musson, 2012). Moreover, the change in global and regional demographics has occurred in a time frame that is short compared to the recurrence

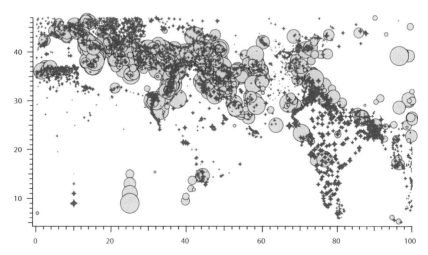

FIGURE 5.1 Earthquake deaths in the past 2,000 years along the southern edge of the EuroAsian Plate compared to present-day centers of population. The shores of the Mediterranean, the River Nile, India, and the Arabian continent can be recognized by their population distributions. Deaths from earthquakes in the past 1,000 years are shown as yellow circles. Figure 5.2 highlights the post-1900 increase in fatality rate east of 47°E.

intervals of earthquakes, especially in the broad collision zones of south Asia, and as a result, the decadal loss of life from earthquakes has significantly increased since the nineteenth century (Figure 5.2), and is likely to continue to increase unless earthquake-resistant construction is not adopted throughout the region.

A disproportionate number of earthquake deaths occur far from the ocean–continent plate boundaries where much of the global earthquake energy release occurs (Figure 5.3). Many of these deaths occur within the wide collision zone that borders the southern edge of the Eurasian plate between Turkey and Burma, amid which numerous midcontinent faults exhibit low

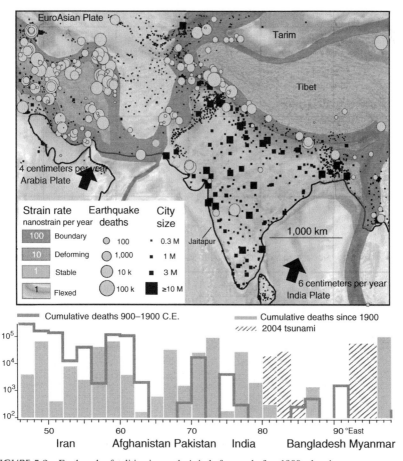

FIGURE 5.2   Earthquake fatalities in south Asia before and after 1900, showing current population demographics *(Updated from Bilham and Gaur, 2013.)*. The figure illustrates plate boundaries and the breadth of the deforming regions of Iran and the Tibetan plateau, and the collisional flexure of the Indian plate responsible for midplate earthquakes in parts of India, Pakistan, and Bangladesh.

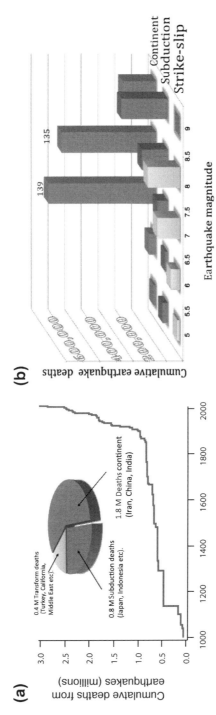

FIGURE 5.3   (a) Two-thirds of earthquake deaths prior to 2012 have not occurred on simple strike-slip or subduction zone plate boundaries where most of the world's seismic energy is released. (b) Most of these deaths occur in $6.5 < M_w < 8$ earthquakes in the broad collisional zones interior to continents.

levels of seismic activity. In the desert regions of Iran, villages are preferentially located near the edges of mountain ranges bordered by these faults, since it is in these regions where water supplies are usually abundant (Jackson, 2006). Similar settings prevail in the villages in Baluchistan (e.g., Quetta destroyed in 1935) and the Pakistan Himalaya (e.g., Taxila near Islamabad destroyed in 55 AD).

5.3 PROBLEMS ASSOCIATED WITH ASSESSMENTS OF SEISMIC HAZARDS

The critical reader should be aware that a number of problems impede a precise knowledge of seismic risk based on traditional methods for evaluating seismic hazards. The following four approaches to assessing seismic hazards are particularly problematic in south Asia.

5.3.1 Instrumental Catalogs

A common assumption in the interpretation of seismic catalogs is that the constants in the Gutenberg Richter relation for a region are invariant with time. Simply stated, in a given period of time, 10 times more earthquakes occur with a magnitude of 6, than with a magnitude of 7, and one-tenth as many with a magnitude of 8. The relationship is stated in the form $\log N = a - bM$, where N is the number of events with magnitude $\geq M$, and a is the measure of seismic productivity (i.e., the total number of earthquakes considered is 10^a). b is often a constant close to 1.0. The utility of this relation is that if it is truly valid, a relatively short (40–100 years) period of instrumental data can be used to evaluate the probability of future less frequent larger damaging earthquakes in a region. The b value is frequently invoked for its importance in estimating large future earthquakes that may be sparsely sampled by an existing regional instrumental or historical record. It can be tested for its ability to forecast the number of moderate earthquakes from the number of minor earthquakes, thus providing a measure of confidence to those that seek to view the future. However, the time invariance of the "a value", which also contributes to this forecast, is rarely questioned. The a value numerically indicates the total number of earthquakes available in a catalog of finite length. In some regions where the historical catalog is sufficiently long, the a value has been shown to vary with time, as for example, in Turkey, where seismic activity alternates between the northern and eastern Anatolian faults over hundreds of years (Ambraseys, 1970, 1971, 1975). In India, a similar time variable trade-off in seismic productivity may occur between earthquakes in Bhutan and those in the Shillong plateau (Vernant et al., in press). Forward projections of seismic catalogs based on the b value alone may thus underestimate (or overestimate) the potential for future damaging earthquakes.

5.3.2 An Inadequate Record of Historical Earthquakes

Although the Middle East, Turkey, and Iran have extensive records of historical earthquakes in the past 2,000 years, an incomplete or inaccurate seismic history is typical of many parts of the world, and in some south Asian countries (Afghanistan, Pakistan, India, Bangladesh, Nepal, Tibet, Bhutan, and Burma), the historical record is surprisingly scant. This may arise from the absence of a historical tradition of archival retention, or an absence in the continuity of governance resulting in the loss of records (wars, malicious destruction of archives, etc.). An example is that many of the written histories of Assam, India, were sought and deliberately destroyed in the eighteenth century, by those in power because in some cases they showed a less-than honorable family history. Thereby the sins of ancestors were erased along with information on destructive earthquakes. A written history of Sikkim was destroyed by an invading Nepalese army that used its wooden tablets as roofing tiles. The Jesuit archives of India's history were destroyed in the Lisbon earthquake, and duplicates in Goa were burned on the dock by the captain sent to retrieve them, to make space for a more lucrative cargo.

More often, the absence of written history is directly related to the decay of written records due to dampness or insect infestation or accidental fire. As a result of this loss of written archives, damaging earthquakes may be missing from catalogs, or the scant surviving records may bias the inferred location of major earthquakes. Place names may have changed and earthquakes assigned to inappropriate coordinates. Some undoubtedly major earthquakes may be omitted or misrepresented because they are recorded at only one location. For example, a major earthquake occurred on January 3, 1519, near the Kunar fault in Afghanistan during the eastward march of Baber's army (Ambraseys and Bilham, 2003a). Its magnitude may have exceeded $M_w \approx 7.5$, because main shock and aftershock shaking is reported to have persisted for half an hour, but because only one report exists, location and magnitude remain speculative, and it cannot be placed with any reliability in parametric catalogs, and hence is omitted by most researchers. In contrast, some early compilations of earthquakes (e.g., Oldham, 1883) have erroneously listed storms or cyclones as earthquakes, as has occurred in Mumbai in 1618 and Kolkata in 1737 (Bilham, 1994; Bilham and Gaur, 2013). These early entries have been adopted uncritically in later compilations and persist in many recent catalogs.

5.3.3 Unmapped Faults

Earthquakes occur on faults, and a profitable investigative approach has been to exhume their surface rupture to determine a history of slip, that in some cases has extended the local seismic record for many tens of thousands of years (e.g., the strike-slip faults of California and Iran) thereby permitting estimates of recurrence intervals and the variability of slip in sequential earthquakes. However, where strain rates are low and faults activated infrequently, faults may be obscured by erosion and deposition rendering their surface expression invisible. Worse still, subsurface faulting may be responsible for earthquakes that leave no local distinct expression of subsurface activity. The existence of unmapped faults in a region may thus render paleoseismic investigations impotent, and bias estimates of seismic risk downward.

5.3.4 Low Strain Rates

In the absence of recent earthquakes or a history of slip on known faults, it is common to invoke measurements of geodetic strain (GPS and Interferometric synthetic aperture radar) to provide an estimate of potential seismic productivity. The underlying physics is that where strain rates are low, earthquakes will occur less frequently. However, low seismic productivity does not imply no seismic productivity. If an earthquake on a fault occurs every 10,000 years and the last one was 10,000 years ago, the earthquake could obviously occur soon. Observations of low geodetic strain rates in isolation may thus misleadingly lower estimates of seismic risk.

Low strain rates prevail in the Indian subcontinent—typically around one nanostrain per year (Paul et al., 1995; Banerjee et al., 2008). Thus, if an earthquake has just ruptured at a failure strain of 100 microstrain, this tiny background strain rate would take 100,000 years to return the rocks in the nucleation zone back to a point close to failure in a future earthquake. But because India's collision with Asia has been ongoing for hundreds of millions of years, stresses are high everywhere. The forces involved are not just the north—south stress of collision, but are imprinted also with a combination of flexural and gravitational forces (Figure 5.4). The flexure comes from the bent northern edge of India as it is depressed under the Tibetan plateau (Bilham et al., 2003), and from the weight of the sediments deposited in the Arabian sea and the Bay of Bengal by the great rivers of India. The gravitational stress arises from the forces resulting from supporting the great weight of the Tibetan Plateau, and the oceanic edges of the Indian continent (Ghosh et al., 2006).

The result of these stresses is that throughout the Indian subcontinent, except where an earthquake has just occurred, the rocks may be close to rupture everywhere. The recent occurrence of $M > 6$ earthquakes there has been attributed to the release of these large stresses by minor perturbations resulting from reservoir impoundment (Seeber et al., 1996; Talwani, 1997). Viewed in terms of our ability to assess future seismic risk, reservoir construction for power and irrigation purposes is responsible for subtly raising the "a value" in the Gutenburg—Richter relationship.

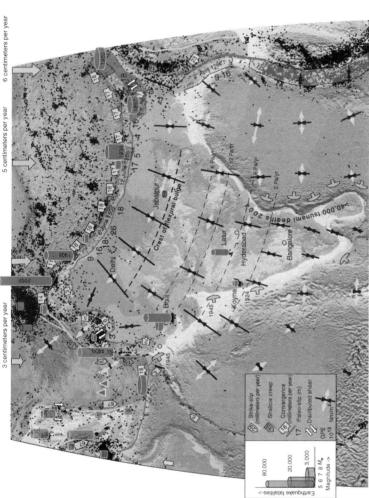

FIGURE 5.4 Earthquakes, death tolls, plate-boundary velocities, and interplate stresses in the Indian plate. Inferred flexural stressing rates (dashed lines in Pa/yr, Bilham et al., 2003) and potential energy stress in the Indian subcontinent (Ghosh et al., 2006), with megaquake rupture zones (green), recent earthquakes (Preliminary Determination of Epicenters (PDE) by the International Seismological Centre), inferred GPS plate-boundary vectors (arrowed tabs with numbers in millimeters per year) (Socquet et al., 2006; Banerjee et al., 2008; Ader et al., 2012; Khan et al., 2008; Szeliga et al., 2012; Drukpa et al., 2012; Gahalaut et al., 2013; and Himalayan paleoseismic slip (black) numbers adjoining collision zones in meters (Lavé et al., 2005; Kumar et al., 2001, 2010; Jayangondaperumal et al., 2011; and Sapkota et al., 2013). Motion of the Ormara plate shaded blue near Karachi is quantified by Kukowski et al. (2000) and the consequences of rotation of the Shillong block (also blue) by Drukpa et al. (2012). Known tsunamis are indicated by a blue hammerhead, with major earthquake death tolls and magnitudes shown as vertical cylinders. Red triangles are volcanoes. Due to India's slow anticlockwise rotation, the velocities of collision increase eastward, but the convergence across the Himalaya (yellow tabs in millimeters per year) remains approximately uniform. Green tabs indicate observed creep rates on surface faults

5.4 A SUMMARY OF EARTHQUAKE HAZARDS IN AND SURROUNDING THE INDIAN PLATE

The highest strain rates measured in the region (>100 nanostrain per year) occur across plate boundaries (Figure 5.2). It is these regions that exhibit the highest seismic productivity for all ranges of earthquake magnitudes. Wrapped around the edges of the Indian plate are three distinct plate-boundary settings where in the past 50 million years the Indian plate has been forcibly inserted into Asia at rates that once exceeded 10 centimeters per year, but have now slowed down to 3–5 centimeters per year. To the west, the Chaman fault system between Karachi and Kabul permits the western edge of the Indian plate to slide past Baluchistan and Afghanistan. To the east, the eastern edge of the Indian plate partly collides and partly slides past Indonesia and the Andaman islands along the Sagaing fault system through Myanmar to Assam. The northern edge of the Indian plate lies hidden beneath the Tibetan plateau several 100 km north of the Ganges, but the northern edge of the Ganges/Brahmaputra plain is truncated by a thrust fault slipping at roughly 2 centimeters per year that marks the start of the Himalayan collision zone, a 2,000-km-long belt of seismicity that has hosted two $M_w \geq 8.4$ earthquakes in the past 100 years, and more are considered overdue. In the following sections, we discuss the history of earthquakes in each of these regions in detail, the structures that underlie these earthquakes, and the populations at risk in each setting.

5.4.1 Western Edge of the Indian Plate; Afghanistan, Pakistan, and Baluchistan

The 1,000-km-long Chaman fault system, a left-lateral boundary between the Indian and EuroAsian plates has an overall slip rate of 29 millimeters per year distributed over a 100- to 400-km-wide zone. It includes numerous branches, few of which are parallel to the slip vector between India and Asia, but several segments exceed 200 km indicating Mmax (Maximum credible earthquake) capabilities exceed $M_w = 7.5$. The largest known earthquakes in the fault system (1935 $M_w = 7.7$ Quetta, 2013 $M_w = 7.7$ Arawan) occur at intervals of approximately 500 years given observed GPS rates. It is clear from Figure 5.5(a) that the historical record is remarkably incomplete prior to 1800, and many major earthquakes are probably missing from early earthquake histories in the region (Ambraseys and Bilham, 2003a,b).

Subparallel faults throughout the fault system share the burden of sinestral slip and partitioned convergence (note the arrows on fold belts orthogonal to India's motion). Convergence evident in Figures 5.1 and 5.5 between the fold belts bordering the Chaman fault and the Indus Valley occurs at rates from 2 to 17 millimeters per year, the fastest rate occurring in the southern Sulaiman range. Right-lateral shear at 11 millimeters per year occurs between the Sulaiman range and the fold belts south of Quetta, which is absorbed by slip on

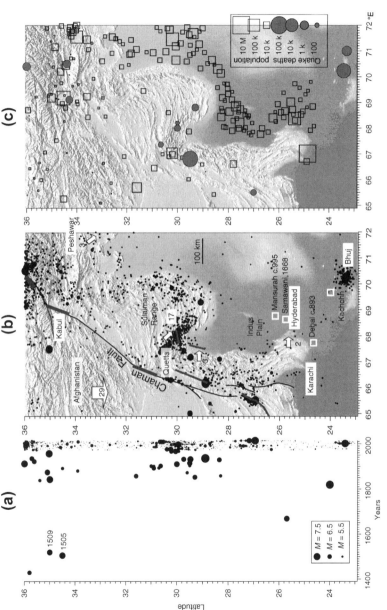

FIGURE 5.5 (a) Historical earthquakes 1400–2014 on the Chaman fault system and the western edge of the Indian plate. (b) Pre-1900 earthquakes (Ambraseys and Bilham, 2003a,b; Bilham and Lodi, 2010) and instrumental earthquakes (PDE and Centennial catalog). (c) Populations and recent deaths from earthquakes. Blue = ruptured faults. Arrows indicate GPS velocities relative to the Indian plate (millimeters per year). The obliquity of the plate boundary to the plate convergence direction (≈ north/south) is responsible for the considerable thrust fault activity to the east of the Chaman fault. Despite two $M_w \geq 7.6$ earthquakes in the Bhuj region of India in the past 200 years, GPS rates in the 200-km region near Bhuj indicate deformation rates of <2 millimeters per year. An east–west GPS convergence rate >2 millimeters per year exists near Hyderabad north of Karachi, which has been the locus of several damaging earthquakes (white squares).

subparallel SW/NE trending sinestral "bookshelf" faults. In 2008, a pair of $M_w = 6.0$ earthquakes occurred near Pishin on two contiguous faults.

A cursory view of Figure 5.5(a) and (b) suggests that a slip deficit exists between 32° and 34°N on the Chaman system. Segments of the fault system north of 32° are believed to be creeping although there is incomplete agreement on the depth to which the Chaman fault creeps. The presence or absence of creep on this part of the Chaman fault is important, because the fault here is a single strand for >200 km, and it could slip >10 m if it were to rupture in a single earthquake ($M_w \geq 7.9$). Surface creep would reduce the slip potential to be released in a future earthquake, which would otherwise have a renewal time ≥ 500 years and would result in considerable damage in Kandahar and Kabul.

Populations are sparse along much of the Chaman fault system, but building styles in villages (unreinforced masonry, adobe etc.,) are particularly vulnerable to damage from quite modest shaking. In Figure 5.5(c), it is clear that historically fatality losses from earthquakes in the region have occurred where populations are thinly distributed. In the 1935 Quetta earthquake, there was a high death toll (35,000) because it was a major town in a sparsely populated desert region that took a direct hit from a nearby earthquake. In 1936, Quetta was unique among Indian cities for being the first city in what was then India, to rebuild with mandatory earthquake resistance. The largest densities of population near the Chaman fault system are Kabul (3.3 million in 2012), Kandahar (0.5M), Quetta (2.8M), and Karachi (with 24 million, Pakistan's largest city), all close to historically damaging earthquakes.

Recent earthquakes in the past century near Karachi have been minor, but segments of the fold belt upon which it is constructed are converging with the Indian plate at a rate of ≈ 2 millimeters per year. Former damaging earthquakes in the Indus Valley within 200 km NE of Karachi have resulted in the historical destruction of several now-abandoned towns (Figure 5.5(c)). Hyderabad (current population 1.2 million) was relocated to its present location in 1768 following an avulsion of the river Indus following the 1668 Samawani earthquake (Bilham et al., 2007; Bilham and Lodi, 2010). Samawani is presently a small village near 25.53°N, 68.58°E. The destruction of Mansurah, the former Arab capital of Sindh, is approximately dated at CE 950 from the time of its abandonment some years after an earthquake. This earthquake (of unknown magnitude or precise location) destroyed major public structures in the city (also known as Brahmanabad, 25.881°N, 68.777°E) and survivors abandoned the city when the Indus shifted its course to the west, compromising its water supply and removing its commercial viability as a river trading station (Bilham and Lodi, 2010). It is likely that Debal (the archaeological site of Bhanbore at 24.751°N, 67.521°E) was also abandoned when the river nearby, the westernmost distributary of the Indus silted up, rendering the town no longer viable as a port. Kovach et al. (2010) interpret a stone tablet discovered in Debal as indicating that the city was reconstructed after an earthquake in CE 893.

The southernmost Chaman fault (the Ornach Nal fault) enters the sea roughly 100 km west of Karachi at a triple junction between the Ormara plate (Kukowski et al., 2000), the Asian Plate, and the Indian plate. No great earthquake is known between the 1945 $M_w = 8.1$ Makran earthquake and the triple junction (Bilham et al., 2007), although there is currently no reason to suppose that this segment of the plate boundary could not have an $M_w \geq 8.0$ earthquake. The 2012 $M_w = 7.7$ Awaran earthquake in Baluchistan released strain at shallow depths, and quite locally close to the fault, and hence, the slip potential from Ormara/Asia plate convergence may currently be high. A great earthquake on this segment of the plate boundary would be potentially damaging in Karachi where earthquake resistance is largely nonexistent, and where several critical facilities would be vulnerable to the resulting tsunami.

5.4.2 Northern Edge of the Indian Plate: The Himalaya from Kashmir to Arunachal Pradesh

The mean convergence rate across the central Himalaya is approximately 18 millimeters per year (Ader et al., 2012) with lower rates observed near the western end of the arc—11 millimeters per year in Kashmir (Schiffman et al., 2013) and slightly higher rates (19 millimeters per year) in eastern Assam (Drukpa et al., 2012; Vernant et al., in press; Figures 5.1 and 5.6). Several great earthquakes ($M \geq 8.0$) have occurred in the past 500 years along the Himalaya but the rupture areas of all these earthquakes, even the most recent in 1950, have not yet been studied in detail. It is certain that the largest earthquakes nucleate beneath the high mountains of the greater Himalaya and rupture the entire 100-km width of a locked decollement. The along-arc rupture lengths of Himalayan earthquakes are not well defined. Even for the 1934 earthquake, whose main shock location is known from instrumental records (Chen and Molnar, 1977; Molnar and Qidong, 1984) and whose surface rupture has been partly exhumed (Sapkota et al., 2013), the mechanisms that initiate and terminate rupture are unknown.

Figure 5.6(b) and (c) demonstrate a striking complimentary asymmetry between sparse populations and widespread seismicity north of the Himalaya, and low seismicity and high population densities south of the Himalaya. The highest seismic productivity and most of the known deaths from earthquakes have occurred within the Himalaya at the confluence of these two extremes, and all of the known high fatality earthquakes have occurred in the past 200 years. The historical database (Figure 5.6(a)) includes a handful of damaging earthquakes prior to 1800, but for these earthquakes, we have no quantitative estimates of magnitude or location. The fatality counts for most of these historical earthquakes are unknown or unreliable. There has been a tenfold increase in population, and a change in building styles towards increased fragility since 1900, which suggests that future earthquakes in the Himalaya will be associated with a significantly greater number of fatalities.

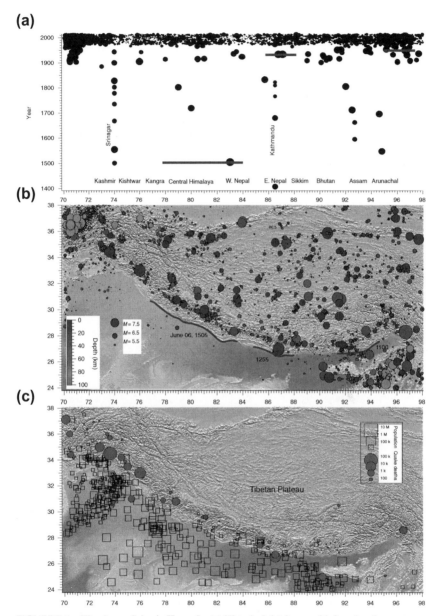

FIGURE 5.6   Himalayan Seismic Hazards. (a) Historical database and inferred major ruptures. (b) Distribution of seismicity with major surface ruptures. Deep earthquakes in green (>100 km) occur at the ends of the arc. (c) Populations are dense in the plains of India, Pakistan, and Bangladesh and sparse in Tibet.

From extant histories that have survived in the principal cities of isolated Medieval Kingdoms (Srinagar and Kathmandu), we know of numerous damaging historical earthquakes, but not their magnitudes or precise locations. Neither Kashmir nor Nepal is potentially higher in seismic productivity than elsewhere in the Himalaya, and Figure 5.6(a) suggests that had similar histories survived elsewhere along the arc the figure would now be densely populated by many more damaging earthquakes. The question of concern in seismic risk studies is the future recurrence of the most damaging earthquakes in the Himalaya. To answer this question, we would need to know the timing and location of, say, future $M_w \approx 7$ earthquakes that now occur at several decade intervals somewhere along the arc, and to know the recurrence interval of infrequent $M_w \geq 8$ earthquakes (every few centuries) along the arc, since these, with their large footprint, can damage the cities in the Ganges plains south of the Himalaya.

An often-neglected constraint in assessing seismic hazards in the Himalaya is that a Himalayan loading rate of 18 millimeters per year means that each 100-km segment of the Himalaya could host a $M_w = 7.8$ earthquake every 100 years (100 km × 100 km × 1.8 m). Thus, in any given century, 20 $M_w = 7.8$ earthquakes could occur on a segment of the Himalayan arc, that is, one every 5 years. The actual recurrence rate exceeds 50 years, that is, one-tenth the anticipated rate. The inescapable conclusion is that Himalayan stress is not being released in these smaller (but nonetheless damaging earthquakes), and is instead building up stress that will be released in larger ones. If we assume that $M_w \approx 8.5$ earthquakes are the norm for the Himalaya (1950 Assam $M_w = 8.6$, 1934 Nepal $M_w = 8.4$ see discussion below), and these occasionally rupture the Himalaya in seven 300-km-long segments, they could do so every 300 years with 5.4 m of slip. We might expect two to three of these great earthquakes to occur every 100 years somewhere along the Himalayan arc. Coincidentally, this is about the rate we have seen in the past century, but from what little we know of the history of these very large earthquakes (Figure 5.6(a)), it appears that no $M_w = 8.5$ earthquakes occurred in the previous 100 years, or even the past 400 years, which fits this pattern. We therefore conclude that some Himalayan earthquakes are much larger than $M_w = 8.5$, and that they occur very infrequently, possibly at intervals of $\geq 1,000$ years.

One such earthquake occurred at dawn on the June 6, 1505 (Ambraseys and Jackson, 2003), and may have had a magnitude of $M_w = 9.0$ (Figure 5.6(b)). Evidence for its magnitude comes not so much from the distribution of recorded intensities, which are sparse and imprecise, but from paleoseismic evidence. Several trenches across the Himalayan frontal thrust between 77° and 84°E revealed offset sediments containing detrital charcoal dated at about 1400 CE (Wesnousky et al., 1999; Kumar et al., 2001, 2010). The causal earthquake must have thus occurred after this date, and many are now of the opinion that these trenches have sampled a single earthquake corresponding to the June 1505 earthquake along at least 550 km of the arc. Although not a certain indication of a single megaquake, it is suggestive, and consistent with the broad felt area of damage in the 1505 earthquake (Joshi et al., 2009). Slip of as much as 23 m has been inferred in some of these trench investigations (Lavé et al., 2005; Kumar et al., 2001, 2010; Yule et al., 2006) and scaling laws suggest that such a large slip requires a correspondingly long along-arc ruptures (Stirling et al., 2014).

In segments of the Himalaya where the slip in meters and the date of the most recent great earthquake and the convergence rate (Figure 5.4) are all known, it is possible to infer the approximate time to the next earthquake (an important assumption is that a future earthquake will slip the same amount as in the previous earthquake). Applying this reasoning to the June 6, 1505, earthquake and a time invariant convergence rate of 18 millimeters per year (Ader et al., 2012), a 1000- to 1300-year time between earthquakes must elapse, and hence, an identical repeat of this earthquake cannot occur for a further 500−700 years. There are inherent observational difficulties that attend this calculation because it is not certain how well point measurements of a paleoseismic slip characterize the mean slip in an earthquake (i.e., are point measurements of >20 m of observed slip representative of the mean slip), or whether the large observed slip exposed in trenches crossing the fault represents the slip in one earthquake, or from several major earthquakes that occurred years or decades apart.

Even if these uncertainties were overcome, the problem with this approach is that we would typically like several historical demonstrations that Himalayan earthquakes indeed recur with the kind of regularity implied by this type of time-predictable renewal process, a regularity that depends upon not just the strain at failure but also upon the friction and strength of the rock-rupture process being replicated in consecutive earthquakes. Examples of repeating earthquakes elsewhere in the world show that this regularity exists only in a statistical sense, with large variations in intervals between earthquakes.

No statistics for the recurrence interval of earthquakes in India exist. In the Himalaya, and in fact throughout India, we know of only two regions where earthquakes have repeated historically—the 1255 and 1934 $M_w > 8$ earthquakes in eastern Nepal (Sapkota et al., 2013) and possibly the 1999 Chamoli and 1803 $M_w < 7$ earthquakes in the central Himalaya. In Gujarat, two nearby $M_w > 7.6$ earthquakes, in 1819 and 2001, did not occur on the same fault and cannot be considered repeats of one another.

5.4.3 Eastern Boundary of the Indian Plate

The disastrous Indian Ocean tsunami resulted from the rupture, in a time span of 10 min, of a 1,600-km-long segment of the eastern plate boundary from 2°N to 15°N (Lay et al., 2004; Chlieh et al., 2007), almost twice as long as the inferred 1505 central Himalayan earthquake discussed above. Previous tsunamigenic earthquakes (e.g., 1881 Ortiz and Bilham, 2003; 1945 Bilham et al., 2005) had been described before the earthquake and evidence for an earlier tsunami have since been documented (Malik et al., 2011; Wang et al., 2013). North of the Andaman Islands, the rupture in 2004 terminated near 15°N where the plate boundary veers to be subparallel to the slip vector. In Myanmar, north of 18°N, the northward slip of the Indian plate is accommodated on the right-lateral Sagaing fault system and subparallel faults, and the east–west component of convergence with the Burma plate is absorbed by a subduction zone whose central segment slipped in a great earthquake in April 1662 (Cummins, 2007; Wang et al., 2014). The shorter segment between 22° and 25°N may have slipped in 1664, for which there are several authentic accounts (Iyengar et al., 1999; Ambraseys, 2004). An earthquake in 1548 is repeated in numerous popular media as affecting Sylhet, and by implication Dhaka and most of Bangladesh, but original accounts of this event are brief, and its location may have been in the Naga Hills bordering the Brahmaputra Valley (Iyengar et al., 1999).

The Indian plate dips eastward beneath the Indo-Burman ranges (Satyabala, 1998) to depths exceeding 100 km with a Benioff zone (Figure 5.7(b)) defined by moderate and occasional major earthquakes (e.g., $M_w = 7.4$, 1954). Earthquakes at shallow depths above this east dipping zone are associated with east–west compression and folding (Steckler et al., 2008; Wang et al., 2014). In the past century, sufficient earthquakes on contiguous segments of the dextral strike-slip Sagaing fault system have occurred to identify gaps where future earthquakes are now probable (Wang et al., 2014). Numerous population centers follow the active faults of south central Myanmar (Figure 5.7(c)), and there is considerable concern that future $M > 7$ earthquakes will occur near major cities (e.g., Wyss, 2008).

5.5 CONSERVATISM AND DENIAL AS AGGRAVATED RISK

In the foregoing sections, I provide a brief summary of the seismic hazards that follow the western, northern, and eastern plate-boundary regions of the Indian plate. These provide the input that have been used by individuals, governments, and global organizations such as Global Seismic Hazard Program (GSHAP) and Global Earthquake Model (GEM) as input to maps of seismic hazards, which in turn form the basis of numerical estimates of seismic risk. A glance at the figures is enough to show that numerous population centers are at risk from being damaged by earthquakes in each region and that the number of historical "hits" represents a small fraction of potential future "hits" from earthquakes. In this context, a "hit" is the occurrence of a major earthquake close to a major population center, an example being the 1935 Quetta earthquake with its 35,000 human fatalities.

The left-hand (a) panels of each figure each reveal the burst of seismic information that occurs after the installation of seismometers around 1900 and the yet greater increment that occurs following the deployment of the standardized seismic network in 1960. The corollary of these data-rich improvements in

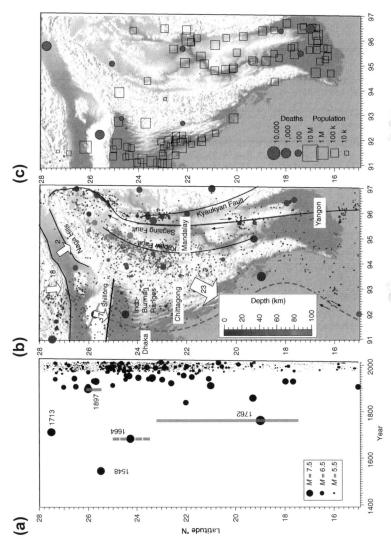

FIGURE 5.7 (a) Historical earthquakes. (b) Earthquake distribution and GPS velocities relative to the Indian plate, and principal surface faults. (c) Populations and fatality losses from historical earthquakes of Bangladesh and Myanmar. Population data are spatially incomplete.

knowledge is the extraordinary incompleteness of historical information about earlier earthquakes. From time to time, new earthquakes are added as a result of the discovery of an as-yet unexamined history of a region, or increasingly, as paleo-seismic investigations of surface faults yield information extending our knowledge of earthquakes to prehistory. It is my opinion, however, that the historical record is likely to remain in its present incomplete state for many years.

I now return to the theme of aggravated seismic risk. This section introduces the notion that seismic risk can be aggravated by interpretative denial. Because historical data are interpreted by so-called experts, it is possible, and indeed expected, that expert opinion will form a diversity of conclusions based on these available data. If a damaging earthquake is deleted from the historical record, it can effectively reduce future seismic risk, thereby relaxing local building codes; if one is added, it will increase the assessments of future hazards, and lead to stronger building codes. Often, these diverse opinions can be incorporated into so-called logic-tree calculations of hazard where extreme views are considered and assigned probabilities. The consequences of these considerations are not academic, they affect human safety, and often have enormous economic consequences. For example, a conservative view of the historical earthquake record in a specific region may permit the construction of a nuclear power plant at half the cost of one that incorporates a more cautious interpretation of these same data.

It is important to understand why some expert opinions would wish to err on the side of a conservative interpretation of earthquake histories.

5.5.1 Unsensational Science versus Banner Headlines

Scientists, in general, know the perils of sensationalizing their results. Unnecessary alarm is counterproductive since it desensitizes the listener to real alarm. Its consequences are exemplified by the story of the young shepherd who cried wolf when there was none, knowing that he would grab the attention of the concerned community, only to be ignored by the community when a real wolf appeared. The problem for the earthquake scientist is that the historical seismic record describes the past imperfectly, and the future only in a probabalistic sense. When findings indicate that the seismic risk of a region has been hitherto underestimated, it is common for scientists to report their findings to each other in the vacuum of academic discourse—at scientific meetings, in learned articles and in conversations in corridors—but it is difficult to convey this information to the public without the information being considered sensational. This is partly due to the substantial reporting differences between scientific and journalistic reporting. Science proceeds by building hypotheses that can be refuted. Journalists write what they consider to be confirmed truths that they hope (if they are honest) are beyond refutation. Words spoken at a scientific meeting sound very different on prime-time news.

Four recent examples of the gap between scientific knowledge and implementation are noted, of several that have occurred in recent years.

- The L'Aquilla disaster in Italy followed from difficulties in communicating the statistical uncertainty of a damaging earthquake. When a fatal earthquake occurred, scientists involved in assessing the significance of foreshocks were accused of providing "incomplete, inaccurate and contradictory" statements (Hall, 2011; Nosenga, 2012).
- A few years prior to the Tohuko 2011 tsunami, scientists knew that a previous tsunami had swept ashore with a similar amplitude and reach (Satake et al., 2008; Tajima et al., 2013), but mechanisms to communicate these new data to those responsible for coastal defenses were unavailable. Attempts to convey this information through media outlets would have been characterized as alarmist.
- In the years preceding the 2010 Haiti earthquake, Eric Calais and his colleagues approached government officials in Haiti with their findings that geodetic strain was close to being renewed following two eighteenth century earthquakes that had destroyed the former city of Port au Prince (Manaker et al., 2008). Officials listened but were unable to act.
- In June 2004, an invited talk was given at a scientific meeting in Bangalore, sponsored in part by the National Science Foundation and National Aeronautics and Space Administration, on the natural hazards of southern India. During the talk, graphical views of potential tsunami runup along the east coast of India were presented, based on the historical tsunami in the Nicobar and Andaman islands (Ortiz and Bilham, 2003). Neither the speaker nor the audience had any intention of releasing these findings to the press, or persuading coastal authorities to implement tsunami warnings. I was the speaker, and the tsunami that occurred six months later was far worse than I had calculated.

The problem scientists face is that the transition from a tentative, or certain, scientific finding to a news release is a quantum leap. There is no smooth transition from abstract hypothesis to headline news. Societal infrastructure responds not to the scientist but to the megaphone of the journalist. Scientific findings may be important to societal planning, but societal action frequently requires the litmus test of banner headlines, a course of action that scientists loathe to invoke.

5.5.2 Denying the Unprecedented—"$M_w = 9.0$ Earthquakes Cannot Occur in the Himalaya"

The $M_w = 9.1$ earthquake that occurred along the submarine southeastern boundary of the Indian plate in 2004 had no historical precedent. The largest known historical earthquake at the time was the 1881 $M_w = 7.9$ Car Nicobar earthquake. The parallels between Himalayan seismicity and the length of the

Himalayan plate boundary prompted speculation that the 1505 earthquake in the central Himalaya may have been a similarly large earthquake (Bilham and Wallace, 2005). To qualify as an $M_w = 9.0$, the earthquake had to satisfy an area and slip combination that was unprecedented in the historical record of the Himalaya—but the arithmetic was, and is, plausible—a rupture length of 600 km, a down-dip width of 100 km and a slip of 20 m yield a magnitude $M_w = 9.0$.

Yet to some scientists, the conclusion is unacceptable. To support their view, a litany of objections can be assembled, similar to the response of the scientific community to the proposition by Alfred Wegener of his theory of continental drift. The arithmetic that leads to the conclusion that $M_w = 9$ earthquakes must be considered possible in the Himalaya is simple, but the three quantities that go into the arithmetic can each be questioned, as can the historical record itself.

A recent article by Srivastava et al. (2013) criticizes the weaknesses of the historical record (Figure 5.6(a)) but then uses it as a template to conclude that the Himalaya are unable to sustain earthquakes larger than $M_w = 8.6$ in the east or $M_w = 7.6$ in the west, with a range of intermediate magnitudes in between, as exemplified by the past century of earthquakes. The authors deny that the Himalaya can sustain approximately 600-km-long ruptures, and propose that the region is too segmented to permit long ruptures to propagate. To overcome the inadequacy of historical earthquakes to keep pace with the known seismic-deficit accumulation rate, they invoke an as-yet unidentified creep process on the Himalayan décollement. Yet, most certainly creep is currently absent (Bilham et al., 1997; Avouac, 2003; Ader et al., 2012). Historical accounts of collapsed temples in Tibet and damage to Agra in 1505 are attributed by Srivastava et al. (2013) to a modest earthquake in the region and to local amplification effects. Paleoseismic slip >20 m reported in trench investigations of the Himalayan Frontal Thrust are dismissed as unreliable. Age dating of detrital carbon is considered questionable. All these arguments are a fundamental component of scientific refutation, but for none are compelling arguments provided that refute the published observations considered unacceptable by Srivastava et al. (2013).

Other authors have questioned the magnitude of Himalayan earthquakes in the past millennium (Rajendran et al., 2013, and previous articles cited therein) based on specific observational data. These authors place considerable emphasis on the survival of Medieval masonry structures as indicative of the absence of great earthquakes. Their findings are of interest since numerous examples of repairs to temples have been discovered that supplement our knowledge of historical earthquakes. However, the findings presented are qualitative in that they provide few numerical estimates for the maximum accelerations and shaking durations implied by the survival and repair of these monuments. The survival of ancient structures is enigmatic in that those that have survived tell us little about those that have succumbed to shaking (Ambraseys, 2009). For example, accelerations in the Andaman islands did not exceed Intensity VIII in the 2004 $M_w = 9.1$ earthquake. In Kashmir, several structures assembled at approximately 800 CE are known to have survived Intensity VII shaking, and possible Intensity VIII shaking (Bilham et al., 2010; Bilham and Bali, 2013), each successive earthquake resulting in incremental damage but incomplete collapse.

5.5.3 National Pride and the Need to Protect the Public from Alarm

The populations at risk from an $M_w \geq 8$ earthquake in the Himalaya are considerable because of the anticipated high accelerations and the long shaking duration of great earthquakes. Scenario studies undertaken to assess the death toll from large earthquakes predict huge losses. For example, Wyss (2006) estimated a fatality count of 29,000−56,000 for a possible future Indian Kashmir earthquake. Hitherto, India's reaction to the recognition that great earthquakes are inevitable in northern India (e.g., Bilham et al., 2001) has been inconsequential; however, in 2012, a series of scenario projections of damage from $M_w = 8$ earthquakes along the Himalaya were initiated by India's National Disaster Management Authority. Early results indicate that a repeat of the 1897 $M_w = 8.1$ earthquake could result in 600,000 deaths in Assam, and death tolls from $M_w \geq 8$ earthquakes elsewhere could reach 800,000. The simulation model for the Mandi region ($\approx 76°-77°$E) is based on a line source model (Sinha and Murty, 2012) and forecasts Intensity IX shaking over a 10,000 km^2 epicentral region. The simulation may be insufficiently conservative because in previous Himalayan earthquakes Intensity IX shaking has been observed in quite limited (≈ 100 km^2) regions (Ambraseys and Bilham, 2003c; Ambraseys and Douglas, 2004).

However, resistance to the notion of an $M_w = 9$ earthquake remains, and one reason is thought to be that such earthquakes are intrinsically "off-scale". The global headlines following the 2004 Andaman and 2010 Tohuko $M_w = 9$ headlines would be unthinkable were they associated with Delhi or Srinagar, or Gauhati, or Agra. Local reactions to an assessment of possible scenarios for future earthquakes in Kashmir by an international team in December 2011 (Jones, 2011) understandably resulted in concern, panic, and denial by local officials. It is possible that denial comes from an unjustified common assumption that earthquake shaking intensity (distinct from shaking duration) rises with magnitude. As mentioned earlier, the recurrence of such a megaquake in the central Himalaya is not anticipated for approximately 700 years when calculated from the renewal time for approximately 20 m of slip. In Kashmir, Mmax has in fact been calculated to be $M_w = 9$ based on the (unlikely) contiguous rupture of a 180-km-wide, 300-km-long segment length between the Kangra 1905 and the Kashmir 2005 earthquakes, one of a dozen scenario ruptures (Schiffman et al., 2013). Although the shaking duration increases with great earthquakes, there is no evidence to suggest that the accelerations during $M_w > 8.5$ earthquakes may be significantly larger than those associated with

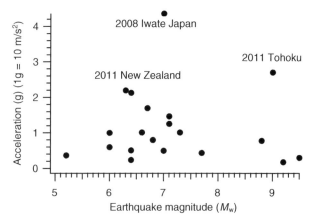

FIGURE 5.8   **Magnitude versus maximum shaking intensity for some recent earthquakes.**

$M_w < 8.0$ earthquakes. In general, the stress at failure, and proximity to the fault rupture, determines local acceleration, but the elastic energy released in an earthquake is stored in a volume of rock surrounding the fault and there is a corresponding increase in mass upon which these accelerations act. Observational data confirm an absence of a linear relationship between M_w and measured peak surface acceleration (e.g., Figure 5.8). As an extreme example consider the 2011 Tohuko $M_w = 9$ earthquake in which accelerations locally exceeded 2 g. In the same year, similar peak accelerations were recorded during the $M_w = 6.3$ and $M_w = 6.4$ Christchurch, New Zealand earthquakes that released 0.01 percent of the energy released in the Tohuko earthquake. During the 2004 $M_w = 9.1$ Indonesian/Andaman earthquake, shaking intensities in the Andaman islands did not exceed MSK VIII (<0.65 g) above the rupture zone. Such considerations are of little comfort to populations at risk in and near the Himalaya, since accelerations of the order of 1 g will prove disastrous to infrastructure and typical constructions in cities in the Ganges plains to their south where shaking durations and amplitudes are expected to be amplified by basin resonances (Hough et al., 2005a,b). The denial of the possibility of the future occurrence of a great earthquake by those responsible for public safety should be accompanied by irrefutable evidence to support such opinions. Unsubstantiated claims aggravate the assessment of seismic risk.

5.5.4 Jaitapur: "All Apprehensions Raised Have Been Considered"

Jaitapur (16.6°N, 73.3°E) is a small village on the west coast of India far from the boundaries of the Indian plate. The Nuclear Power authority of India has been advised that it is an ideal location to site the largest nuclear power station in the world.

This apparent perfection, however, is marred by the observation that a significant fault scarp is mapped near the planned power plant. No recent earthquake has been recorded instrumentally closer than 44 km, but two earthquakes in the past century at a similar latitude indicate that stresses in the Indian plate are sufficiently high to sustain moderate earthquakes: in 1967, an $M_w = 6.7$ earthquake occurred at Koyna (17.4°N, 73.8°E) 110 km to the north, and in 1993, an $M_w = 6.3$ earthquake occurred at Latur (18.1°N, 76.5°E) to the NW (Figure 5.4). The origin of the stresses released by these earthquakes is related to India's collision with southern Tibet, which flexes the Indian plate into several hundred kilometers of undulation and compresses it in a north–south sense (Lyon Caen and Molnar, 1983; Bilham et al., 2003), and to the gravitational potential energy of the >9 km range in topography, from the sea floor to the 5 km mean elevation of the Tibetan plateau (Gosh et al., 2006). The existence of a substantial deviatoric stress in the Indian plate accounts for the faint background of microseismicity manifest throughout India in small and sometimes damaging earthquakes. The larger earthquakes occur on faults with dimensions of tens of kilometers, at intervals of many hundreds if not many thousands of years. Hence, the existence of any fault near Jaitapur, given the history of moderate earthquakes at these latitudes, is of concern to characterizing seismic hazards in the region (Bilham and Gaur, 2011; Gaur and Bilham, 2012).

Usually, the faults of southern India are subtly expressed, and many of them are concealed in the subsurface, but a glance at the Google imagery south of Jaitapur reveals a 35-km-long, 20-m-high step in the marine terrace on which the power plant will be constructed. There is no disagreement among geologists that this is the scarp of a tectonic fault (the Vijaydurg fault), and were this in Europe, Japan, or the USA, it would have been the subject of extensive paleoseismic trenching investigations to determine when it last slipped. Its on-land length exceeds 35 km, and its NNW projection offshore passes within 10 km of the site an unknown distance to the north. The dimensions of the fault suggest that if it slipped in a single earthquake, it could do so with $6.5 <$ Mmax < 7.5 depending on its offshore extent, which is currently unknown. The site report and a recent article (Rastogi, 2012) asserts that the fault is inactive—incapable of slipping in a future earthquake—based on the surface inspection of the fault scarp. The site report (not released publicly as of January 2014) indicates that no trenching has been undertaken and that no offshore mapping has been undertaken. In the USA, a capable fault is defined as one that has slipped once in the past 35,000 years, or slips with a recurring nature in the past 500,000 years. Such questions have not been addressed at Jaitapur.

The issue is not so much that a local fault may potentially slip in an earthquake near a planned nuclear power plant, since a power plant can be built to withstand accelerations of ≥ 1 g (although the design of a nuclear plant potentially 10 km from the epicenter of an $M_w = 7.5$ earthquake would be challenging and costly). At issue is the absence of transparency in the site review process, and the apparent absence of many of the geological investigative techniques currently available to quantify fault activity both onshore and offshore (Gaur and Bilham, 2012). Perhaps of greatest concern is the overt attempt to stifle scientific discussion concerning seismicity near Jaitapur (Bagla, 2012). Government suppression of the discussion of seismic hazards must surely aggravate objective assessments of future seismic risk. It is of little value claiming that "all apprehensions raised have been considered" (Rastogi, 2012), without quantitative and transparent evidence supporting such claims.

5.6 EARTHQUAKE KNOWLEDGE AND ITS APPLICATION

In the above account, I consider the aspects of earthquake hazards, an imperfect knowledge of which in most cases underestimates earthquake risk. I avoid the presentation of seismic hazard maps, since these are maps of an imperfect earthquake history, not of a possible seismic future (Nikolic-Brzev et al., 1999). I also omit a discussion of the extensive probabilistic calculation of earthquake risk from the best available information of seismic hazards. The remainder of this chapter assumes that accurate estimates of seismic hazard are available, and focuses on the imperfect transfer of this knowledge to the user community.

Earthquake risk has been discussed for >100 years, and the developed nations have responded by incrementally improving the guidelines and reach of recommendations for earthquake-resistant construction. In wealthy industrial countries such as Japan and in regions of the USA where earthquakes occur, earthquake risk is recognized as a problem with a simple solution—quantify the risk and design structures accordingly (FEMA, 1995). Earthquake-resistant design is typically considered in a cost—benefit analysis, and for civic structures is mandatory where the hazard has been quantified. Earthquake resistance in historical public structures (retrofits) is introduced where existing earthquake resistance is considered inadequate, but is left to the home owner in older housing. New housing starts in California must comply with earthquake-resistant design recommendations before a planning permit is awarded, and at each step in construction, compliance is verified before the next stage in construction can be undertaken. Thus, to a large degree, the public in developed nations is protected from unexpected home fragility, or the collapse of civic structures such as schools, fire stations, and hospitals, or from earthquake damage to city infrastructure such as harbors, pipelines, bridges, and roadways.

In the developing world where most of the population increase in the past century has occurred, earthquake resistance has not been an important factor in the assembly of dwellings (Jain, 2005). Earthquake resistance has been ignored through ignorance or poverty. In some cases, earthquake resistance has been deliberately avoided in order to maximize construction profits.

Adobe mud buildings with thick mud roofs are still common in Iran (Berberian, 1979; Manafpour, 2008). The use of undressed field stones cemented with mud to assemble walls proved fatal in the Latur earthquake in India. In contrast, some traditional construction methods are inherently resistant to earthquake shaking. In the Kachchh region of India, adobe mud walls surmounted by thatch survived the Bhuj earthquake. In Kashmir, Sikkim, and Ladakh traditional stone and masonry buildings are intercalated with wooden beams that absorb shaking energy (Langenbach, 2009; Tessman, 2012). In Assam and in villages in Kachchh, wattle-and-daub or bamboo lightweight dwellings (Das et al., 2012) once common and resistant to earthquake shaking are now shunned by those who can afford concrete and steel structures. In some case, villagers spurn traditional construction methods in favor of the perceived prestige attending the adoption of buildings that have adopted new methods of construction.

The demise of wooden construction in most parts of Asia due to the depletion of forests means that some traditionally earthquake resistant buildings can no longer be considered by villagers. Increasingly, the only materials available for new construction consist of bricks, cement, and steel. In nearly every settlement, village, town, and city, new construction is based on these three materials. When they are of adequate quality and correctly assembled, they can provide safety from shaking in an earthquake. When the bricks or cement are weak, or if the steel is brittle, or of an insufficient quantity with inadequate internal connectivity, the assembled buildings are prone to collapse in an earthquake. This is particularly the case in cities where space is at premium and the need to house many families requires buildings with three or more storys, resulting in higher loading stresses during earthquakes.

In the major cities and megacities of south Asia, the current building stock is fragile and is an easy target for earthquake damage (Bilham and Gaur, 2013). A measure of the fragility of structures in south Asia is the frequent occurrence of buildings that collapse spontaneously, but the percentage of substandard construction is unknown. A well-known example is the collapse of a garment factory on May 24, 2013, in Dhaka that killed 1,123 people, and injured 2,500 others. Although the largest of these disastrous collapses are reported internationally, local news reports of building collapse suggests that it is far more common than realized. During the year 2012 in India, 2,737 instances of structural failure (including bridges, walls and buildings) resulted in 2,651 dead and 850 injured (Chalabi, 2013). In Pakistan, the statistics are yet to be gathered, but a cursory survey of collapse reports in Karachi, Islamabad, and Hyderabad reveals that collapse in "zero intensity" shaking is not uncommon. Almost on an annual basis a building collapses in Karachi killing or injuring occupants. After 21 were killed in August 2011, the Sindh Building Authority indicated that they had issued notices to occupants to leave 200 buildings considered in a dangerous condition. Despite warnings, occupants rarely abandoned these condemned buildings, and in 2013, Karachi witnessed the spontaneous collapse of buildings in January, April, and August.

5.6.1 Wealth, Ignorance, and Corruption

An essential feature of civic institutions in cities with a moderate population density is the presence of an engineering/planning facility responsible for ensuring the wise design of city infrastructure and the safe assembly of public structures and private dwellings. The collapse of buildings due to shoddy construction thus speaks to a breakdown in the responsibilities of these planning departments. To varying degrees, the fragility of construction methods in a city is related to dysfunction in the inspection process and Ambraseys and Bilham (2011) hypothesize that this weakness is in part related to the prevalence of corruption in a country (Figure 5.9). Corruption statistics (CPI = Corruption Perception Index) are taken from Transparency International (2013). The association is not simple because corruption, income levels, and education are to some degree related to each other. But although poor countries are generally more corrupt than are rich countries, some countries are more corrupt than others with similar income levels, and this has been quantified numerically as an *expectation index*. The striking conclusion is that more than 80 percent of global earthquake fatalities have occurred in nations that are more corrupt than predicated by their level of poverty (Figure 5.10).

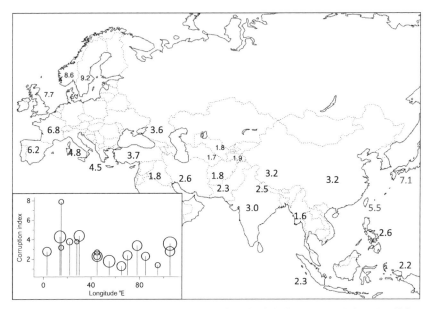

FIGURE 5.9   Perceived Corruption Index since 2002–2012 declines eastward along the Alpine Himalayan collision zone to low values between Iran and Myanmar. Ten is not corrupt (transparent with no bribes) and 1 is very corrupt (opaque with rampant bribery). The inset shows a weak relationship between the corruption index and cumulative fatalities (circle size is proportional to log cumulative earthquake deaths since 1900).

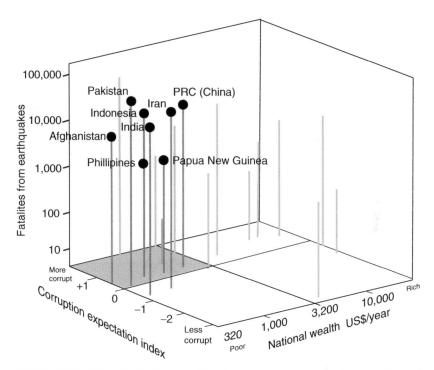

FIGURE 5.10   Global deaths from earthquakes as a function of national wealth and corruption expectation index. Selected Asian countries are highlighted with bold circles. The expectation index is a measure of the observed corruption perception index (CPI) above or below the mean CPI anticipated from a nation's average-income level (Ambraseys and Bilham, 2011). Of the 20 countries in this plot, one-third lie in south Asia. The gray area on the NW side of the base of this 3D plot indicates low average-income countries with higher than expected levels of perceived corruption, where >80 percent of all deaths from earthquakes have occurred in the past century.

The difficulties associated with examining the intertwined global statistics of corruption and poverty are addressed by Asquer (2011).

5.6.2 Corruption and Its Many Forms in the Building Industry

The approximately $10 trillion ($10^{13}$) per year building industry is the most lucrative component of the global economy: roads, dams, bridges, civic structures, schools, banks, business centers, malls, parking lots, and housing. The basic ingredients of construction concern real estate, steel, and cement. Their manipulation requires the participation and skillful organization of numerous engineering trades: architects, civil engineers, demolition experts, excavation and soil engineers, transport planners, assembly contractors, water and waste engineering, electrical distribution, and public safety.

The enormity of the funds available and the complexities of completing construction projects over periods of several months or years offer many opportunities for dishonest elements of the building industry to participate and to maximize profits where possible. The most common form of dishonest practice is to use substandard materials and/or assembly methods, and to conceal these shortcuts invisibly within the structure. The term *cover-up* originates in the building industry. At few points during the incremental progression from planning, to foundation, to structure, and to a final coat of paint, are retroactive tests of structural integrity possible; hence, in the industrial nations, a rigid sequence of building inspections has been introduced.

It is important to distinguish between engineered and nonengineered construction. Engineered construction includes the official construction of civic structures and major buildings that require architectural and engineering design and are subject to mandatory supervision by experts during assembly. In contrast, nonengineered construction applies to buildings that are assembled by contractors and home owners without supervision by engineers. Many dwelling units in the large cities of south Asia are nonengineered structures. Both forms and construction are nominally subject to inspection, and both forms of construction are, or should be, required to adhere to local earthquake-resistant construction guidelines.

Although it is self-evident that earthquake resistance will be considered essential by an engineer aware of local code requirements in an industrial nation, a rural contractor in a developing nation may be unaware of earthquake-resistant codes, or may consider them irrelevant unless an earthquake has occurred recently nearby. The city building inspector will expect adherence to earthquake-resistant codes, but the rural building inspector, even when aware of earthquake-resistant codes, may judge them inappropriate especially when persuaded by visible hardship, or a small payment to turn a blind eye to their absence.

The circumvention of inspections during the construction process is the most common form of corruption in the building industry in south Asia (Bilham, 2009, 2012). Permits are required to initiate construction and the issuance of these permits can often be influenced by bribes. Code avoidance can occur at the planning stage (construction in environmentally protected land), at the foundation level (inappropriate location in regions of unstable ground), at the design stage (avoidance of earthquake resistant elements), in the construction stage (avoidance of appropriate materials or incorporation of substandard assembly practices), and at the final approval stage (ignoring known noncompliant elements in the building). Each of these corrupt practices enhanced construction profits for the builder, and income for the inspectors who accepted these bribes.

A substantial number of structures are assembled without any kind of planning or construction oversight by user/owners, some on illegally occupied lots. Once built, the structure becomes part of the landscape and other illegal structures follow in time encroaching on contiguous areas. Occasionally, authorities will react and demolish such structures, but all too often, they remain and become part of a city's building stock. A remarkable example of corrected encroachment recently occurred in Kathmandu when authorities ordered all buildings encroaching on main thoroughfares to be truncated at their legal lot lines. The truncated buildings exposed half rooms and severed hallways and were considerably weakened by the removal of shear walls, requiring substantial retrofits to retain building integrity.

5.7 DISCUSSION—WHO GAINS, WHO LOSES

A curious feature of worldwide housing is that few buildings can be purchased or rented with any kind of a guarantee of their structural integrity, or resistance to shaking, yet it would be unthinkable to purchase a motor vehicle that did not adhere to the most stringent safety criteria. While this is unlikely to change in the foreseeable future, earthquake risks remain a societal problem that can be managed only by the resolve of the State. The collapse of hundreds of schools in the 2005 Pakistan earthquake, for example, reveals that the State itself may be unaware of its responsibilities for code awareness and enforcement. Home owners and apartment renters are not seismologists, nor are city planners, contractors, developers, and engineers intrinsically aware of earthquake risks. In the absence of a recent earthquake, knowledge that buildings must be assembled to resist occasional strong shaking must come from the collective wisdom of a small group of historians, geologists, and seismologists familiar with the geological structure and seismic history of a region. A knowledge of these earthquake hazards must be transferred to, and acted on, by the State if it is to effectively reduce earthquake risk to its citizens. Numerous societal impediments, however, act to prevent the successful application of this knowledge (Bilham, 2009). Seismologists are adept at describing hazard and at quantifying risk, but are ill equipped to see the conversion of quantified risk to safe housing. The "unkindest cut of all" is the circumvention of the collective knowledge of seismology, engineering, and state laws during the assembly of a structure by an ignorant or corrupt contractor (Bilham et al., 2012).

The cost of earthquakes to society is considerable. In recent earthquakes, the costs of reconstruction have exceeded \$500 billion (\$5 \times 10^{11}$) in the industrial nations and deaths have exceeded 0.5 million in the developing nations (Figure 5.11). A line approximately divides the poor nations from the industrial nations with a slope of 100 deaths per million dollars. "Transparency International" has estimated that the cost of corruption in the building industry results in worldwide construction costing the public 15 percent more than it would were corruption absent. Engineers indicate that the incorporation of earthquake resistance to a building may add less than an additional 10 percent to its total cost. Global earthquake resistance would appear to be less expensive than the costs to society of corruption.

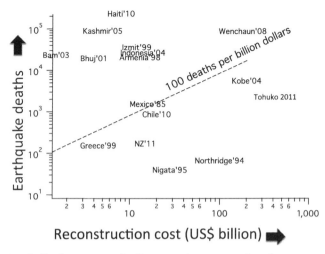

FIGURE 5.11   **Dollar losses versus fatality counts in recent earthquakes.**

The doubling and redoubling of SE Asia building inventories in the past four decades represents a lost opportunity to construct safer cities. Instead, the building boom was undertaken at minimal cost, leaving many cities intrinsically fragile. Construction was undertaken largely with indifference to earthquake resistance, and in many cases without construction code adherence of any kind. The true state of building fragility in a region is often revealed only during earthquakes, and numerous examples have occurred in the past decade to remind us that deaths in urban earthquakes can exceed 12 percent of an urban population. Clearly, this state of affairs should not continue, but it is unlikely to change if the disconnect between earthquake hazards/risks and safe construction is not firmly addressed. In many countries, national earthquake-resistant codes exist, but their implementation is applied without rigor. The challenge for future seismologists and engineers is perceived to be one of education: to educate the State of the certainty of future earthquake disasters given current policies, and to educate the public that construction shortcuts by contractors who assemble their dwellings are unacceptable.

Several mechanisms are available to politicians and developers who are willing to place progress ahead of considerations of safety. In the case of the Tehri Dam, disagreements among scientists with different views were exploited to demonstrate a lack of consensus, and to select a result favorable to government designs (Gaur and Valdiya, 1993). In the case of the Jaitapur nuclear power plant investigation, and estimates of Mmax for Kashmir, it was considered necessary for the Indian Government to banish from India the US coauthor of studies attempting to quantify these hazards (Bagla, 2012). A trickle-down effect also exists that goes far beyond initial conflicts of interest. Once a scientific study has been touched by political controversy, young scientists are discouraged from investigating the issues raised, or similar issues, for fear of government reprisals in the form of restrictions in future research funding, or threats to job security and future promotion.

5.8 CONCLUSIONS

The implementation of earthquake-resistant construction in south Asia depends not only on the objective evaluation of the history and probable future of earthquakes in the region but also on the transfer of this knowledge to an appropriate user community. The seismic history of India's plate boundaries are incomplete, but in some locations, they are adequate to form probabilistic estimates of future risk for regions of high seismic productivity in India, Pakistan, Bangladesh, Nepal, Bhutan, and Myanmar. In contrast, the short historical record of intraplate earthquakes renders the characterization of seismicity in central and southern India very uncertain. The probability of supplementing the written historical record is low everywhere, but there remains hope that paleoseismic studies of surface faults will extend the record of major earthquakes back several thousands of years along plate boundaries, continental margins, and in the interior of the plate.

The development of a reliable historical database and from it estimates of seismic risk and regional code guidelines is considered insufficient to guarantee that building codes are applied ubiquitously to civic and private structures. Corrupt practices prevail throughout the building industry that can prevent engineered structures from performing according to design. Where private dwellings are assembled by contractors, a mix of corrupt practices and ignorance can lead to structures that are unsafe even without earthquake shaking.

Despite competence in the geophysical and engineering communities of south Asia, numerous societal pressures are at work in India that in the past have tended to downplay seismic hazards. Instead of being seen as a necessary prelude to safe engineering design, studies of seismic hazards are often perceived to constitute a threat to progress, with the potential to delay or cancel plans for major engineering works. It is almost impossible to believe that any public organization would justify attempting to influence objective assessments of seismic hazards, but circumstantial evidence demonstrate that this indeed occurs. Political influence of the scientific method is a form of high-level corruption.

Various forms of aggravated seismic risk are identified. Seismic risk is high in many locations in south Asia due to large populations located on the wide collision zone that follows the southern margin of the EuroAsian plate. Political pressures to bias seismic hazards to a level favorable for inexpensive construction of major engineering works do not reduce seismic risk. Instead, they aggravate the potential for future disaster. The development of building code guidelines based on seismic hazard mapping is successful where seismic productivity is high, largely because the historical database is short in south Asia and is best sampled where earthquake recurrence intervals are short. In other regions, building code guidelines are likely to be misleading because maps of seismic hazard reflect recent earthquakes, which may not at all be where future earthquakes will occur. Earthquake zoning maps in some mid-continent regions may aggravate seismic risk, since they lull design engineers into a false sense of seismic stability. Finally, and perhaps most importantly, the combined efforts of seismologists, historians, and engineers to implement safe dwellings and civic structures to protect future societies from harm in earthquakes may be failing to reach those most in need of this protection—in the low cost housing in villages and cities in south Asia where more than 2 billion people live.

REFERENCES

Ader, T., Avouac, J.-P., Liu-Zeng, J., Lyon-Caen, H., Bollinger, L., Galetzka, J., Genrich, J., Thomas, M., Chanard, K., Sapkota, S.N., 2012. Convergence rate across the Nepal Himalaya and interseismic coupling on the Main Himalayan Thrust: Implications for seismic hazard, J. Geophy. Res. 117 (B4). http://dx.doi.org/10.1029/2011JB009071.

Ambraseys, N., 1970. Some characteristic features of the Anatolian fault zone. Tectonophysics 9, 143−165.

Ambraseys, N., 1971. Value of historical records of earthquakes. Nature 232, 375−379.

Ambraseys, N., 2009. Earthquakes in the Mediterranean and Middle East: A Multidisciplinary Study of Seismicity up to 1900. Cambridge University Press, pp. 968.

Ambraseys, N., 1975. Studies in Historical Seismicity and Tectonics, Geodynamics Today. The Royal Society, London, pp. 7−16.

Ambraseys, N., 2004. Three little known earthquakes in India. Curr. Sci. 86, 506−508.

Ambraseys, N., Bilham, R., 2003a. Earthquakes in Afghanistan. Seismol. Res. Lett. 74 (2), 107−123.

Ambraseys, N., Bilham, R., 2003b. Earthquakes and crustal deformation in northern Baluchistan. Bull. Seismol. Soc. Am. 93 (4), 1573−1605.

Ambraseys, N., Bilham, R., 2003c. MSK Isoseismal intensities evaluated for the 1897 Great Assam Earthquake. Bull. Seismol. Soc. Am. 93 (2), 655−673.

Ambraseys, N., Bilham, R., January 13, 2011. Corruption kills. Nature 469, 143−145.

Ambraseys, N., Jackson, D., 2003. A note on early earthquakes in northern India and southern Tibet. Curr. Sci. 84, 570−582.

Ambraseys, N.N., Douglas, J., 2004. Magnitude calibration of North Indian earthquakes. Geophys. J. Int. 158, 1−42. http://dx.doi.org/10.1111/j.1365-246X.2004.02323.

Asquer, R., 2011. The Deadly Effects of Corruption: A Cross National Study of Natural Disasters, 1980−2010. Unpublished paper UCLA Dept. of Political Science. https://www.academia.edu/845400/The_Deadly_Effects_of_Corruption_A_Cross-National_Study_of_Natural_Disasters_1980-2010.

Avouac, J.P., 2003. Mountain building, erosion, and the seismic cycle in the Nepal Himalaya. Adv. Geophys. 46, 1−80.

Bagla, P., 2012. India barred entry to U.S. author of seismic review. Science 338, 1275.

Banerjee, P., Bürgmann, R., Nagarajan, B., Apel, E., 2008. Intraplate deformation of the Indian subcontinent. Geophys. Res. Lett. 35, L18301. http://dx.doi.org/10.1029/2008GL035468.

Berberian, M., 1979. Tabas-e-Golshan (Iran) earthquake of 16 September 1978; a preliminary field report. Disasters 2, 207—219.

Bilham, R., 1994. The 1737 Calcutta earthquake and cyclone evaluated. Bull. Seismol. Soc. Am. 84 (5), 1650—1657.

Bilham, R., Bendick, R., Wallace, K., 2003. Flexure of the Indian plate and intraplate earthquakes. Proc. Indian Acad. Sci. (Earth Planet. Sci.) 112 (3), 1—14.

Bilham, R., Larson, K., Freymueller, J., Project Idylhim members, 1997. GPS measurements of present-day convergence across the Nepal Himalaya. Nature(Lond) 386, 61—64.

Bilham, R., Wallace, K., 2005. Future $M_w > 8$ earthquakes in the Himalaya: implications from the 26 Dec 2004 $M_w = 9.0$ earthquake on India's eastern plate margin. Geol. Surv. India Spec. Publ. 85, 1—14.

Bilham, R., Singh, B., Bhat, I., Hough, S., 2010. In: Sintubin, M., Stewart, I.S., Niemi, T.M., Altunel, E. (Eds.), Historical Earthquakes in Srinagar, Kashmir: Clues from the Shiva Temple at Pandrethan, GSA Special Paper 471 on Ancient Earthquakes, ISBN 9780813724713.

Bilham, R., Bali, B.S., 2013. A ninth century earthquake-induced landslide and flood in the Kashmir valley, and earthquake damage to Kashmir's medieval temples. Bull. Earthquake Eng. 11 (4), 1—31. http://dx.doi.org/10.1007/s10518-013-9504-x.

Bilham, R., Lodi, S., 2010. In: Sintubin, M., Stewart, I.S., Niemi, T.M., Altunel, E. (Eds.), The Door Knockers of Mansurah: Strong Shaking in a Region of Low Perceived Seismic Risk, Sindh, Pakistan. GSA Special Paper 471 on Ancient Earthquakes, ISBN 9780813724713.

Bilham, R., Lodi, S., Hough, S., Bukhary, S., Khan, A.M., Rafeeqi, S.F.A., 2007. Seismic hazard in Karachi, Pakistan: uncertain past, uncertain future. Seismol. Res. Lett. 78 (6), 601—631.

Bilham, R., Engdahl, E.R., Feldl, N., Satyabala, S.P., 2005. Partial and complete rupture of the Indo-Andaman plate boundary 1847—2004. Seismol. Res. Lett. 76, 299—311. http://dx.doi.org/10.1785/gssrl.76.3.299.

Bilham, R., 2012. Societal and observational problems in earthquake risk assessments and their delivery to those most at risk. Tectonophysics 584, 166—173. Active Tectonic Deformation of the Tibetan Plateau and Great Earthquakes. http://dx.doi.org/10.1016/j.tecto.2012.03.023.

Bilham, R., 2009. The seismic future of cities, Twelfth annual Mallet Milne Lecture. Bull. Earthquake Eng. 7 (4), 839—887. http://dx.doi.org/10.1007/s10518-009-9147-0.

Bilham, R., Gaur, V.K., Molnar, P., 2001. Himalayan seismic hazard. Science 293 (5534), 1442—1444.

Bilham, R., Gaur, V.K., 2011. Historical and future seismicity near Jaitapur, India. Curr. Sci. 101 (10), 1275—1281.

Bilham, R., Gaur, V., 2013. Buildings as weapons of mass destruction: earthquake risk in South Asia. Science 341, 618—619.

Chalabi, M., September 3, 2013. How many die in building collapses in India? Interactive map. The Guardian. http://www.theguardian.com/news/datablog/interactive/2013/sep/03/india-falling-2651-deaths-structural-collapse-architecture-housing.

Chen, W.-P., Molnar, P., 1977. Seismic moments of major earthquakes and the average rate of slip in Central Asia. J. Geophys. Res. 82, 2945—2969.

Chlieh, M., et al., 2007. Coseismic slip and afterslip of the great M_w 9.15 Sumatra—Andaman earthquake of 2004. Bull. Seismol. Soc. Am. 97 (1A), S152—S173. http://dx.doi.org/10.1785/0120050631.

Cummins, P.R., 2007. The potential for giant tsunamigenic earthquakes in the northern Bay of Bengal. Nature 449 (7158), 75—78.

Das, P., Korde, C., Sudhakar, P., Satya, S., 2012. Traditional Bamboo houses of NE region: a field study of Assam and Mizoram. Key Eng. Mater. 517, 197—202.

Drukpa, D., Pelgay, P., Bhattacharya, A., Vernant, P., Szeliga, W., Bilham, R., December 2012. GPS constraints on Indo-Asian convergence in the Bhutan Himalaya: segmentation and potential for a $8.2 < M_w < 8.8$ earthquake. HKT meeting, Kathmandu, Nepal. J. Nepal Geol. Soc. 45 (special issue), 43–44.

England, P., Jackson, J., 2011. Uncharted seismic risk. Nat. Geosci. 4, 348–349. http://dx.doi.org/10.1038/ngeo1168.

FEMA, 1995. Seismic Considerations for Communities at Risk, Program on Improved Seismic Safety Provisions, Federal Emergency Management Agency, Building Seismic Safety Council Washington. US Government Printing Office, pp. 114.

Gahalaut, V.K., Khundu, D., Laishram, S.S., Catherine, J., Kumar, A., Singh, M.D., Tiwari, R.P., Chadha, R.K., Samanta, S.K., Ambikapathy, A., Mahesh, P., Bansal, A., Narsaiah, M., 2013. Aseismic slip boundary on the Indo-Burmese wedge, northwest of the Sunda Arc. Geology 41, 235–238.

Gaur, V.K., Valdiya, K.S., 1993. Earthquake Hazard and Large Dams in the Himalaya. Indian National Trust for Art and Cultural Heritage, New Delhi, ISBN 81-900281-2-X.

Gaur, V.K., Bilham, R., December 10, 2012. Discussion of seismicity near Jaitipur. Curr. Sci. 103 (11), 1273–1278.

Ghosh, A., Holt, W.E., Flesch, L.M., Haines, A.J., 2006. Gravitational potential energy of the Tibetan Plateau and the forces driving the Indian plate. Geology 34 (5), 321–324. http://dx.doi.org/10.1130/G22071.1.

Hall, S.S., 2011. Scientists on trial: at fault. Nature 477, 264–269.

Holzer, T.L., Savage, J.C., 2013. Global earthquake fatalities and population. Earthquake Spectra 29 (1), 155–175.

Hough, S.E., Bilham, R., 2006. After the Earth Quakes. Oxford University Press pp. 321.

Hough, S.E., Bilham, R., Ambraseys, N., Feldl, N., 2005a. Revisiting the 1897 Shillong and 1905 Kangra earthquakes in northern India: site response, Moho reflections and a triggered earthquake. Curr. Sci. 88 (10), 1632–1638.

Hough, S.E., Bilham, R., Ambraseys, N., Feldl, N., 2005b. The 1905 Kangra and Dehra Dun earthquakes. Geol. Surv. India Spec. Publ. 85, 15–22.

Iyengar, R.N., Sharma, D., Siddiqui, J.M., 1999. Earthquake history of India in medieval times. Indian J. Hist. Sci. 34 (3), 181–237.

Iyengar, R.N., 1999. Earthquakes in ancient India. Curr. Sci. India 77 (6), 827–829.

Jackson, J., 2006. Fatal attraction: living with earthquakes, the growth of villages into megacities, and earthquake vulnerability in the modern world. Phil. Trans. R. Soc. 364, 1911–1925. http://dx.doi.org/10.1098/rsta.2006.1805.

Jain, S.K., 2005. The Indian earthquake problem. Curr. Sci. 89, 1464–1466.

Jaiswal, K., Wald, D., 2010. An empirical model for global earthquake fatality estimation. Earthquake Spectra 26, 1017–1037.

Jayangondaperumal, R., Wesnousky, S.G., Choudhuri, B.K., 2011. Near-surface expression of early to late Holocene displacement along the northeastern Himalayan frontal thrust at Marbang Korong Creek, Arunachal Pradesh. India Bull. Seismol. Soc. Am. 101 (6), 3060–3064.

Jones, N., December 8, 2011. Western Himalayan region faces big quake risk. Nat. Newsblog.

Joshi, D.D., John, B., Kandpal, G.C., Pande, P., 2009. Paleoliquefaction features from the Himalayan frontal belt, India and its implications to the status of 'Central Seismic Gap'. J. South Asia Disaster Stud. 2, 139–154.

Khan, M.A., Bendick, R., Bhat, I.M., Bilham, R., Kakar, D.M., Khan, S.F., Lodi, S.H., Qazi, M.S., Singh, B., Szeliga, W., Wahab, A., 2008. Preliminary geodetic constraints on plate boundary deformation on the western edge of the Indian Plate from TriGGNnet. J. Himalayan Earth Sci. 41, 71−87.

Kovach, R.L., Grijalva, K., Nur, A., 2010. Earthquakes and civilisations of the Indus Valley: a challenge for archaeoseismology. Geol. Soc. Amer. Sp. Paper 471, 119−127.

Kukowski, N., Schillhorn, T., Flueh, E.R., Huhn, K., 2000. Newly identified strike-slip plate boundary in the northeastern Arabian Sea. Geology 28, 355−358.

Kumar, S., Wesnousky, S.G., Jayangondaperumal, R., Nakata, T., Kumahara, Y., Singh, V., 2010. Paleoseismological evidence of surface faulting along the northeastern Himalayan front, India: timing, size, and spatial extent of great earthquakes. J. Geophys. Res. 115, B12422. http://dx. doi.org/10.1029/2009JB006789.

Kumar, S., Wesnousky, S.G., Rockwell, T.K., Ragona, D., Thakur, V.C., Seitz, G.G., 2001. Earthquake recurrence and rupture dynamics of Himalayan frontal thrust, India. Science 295 (5550), 2328−2331.

Langenbach, R., 2009. Dont Tear it Down! Preserving the Earthquake Resistant Vernacular Architecture of Kashmir. UNESCO, India.

Lavé, J., Yule, D., Sapkota, S., Basenta, K., Madden, C., Attal, M., Pandey, M.R., 2005. Evidence for a great medieval earthquake (1100 A.D.) in the Central Himalayas, Nepal. Science 307, 1302−1305.

Lay, T., Kanamori, H., Ammon, C., Nettles, M., Ward, S., Aster, R., Beck, S., Bilek, S., Brudzinski, M., Butler, R., DeShon, H., Ekström, G., Satake, K., Sipkin, S., December 26, 2004. The great Sumatra−Andaman earthquake of. Science 308, 1127−1133.

Lyon Caen, H., Molnar, P., 1983. Constraints on the structure of the Himalaya from an analysis of gravity anomalies and a flexural model of the lithosphere. J. Geophys. Res. 88 (B10), 8171−8191.

Malik, J.N., Shishikura, M., Echigo, T., Ikeda, Y., Satake, K., Kayanne, H., Sawai, Y., Murty, C.V.R., Dikshit, O., 2011. Geologic evidence for two pre-2004 earthquakes during recent centuries near Port Blair, South Andaman Island, India. Geology 39, 559−562.

Manafpour, A.R., 2008. Bam Earthquake, Iran: lessons on the seismic behaviour of building structures. In: The 14th World Conference on Earthquake Engineering October 12−17, 2008, Beijing, China.

Manaker, D.M., Calais, E., Freed, A.M., Ali, S.T., Przybylski, P., Mattioli, G., Jansma, P., Prepetit, C., De Chabalier, J.B., 2008. Interseismic plate coupling and strain partitioning in the northeastern Caribbean. Geophys. J. Int. 174, 889−903.

Molnar, P., Qidong, D., 1984. Faulting associated with large earthquakes and the average rate of deformation in central and eastern Asia. J. Geophys. Res. 89, 6203−6227.

Musson, R., 2012. The Million Death Quake: The Science of Predicting Earth's Deadliest Natural Disaster. Palgrave Macmillan, NY, ISBN 978-0-230-11941-3, pp. 255.

Nikolic-Brzev, S., Green, M., Krimgold, F., Seeber, L., 1999. Development of an earthquake hazard map for Maharashtra, 81−89. In: Lessons Learned over Time. Learning from Earthquakes Series. Innovative Earthquake Recovery in India, vol. 2.

Nosenga, N., 23 Oct 2012. Italian court finds seismologists guilty of manslaughter. Nature News. http://dx.doi.org/10.1038/nature.2012.11640.

Oldham, T., 1883. A Catalogue of Indian earthquakes. In: Oldham, R.D. (Ed.), Mem. Geol. Surv. India, vol. 19, pp. 163−215. Geol. Surv. India, Calcutta.

Ortiz, M., Bilham, R., April 23, 2003. Source area and rupture parameters of the 31 Dec. 1881 M_w 7.9 Car Nicobar earthquake estimated from Tsunamis recorded in the Bay of Bengal. J. Geophys. Res. 108 (B4) [2002JB001941RR 2003].

Paul, J., et al., March 1995. Microstrain stability of Peninsula India 1864–1994. Proc. Indian Acad. Sci. (Earth Planet. Sci.) 104 (1).

Rastogi, B.K., 2012. Seismicity near Jaitapur India. Curr. Sci. 103, 130–131.

Rajendran, C.P., Rajendran, K., Sanwal, J., Sandiford, M., 2013. Archeological and historical database on the medieval earthquakes of the Central Himalaya. Ambiguities Inferences Seismol. Res. Lett. 84, 1098–1108. http://dx.doi.org/10.1785/0220130077.

Sapkota, S.N., Bollinger, L., Klinger, Y., Tapponier, P., Gaudemer, Y., Tiwari, D., 2013. Primary surface ruptures of the great Himalayan earthquakes in 1934 and 1255. Nat. Geosci. 6, 71–76.

Satake, K., Namegaya, Y., Yamaki, S., 2008. Numerical Simulation of the AD 869 Jogan Tsunami in Ishinomaki and Sendai Plains. In: Annual Report on Active Fault and Paleoearthquake Researches, vol. 8, pp. 71–89 (in Japanese with an English abstract).

Satyabala, S., 1998. Subduction in the IndoBurma region: Is it still active? Geophys. Res. Lett. 25 (16), 3189–3192.

Schiffman, C., Bali, B.S., Szeliga, W., Bilham, R., 2013. Seismic slip deficit in the Kashmir Himalaya from GPS observations. Geophys. Res. Lett. 40, 5642–5645. http://dx.doi.org/10.1002/2013GL057700.

Seeber, L., Ekstrom, G., Jain, S.K., Murthy, C.V.R., Chandak, N., Ambruster, J.G., 1996. The Killari earthquake in central India: a new fault in Mesozoic basalt flows? J. Geophys. Res. 101 (B4), 8543–8560.

Sinha, R., Murty, C.V.R., Goyal, A., Krishna, C., Vishnoi, P., May 14, 2012. $M_w = 8.0$ Mandi Earthquake Disaster Scenario for Disaster Risk Management. Report Number 2. http://revenueharyana.gov.in/html/disastermgt/Report_scenario_IITbombay.pdf.

Socquet, A., Vigny, C., Chamot-Rooke, N., Simons, W., Rangin, C., Ambrosius, B., 2006. India and Sunda plates motion and deformation along their boundary in Myanmar determined by GPS. J. Geophys. Res. 111, B05406. http://dx.doi.org/10.1029/2005JB003877.

Srivastava, H.N., Bansal, B.K., Verma, M., 2013. Largest earthquake in Himalaya: an appraisal. J. Geol. Soc. India 82, 15–22.

Stirling, M., Goded, T., Berryman, K., Litchfield, N., 2014. Selection of earthquake scaling relationships for seismic-hazard analysis. Bull. Seismol. Soc. Am. 103, 2993–3011.

Steckler, M., Akhter, S.H., Seeber, L., 2008. Collision of the Ganges–Brahmaputra Delta with the Burma arc. Earth Planet. Sci. Lett. 273, 367–378.

Szeliga, W., Bilham, R., Kakar, D.M., Lodi, S.H., 2012. Interseismic strain accumulation along the western boundary of the Indian subcontinent. J. Geophys. Res. 117, B08404. http://dx.doi.org/10.1029/2011JB008822.

Tajima, F., Mori, J., Kennett, B.L.N., 2013. A review of the 2011 Tohoku-Oki earthquake (M_w 9.0): large-scale rupture across heterogeneous plate coupling. Tectonophysics 586 (2013), 15–34.

Talwani, P., 1997. Seismotectonics of the Koyna–Warna area, India. Pure Appl. Geophys. 150, 511–550.

Tessman, R., 2012. Earthquake Damage Assessment—Vulnerability of Sikkim's Built Heritage. INTACH- Indian National Trust for Arts and Cultural Heritage, 1-79, New Delhi. Transparency International 2013. http://www.transparency.org/cpi2013/results.

Vernant, P., Bilham, R., Szeliga, W., Drupka, D., Skalita, S., Bhattacharyya, A., Gaur, V.K., Clockwise Rotation of the Brahmaputra valley: tectonic convergence in the eastern Himalaya, Naga Hills and Shillong Plateau, J. Geophys. Res. In press.

Wang, Y., Bruce, J., Shyu, H., Sieh, K., Chiang, H.-W., Wang, C.-C., Aung, T., Nina Lin, Y.-N., Shen, C.-C., Min, S., Than, O., Lin, K.K., Tun, S.T., 2013. Permanent upper-plate deformation in western Myanmar during the great 1762 earthquake: implications for neotectonic behavior of the northern Sunda megathrust. J. Geophys. Res. Solid Earth 118. http://dx.doi.org/10.1002/jgrb.50121.

Wang, Y., Sieh, K., Tun, S.T., Lai, K.-Y., Myint, T., 2014. Active tectonics and earthquake potential of the Myanmar region. J. Geophys. Res. http://dx.doi.org/10.1002/2013JB010762.

Wesnousky, S.G., Kumar, S., Mohindra, R., Thakur, V.C., 1999. Uplift and convergence along the Himalayan Frontal Thrust of India. Tectonics 18, 967—976.

Wyss, M., 2006. The Kashmir M7.6 Shock of 8 October 2005 Calibrates estimates of losses in future Himalayan earthquakes. In: Van de Walle, B., Turoff, M. (Eds.), Paper Presented at Proceedings of the Conference of the International Community on Information Systems for Crisis Response and Management, p. 5. Newark.

Wyss, M., Wang, R., Zschau, J., Xia, Y., 2006. Earthquake loss estimates in near real-time. EOS, 477—479.

Wyss, M., 2008. Estimated human losses in future earthquakes in central Myanmar. Seismol. Res. Lett. 79, 520—525.

Yule, D., Dawson, S., Lave, J., Sapkota, S., Tiwari, D., 2006. Possible evidence for surface rupture of the main frontal thrust during the Great 1505 Himalayan Earthquake, far-western Nepal, AGU Fall Meeting, abstract #S33C-05.

Chapter 6

Ten Years of Real-time Earthquake Loss Alerts

Max Wyss

University of Alaska; International Centre for Earth Simulation, Geneva, Switzerland

ABSTRACT

The priorities of the most important parameters of an earthquake disaster are number of fatalities, number of injured, mean damage as a function of settlement, and expected intensity of shaking at critical facilities. The requirements to calculate these parameters in real time are (1) Availability of reliable earthquake source parameters within minutes. (2) Capability of calculating expected intensities of strong ground shaking. (3) Data sets on population distribution and conditions of building stock as a function of settlements. (4) Data on locations of critical facilities. (5) Verified methods of calculating damage and losses. (6) Personnel available on a 24/7 basis to perform and review these calculations. Three services are available that distribute information about the likely consequences of earthquakes within about half an hour of the event. Two of these calculate losses, one gives only general information. Although, much progress has been made during the last 10 years in improving data sets and calculating methods, much remains to be done. Data sets are only first-order approximations and the methods bare refinement. Nevertheless, the quantitative loss estimates after damaging earthquakes in real time are generally correct in the sense that they allow one to distinguish disastrous from inconsequential events.

6.1 INTRODUCTION

Ten years ago, there existed no service to estimate losses due to earthquakes worldwide in real time. Today, three services exist that comment on the consequences within about 30 min. Email and Short Message Service loss alerts are distributed by the World Agency for Planetary Monitoring and Earthquake Risk Reduction (WAPMERR, Geneva, Switzerland), and the US Geological Survey (USGS; PAGER, Golden, Colorado, USA). The Joint Research Center (JRC, Ispra, Italy) distributes information about the number of population and critical facilities in the vicinity. Only WAPMERR has

Earthquake Hazard, Risk, and Disasters. http://dx.doi.org/10.1016/B978-0-12-394848-9.00006-7

distributed loss estimates for 10 years, PAGER started its activity 3.5 years later. Therefore, this report is mostly about the 10-year experience gained by WAPMERR.

The first requirement for making such services possible is the rapid determination of hypocenters and magnitudes worldwide. Based on these, one can calculate the expected ground accelerations as a function of distance from the epicenter. Next, one needs to know the distribution of the population and the properties of the building stock in the area of strong shaking. The fact that such data sets have been collected during the last 10 years made these services possible.

The global population and building data sets are still approximate, and most models are primitive, but many research groups continue to refine and add to these data. Thus, real-time loss estimates and a logical extrapolation for loss estimates of future possible earthquakes become increasingly reliable.

The term *losses*, as used here, includes damage to the built environment, and casualties. *Casualties* are the sum of *fatalities* and *patients*, the latter meaning people who are *injured* to a degree that requires treatment.

The purpose of this chapter is to briefly review how far we have come with real-time earthquake loss estimates to date and what is needed to improve these services.

6.2 BRIEF REVIEW OF THE METHODS

6.2.1 Calculating Strong Ground Motions

Once the epicenter and magnitude are known, one can calculate the displacement, the velocity, and the acceleration that any point on the Earth's surface is likely to experience due to the elastic waves emanating from the earthquake rupture. These amplitudes decrease as a function of distance. First, due to geometrical spreading of the energy, the wave energy that flows through a small circle (or surface) near the origin is distributed over an ever-widening circle (or surface). Second, the rocks through which the waves travel are not perfectly elastic; hence, some of the wave energy is absorbed by internal friction called attenuation and turns into heat. In tectonically stable parts of continents, such as the Eastern United States, the transmission properties for elastic waves are excellent, whereas they are often relatively poor in tectonically active regions, such as in California. These contrasts give rise to the first issue in estimating losses: Which attenuation function should be used for which part of the globe? In regions where dense seismograph networks exist with station separations of about 10–50 km, attenuation properties are well known, but not in most seismically active areas because they are not sufficiently instrumented.

Many proposals for attenuation functions exist, but not all are equally reliable. Currently, the most prudent choice of attenuation function is to follow the recommendations of the Next Generation Attenuation Models (https://www.eeri.org/products-page/next-generation-attenuation-models/).

The units in which the strong motions are mapped in a scientific, relatively precise way are acceleration in meters per second squared. However, intensity maps are commonly used in practice because the macroseismic intensity (I) scale (Wood and Neumann, 1931) is derived from felt and damage reports, giving the user an idea what happened to the built environment. Using the 12-point scale of intensity for mapping strong ground motions is less precise because the units are coarse, but such maps are what first responders need in the field because the intensity scale is more informative than acceleration values. An example of a map showing calculated intensities by settlement is shown in Figure 6.1(a).

This figure illustrates how Quake Loss Assessment for Response and Mitigation (QLARM) maps look like for great earthquakes, distributed in email alerts. At the same time, it demonstrates how future or past earthquakes can be modeled, once confidence has been established by real-time alerts that loss results are correct to an acceptable degree in the region in question (Wyss 2005, 2006). This is a model for a repeat of the Bihar earthquake for which 10,700 fatalities were reported (Utsu, 2002). With the Indian population having increased since 1934 by an approximate factor of 4, one would expect about 43,000 fatalities due to the increase, assuming no change in building resistance to shaking. The QLARM estimate of fatalities for a repeat in 2010 ranges from 8,000 to 50,000, agreeing with this expectation. The number of settlements for which QLARM estimates $I \geq IV$ is 9,550 in this case.

6.2.2 Calculating Damage Inflicted

An informative way of conveying the state of a settlement to first responders is to present maps on which the estimated mean damage in a settlement is shown (Figure 6.1(b)). Settlements are composed of a range of building types (Jaiswal et al., 2010) with a range of resistance to strong shaking. In order to estimate the degree of damage a building of a given type is likely to sustain, a set of fragility curves have been constructed that describe the probability of collapse (for example) as a function of the intensity of shaking. The European Macroseismic Scale, EMS98 (Grünthal, 1993), recognizes classes ranging from A, the weakest, to F, the most resistant building.

The distribution of damage states is calculated for each building type present in a settlement, based on the intensity expected. The mean damage state of the whole settlement is then derived by considering all the expected states of damage, given the percentages of each of the building types present. A scale ranging from 0, no damage, to 5, near total destruction, is generally used.

6.2.3 Calculating Casualties

A casualty matrix gives the distribution of occupants into the three possible classes of dead, injured, or unscathed in a building that sustains a certain level

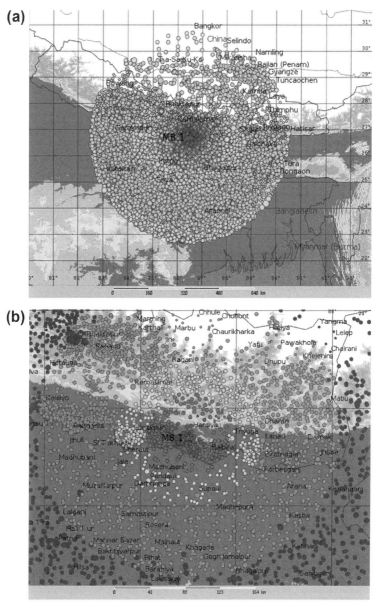

FIGURE 6.1 Maps of settlements that would be affected by a repeat of the M8.1 Bihar earth-quake of 1934 (Pandey and Molnar, 1988) modeled as a point source. Dot diameters are pro-portional to population. (a) Intensity map with colors representing I = IV at the periphery (blue) to I = IX near the epicenter (red). (b) Mean damage per settlement from green = 1, to read = 4, on a scale of 5.

of damage. This matrix varies strongly between countries with different building practices. At a given ground acceleration, few fatalities occur, although injuries will can happen in regions where modern building codes are strictly enforced, whereas in regions with poorly constructed buildings sometimes more occupants are killed than injured (Wyss and Trendafiloski, 2011).

6.3 BRIEF REVIEW OF THE DATA SETS

6.3.1 Models for the Built Environment

Ideally, one would like to know the building type, size, age, height, shape, and use for every building in the approximately two million settlements known by name on our planet. In some of the most developed nations, much information on individual buildings exists, but in the vast majority of the world, even general information is hard to come by. Much information about common building types and occupancy rate has been collected by the World Housing Encyclopedia (WHE) (http://www.world-housing.net/), especially with the WHE/PAGER project. In some countries, including in Africa, limited information on housing types is contained in census reports.

Given the absence of detailed housing information, one has to resort to approximate models. For some countries, different housing inventories are given for two classes: urban and rural. This distinction is necessary because villages contain fundamentally different compositions of structures than do cities. However, distinguishing only two classes of settlements is a strong simplification of a spectrum of housing inventories as a function of size and of character of each settlement. Settlements with the same population numbers may have different building inventories because they may be predominantly administrative, agricultural, touristic, or industrial. In the data set for the computer code QLARM (Trendafiloski et al., 2011), we have modeled all countries in three size classes of settlements, <2,000, between 2,000 and 20,000, and >20,000. These boundaries are likely to vary between countries, but there is currently no reference we could find that gives differentiated information; therefore, we follow Satterthwaite (Satterthwaite, 2006), who proposes that the aforementioned limits are valid for most developing countries.

In the struggle to construct models for building types in all countries of the world, the neighborhood rule is often invoked. The idea is that neighboring countries, such as Germany, Austria, Switzerland, and Liechtenstein can be assumed to have the same distribution of building types as a first approximation. In such a group, the country with the most detailed information on buildings is used as the lead country, and the others are modeled after it. Far from being satisfactory, this approximation is necessary in all parts of the world.

6.3.2 Models for the Population Distribution

Some of the population resides in clearly defined settlements, whereas a smaller percentage is distributed in a diffuse way. In some countries, detailed census data exist. For most countries, however, census data are not available down to the smallest settlements, but only for large cities and for districts. Satellite images provide a way to map the distribution of the population without counting them by surveys on the ground (Taubenböck et al., 2014). Social, economic, and demographic indicators, such as the brightness of lights at night (http://www.ornl.gov/sci/landscan/), are used to derive the local population density. High-resolution satellite images allow the construction of population maps with units of 1-km resolution.

For first responders, it may be more useful to carry maps on which the population is represented in settlements with names, coordinates, and size, rather than in pixels that have coordinates and population, but no names. With a known size of settlements, it is also easier to characterize the built environment because the latter varies as a function of settlement size. Believing that population data by settlement are the most informative for disaster responders, WAPMERR has constructed a world data set from the following sources, prioritized according to the list below:

1. Census (does not exist by settlement in detail for most countries).
2. World Gazetteer (contains information on all countries, but only the larger cities with name, coordinates, and population) (http://world-gazetteer.com/).
3. Geonames (an extensive list of settlements with coordinates and names, but no population numbers) (http://www.geonames.org/).
4. National Geographical Intelligence Agency (NGA) (data set similar to Geonames) (http://earth-info.nga.mil/gns/html/namefiles.htm).

For most countries, the sum of the population in settlements listed in the primary sources (Census and World Gazetteer) is substantially smaller than the total population as given by the Central Intelligence Agency (CIA) on their website World Factbook (https://www.cia.gov/library/publications/the-world-factbook/geos/do.html). For countries for which the total population was not given in its census, we accepted the value given by the CIA. To reach the CIA target value for the total population, we distributed the population missing in the Census and World Gazetteer evenly into the settlements with known names extracted from Geonames and NGA. The number of people necessary to fill in these additional settlements ranged from 5 (e.g., Iceland) to approximately 5,000 (e.g., Egypt) per settlement. As a result, the QLARM data set contains nearly two million settlements, and the sum of the population in all countries is in agreement with the CIA target value to within <3 percent. For differences smaller than this value, WAPMERR could not afford the effort necessary to adjust the population more accurately.

The difference between mean damage calculations based on Landscan urban mask and QLARM is shown in Figure 6.2. This example is based on an earthquake that did happen, and the country is selected as being representative for developing nations for which the information form satellite images may be

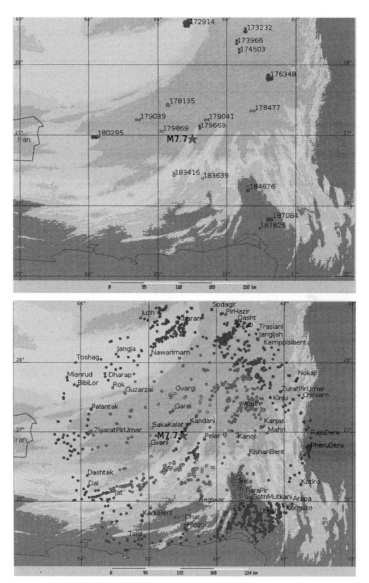

FIGURE 6.2   Map of settlements in southwest Asia estimated to have been damaged by the M7.7 earthquake of April 16, 2013. Colors indicate mean damage on a scale of 0 to 5, with 5 meaning total destruction. On top the urban centers from the Landscan urban mask, at the bottom the settlements in the QLARM data set.

expected to be superior to that based on the QLARM method, which relies on information generated in the country in question. It turns out that the contrary is the case. It is evident that the Landscan urban mask data on urban centers miss numerous settlements for which names and locations are known. Also, the Landscan urban mask data contain only about 5 percent of the population believed to be present in this area based on QLARM data.

6.3.3 World Data for Airports

QLARM contains a world data set for airports. Real-time alerts deliver a list of the 10 airports closest to the epicenter with the intensity of strong ground motion estimated for those that are close enough for this parameter being of interest. This data set contains 9464 airports for which the co-ordinates and names are known and which are distributed in 207 countries. More than 100 and >50 airports each are known for 14 and 31 countries, respectively. The average and median numbers of airports in 41 countries judged to be the most earthquake prone are 44 and 30, respectively. In these statistics, the USA and Russia are excluded, as well as islands smaller than Iceland.

6.4 BRIEF REVIEW OF THE SERVICES PROVIDED

This review is based on the alerts distributed for a set of potentially damaging earthquakes. Because it makes no sense to flood the email lists of people interested in receiving earthquake alerts in real-time, the Swiss Seismological Service (SED) (collaborating with WAPMERR in estimating alert parameters and in distributing these) has designed a filter of minimum magnitude and maximum depth for events to be analyzed (Figure 6.3). The information given in the following for all three agencies providing real-time earthquake alerts is derived from the *potentially damaging earthquakes*, as defined by the SED in Figure 6.3. In the case of the PAGER (Prompt Assessment of Global Earthquakes for Response) alerts, events smaller than M5.5 are not considered, and for the JRC in Ispra, Italy, which furnishes information to the Global Disaster Alerts and Coordination System of the United Nations only alerts with level orange or red (according to the JRC) are considered.

6.4.1 Delay Times

The WAPMERR service started about three years before the other services (Table 6.1). All WAPMERR alerts are calculated and distributed manually because predicting the numbers of fatalities is sensitive. All JRC alerts are calculated and distributed automatically, whereas it is a mixture of the two modes in the case of PAGER. All three agencies cover the entire globe with their service.

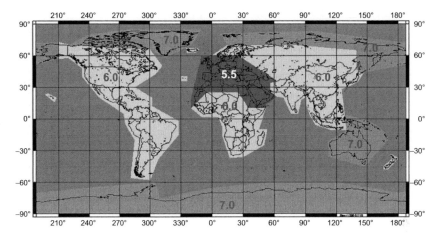

FIGURE 6.3   Map of the magnitude thresholds for shallow earthquakes for which a loss estimate is required in real time by the SED and the rescue team of the Swiss Agency for Development and Cooperation.

TABLE 6.1 Delays of Real-time Earthquake Alerts

Agency	Start Date	Type	Average Delay (min)	Median Delay (min)	Numbers Total
WAPMERR	31/10/2002	Manual	36	30	754
PAGER	11/04/2006	Automatic and manual	80	29	294
JRC	26/09/2005	Automatic	38	21	143

The median delay times between the earthquake occurrence time and the loss alerts (Table 6.1) are about 30 min for WAPMERR and PAGER because all and many are issued manually, by the two agencies, respectively. By skipping the review process, the JRC gains 9 min, median. These delay times are based on the time the email alerts are registered in my email server recipient as a regularly subscribing user to all three services.

The total number of alerts listed in Table 6.1 are those I received as a user, subject to the previously explained filters. These may not exactly equal the number of alerts issued because my email server may have been down over short periods.

The total delay time (Table 6.1) is composed of two parts: The time it takes for the earthquake source parameters (latitude, longitude, depth, and magnitude) to be calculated and disseminated by other agencies, plus the time to calculate the estimated losses by the three services. The time it takes for reviewed source parameters to be distributed by the USGS worldwide has been 18 min, median, considering all events in the test set since October 2002. Thus, it takes WAPMERR and PAGER about 12 min to calculate and distribute loss estimates. The Geoforschungszentrum (GFZ), Potsdam, distributes high-quality automatic source parameter determinations within 8 min of worldwide events since May 2008. This gives a useful additional 10 min to think about possible error sources that may affect the loss estimates. The Tsunami Warning Center distributes hypocenters and magnitudes for earthquakes in the Pacific and neighboring oceans within 13 min, median. WAPMERR considers all three of these sources for calculating losses. The European Mediterranean Seismological Center distributes earthquake parameters after 27 min (median), and does not cover the globe on a consistent basis, concentrating on the Mediterranean. Thus, their information arrives too late for consideration by WAPMERR.

6.4.2 Alert Contents

The contents of the three real-time alert services are summarized in Table 6.2. The strengths of the QLARM messages are that mean damages of hundreds to thousands settlements are shown on a map and that the list of these can be obtained on request. Also, the number of fatalities, as well as the number of those injured, is estimated. As a new feature, the expected intensities of shaking at regional airports are provided. The QLARM alerts are prepared manually, including consideration of possible input errors such as hypocenter location and magnitude. The weaknesses of QLARM alerts include the fact that there are no critical facilities considered and economical losses are not calculated.

The strengths of the PAGER messages are that economic losses are estimated and historical earthquakes are reported. The weaknesses include that the number of injured is not calculated, small settlements are not included on the map, and the mean damage is not presented. The PAGER messages are reviewed for important cases.

The strength of the JRC messages is that distances of nuclear plants, hydrodams, airports, and ports are listed. The chief weakness is that the strength of shaking (intensity) is not calculated. This leads to the inability to estimate damage, fatalities, injured, and economic losses. Another weakness is that the alerts are distributed automatically. This leads to errors, such as (1) On January 15, 2009, a red alert was issued for an earthquake located offshore the Kurile Islands, which affected zero population. (2) On September 8, 2008, a tsunami alert was issued for an earthquake with a depth of 175 km. An earthquake at this depth cannot generate a tsunami. (3) On December 9, 2012, an orange alert was issued for an earthquake in the Philippines with M5.6 and at depth of 56 km; an event that generated a maximum Intensity of 4, which

TABLE 6.2 List of Alert Contents by the Three Agencies Distributing Them

Information	QLARM	PAGER	JRC
Alert level on a color scale	Yes	Yes	Yes
Earthquake source parameters	Yes	Yes	Yes
Estimated fatalities	Range	Probability distribution	No
Estimated injured	Range	No	No
Estimated economic losses	No	Probability distribution	No
Population exposed as a function of intensity	Yes	Yes	No
Map of intensity	Settlements	Contours only	No
Map of mean damage	Upon request	No	No
List of a few selected settlements	Yes	Yes	Yes
Description of local building stock	No	Yes	No
Historical earthquake list of the region	No	Yes	No
Airports	Yes	No	Yes
Critical facilities	No	No	Yes
Comments on socioeconomy and landslides	No	No	Yes

means that the perceived shaking is light and that no structures are expected to be damaged. These errors were avoidable; below I will report on some unavoidable errors.

6.5 ERROR SOURCES

6.5.1 Epicenter Errors

For practically all worldwide loss estimates, the real-time earthquake parameters are based on teleseismic recordings. This means that the seismic waves have to travel to distant seismographs, making it impossible to estimate source parameters within $<10 \pm 2$ min. As the seismic waves progress and more records come online, the parameters displayed on the GFZ Geophone web page are refined (http://geofon.gfz-potsdam.de/eqinfo/list.php). Nevertheless, the parameters distributed during the first few hours to days are less accurate than the eventual ones.

For example, the epicenter estimates rapidly distributed by the USGS differ by 25 km, on average, from local, accurate final estimates in Japan and Taiwan (Wyss et al., 2011). In the listings of the Preliminary Determination of Epicenters (USGS, which is the final list in spite of its name), this error is reduced by 9 km, on average. Depending on the absence or presence of population centers and the quality of buildings in them, the difference in estimates of fatalities can range between 0 and 10,000, due to a 25-km error in the epicenter determination (Wyss et al., 2011).

6.5.2 Depth Errors

Without a station close to the epicenter, the standard teleseismic depth estimates are uncertain by ± 25 km in the top 60 km. If a population center is close by, the resulting error of the estimate of the extent of the disaster is similar to that for epicenter errors. If the nearest population center is farther away, the difference in loss estimates diminishes.

6.5.3 Magnitude Errors

For 150 earthquakes with $M \geq 5.5$ that occurred since January 2011 and were subject to the aforementioned filters, the absolute difference between the initial magnitudes distributed by the USGS and GFZ is 0.20, median. This is a reasonably small value that contributes moderately to errors in loss estimates.

However, for great earthquakes ($M \geq 8$), the error in the first few hours is usually very large because the first pulses recorded always yield values that are far too low. There are two physical reasons for this. First, large ruptures are usually multiple events, with the first energy radiated coming from a limited source, while the rupture has not finished propagating, as in the case of the great Alaska earthquake of 1964 (M9.2), where the initial event measured M6.8 (Wyss and Brune, 1967). In addition, the standard magnitude is determined from amplitudes of waves with wavelengths of ≤ 100 km, which do not carry the full information about a source that is much longer. Therefore, the standard magnitude measurements saturate at about $M = 8$ (rupture length ~ 200 km). The underestimates of losses in cases of great earthquakes for which the magnitude is underestimated by 1 ± 0.5 units can reach tens of thousands of fatalities.

The magnitude of the Wenchuan earthquake of May 12, 2005, was first estimated as M7.5, which led to a fatality estimate of $2,500 \pm 1,500$. However, based on information by telephone that the magnitude was probably M7.9 (personal communication, Z. L. Wu, 12 May, 2005), WAPMERR revised the estimate of fatalities to 20,000–90,000. After a week of uncertainty, this estimate, issued 5½ h after the event, turned out to contain the result within its error limits (Figure 6.4). The large uncertainty is composed of formal errors arising from the probability distribution of the fragility curves, plus possible

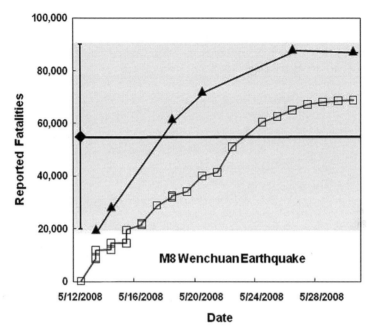

FIGURE 6.4   Fatalities reported by the Chinese media (squares) and the sum of fatalities plus missing (triangles) as a function of date after the 2008 M8 Wenchuan earthquake. The diamond with error bars shows the WAPMERR estimate of $55,000 \pm 35,000$, based on the revised magnitude of 7.9, 100 min after the event. The true extent of this disaster was not known to Chinese authorities for nearly two weeks.

influences of unknown attenuation properties and magnitude uncertainties estimated by expert judgment.

6.5.4 Incorrect Properties of Building Stock

In countries such as Iran, where many recent earthquakes have caused fatalities, the loss calculations have been verified repeatedly. Therefore, the fatality counts continue to fall within the range of the real-time estimates. However, in Haiti, where there have not been damaging earthquakes before the M7 event of January 12, 2010, the buildings turned out to be far less resistant than was assumed previously and the disaster was underestimated by WAPMERR by an order of magnitude in real time. Nevertheless, the event was classified as a disaster, and the Swiss rescue team was alerted by telephone within less than an hour.

6.5.5 Unknown Extent of Rupture

Most damaging earthquakes have $M \geq 6.5$, which means that their rupture length $L \geq 25$ km. In real time, the extent and direction of the rupture are not known. The coordinates of the initiation point of the rupture may be located quite far from a population center, leading to a low estimate of the number of fatalities. If the rupture runs toward a major population center releasing energy far closer to the latter than expected, based on the epicenter position, then a disaster will be underestimated. This was the case at Yogyakarta where on May 27, 2006, an M6.2 earthquake, at first incorrectly located offshore at a distance of 30 km from the city, was actually located on shore and broke along a fault toward the major population center of about half a million, killing 5,749. This event was not classified in real time by WAPMERR as a disaster.

6.5.6 Additional Errors

These can be introduced in loss estimates if an inappropriate attenuation function is used, if the earthquake emanates an unusual amount of high-frequency waves, and by unknown soil conditions below a population center that may amplify the strong ground motion.

6.6 CITY MODELS

For villages, it is an appropriate model to place the entire population in a single coordinate point. For medium and large cities, the approximation is inadequate. However, with tens of thousands of large cities being at risk, the cost of constructing detailed models is enormous. Besides efforts to construct 3D models using crowd sourcing (Google, Open Street Map), a highly professional industry exists serving telecommunications companies to construct 3D models of most large cities in order to place their antennas such that in the canyons of the downtowns communication can be guaranteed. If funds are available, 3D models in which each building is just a prism (Figure 6.5(a)) can

(a)   **(b)**

FIGURE 6.5   (a) Example of the type of 3D model constructed routinely by telecommunication companies for large cities. In this model of Bucharest, neighborhoods of relatively homogeneous composition are discernible. (b) Example of a 3D city model for Dubai constructed from a high-resolution satellite image, with facades photographed from the street level added. In this model, not only the building properties but also the businesses located in the office buildings are known.

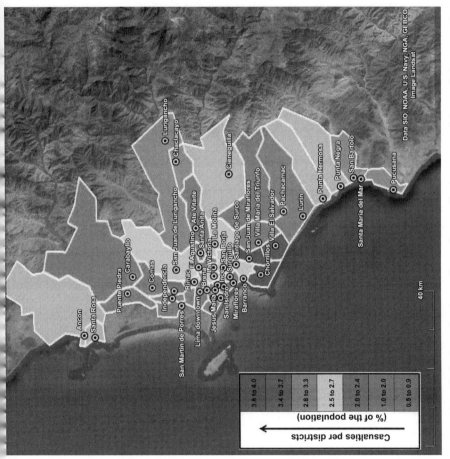

FIGURE 6.6 Map of Lima, Peru, with its 42 districts colored according to the mortality (percentage of the population expected to die) in the case of an M8.5 earthquake along the plate boundary offshore. Expected amplifications due to soil conditions are considered in this case.

be beautified by adding photographs of the facades taken from the street level from vehicles (Figure 6.5(b)). From the frontal appearance, a building may be classified and assigned a fragility class.

Construction of city models in which neighborhoods of similar buildings are mapped from satellite images is less expensive than the construction of detailed models as in Figure 6.5(b). Some parts of most cities are quite uniform in the building types they contain, as can be seen in Figure 6.5(a). Great progress has been made to construct open-source software with which non-specialists can map relatively homogeneous neighborhoods based on medium- and high-resolution satellite images. Such a multisource imaging approach was tested for the city of Bishkek, Kyrgyzstan, where medium-resolution satellite images, together with a limited use of photographs taken by omnidirectional cameras on the ground, were used to inexpensively construct the type of city model needed for loss assessments: A map of relatively homogeneous neigh-borhoods with the predominant building types probabilistically classified on the EMS98 scale (Wieland et al., 2012; Pittore and Wieland, 2012).

The most cost-effective city model can be constructed from a census of the districts. The population of each district of a large city is usually known, and sometimes, information on building types is also available. In addition, in-formation exists on the soil conditions in some cities. So far, QLARM cal-culates losses for about 100 cities by district, with consideration of the soil conditions that can amplify strong ground motion (Figure 6.6).

6.7 BASIC CONCEPTS

Not all basic concepts in estimating earthquake-generated losses are well understood by nonpractitioners. Thus, some of these are briefly reviewed in this section.

6.7.1 The Law of Averages

It does not make sense to attempt calculating the damage state of a single building, in general. In the absence of blue prints for the building, without a knowledge of the degree of enforcement of the building code, without infor-mation on the soil conditions, without a knowledge of the age of the building and its state of maintenance, its response to strong shaking cannot be reliably calculated. However, if one considers numerous buildings together, calculating their average response, some of these unknowns will average out and the loss estimate becomes acceptably reliable. Thus, it is not advisable to divide cities into very small neighborhoods for loss calculations.

6.7.2 Occupancy Rate as a Function of Time

The occupancy rate of buildings being larger at night than during the day is modeled in QLARM as a simple function assumed to be valid for all buildings.

However, it would be desirable to model movements of the population from dwellings into office and factory buildings in detail. Also, seasonal changes due to tourism should be modeled because the population of beach resorts can increase by an order of magnitude.

6.7.3 Population Shifts over the Years

Populations of most countries increase, and for some, they decrease with time. In addition, the population of many countries tends to gravitate to the urban centers, while the rural population decreases. Unfortunately, it takes considerable resources to keep up with these changes annually. Luckily, the averaging effect helps when summing up losses over urban and rural areas.

The annual population increase is measured in units of ten and hundred thousand for some of the largest population centers. If the location of the newcomers can be mapped, then loss estimates could be improved usefully. However, without knowing the exact whereabouts of the added population, the average sum of the human losses due to an earthquake changes only by a few percent the amount of the increase in population. An error of a few percent, ranks as a relatively small error compared to the other possible errors discussed above.

6.7.4 Preoccupation with Megacities

Much ado is made of disasters that are possible in megacities. Striving to reduce the earthquake risk in very large population centers makes sense. However, focusing only on the largest city of a country may be too simple a strategy for planning responses and mitigation. Many cases exist where the ratio of the rural to the urban population is far >1. For example, if a repeat of the earthquake of 1356 in Basel (estimates of magnitude range from 6.6 to 6.9) would occur, the sum of the fatalities and the injured in the surrounding settlements would exceed those in the city by a factor of three to four (Figure 6.7). In such cases, response and mitigation plans must include the

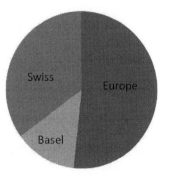

FIGURE 6.7 Comparison of the distributions of fatalities in the city of Basel, Switzerland, to those in the Swiss surroundings, and those in the European Union (Germany and France), illustrating the need in some cases to not only consider the largest city affected but also the surrounding population for mitigation and preparation measures.

countryside, and not only the major city, although the losses in the latter are the largest of any single settlement affected.

6.8 THE FUTURE OF REAL-TIME ESTIMATES OF LOSSES DUE TO EARTHQUAKES

The future of real-time earthquake loss estimates is bright, although a thorny path leads to the necessary improvements. Many groups are working on improving the data sets and the methods. I am summarizing only a few of them that seem to be the most obvious.

6.8.1 Improvements in Rapid Earthquake Source Parameter Estimates

For the parts of the world where advanced seismograph networks exist, the errors introduced by teleseismic epicenter and depth uncertainties could be greatly reduced, if the very accurate local hypocenter solutions would be made available for estimating losses. Good reasons do not seem to exist for the operators of these networks not to disseminate their results as soon as they are available, that is, within <1 min. However, for most of the world, no local information exists in real time, so one may ask how teleseismic estimates of source parameters may be improved.

Even for regions without seismographs, epicenter accuracies could be vastly improved based on InSAR satellite images, if they became available within a few hours of an earthquake. It has been shown that for shallow earthquakes, the surface deformations measured by InSAR can be inverted to yield a location of the energy release within 1—2 km and a highly accurate model of the rupture plane and the displacement on it (Wyss et al., 2006).

One of the parameters that can be improved without too much effort is the depth estimate. When the depth exceeds about 60 km, the direct pulse and the pulse reflected from the surface near the epicenter are usually recognizable as distinct phase arrivals, and from the delay between them the depth can be calculated. However, for shallower depths, the two pulses are not clearly separated. By filtering the signals, the pulses could be made recognizable, but this process may not easily be automated.

The fact that the magnitude scale saturates near M8 has been recognized decades ago, and various approaches to overcome this problem have been proposed (Di Giacomo et al., 2008; Kanamori and Rivera, 2008), but none of them have been implemented in real-time magnitude-estimating procedures.

6.8.2 The Direction of Propagation

Moment tensor solutions come online rapidly, for example, on the web page of GFZ. These define the orientations of two possible fault planes and the slip directions on them. By expert choice, one could define which is most likely the actual fault plane, and model an extended rupture along this plane. However, the direction in which the rupture propagated from the hypocenter is unknown. In many cases, the direction of propagation could be derived without great effort from the ratio of amplitudes as a function of the azimuth. Amplitudes in the forward direction of propagation are larger than those in the backward direction.

6.8.3 Improvements in Population Data

These are necessary as is evident from the above discussion. New census data will doubtlessly become available. They would be of great help, if these counts, which necessarily start in the smallest settlement, would be made available fully and not only by district. Techniques to derive grids of population from satellite data are also likely going to improve (Taubenböck et al., 2014).

6.8.4 Improvements in Building Stock Data

The web page of the WHE continues to make more data available on buildings worldwide. The Global Earthquake Model (GEM, http://www.globalquakemodel.org/) is making great strides to accumulate better data on the built environment and also on methods to gather these data.

6.8.5 Improvements in City Models

The UNISDR (United Nations Office for Disaster Risk Reduction, http://www.unisdr.org/), GEM, the World Bank and other agencies are funding projects to gather data on population and building stock in large population centers. If these data become available as open source, they will be very helpful for more detailed loss assessments in large cities in real time.

6.8.6 Dissemination of Loss Estimates in Real-time

The mass media do a great disservice to injured people who need help to get out from underneath the rubble of their dwellings and into a hospital, by announcing regularly "two people reported killed by earthquake", this at a time when WAPMERR has issued an alert estimating that the fatality and injured numbers would likely be 1000 and 3000, respectively. Actually, the final death count was 631. A statement by the mass media prompted the responsible civil protection official of the country affected to refuse international help offered within 2 h. Twenty hours later, when it was too late, the country finally realized that a disaster had happened and invited international help. To remedy this counterproductive situation, an international body should be formed that assesses earthquake disasters in real time. This body should include loss estimating experts, seismologists, engineers, civil defense officials, and media representatives.

6.8.7 Critical Facilities

These can become the source of secondary disasters, such as fires, pollution, and flooding in the case of a dam rupturing. Calculation of the amplitude of strong ground motion at such locations would be helpful. Presently, only the JRC has a data set of the locations of critical facilities, but these data are proprietary, and the intensity of shaking that these facilities may experience is not calculated by the JGR.

6.8.8 Building Stock can Vary Strongly in a Single Country

Currently, models for the built environment are mostly applied to entire countries. For example, in Myanmar, the typical dwelling in the tropical low lands is a bamboo hut with palm leaves as a roof, whereas the buildings in the mountain ranges (e.g., >1000 m) are necessarily masonry structures. Clearly, buildings in sections of countries should be defined, instead of employing the current approach (dictated by necessity) of assuming that buildings are the same within one country and in neighboring countries as well.

6.8.9 Calibration and Verification

Calibration and verification of the data sets and methods used will continue, as the observed effects of damaging earthquakes can be compared to the estimated effects calculated in real time without input from the affected area. If strong differences exist (larger than a factor of two), the cause of the difference should be identified and the data or methods to calculate can be improved.

6.8.10 The Benefits of Improving the Input Data

Input errors that can cause errors of an order of magnitude in casualty estimates include (1) Underestimates of magnitudes of great earthquakes by ≥ 0.5 units. (2) Errors in the hypocentral depth >20 km. (3) Strong overestimate of the building stock's resistance to shaking. (4) Direction of rupture propagation of an extended source.

If these input errors were removed, the accuracy of some real-time loss estimates could be significantly improved (examples discussed above are Haicheng, Yogyakarta, and Haiti). Strong efforts are made by numerous organizations to collect better information on the built environment, which will reduce error (3). The other errors are fairly well-understood seismological problems that could be solved, if the will to do this existed.

6.9 DISCUSSION AND CONCLUSIONS

The purpose of real-time estimates of losses due to earthquakes is to distinguish disastrous from inconsequential events. This information is useful to responders because they do not need to worry whether or not to mobilize in case a large-magnitude event is reported.

As earthquakes of M5.5 can cause damage and even casualties in some countries, one needs loss estimates for such small events for some regions. Because more than one such earthquake occurs per day on average worldwide, one needs a filter to select only those for preparing a loss report for which the potential for damage exists. The Swiss Seismological Service (SED) has defined such a filter that works reasonably well and leads to about 70 alerts per year, a little more than one per week. Most of these alerts indicate that most likely no fatalities have occurred.

The number of fatalities is probably the most useful measure of the degree of an earthquake disaster because they are always reported, whereas the number of injured and economic losses is often unknown. However, the number of fatalities can also be unreliable. In remote areas, they may simply not be known. In some countries, they are exaggerated to obtain aid, and in others, they are purposely underestimated to hide the dilapidated condition of the built environment.

From a scientific point of view, a real-time estimate of fatalities that comes within a factor of two of the eventual count can be considered a successful result, in case of large numbers. If the numbers are small, say <400, agreement within about 100 fatalities can be considered a success. However, from an operational point of view, simply recognizing the event as a disaster is satisfactory to most international responders because they will offer help regardless of the precise number of fatalities.

Whereas the sum of fatalities may be correct, the detailed distribution may be a poor approximation to the truth. Individual settlements may be more or less affected than estimated and information on different districts of large population centers is not available most of the time.

Loss estimates of past earthquakes serve to verify the methods and data sets used. In countries where damaging earthquakes occur frequently, as in Iran, loss estimates are therefore fairly reliable. In countries where large earthquakes are possible, but have not happened for centuries, uncertainties of real-time loss estimates may be large.

Error sources that affect loss estimates are numerous, and their influence is not known quantitatively for all. Thus, it is unavoidable that some events are classified as disastrous in real time when they are not and vice versa. By reviewing real-time alerts before distributing them, errors that can be avoided are usually eliminated.

Improvements are possible for almost all the error sources, so one may expect an increase in reliability of loss estimates in the future. For population distribution and building stock properties, strong progress is expected thanks to the GEM project. Improving the earthquake parameter input may not happen for a long time because there seems to be no interest in applying techniques that are actually known, but it would require some work to be implemented.

Lack of appreciation of the expert real-time loss estimates by news media and local civil defense officials is a significant obstacle for the alerts to become useful. No means to change that unfortunate situation seems to exist.

Funding for quantitative real-time loss estimates exists only for the USGS PAGER group. This is an undesirable situation because several groups should be working on improving the necessary methods and data sets because different approaches exist concerning these complex questions, none of which are necessarily recognizable as being the most effective.

ACKNOWLEDGMENTS

I thank M. Wieland for helpful criticism of the manuscript and the JTI Foundation for partially supporting this study.

REFERENCES

Di Giacomo, D., Parolai, S., Saul, J., Grosser, H., Bormann, P., Wang, R., Zschau, J., 2008. Rapid determination of the energy magnitude Me, paper presented at European Seismological Commission 31st General Assembly, Hersonissos.

Grünthal, G., 1993. European Macroseismic Scale 1992 (Up-dated MSK-scale). Conseil de l'Europe, Luxembourg, 79 pp.

Jaiswal, K., Wald, D.J., Porter, K., 2010. Creating global building inventory for earthquake loss estimation and risk management. Earthq. Spectra 26 (3), 731−748.

Kanamori, H., Rivera, L., 2008. Source inversion of W phase: speeding up seismic tsunami warning. Geophys. J. Int. 175, 222−238.

Pandey, M.R., Molnar, P., 1988. The distribution of intensity of the Bihar−Nepal earthquake of January 15, 1934 and bounds on the extent of the rupture zone. J. Geol. Soc. Nepal 5, 22−44.

Pittore, M., Wieland, M., 2012. Toward a rapid probabilistic seismic vulnerability assessment using satellite and ground-based remote sensing. Nat. Hazards 68 (1), 115−145.

Satterthwaite, D., 2006. Outside the Large Cities; the Demographic Importance of Small Urban Centres and Large Villages in Africa, Asia, and Latin America. International Institute for Environment and Development, London.

Trendafiloski, G., Wyss, M., Rosset, P., 2011. Loss estimation module in the second generation software QLARM. In: Spence, R., et al. (Eds.), Human Casualties in Earthquakes: Progress in Modeling and Mitigation. Springer, pp. 381−391.

Taubenböck, H., et al., 2014. The capabilities of earth observation to contribute along the risk cycle. Elsevier, Waltham, Massachusetts. In: Shroder, J., Wyss, M. (Eds.), Earthquake Hazard, Risk, and Disasters. Elsevier, London, pp. 25−53.

Utsu, T., 2002. A list of deadly earthquakes in the world: 1500−2000. In: Lee, W.K., et al. (Eds.), International Handbook of Earthquake Engineering and Seismology. Academic Press, Amsterdam, pp. 691−717.

Wieland, M., Pittore, M., Parolai, S., Zschau, J., Moldobekov, B., Begaliev, U., 2012. Estimating building inventory for rapid seismic vulnerability assessment: Towards an integrated approach based on multi-source imaging. Soil Dyn. Earthq. Eng. 36, 70−83.

Wood, H.O., Neumann, F., 1931. Modified Mercalli intensity scale of 1931. Bull. Seismol. Soc. Am. 21 (4), 277−283.

Wyss, M., Brune, J.N., 1967. The Alaska earthquake of March 28, 1964: a complex multiple rupture. Bull. Seismol. Soc. Am. 57 (5), 1017−1023.

Wyss, M., 2005. Human losses expected in Himalayan earthquakes. Nat. Hazards 34, 305−314.

Wyss, M., Wang, R., Zschau, J., Xia, Y., 2006. Earthquake loss estimates in near real-time. EOS Trans AGU, 477−479.

Wyss, M., 2006. The Kashmir M7.6 shock of October 8, 2005 calibrates estimates of losses in future Himalayan earthquakes, paper presented at proceedings of the conference of the International Community on Information Systems for Crisis Response and Management, Newark.

Wyss, M., Elashvili, M., Jorjiashvili, N., Javakhishvili, Z., 2011. Uncertainties in teleseismic epicenter estimates: implications for real-time loss estimate. Bull. Seismol. Soc. Am. 101 (3), 1152−1161.

Wyss, M., Trendafiloski, G., 2011. Trends in the casualty ratio of injured to fatalities in earthquakes. In: Spence, R., et al. (Eds.), Human Casualties in Natural Disasters: Progress in Modeling and Mitigation. Springer, London, pp. 267−274.

.

Chapter 7

Forecasting Seismic Risk as an Earthquake Sequence Happens

J. Douglas Zechar[1], Marcus Herrmann[1], Thomas van Stiphout[2] and Stefan Wiemer[1]

[1]*Swiss Seismological Service, ETH Zurich, Zurich, Switzerland,* [2]*Independent Researcher, Zurich, Switzerland*

ABSTRACT

We describe a model-based approach to forecast seismic risk during an earthquake sequence, emphasizing building damage and human injuries. This approach, which could also be used to forecast financial losses, incorporates several models of varying complexity, but it is intuitive and its output can be succinctly described with simple mathematics. We describe two applications of this approach—one in the wake of the destructive 2009 L'Aquila, Italy earthquake, and another with a hypothetical $M_w6.6$ earthquake in Basel, Switzerland. We discuss the challenges of short-term seismic risk forecasting and suggest potential improvements.

7.1. INTRODUCTION

When we think of seismic risk, we have in mind a place, a period, and consequences: probablilistic seismic risk quantifies the probability of those consequences at that place within that period. Because earthquakes do not occur constantly, seismic hazard and, as a consequence, risk vary from moment to moment, especially when earthquake activity increases. Such an increase might follow a large earthquake, when many smaller earthquakes occur nearby. Sometimes, small or moderate events precede a large earthquake and seem, in retrospect, to have signaled the impending large event. The preceding and following earthquakes are colloquially referred to as foreshocks and aftershocks, respectively, of the mainshock. Sometimes, several events of about the same size happen and there is no mainshock; this is called a *swarm*. Swarm sequences and foreshock—mainshock—aftershock sequences increase seismic risk, and in this chapter, we consider how we can forecast seismic risk as an earthquake sequence happens.

Earthquake Hazard, Risk, and Disasters. http://dx.doi.org/10.1016/B978-0-12-394848-9.00007-9
167

Forecasting seismic risk is subtly different from *assessing* seismic risk as discussed in Chapter 6: the goal of that type of assessment is to estimate the impact of earthquakes that have occurred, while the goal of forecasting is to predict both the distribution of future earthquakes and the corresponding effects. In this chapter, we emphasize time-varying risk forecasts that respond to the space-time clustering of seismicity. Our motivations for doing this are so intuitive that you likely already know them: to advise emergency managers and respond to public concerns, thereby reducing losses. Conceptually, our approach is also intuitive: we combine an earthquake occurrence model, a hazard model, a damage model, and a loss model to make corresponding probabilistic predictive statements about future seismicity, shaking, damage, and losses. In combining these elements, we synthesize concepts from the chapters on earthquake prediction (Wu, Chapter 16; Sobolev, Chapter 17; Kossobokov, Chapter 18), forecast testing (Schorlemmer & Gerstenberger, Chapter 15), seismic risk (Wyss, Chapter 6), and losses (Michel, Chapter 21). Like those chapters, we have in mind tectonic earthquakes, not induced seismicity. And like Wyss (Chapter 6), we use "loss" to refer to either loss of human life or financial loss incurred from damage to the built environment.

In the following sections, we introduce notation and suggest a mathematical formulation of the problem; illustrate our approach using data from the L'Aquila earthquake sequence of 2009; demonstrate an extension using a spatially-varying seismicity model for Switzerland's SEISMO-12 preparatory exercise; and discuss shortcomings of our implementation and directions for future research.

7.2. SEISMIC RISK

What do we need to assess seismic risk and predict the resulting losses? We are interested in building damage, so we need to know about the earthquakes that may have caused damage and the buildings that may have been damaged. Specifically, we need to know where the earthquakes occurred and how big they were, so we can estimate the shaking they caused. And we need to know something about the strength of the potentially affected buildings, so we can estimate the damage caused by such shaking. We are also interested in casualties, so we need to know about the people that were potentially affected by building damage: how many people were affected, and where were they?

Consider the first-person perspective of risk assessment: imagine that a set of earthquakes, E, is going to happen and you want to know the probability that you are injured—in other words, that you reach a casualty degree C. We can express this risk, $\Pr(C|E)$, as a product of several (mostly conditional) probabilities:

$$\Pr(C|E) = \sum_{e \in E} \sum_{n=0}^{5} \sum_{k=1}^{12} \sum_{j=A}^{F} \Pr(C|D_n, V_j)\Pr(D_n|I_k, V_j)\Pr(I_k|e)\Pr(V_j) \quad (1)$$

Let's step through the product terms left to right. The first denotes the probability that casualty degree C is reached conditional on you being in a building with vulnerability V_j that reached damage state D_n. The next term is the probability that the damage state was reached, conditional on the vulnerability and a given intensity I_k. This is analogous to an earthquake engineering fragility function. The third term is the probability that the earthquake e produced the given intensity, something we might call the conditional seismic hazard. And the final term describes the distribution of vulnerability. Note that we treat building vulnerability, intensity, damage states, and casualty degrees as categorical data. In fact, when we write risk in this way, we have particular discretizations in mind: European Macroseismic Scale (EMS-98, Grünthal, 1998) for vulnerability, intensity, and damage grade, and HAZUS (FEMA, 2010) for casualty degree. But this approach to assessing risk can be generalized to other discretizations, or to continuous data. For example, I could be given in units of Modified Mercalli Intensity, peak ground velocity, peak ground acceleration, and so on.

When we *forecast* seismic risk, we are not interested in the effect of earthquakes that have already happened, but rather of those that may happen in some specific future period. This puts us in the land of models; we need the following:

1. An earthquake occurrence model to forecast when and where future earthquakes will occur, and how big they will be.
2. A ground motion model, including site effects, to forecast the resulting shaking.
3. A building stock model to represent the built environment.
4. A damage model to forecast the damage to the building stock caused by the shaking.
5. A loss model to forecast human casualties (which requires the loss model to include a model of the space-time distribution of people) and/or financial losses (which requires a model of costs for potential loss events).

Frequently, the first two models are combined to form a seismic hazard model, where hazard can be expressed as the probability of exceeding some ground motion. Then we can write the time-varying, first-person-perspective risk as a function of hazard:

$$\Pr(C, t) = \sum_{n=0}^{5} \sum_{k=1}^{12} \sum_{j=A}^{F} \Pr(C|D_n, V_j)\Pr(D_n|I_k, V_j)\Pr(I_k, t)\Pr(V_j) \quad (2)$$

Note the explicit time dependence in the risk forecast and the hazard model $\Pr(I, t)$: in subtle contrast to Equation 1, Equation 2 describes the probability that you will experience a casualty degree C in the future period t. In principle, the vulnerability distribution could also vary in time, allowing one to account for progressive damage, but this is not yet done in practice. When an earth-quake sequence is ongoing, we can update our risk forecast by updating our earthquake occurrence model (and thus our seismic hazard model). This is analogous to day-by-day weather forecasting: based on new observations, the weekend weather forecast may change throughout the week.

In the following two sections, we demonstrate this approach to forecasting short-term risk in Italy and Switzerland and give concrete examples of the terms in the seismic risk forecast equation.

7.3. FORECASTING SEISMIC RISK DURING THE L'AQUILA SEQUENCE

On April 6, 2009, an $M_w6.3$ earthquake devastated L'Aquila, Italy, causing widespread damage and killing 299 residents. In the weeks leading up the deadly earthquake, many smaller events occurred near L'Aquila; with the benefit of hindsight, we can say that those events were foreshocks. Many authors have addressed the controversy surrounding the L'Aquila sequence, and we will not rehash the details (see for example Jordan et al., 2011; Marzocchi, 2012). For this chapter, we are only interested in the L'Aquila sequence as a set of scientific circumstances, and we will ignore the legal, philosophical, and political ramifications, despite the fact that they may have a more important legacy than the earthquakes themselves.

In the aftermath of the L'Aquila sequence, van Stiphout et al. (2010, hereafter vS) recognized that seismologists must be able to effectively communicate time-varying seismic risk to emergency managers. They pioneered the model-based approach described in the previous section, and they conducted an experiment to forecast seismic risk as the L'Aquila sequence progressed. Note that they did this after the fact, not in real time, so the experiment was a proof of concept that used observed data.

For the earthquake occurrence model, vS used the RJ (Reasenberg and Jones, 1994) model, which is founded on two first-order observations of earthquakes:

1. The distribution of magnitudes is well-approximated by the Gutenberg–Richter relation (i.e., an exponential function, Gutenberg and Richter, 1944).
2. The rate of events occurring nearby and soon after a large earthquake is well approximated by the Omori–Utsu relation (i.e., an exponential decay function, Utsu, 1961).

The RJ model aims to reproduce the temporal clustering and size distribution of earthquakes and is a stochastic triggering model: it estimates the rate and sizes of future earthquakes triggered by observed events. For the application to the L'Aquila sequence, vS used RJ parameter values that Lolli and Gasperini (2003) estimated by fitting the model to earthquake sequences in Italy.

For the ground motion model, vS combined the ground motion prediction equation of Akkar and Bommer (2007) with the ShakeMap method using Italy-specific parameter values (Michelini et al., 2008), and they included a site amplification factor of 1.25 intensity units to account for local soil conditions in L'Aquila. For the building stock model, damage model, and the loss model, vS used the module developed for the QLARM software (Trendafiloski et al., 2011). To estimate the distribution of damage, QLARM employs the European Macroseismic Method (EMM) (Giovinazzi, 2005), which maps intensity and vulnerability class to a probability mass function of damage. In other words, this permits the possibility that two buildings with the same vulnerability are subject to the same shaking intensity yet experience different damage. For the building stock model, QLARM uses a generic model in regions where detailed models are not available; in the vS study, QLARM assigned 30 percent of the L'Aquila buildings to EMS-98 vulnerability class A, 30 percent to B, 30 percent to C, 10 percent to D, and none to E and F (class A is the most vulnerable to shaking and F, the least vulnerable). The QLARM loss model maps the estimated building damage to HAZUS casualty degree using an empirical approach calibrated using previous events (Trendafiloski et al., 2011). For the loss model component that describes the distribution of people, QLARM assumes that the distribution of inhabitants matches the building stock distribution: 30 percent of the inhabitants were in class A buildings, 30 percent in B, 30 percent in C, and 10 percent in D.

For their study of the L'Aquila sequence, vS analyzed regional earthquakes with magnitude greater than 2.5 between November 1, 2008 and May 1, 2009 (weg damit Figure 7.1). They made 24 h forecasts that were updated every 3 h or whenever a new earthquake happened, whichever came first. (Three hours *without* any earthquake is an observation that can be used to update the earthquake occurrence model, and therefore, the seismic risk forecast.) Because this forecast approach is probabilistic, we can look at the risk in several different ways: for example, imagine that you are an emergency manager and you are concerned not about your own risk but the risk of all people in the region—call this a third-person risk perspective. Then you would be interested in a loss curve that shows the probability of exceeding some number of fatalities (Figure 7.2). By integrating the curve and dividing by the population, we can recover the first-person risk forecast emphasized in the previous section. We can also visualize the temporal evolution of the seismic risk forecast as the sequence progresses, as shown in Figure 7.3. Figure 7.3 illustrates the effect of the ongoing sequence on the risk forecast: whenever a new event—a possible foreshock—occurs, the risk suddenly increases, and as time passes and no events occur, risk gradually decreases.

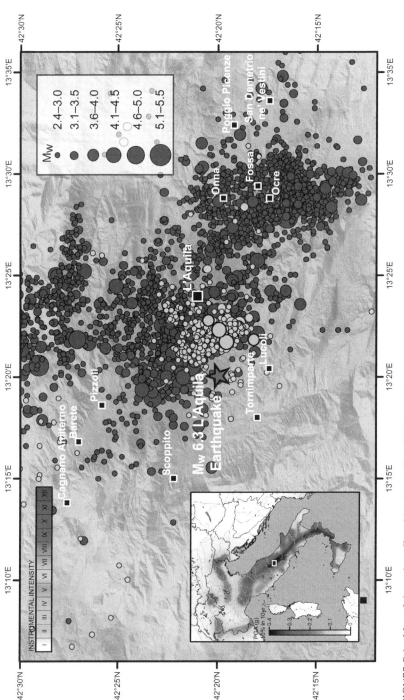

FIGURE 7.1 Map of the region affected by the April 6, 2009 L'Aquila $M_w6.3$ earthquake (red star), including the predicted ground motion, the foreshocks between 1 November and 6 April (yellow), aftershocks between 6 April and 1 May (gray), and the settlements (black squares). Inset shows the national seismic hazard map (Meletti et al., 2008) with the white box indicating the region in the main panel.

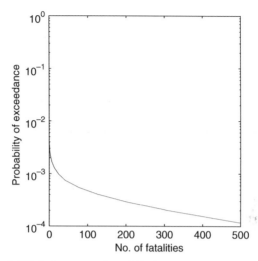

FIGURE 7.2   Probabilistic loss curve for EMS-98 building class type A in L'Aquila on April 6, 2009 at 2 a.m. local time, for the following 24 h.

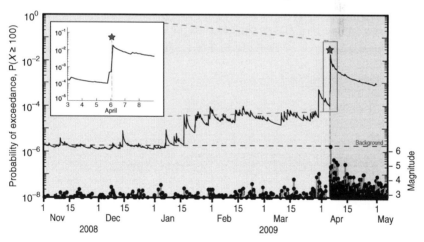

FIGURE 7.3   Probability of exceeding 100 fatalities in the next 24 h, updated after each earthquake or every 3 h (black). The time of the mainshock is indicated by a red star and the background probability of exceeding 100 fatalities with the next 24 h based on Meletti et al. (2008) is shown with the dashed blue line. The inset shows details of the risk forecast immediately before and after the occurrence of the mainshock. Right axis: earthquake magnitudes as a function of time. Note: the probability is based on the seismicity within a box 25 by 25 km centered at L'Aquila.

7.4. FORECASTING SEISMIC RISK FOR THE SEISMO-12 SCENARIO SEQUENCE

In May 2012, the Swiss Federal Office for Defense, Civil Protection, and Sport conducted an earthquake exercise involving a hypothetical M_w6.6 event in Basel, Switzerland. This exercise, called SEISMO-12, was designed to explore how authorities might react in case of a "repeat" of the 1356 Basel earthquake (Meghraoui et al., 2001; Gisler et al., 2008; Fäh et al., 2009). Rather than isolating the scenario mainshock, SEISMO-12 participants were asked to respond to an entire sequence. The SEISMO-12 sequence combined events from an automated simulation of the Epidemic Type Aftershock Sequence (Ogata, 1988) model (specifically, the implementation of Hainzl et al., 2008) and some manually inserted earthquakes: five foreshocks (including an M_w5.1 20 min prior to the M_w6.6 event) and several large aftershocks (M3.4–M5.9). Figures 7.4 and 7.5 show the temporal and spatial distribution, respectively, of the SEISMO-12 sequence.

Herrmann (2013, hereafter H13) extended the approach of vS to forecast risk for the SEISMO-12 sequence. Rather than using the RJ earthquake occurrence model, H13 used the Short-Term Earthquake Probability model (STEP) (Gerstenberger et al., 2005), which is an extension of the RJ model. In particular, STEP extends the RJ model by incorporating spatial information in so-called *aftershock zones*. With these aftershock zones, STEP aims to include an additional first-order observation: earthquakes cluster in space. As more events are observed, STEP gradually increases the resolution of the aftershock zone. One of the ways that STEP attempts to reproduce spatial clustering is to estimate the geometry of the fault that ruptured in large earthquakes.

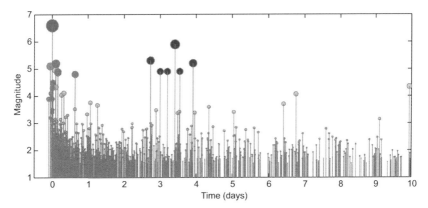

FIGURE 7.4 Time-magnitude distribution of the SEISMO-12 sequence for the 10 day duration of the exercise, with the M_w6.6 event at $t = 0$ days. Events from the ETAS simulation are shown in gray, the M_w6.6 event is red, manually inserted foreshocks are orange, and manually-inserted aftershocks are green and blue. The blue circles denote larger aftershocks occurring after $t = 3$ days.

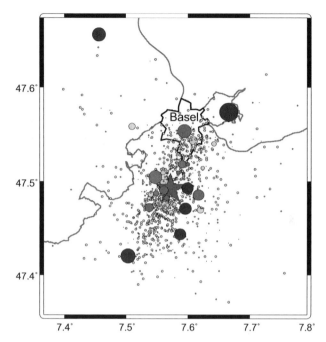

FIGURE 7.5   Map view of the SEISMO-12 sequence. The colors of the symbols are the same as in Figure 7.4.

In the original STEP implementation, a fault was represented as two line segments with an endpoint at the epicenter of the large event, and the length and direction of the two line segments were determined by the spatial extent of aftershocks. But H13 noted that this fault identification algorithm was sensitive to outliers and yielded counterintuitive results for the SEISMO-12 sequence, and he therefore implemented an improvement that emphasizes the regions with the greatest aftershock density. Figure 7.6 shows a comparison of the fault approximation methods for the SEISMO-12 sequence.

As shown in Figure 7.6(a) and (b), STEP generates spatially-varying earthquake occurrence forecasts; for the ground motion model, H13 used such forecasts as input to the intensity prediction equation of Allen et al. (2012) and summed the results with regionally appropriate site amplifications from P. Kästli (written communication). Rather than using the generic building stock model in QLARM, H13 benefited from being able to use a superior building stock model for the Basel region. Because of a geothermal project, a detailed risk assessment of Basel had been conducted and the corresponding report was published, including building inventory for 19 districts of Basel and 60 surrounding settlements (Baisch et al., 2009); see Figure 7.7 for a map-view representation of these data. The report also contained population estimates for each settlement and the district of Basel. Like the building stock model, the population dataset

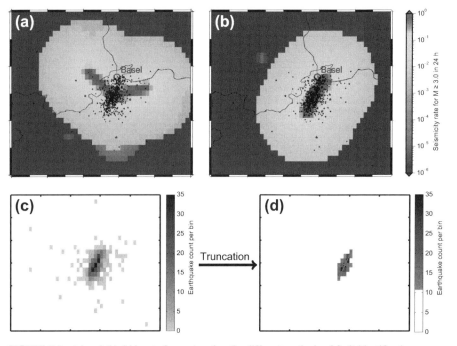

FIGURE 7.6 (a) and (b) 24 h rate forecasts using the different methods of fault identification. (a) The original STEP implementation. (b) The H13 approximation. Both forecasts are generated 18 h after the mainshock, with approximately 530 aftershocks having already occurred. Using the same data, (c) and (d) demonstrate how the H13 fault approximation works. (c) is a 2D spatial histogram of the aftershock, (d) is the same after filtering, and the red line denotes the inferred fault.

offers a higher resolution than that of the default data in QLARM (which were used by vS).

Following vS, H13 also used QLARM for the damage model and loss model. Also following vS, for the loss model H13 assumed that the distribution of inhabitants weg damit matched the local building stock distribution weg damit.

To estimate seismic risk during the SEISMO-12 sequence, H13 analyzed those earthquakes shown in Figures 7.4 and 7.5 and generated 24 h earthquake occurrence forecasts, seismic hazard forecasts, and seismic risk forecasts. Figure 7.8 is analogous to Figure 7.3—it shows, for the SEISMO-12 sequence, the temporal evolution of hazard and risk. Figures 7.9 and 7.10 emphasize the spatial information available in these risk forecasts—they are risk maps showing the space-time variation of risk that results from seismicity that varies in space and time, as well as vulnerability and population that varies spatially. Figure 7.10(b) shows a third-person risk perspective that is normalized by population, a view that lets you identify relative risk across the region.

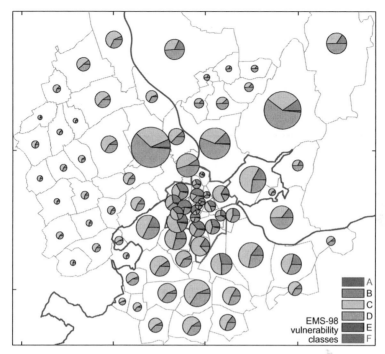

FIGURE 7.7   Map view of the EMS vulnerability class distribution for each settlement. The area of each pie chart is proportional to the number of buildings in the corresponding settlement.

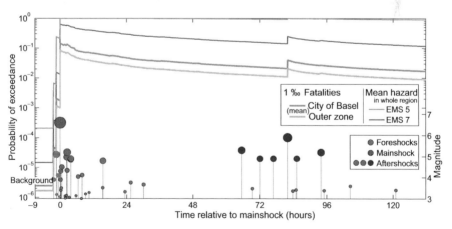

FIGURE 7.8   Time-varying probability of exceeding loss (orange and blue) and hazard thresholds (gray to black) for a few hours before and the first 6 days after the Basel scenario earthquake. The loss threshold is set to 1 per mill (‰) fatalities; the hazard is illustrated by distinct intensity levels. Thicker lines represent the mean probability of zone-related settlements (see legend), whereas the dashed lines show the maximum and minimum loss probabilities as observed in the two zones separately. The loss and hazard forecasts were issued each hour and refer to the following 24 h. To track the seismicity during this time, stems at the bottom represent earthquakes above M3 with the same color scheme as in Figure 7.4.

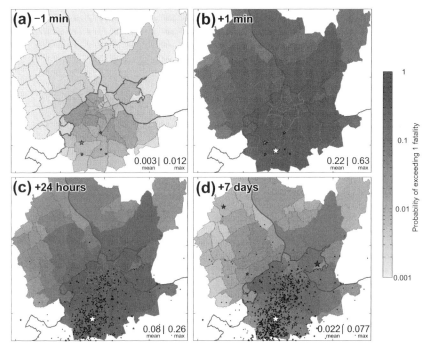

FIGURE 7.9 Snapshots of the regional seismic risk forecast for the following 24 h, issued at the time (relative to the $M_w6.6$ event) given in the top-left corner. The color of each settlement denotes the probability of having one or more fatalities. The average and maximum probabilities for each map are reported in the bottom-right corner of each snapshot. The earthquake symbols were assigned the same color scheme as in Figure 7.4 (except the mainshock, which is white in these maps).

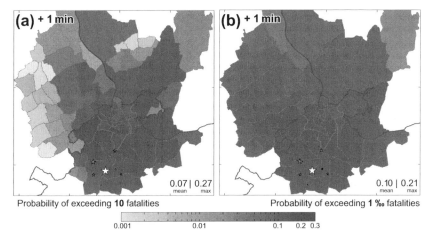

FIGURE 7.10 Short-term risk forecast for the next 24 h, 1 min after the $M_w6.6$. Probabilities of exceeding two different fatality thresholds are presented. Note that the color scale has changed compared to Figure 7.9. (a) Probability of more than 10 fatalities in each settlement. (b) Normalized by settlement population: Probability that more than 1 per mill (‰) of the population in each settlement dies.

7.5. DISCUSSION

As Wyss (Chapter 6) highlighted, the practice of routinely estimating seismic losses in the wake of a large earthquake is relatively new. The practice of forecasting seismic risk as an earthquake sequence happens is even newer—vS were pioneers—and forecasting risk is inherently a more difficult problem. Wyss (Chapter 6, Section 8) mentioned some of the unsolved problems for estimating losses in real time; they also apply to forecasting short-term risk, especially those related to data quality and availability.

Nearly all short-term earthquake occurrence models use a catalog of recent events as input. Because the models we have in mind are founded on the belief that small earthquakes can trigger large earthquakes, including small earthquakes in the catalog is important. Ironically, detecting and analyzing small earthquakes is hardest exactly when they interest us most—when seismic activity is higher than normal. This problem of time-varying catalog completeness is well known and affects all of statistical seismology (Mignan and Woessner, 2012). Some researchers have proposed partial solutions to this problem (e.g., Peng et al., 2006), but the solutions have not been implemented in routine seismic network processing. Of course, errors in catalog data—for example, incorrect magnitude or location estimates, or records of earthquakes that did not happen—will also result in inaccurate seismic risk forecasts. Producing very accurate and very complete catalogs in real time is difficult primarily because the producers lack resources, computational or otherwise. This is a pragmatic problem that is probably best addressed by stressing the importance of high-quality catalog data to funding agencies.

Even if data quality does not improve dramatically, there are steps we should take to improve seismic risk forecasts. For example, in this chapter, we have neglected the topic of uncertainty, and this should be addressed. Uncertainty in earthquake source parameters could be propagated from the catalog into the occurrence model and on to the loss model—in other words, from beginning to end. Beyond data uncertainty, we also cannot be sure that we are using the best model for each component of the seismic risk forecast. Although the examples in the previous two sections emphasized STEP as an earthquake occurrence model, we could use more sophisticated models and/or models that have been developed for a particular region or tectonic setting. Or we could directly account for our uncertainty in selecting an earthquake occurrence model by using an ensemble, aiming to offset the weaknesses of any individual model with the strengths of others (Marzocchi et al., 2012). The same principle applies to ground motion, damage, and loss models.

In addition to improving the earthquake occurrence model, we should leverage the numerous recent advances made by the ground motion modeling community. For example, we could apply methods intended to help rank ground motion prediction equations (Scherbaum et al., 2004, 2009; Kaklamanos and Baise, 2011; Kale and Akkar, 2013), and we can also apply the findings of impending studies based on ground motion data sets from around the world (e.g., the Pacific Earthquake Engineering Research Center ground motion database). We should move away from empirical, qualitative measures such as intensity and toward physical measurements such as peak ground velocity, peak ground acceleration, and spectral acceleration. These quantities can be used in more sophisticated, machanics-based damage models (e.g., Borzi et al. 2008). Such models also permit the possibility of time-varying building stock models. In other words, damage models could feed back into the risk forecast as the sequence progresses, and we could account for progressive damage.

We could already crudely model progressive damage with the simple damage model that vS and H13 used. In those examples, the building stock model and distribution of people were treated as static entries throughout the earthquake sequence. But the damage forecasts themselves implying time-varying vulnerability: to be internally consistent, the fractions of the buildings that were forecast to be destroyed should be removed from the building stock model, and the remaining fraction should be updated. Moreover, as the buildings are damaged, the number and distribution of affected people would change, even if a complete evacuation were not economically justified (as vS claimed regarding the L'Aquila sequence). Future seismic risk forecasts should account for progressive damage and people's movements during an earthquake sequence; ideally, these systematic changes would be estimated from field observations.

Although the previous two sections emphasized QLARM for loss estimation, Wyss (Chapter 6) mentioned a few alternatives, including Prompt Assessment of Global Earthquake Risk (PAGER) (Jaiswal and Wald, 2010). Indeed the approach we described in this chapter is flexible—it is not tied to any particular model. More generally, existing loss models only estimate losses due to earthquake shaking, ignoring the potential impact of tsunamis, landslides, and fires caused by earthquakes. This is a fundamental problem and one for which we have even fewer data available to build empirical models.

Perhaps you have noticed the elephant in the room: how do we use short-term seismic risk forecasts? And how do we effectively communicate results? We don't know the answer, but we can say that it depends on the target user, and in general, we should try to open the lines of communication between seismologists, emergency managers, and social scientists (the latter group is particularly important if our goal is to communicate with the public). In terms of communication, vS discussed cost-benefit analysis, and H13 presented a traffic light system, but both approaches were developed without interaction with risk experts, and in that sense, they are only a guess as to what forecast products might be useful. In this chapter, we emphasized a scientific, model-based method to forecast seismic risk during an earthquake sequence, but it is easy to imagine that this is only the beginning; and the hardest, and most exciting, work—figuring out how to apply such methods effectively to benefit society—lies ahead.

ACKNOWLEDGMENTS

This work was partially supported by the NERA (Network of European Research Infrastructure for Earthquake Risk Assessment and Mitigation) project from the 7th Framework Program by the European Commission.

REFERENCES

Akkar, S., Bommer, J.J., 2007. Empirical prediction equations for peak ground velocity derived from strong-motion records from Europe and the Middle East. Bull. Seismol. Soc. Am. 97, 511−530.

Allen, T.I., Wald, D.J., Worden, C.B., 2012. Intensity attenuation for active crustal regions. J. Seismol 16, 409−433. http://dx.doi.org/10.1007/s10950-012-9278-7.

Baisch, S., Carbon, D., Dannwolf, U., Delacou, B., Devaux, M., Dunand, F., Jung, R., Koller, M., Martin, C., Sartori, M., Secanell, R., Vörös, R., 2009. SERIANEX − deep heat mining Basel. Tech. Rep. (in German).

Borzi, B., Crowley, H., Pinho, R., 2008. Simplified Pushover-Based Earthquake Loss Assessment (SP-BELA) Method for Masonry Buildings. Int. J. Archit. Herit 2, 353−376. http://dx.doi.org/10.1080/15583050701828178.

Fäh, D., Gisler, M., Jaggi, B., Kästli, P., Lutz, T., Masciadri, V., Matt, C., Meyer-Rosa, D., Rippmann, D., Schwarz-Zanetti, G., Tauber, J., Wenk, T., 2009. The 1356 Basel earthquake: an interdisciplinary revision. Geophys. J. Int. 178, 351−374.

FEMA, 2010. Hazus − MH MR5, Earthquake Model − Technical Manual.

Gerstenberger, M.C., Wiemer, S., Jones, L.M., Reasenberg, P.A., 2005. Real-time forecasts of tomorrow's earthquakes in California. Nature 435, 328−331.

Gisler, M., Fäh, D., Giardini, D., 2008. Nachbeben −- Eine Geschichte der Erdbeben in der Schweiz, first ed. Haupt Verlag, Bern. p. 187 (in German).

Giovinazzi, S., 2005. The Vulnerability Assessment and the Damage Scenario in Seismic Risk Analysis (Doctoral dissertation). Department of Civil Engineering Technical University Carolo-Wilhelmina, Braunschweig, Germany.

Grünthal, G. (Ed.), 1998. European Macroseismic Scale 1998. European Seismological Commission, Luxembourg.

Gutenberg, B., Richter, C., 1944. Frequency of earthquakes in California. Bull. Seismol. Soc. Am. 34, 185−188.

Hainzl, S., Christophersen, A., Enescu, B., 2008. Impact of earthquake rupture extensions on parameter estimations of point-process models. Bull. Seismol. Soc. Am. 98, 2066−2072.

Hermann, M., 2013. Forecasting Losses Caused by a M6.6 Scenario Earthquake Sequence is Basel, Switzerland. Institut für Geophysik und Geoinformatik, TU Bergakademie Freiberg. Master Thesis.

Jaiswal, K.S., Wald, D.J., 2010. An empirical model for Global earthquake fatality estimation. Earthquake Spectra 26, 1017−1037.

Jordan, T.H., Chen, Y.-T., Gasparini, P., Madariaga, R., Main, I., Marzocchi, W., Papadopoulos, G., Sobolev, G., Yamaoka, K., Zschau, J., 2011. Operational earthquake forecasting: state of knowledge and guidelines for implementation. Ann. Geophys. 54, 315−391.

Kaklamanos, J., Baise, L.G., 2011. Model validations and comparisons of the next generation attenuation of ground motions (NGA−West) project. Bull. Seismol. Soc. Am. 101, 160−175.

Kale, O., Akkar, S., 2013. A new procedure for selecting and ranking ground-motion prediction equations (GMPEs): the Euclidean distance-based ranking (EDR) method. Bull. Seismol. Soc. Am. 103, 1069—1084.

Lolli, B., Gasperini, P., 2003. Aftershocks hazard in Italy Part I: Estimation of time magnitude distribution model parameters and computation of probabilities of occurrence. Journal of seismology, 235—257. http://dx.doi.org/10.1023/A:1023588007122.

Marzocchi, W., 2012. Putting science on trial. Phys. World 25, 17—18.

Marzocchi, W., Zechar, J.D., Jordan, T.H., 2012. Bayesian forecast evaluation and ensemble earthquake forecasting. Bull. Seismol. Soc. Am. 102, 2574—2584.

Meghraoui, M., Delouis, B., Ferry, M., Giardini, D., Huggenberger, P., Spottke, I., Granet, M., 2001. Active normal faulting in the upper Rhine graben and paleoseismic identification of the 1356 Basel earthquake. Science 293, 2070—2073.

Meletti, C., Galadini, F., Valensise, G., Stucchi, M., Basili, R., Barba, S., Vannucci, G., Boschi, E., 2008. A seismic source zone model for the seismic hazard assessment of the Italian territory. Tectonophysics 450, 85—108.

Michelini, A., Faenza, L., Lauciani, V., Malagnini, L., 2008. Shakemap implementation in Italy. Seismol. Res. Lett. 79, 688—697.

Mignan, A., Woessner, J., 2012. Estimating the Magnitude of Completeness for Earthquake Catalogs. Community Online Resource for Statistical Seismicity Analysis. http://dx.doi.org/ 10.5078/corssa-00180805. Available at: http://www.corssa.org.

Ogata, Y., 1988. Statistical models for earthquake occurrences and residual analysis for point processes. J. Am. Stat. Assoc. 83, 9—27.

Peng, Z., Vidale, J.E., Houston, H., 2006. Anomalous early aftershock decay rate of the 2004 Mw6.0 Parkfield, California, earthquake. Geophys. Res. Lett. 33, L17307. http://dx.doi.org/10. 1029/2006GL026744.

Reasenberg, P.A., Jones, L.M., 1994. Earthquake aftershocks: update. Science 265, 1251—1252.

Scherbaum, F., Cotton, F., Smith, P., 2004. On the use of response spectral-reference data for the selection and ranking of ground-motion models for seismic-hazard analysis in regions of moderate seismicity: the case of rock motion. Bull. Seismol. Soc. Am. 94, 2164—2185.

Scherbaum, F., Delavaud, E., Riggelsen, C., 2009. Model selection in seismic hazard analysis: an information—theoretic perspective. Bull. Seismol. Soc. Am. 99, 3234—3247.

Trendafiloski, G., Wyss, M., Rosset, P., 2011. Loss estimation module in the second generation software QLARM. In: Spence, R., So, E., Scawthornp, C. (Eds.), Human Casualties in Natural Disasters: Progress in Modeling and Mitigation. Springer, Dordrecht, Heidelberg, London, New York, pp. 95—104.

Utsu, T., 1961. A statistical study of the occurrence of aftershocks. Geophys. Mag. 30, 521—605.

van Stiphout, T., Wiemer, S., Marzocchi, W., 2010. Are short-term evacuations warranted? Case of the 2009 L'Aquila earthquake. Geophys. Res. Lett. 37, 1—5.

Chapter 8

How to Render Schools Safe in Developing Countries?

Amod M. Dixit, Surya P. Acharya, Surya N. Shrestha
and Ranjan Dhungel
National Society for Earthquake Technology — Nepal (NSET)

ABSTRACT

The schools in seismic regions of developing countries, including Nepal, face huge risks from earthquakes. Significant proportions of school buildings were destroyed, and thousands of school children and teachers were killed or injured in recent earthquakes in different parts of the region. A recent study conducted by the National Society for Earthquake Technology—Nepal revealed that out of approximately 82,000 school buildings belonging to 34,000 private and public schools in Nepal, >75 percent are highly vulnerable to earthquakes, requiring immediate intervention; of these, 15 percent are dangerous to occupy and need demolition and reconstruction. In the case of intensity IX Modified Mercalli Intensity level of shaking, approximately 111,000 school population (10 percent of the total affected school population) could lose their lives, and another 85,000 could suffer serious injury.

The main reasons behind the high vulnerability and risk of schools are due to (1) poor building construction practices resulting from the use of poor materials and construction technology, lack of skilled and trained construction manpower, and lack of proper supervision and quality control; (2) poor performance of nonstructural elements and systems; and (3) vulnerability caused by locational and physiographic factors. Lack of financial resources of the schools and lack of awareness and knowledge about the underlying risks and possibility of mitigation measures have been the root cause of this ever-increasing earthquake risk.

To address this problem, Nepal has been implementing a School Earthquake Safety Program under the leadership of the Department of Education of the Government of Nepal. The program consists of assessment of seismic vulnerability of school buildings and systems; strengthening of buildings and nonstructural components; training and education of school children, teachers, and parents on aspects of earthquake safety; enhancing earthquake preparedness of schools; and training of local masons on safer earthquake construction technology. Promotion of locally available materials with improved technology, involvement of the community in all program activities, emphasis on awareness raising and capacity building, and transparency in all activities have been the key elements for the success of the

Earthquake Hazard, Risk, and Disasters. http://dx.doi.org/10.1016/B978-0-12-394848-9.00008-0
183

program. The approaches adopted can be useful and replicable in other countries with similar building typologies, school management systems, and awareness levels; although the technical methods for addressing structural and nonstructural vulnerabilities may differ in different contexts.

8.1 INTRODUCTION

Schools in seismic regions of developing countries face huge risks from natural hazards such as earthquakes, cyclones, floods, and landslides mainly as a result of locational, structural, and nonstructural vulnerabilities, as well as inadequate levels of preparedness. Among the various natural hazards, earthquakes have the potential to inflict huge destruction to a large number of school buildings in a single event. For example, the 7.6 magnitude 2005 earthquake destroyed almost all education facilities in Azad Jammu & Kashmir of Pakistan, causing death to >18,000 students and injury to many more. More than 10,000 children lost their lives, and thousands more were injured in the Sichuan China earthquake of 2008. A majority of the school buildings within the earthquake affected areas of China (2008), India (2001), Pakistan (2005), Iran (2003), and Haiti (2010) suffered heavy damage—almost all school buildings in the areas of shaking intensity IX or more either collapsed or were damaged beyond repair.

In Nepal, the 1988 Udayapur earthquake (magnitude 6.5) caused the destruction of >6,000 classrooms, disrupting the education system—approximately 300,000 children were not able to properly attend schools for several months after the event (Thapa, 1989; Dixit, 1991). Fortunately, the earthquake occurred during nonschool hours; otherwise, the casualty rate would have been higher. More recently, in September 2011, the moderate 6.9 magnitude earthquake severely damaged >2,000 school buildings of Eastern Nepal, despite the fact that the maximum shaking intensity observed was of only VII on the Modified Mercalli Intensity (MMI) scale (NSET, unpublished c).

In the context, ensuring the safety of schools from earthquake risks is emerging as one of the most urgent tasks to be addressed by governments, international financial institutions, development partners, and parents. This chapter aims to encourage national and international stakeholders to invest in making schools safe against earthquake risks. It provides evidence of successful cases of cost-effective methods of enhancing seismic performance of schools in Nepal. Although the technical methods for addressing structural and nonstructural vulnerabilities may differ from country to country, the approach is expected to be useful everywhere and several of the intervention techniques could also be replicated in conditions of similar building typologies and school management.

8.2 EARTHQUAKE RISK OF NEPAL

Nepal sits astride the boundary between Indian and Tibetan plates along which a relative strain of approximately 2 cm per year has been estimated (Dixit, 1994). The existence of the Himalayan Range, with the world's highest peaks, is evidence of the continued tectonic activity beneath the country (Bilham, 2014). In fact, Nepal has a long history of destructive earthquakes. The earliest recorded event in the most comprehensive catalog to date occurred in 1255. Earthquakes with great damage occurred in 1833, 1934, 1960, and 1988. In the past century alone, >11,000 people have lost their lives in four major earthquakes. The 1934 earthquake produced an intensity of IX–X on the MMI scale in the Kathmandu Valley, destroying 20 percent and damaging 40 percent of the valley's building stock. In Kathmandu itself, one quarter of all homes were destroyed. Many of the temples in Bhaktapur were destroyed as well. A possible repeat of the 1934 level shaking in modern day Kathmandu is estimated to cause deaths of up to 100,000 and injury to 300,000 people (NSET, 1998 modified by NSET, unpublished report a; Dixit et al., 2012).

8.2.1 Unsafe Buildings are the Root of Earthquake Risk

Although rampant poverty, rapid population growth, illiteracy, and lack of awareness are believed to be responsible for high levels of earthquake risk, poor building performance has been singled out as the most important constituent source of the ever-increasing earthquake risk in Nepal (GESI, 2001). Therefore, improving seismic performance of new, as well as existing, buildings should become one of the main thrusts toward earthquake safety in Nepal. Promoting safer building construction is an objective necessity for developing nations where urban populations seem to be doubling every 10–15 years. Such rapid urban population growth demands a high rate of building production, which, in the absence of a proper building permit process, and a general lack of knowledge and skills for earthquake-resistant construction, end up in shanty constructions that are extremely vulnerable to earthquake.

To compound the problem, no system exists for controlling the professional standards of engineers/designers through reference to professional qualifications/membership, peer review processes, or by legal means. Further, the owner builders, who follow the advice of local craftsmen and mason leaders, build a significant proportion of the buildings in Nepal. Neither the owner builder nor the crafts persons are aware of the possible disastrous consequences of an imminent earthquake. They do not have adequate access to information related to safer building practices and incorporation of simple earthquake-resisting features at nominal extra costs. Even building construction projects funded by national and multilateral agencies generally do not spell out adequate requirements related to seismic safety in their terms of reference to their consultants (Dixit, 2005).

8.3 SEISMIC VULNERABILITY OF SCHOOLS IN NEPAL

8.3.1 Structural Vulnerability

Nepalese schools, both their buildings and occupants, face extreme earthquake risks. This is because the majority of school buildings, even those constructed in recent years, are generally not constructed with the input of engineers trained in seismic design or construction supervision. Management of public schools is largely the responsibility of the local community—government provides only the curriculum and minimum financial support. Characteristically, only low annual budgets are available to the school management system, which increases the likelihood that poor materials or workmanship are used in the construction of the school buildings, making them structurally vulnerable to earthquakes.

A 1997 study by the National Society for Earthquake Technology—Nepal (NSET) in Kathmandu Valley revealed that >66 percent of the valley's public schools are likely to collapse if the valley experiences MMI intensity IX level of earthquake shaking. Such shaking during school hours could kill >29,000 students and teachers, and injure an additional 43,000 occupants of these schools (NSET, 2000; Dixit et al., 2002). The results obtained from a seismic assessment of existing school buildings in Kathmandu Valley, Lamjung, Nawalparasi, Humla, and other districts constitute the following vulnerability scenario (NSET, unpublished b).

Extrapolation of the above findings to the national level indicates that >75 percent of the existing buildings are highly vulnerable to earthquake risk (Figure 8.1), and require immediate intervention. Of these, 15 percent are too dangerous to occupy on a daily basis, and must be dismantled and reconstructed in an earthquake-safe method immediately.

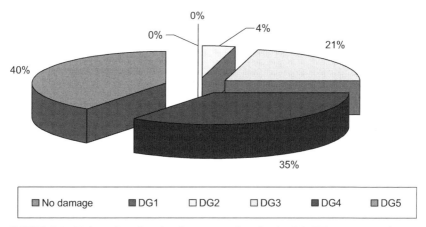

FIGURE 8.1 **Estimated nationwise damage grades of school buildings at Intensity IX Modified Mercalli Intensity.** DG, damage grades.

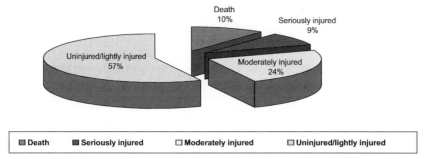

FIGURE 8.2   **Estimated casualty in schools at Intensity IX Modified Mercalli Intensity.**

In Nepal, approximately 82,000 school buildings belong to about 34,000 private and public schools. NSET has identified that approximately 49,000 buildings require seismic strengthening and 12,000 buildings must be demolished and reconstructed earthquake-resistant. If the problem is not addressed, a rough estimation reveals that in the case of an earthquake of intensity IX during school hours, in any part of the country, approximately 111,000 students (10 percent of the total affected population) could lose their lives and another 85,000 could suffer serious injury (NSET, unpublished b) (Figure 8.2).

8.3.2 Nonstructural Vulnerability and Earthquake Preparedness

Lack of earthquake awareness and earthquake response planning in schools is conspicuous. School syllabi educate students on the causes of earthquake phenomenon, but do not teach earthquake evacuation and survival. Earthquake evacuation plans and periodic drills are not practiced in almost all schools, both public and private.

8.4 REASONS FOR HIGH SEISMIC VULNERABILITY OF SCHOOLS

In order to understand the genesis of the seismic vulnerability of schools, one has to understand the school system of Nepal. Broadly, schools belong to one of the following five categories in terms of ownership and funding:

1. *Government-aided community schools*: full funds provided by and managed by government;
2. *Community-managed schools*: government provides funds for the salary of a prescribed number of teachers and other resources, all management responsibility taken by the community;
3. *Community schools, government-unaided*: limited to no funding or management support from the government;

4. *Schools managed by faith-based institutions*: limited governmental support; and
5. *Private/institutional*: privately funded; supported by fees and trustees.

Excepting a few Kathmandu-based schools that fall into category A, most public schools in Nepal fall under categories B, C, and E. Local communities have established, and continue to run the majority of Nepali schools. Government responsibility lies only in controlling curricula, teaching standards, and examinations, as well as providing limited financial resources to cover mainly the salaries of a minimum number of teachers. The community manages all remaining expenses, relying mainly on donations, as tuition fees are kept minimal lest the parents should be discouraged to send their children to school. In this process, the school buildings are constructed with limited resources employing building construction practices and using construction materials prevalent in the community. Due to the lack of appropriate regulation related to physical facilities and safety provisions, the buildings are weak, and the schools are operated in an unsafe and unhealthy environment.

8.4.1 Vulnerability Due to the Use of Poor Building Materials and Technology

More than 95 percent of the school buildings are believed to be constructed using informal processes and are nonengineered. Although the Department of Education (DOE) has prescribed standard designs for school buildings, these standards are not adhered to: the actual construction even of newly constructed school buildings differs totally from the design in terms of types, materials, size, and quality of structural elements, and the size of buildings. No standard supervision or quality assurance processes are maintained for community-constructed schools. The remaining 5 percent of engineered school buildings are in urban areas. The buildings and ancillary structure in a school made of traditional construction materials such as bricks, adobe, and stone have inherent weaknesses and are susceptible to damage due to the lateral loads of earthquake forces.

8.4.2 Vulnerability Due to Poor Building Construction Practices

Building construction practice relies heavily on untrained local masons who rely on skills handed down over the generations. Unfortunately, the traditional wisdom and skills of earthquake-resistant elements of building construction are gradually being lost due to the advent of modern construction materials, such as the reinforced cement concrete (RCC), that require the use of defined procedures such as a controlled water-to-cement ratio in concrete.

Local masons are generally not familiar with the properties, limitations, and proper techniques of modern construction materials such as cement, steel, reinforcement bars in different components, and the required quality of such

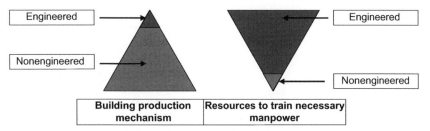

| Engineered | | Engineered |
| Nonengineered | | Nonengineered |

| Building production mechanism | Resources to train necessary manpower |

FIGURE 8.3   **Resource distribution in comparison to construction mechanism.**

materials. The masons do not have formal training on the proper techniques of bending and placing of reinforcing bars, grade of concrete demanded, required cover to reinforcement and curing (Dixit, 2004). Most of the buildings in Nepal are constructed with the technical inputs from local masons and contractors. However, the level of investment for training them is extremely low as compared to the investment to develop engineers (Figure 8.3). Therefore, the local craftpersons have limited knowledge and skills on earthquake-safer construction of buildings. This leads to increasing vulnerability, even in new constructions with the use of modern materials. Further, the institutional and technical capacity of local authorities is generally very poor to properly implement even the standard designs prescribed by the DOE under the Ministry of Education (MoE), let alone adoption of emerging technologies.

In most of the cases, local masons act as technical advisors to building owners. Consequently, most school buildings, even in urban environments do not incorporate sufficient seismic design for structural components.

8.4.3 Supervision and Quality Control

Generally, no formal mechanism exists for on-the-site quality control or construction supervision for the majority of community-managed schools. School management committees (SMCs), school construction committees, and local community leaders supervise construction. Technical staffs at the District Education Office sometimes conduct casual supervision of the school construction works, and only for some schools nearer to district headquarters.

8.4.4 Vulnerability Due to Poor Performance of Nonstructural Elements

Nonstructural vulnerability equally contributes to increased seismic risks in schools. Such seismic vulnerability is caused by nonstability of goods inside the buildings that do not contribute to the structural integrity of the buildings, such as cupboards containing chemicals in a science laboratory, classroom furniture, staircases, architectural ornamentations, and other objects that could fail or fall during earthquake shaking. Movements or failure of these elements and/or obstruction to evacuation may cause death and injury during an earthquake.

8.4.5 Location Vulnerabilities

Schools are generally constructed on a lot of land donated by the community or public land, which usually are marginal. These lands are riverbanks vulnerable to flooding/flash flood or at the foot of an unstable hill, susceptible to landslide, liquefaction, and erosion. Furthermore, in an effort to minimize the costs, many schools, for example, have light galvanized iron sheeting as the roofing that is very hot in the southern plains of Nepal in summer and extremely cold in winter, especially in the high mountains.

8.4.6 Physiographic Factors of Vulnerability

As the types of construction materials used, as well as the level and use of accessible knowledge differs significantly based on geographic location, there is a distinct variation in the distribution of vulnerability factors among schools located across different physiographic regions of Nepal. The mountainous parts of the country largely lack road networks, so import is expensive and limited, and hence, people continue to construct buildings with locally available construction materials. School buildings were constructed with adobe with irregular timber reinforcement. Indigenous technology in safer construction is almost extinct, and new construction practice is now being employed with very limited knowledge or skill on proper use.

8.4.7 Knowledge Availability

Knowledge in the use of modern construction materials and safer construction techniques in rural areas is limited. The national building code is not mandatory in university engineering curricula; consequently, few engineers understand seismic performance of buildings or the need for design detailing. Thus, standard engineering procedures as discussed by Tolis (2014) are not followed.

8.5 MAKING SCHOOLS SAFE AGAINST EARTHQUAKES

8.5.1 Concept of Seismic Safety of Schools

School earthquake safety should mean that the entire education system is safe against earthquake risks: structural as well as nonstructural elements of school infrastructure built or repaired to acceptable seismic performance, adequate preparedness exists for proper earthquake response, as do adequate knowledge and skills within the school system to ensure safety of the populace in case the school is used as emergency shelter after an earthquake. Earthquake safety of schools should go beyond the concept of safe building to encompass resilience of the school system and the stakeholders it serves.

The following are the main elements of seismic safety of schools.

1. *Structural safety*—all buildings and physical infrastructure of education system must be safe for immediate occupancy after an earthquake. The concept of safer construction should be incorporated in the annual plans and programs of the related authority, allowing construction only in locations that are safe against all hazards, an adequate distance from high-voltage electricity distribution lines and electric substations.
2. *Nonstructural safety*—all nonstructural elements, including classroom partition walls, cupboards, and laboratory chemicals are secured and students' seating arrangements are safe and facilitate an easy emergency evacuation process.
3. *Curriculum on earthquake safety*—disaster safety education is well incorporated in the formal curriculum of relevant subjects of different grades. The curriculum should enable children to understand not only the genesis of earthquakes but also to impart knowledge and skills on protecting themselves during emergencies.
4. *Schools as community learning centers*—in developing countries, local communities are strongly attached to schools because people have often contributed significantly to the establishment of those schools; they are continuously involved in developing physical facilities and in managing its operation. The community easily accepts new ideas, technology, and knowledge introduced in schools. Propagation of knowledge on earthquake safety should take place from schools by means of teachers to children, children to parents, and parents to the community. Further, the learning or practice of disaster education by young brains remains in the society for a longer period of time and is easily propagated to friends and family over the entire life of young students. In this way, a school in a developing country can function as community learning centers for knowledge management, skill transformation, and societal change.

8.5.2 Model Steps of a School Earthquake Safety Program

The following comprises the methodology of a School Earthquake Safety Program (SESP) in Nepal.

1. Inventory of schools, collection of basic physical and demographic parameters;
2. Mapping of seismic hazards, earthquake risk assessment, and identification of appropriate intervention options for vulnerability reduction;
3. Provision of technical assistance to school management and the local government authority in planning, design, and implementation for new building construction, demolition and reconstruction, or seismic retrofitting of existing buildings;

4. Build capacity at the local level on safer construction technology through training of local engineers, architects, masons, and contractors;
5. Provision of technical assistance to school management to develop and implement a school disaster preparedness plan; skills development training for teachers and students on earthquake preparedness; and earthquake evacuation drills including light search and rescue and first aid;
6. Advocacy for policy strengthening on disaster risk reduction (DRR) education;
7. Incorporation of earthquake education in formal school curriculum; and
8. Activities on enhancing earthquake awareness, building/enhancing capacities, and institutionalization of DRR in the school and school management system nationally.

SESPs have been successful in developing appropriate technical methodologies and practices for community-based implementation. The initiative has demonstrated the technical, economic, political, and sociocultural feasibilies of enhancing the earthquake safety of schools in Nepal.

8.6 IMPLEMENTATION OF AN SESP

8.6.1 Interventions

Typical activities under an SESP are

1. Incorporate safety measures against earthquake and other natural hazards into the construction of new buildings, including both structural and nonstructural elements.
2. Implement seismic retrofitting of existing school buildings that have been identified as unsafe for earthquakes, including both structural and nonstructural elements.
3. Develop an Earthquake Preparedness Plan and establish a system of implementation and periodic drills.
4. Conduct earthquake awareness activities, including orientation and training of parents, community members, members of the SMC, and local government officials; airing of public service announcements via local media, workshops, seminars, etc.
5. Conduct target-oriented training programs for engineers, technicians, masons, owner builders, materials traders, and suppliers.
6. Construct a demonstration earthquake-resistant building in the settlement.
7. Promote the SESP through other activities such as Annual Earthquake Safety Day, annual Students' Summit for Earthquake Safety, and learning and exchange visits.

8.6.2 Addressing Seismic Performance of New Constructions

More than 6000 new classrooms are constructed every year in Nepal. It is necessary to ensure that these new constructions are made earthquake resistant by following the national building code with adoption of suitably high safety and importance factors for schools.

A majority of school buildings are elongated, with length more than three times that of their width, and built in an inappropriate shape such as L, T, E, and U. Most columns are weaker than the beams and not laid out on a grid, there is no direct vertical load path, foundations are commonly of an insufficient size, and reinforcing bars are often placed and detailed inappropriately. To achieve the objective of earthquake-resistant building, designs should comply with the following standard:

1. Regular shape of building
2. Length to be less than three times the width ($L < 3B$)
3. Height to be less than three times the width ($H < 3B$)
4. Load path should be strictly vertical—no offsets to be allowed between wall or column footprints on different floors
5. At least three load-bearing structural walls in each direction for masonry buildings
6. In masonry buildings, the unsupported length of wall should be restricted and the wall should have adequate thickness
7. Seismic bands for resisting lateral forces to be placed all around masonry buildings at foundation, plinth, sill, lintel, and roof levels—this provides integrity to different orthogonal walls, acting as a box during earthquakes
8. Strengthen the joints of masonry buildings at the corners and T-joints with stitches
9. Provide through-stones at an interval of not less than 2 ft for stone masonry buildings
10. Ensure structural integrity of buildings from the foundation to the roof level by tying all individual elements to each other
11. Provide at least three columns in perfect grid in each direction for framed structures
12. Follow the principle of "strong columns, weak beams"
13. Provide adequate size of structural elements, such as columns and beams
14. Ensure proper bar detailing and anchorage at beam column joints
15. Ensure high quality of materials and workmanship
16. Ensure proper batching, mixing, placing, compacting, and curing of concrete

All new construction should comply with the minimum requirements of seismic safety. Considering that a significant number of school buildings are constructed every year through different funding sources, including community resources, certain guidelines must be developed to address seismic safety requirements and should be made mandatory to all school construction irrespective of the origin of resources for construction. Additionally, there should be a proper mechanism for supervision of construction work to ensure safety standards are adhered to in construction at the field level. Trained midlevel technicians should be employed for supervision and quality assurance of the work.

8.6.3 Demolition and Seismic Reconstruction of Highly Vulnerable School Buildings

About 15 percent of the existing building stock is found to require demolition and reconstruction. The decision whether to demolish and reconstruct earthquake resistant or implement a seismic retrofit is taken on an objective basis. Both are expensive processes. It is necessary to determine an appropriate threshold of when to retrofit and when to demolish and reconstruct. NSET's past practice indicates that if the cost of retrofitting is <30 percent of the cost of a new construction of the same type and size of building, the building should be retrofitted. The level of safety achieved per unit expense justifies retrofitting. The cost per unit safety is not justified if the estimated cost for seismic retrofitting is >30 percent of the building replacement cost.

8.6.4 Seismic Retrofitting

Retrofitting is a vulnerability reduction technique, with the primary objective of enhancing the seismic performance of an existing, vulnerable building, in order to protect property and lives of school occupants during an earthquake. It requires technical input and skilled human resources.

The majority of school buildings subject to seismic retrofitting are either stone or brick masonry structures. Therefore, Nepal's experience is largely on the retrofitting of masonry buildings. Structural interventions target enhancing structural integrity of the building elements, augmentation of the joints of elements, and enhancing the strength of individual structural elements. Methods employed include adding vertical splints, horizontal bandages, and jacketing of walls and columns. Often, the approach also consists of strengthening walls by adding more area, and reducing the sizes of openings. In the case of an RCC structure, retrofitting consists largely of strengthening the columns and column-beam joints, strengthening of the foundation, and enhancing structural connection between infill walls and columns to avoid out of plane failure.

Past experiences demonstrate that retrofitting is technically and financially feasible in Nepal. Different types of buildings in different regions have been successfully retrofitted, and the efforts have demonstrated good results in terms of social acceptance.

8.6.5 Reduction of Nonstructural Vulnerabilities

Failure of nonstructural elements also account for a significant proportion of deaths in schools due to earthquakes. Strengthening cantilevered walls, fixing of cupboards and other gadgets into walls or ceilings, removal of obstructing materials from narrow corridors, refitting of door hinges to allow them to open outside, and placement of signage are among the methods of nonstructural vulnerability reduction. In Nepal, the cost of such nonstructural vulnerability reduction is low, and hence very attractive. In many cases, the SESP allocates a token budget for implementing small-scale mitigation, such as strengthening the slope near the school or repairing the iron sheet of the roof.

8.6.6 Emergency Response Plan and Earthquake Drill

Nepal has learned the usefulness of preparing Earthquake Preparedness and Response (EPR) plans and conducting earthquake drills (DROP, COVER, and HOLD ON) periodically. While EPR is not yet mandatory, an increasing number of schools are developing it, as well as strengthening their structures under SESP. The DOE is considering making emergency plans and regular drills an integral part of school management by incorporating the following:

1. Disaster Emergency Preparedness and Response planning and implementation mandatory for every school;
2. Build capacity of resource persons and teachers on emergency planning and drill procedure;
3. Build the capacity of teachers, SMC members, and parent representatives in first aid and light search and rescue techniques;
4. Allocate a certain budget for prepositioning light search and rescue and first aid items;

8.6.7 Earthquake Awareness

Awareness is a key to internalizing risk and achieving change in peoples' attitudes toward risk reduction. It is an integral part of the SESP in Nepal. The following are some of the most widely employed methods of earthquake awareness raising activities implemented under the SESP, for both schools and communities.

1. Community lectures, video documentaries and field visit for students, street drama, and simulation exercises are excellent tools. Annual Earthquake Safety Day, on January 15, is a great awareness activity that has been institutionalized in Nepal.
2. Awareness activities for the engagement of parents and community members include orientation lectures, distribution of reading materials, video shows, street dramas, and opportunities to participate in activities conducted in schools. Earthquake walks through city cores with an explanation of the seismic vulnerabilities and their dynamics is a strong motivation tool.
3. Development of community learning centers as training centers for awareness, preparedness, and risk reduction measures are important.

A combination of a series of activities on earthquake awareness, nonstructural vulnerability reduction, and training of teachers and students

FIGURE 8.4 Laying of rebar mesh for jacketing of the brick masonry wall in mud mortar by microconcrete.

using the system of training of trainers has been found to be extremely effective, especially in reaching a large number of students within a short span of time, at nominal costs. This approach is useful when no resources are available for implementing seismic retrofitting of the school building structure (Figures 8.4 and 8.5).

FIGURE 8.5 Student initiated Earthquake Awareness rally.

8.6.8 Capacity Building and Training

The DOE is gradually incorporating the SESP approach to the increasing number of schools in the country. The following are the typical activities conducted for the purpose of building capacities in the system:

1. Establishing a system of disaster preparedness, periodic drills, and organization of disaster reduction activities including earthquake awareness and training, as well as vulnerability assessment and retrofitting of schools in terms of structural and nonstructural safety.
2. Formal training for engineers/designers including vulnerability assessment of existing buildings, design of earthquake-resistant school building, and retrofit design.
3. Formal training of supervisors/junior engineers on construction implementation, including supervision, and quality control of materials and construction processes.
4. Theoretical and hands-on training and practical exercises to masons and petty contractors on each and every step of the construction process for new buildings or seismic retrofitting (Schacher, 2014).

A suite of about 20 curricula have been developed and tested for conducting training for end users as well as training of trainers, incorporating adult learning approaches. Training of trainers has become a regular activity to address the training needs of different stakeholders nationally.

8.6.9 Extracurricular Activities

The SESP actively promotes extracurricular activities for promoting a culture of safety in schools. These activities are designed to support efforts toward enhancing the structural, nonstructural, and functional safety of the school system. The program also encourages students to form disaster Safety Clubs in the school or add DRR activities into existing student clubs. Examples of such extracurricular activities include essay, song, and drama writing and singing; art and quiz competitions, organization of Earthquake Safety Day, and organization of networking and exchange visits such as the annual Student Summit.

8.7 LESSONS LEARNED

8.7.1 Institutionalization of the SESP Demanded a Long Wait

Enthusiasm at the initial success of the SESP instilled a naive belief that the Government of Nepal would willingly adopt the program to cover all schools in Kathmandu Valley and will easily propagate it to other parts of Nepal. The NSET submitted to the MoE a draft program for improving the seismic performance of all 1100 school buildings in the Valley, at an estimated cost of about 10 million US dollars in 1999. The Government of Nepal's MoE adopted the SESP as a national program only in 2010, after the start of a unique initiative called the Nepal Risk Reduction Consortium that advanced the SESP as one of the five flagship programs, and estimated the required total of 50 million US dollars for retrofitting the schools in the Valley. Persistent work spanning more than a decade made the SESP attractive to stakeholders and international development partners of Nepal.

8.7.2 Retrofitting a School as an Important Awareness Raising Opportunity

The most important lesson was not simply that a school could be retrofitted, but that, for an additional 15,000–25,000 dollars, local masons could be trained while retrofitting the school and the villagers could have their earthquake awareness raised. Strengthening the school was important and attractive, but a more attractive outcome of the initiative was retrofitting the school, AND training the masons, AND convincing the masons that the good techniques are better than the poor techniques AND raising the awareness of the villagers, AND teaching the children and teachers what to do during and before an earthquake. All these extras come for a small relative increase in the cost of retrofitting, and they could be possible only because of the retrofitting.

8.7.3 "What is Accepted by the Community" is More Important than "What is Necessary"

The knowledge, program, technology, and training to be given to communities for disaster risk management should be compatible to what they accept, and practice. This implies that the community-based (or community-managed) disaster risk management program should start from low-cost and low-technology options for mitigation and preparedness, to ensure not only that people understand the logic but also accept and use it. The SESP consistently adopted simple technical approaches, which made the initiatives cost effective and understandable to the laypersons. It also helped to focus the project on implementation of risk reducing actions, our major aim. The SESP was accepted and also replicated by the related community also because of the comprehensiveness of approach. Efforts on physical improvement were combined with intensive interaction and two-way dialog with the community, training and education, and establishment of mutual trust and confidence (Figures 8.6 and 8.7).

8.7.4 Community-Based Approach is Key to Risk Management Efforts

Despite the traditional fatalistic outlook, issues of disaster management are becoming popular Traditional attitudes of government bearing sole responsibility for relief and prevention works are being replaced by realizations

FIGURE 8.6   **Earthquake Drill under the SESP program.**

of the need to work at the community level. However, DRR is not the highest priority of the people in view of more pressing needs, such as infrastructure, sanitation, health, education, and environment. Moreover, most communities lack enough financial resources. Therefore, making DRR programs self-

FIGURE 8.7   **Conduction of Community level Search and Rescue Training.**

sustaining is challenging, and requires innovative thinking. At the same time, considering that the benefit—cost ratio is very high in view of the prevailing low level of preparedness, methods to initiate and support community-based disaster management programs must be identified.

8.7.5 Transparency Pays

Low levels of awareness of earthquake risks resulted in criticism during the initial stage of program implementation. NSET's status as a nongovernmental organization (NGO) further increased initial apathy, due to the tarnished image of several NGOs. However, open financial policies and transparency were exercised, limiting the role of the NSET to technical/management assistant, and granting local SMCs' decision-making responsibilities, aiding the development of mutual trust, and the program ultimately received cooperation from all concerned.

School Earthquake Safety Advisory Committees, headed by the Chairman of the District Education Committee, including representatives of local village level government, prominent individuals and members of the SMC, were formed to provide necessary advice. All aspects of reconstruction activities, including finance and other problems, were discussed in the fortnightly meetings of the advisory committee. These committees provided necessary political support to the project, helping to increase its outreach.

A five-member School Retrofitting Committee, consisting of the chairman of the SMC, the headmaster, the NSET engineer, and two local representatives, implemented the project in each school. Committee decisions were made with NSET guidance. The committee operated a bank account with the headmaster and the NSET engineer as the joint signatories. This committee helped achieve transparency, to optimize the limited resources, and foster mutual trust.

The whole process was conducive to peoples' participation. While the local community provided 25 percent of labor as an in-kind contribution, the reconstruction was done completely by the local community with the NSET providing only technical inputs and providing fund-raising assistance.

8.7.6 Training Program for Masons is Essential for a Successful School Earthquake Safety

The training program helped a lot to convince the local masons on the affordability and possibility of constructing earthquake-resistant buildings using slight improvements in the locally employed methods of construction.

8.7.7 Small Dispersed Infrastructures are Better Candidates for Initial Mitigation Investment

A significant proportion of public schools in developing countries are small, simple structures, constructed with traditional materials, offering mitigation specialists excellent objects to work and learn from. When the SESP was started, the NSET was criticized for excessive focus on public schools and neglect of other facilities. Many people questioned why hospitals, a critical facility for postearthquake response, were not chosen. Additionally, people asked why cinemas, private schools, and colleges were not examined. The project team's justification and continued explanation for its focus on school did not quell the criticism. However, given the limited resource availability, Kathmandu Valley Earthquake Risk Management Project continued to focus on schools, noting that the work on schools was building NSET's capacity to evaluate the vulnerability of other systems in the future. The school survey examined many previously unknown activities: the costs of conducting a survey of building vulnerability, the technical expertise required for this type of survey, the costs involved in strengthening existing vulnerable buildings, the types of techniques to use for strengthening typical Nepalese structures, the interest of the community in strengthening buildings, the ability to attract funds (local and international) for this type of work, and the levels of earthquake risk acceptable in Nepalese society.

8.8 CONCLUSION

The decade-long experience of the SESP in Nepal has demonstrated that the safety of schools can be enhanced significantly through organized approaches and the combined efforts of concerned stakeholders. The program is cost effective, and the technical skills can be developed within the country. There are challenges, but they are not difficult to overcome. Successful implementation of an SESP in a country with a weak economy, such as Nepal, convincingly demonstrates that the program is replicable in other parts of the region with suitable adaptation and contextualization.

REFERENCES

Bilham, R., 2014. Aggravated earthquake risk in South Asia: engineering vs. human nature. In: Shroder, J., Wyss, M. (Eds.), Earthquake Hazard, Risk, and Disasters. Elsevier, London, pp. 103−141.

Dixit, A.M., Parajuli, Y.K., Shrestha, S.N., 24−28 September 2012. Preparing for a major earthquake in Nepal: Achievements and lessons. In: 15th World Conference on Earthquake Engineering (15WCEE). Lisbon, Portugal.

Dixit, A.M, 2005. International symposium on "Building safer communities against disaster", Safety on Non-engineered Constructions: A Public Forum Event during the UN World Conference on Disaster Reduction, 18−22, 2005, Kobe, Japan.

Dixit, A.M., August 1−6, 2004. Promoting safer building construction in Nepal. In: Proceedings of 13th World Conference on Earthquake Engineering, Vancouver, B.C., Canada.

Dixit, A.M., Pradhanang, S.B., Bothara, J.K., Guragain, R., Shrestha, S.N., Tucker, B., Dwelley, L.S., Parajuli, Y.K., 2002. Experiences of KVERMP: Promoting Safer Building Construction, Regional Workshop on best Practices in Disaster Mitigation: Lessons Learned from the Asian Urban Disaster Mitigation Program and Other Initiatives, 24−26 September 2002, Bali, Indonesia.

Dixit, A.M., 1994. Status of seismic hazard and risk management in Nepal. In: Meguro, K., Katayama, T. (Eds.), WSSI Workshop on Seismic Risk Management for Countries of the Asia Pacific Region; 1993 Feb. 8; Bangkok, Thailand, Tokyo. INCEDE, Inst. Industrial Sc., Uni. Tokyo, 1994; INCEDE Rep. 1994-02, Sr. No. 5: pp. 133−145.

Dixit, A.M., 1991. Geological effects and intensity distribution of Udayapur earthquake of August 20, 1988. J. Nepal Geol. Soc. 7, 1−17.

GESI, June 2001. Final Report − Global Earthquake Safety Initiative (GESI) Pilot Project, GeoHazards International (GHI) and United Centre for Regional Development (UNCRD).

NSET, 1998. The Kathmandu Valley Earthquake Risk Management Action Plan, published by NSET-Nepal based on the outcome of the Kathmandu Valley Earthquake Risk Management Project (KVERMP) of the Asian Urban Disaster Mitigation Program (AUDMP) implemented by the Asian Disaster Preparedness Center (ADPC) for the US Office of Foreign Disaster Assistance (OFDA), Kathmandu.

NSET, 2000. Seismic Vulnerability of the Public School Buildings of Kathmandu Valley and Methods for Reducing It: A Report of the School Earthquake Safety Program of Kathmandu Valley Earthquake Risk Management Project; National Society for Earthquake Technology-Nepal (NSET), Kathmandu.

NSET. Update of Kathmandu Valley earthquake damage scenario, in-house report of the National Society for Earthquake Technology − Nepal, Kathmandu, unpublished report a.

NSET. Draft National Strategy for Making Schools Safe against Earthquakes in Nepal: a draft prepared by National Society for Earthquake Technology Nepal (NSET) based on a research and seismic retrofitting works implemented in six schools in Nawalparasi and Lamjung districts in 2008−2010 under the GFDRR/WB Grant, Kathmandu, unpublished b.

NSET. A Report on the M ∼ 6.9 Himalayan (Sikkim) Earthquake of September 18, 2011; National Society for Earthquake Technology − Nepal (NSET); Kathmandu, unpublished c.

Schacher, T., 2014. Disaster risk reduction through the training of masons and public information campaigns: experience of SDC's competence centre for reconstruction in Haiti. In: Shroder, J., Wyss, M. (Eds.), Earthquake Hazard, Risk, and Disasters. Elsevier, London, pp. 55−69.

Thapa, N., 1989. Bhadau Panch Ko Vookampa 2045 (The Earthquake of 5th Bhadra, 2045 B.S. in Nepali). Niranjan Thapa, Kathmandu.

Tolis, S.V., 2014. To what extent can engineering reduce seismic risk? In: Shroder, J., Wyss, M. (Eds.), Earthquake Hazard, Risk, and Disasters. Elsevier, London, pp. 531−541.

Chapter 9

The Socioeconomic Impact of Earthquake Disasters

James E. Daniell

*Center for Disaster Management and Risk Reduction Technology; Geophysical Institute,
Karlsruhe Institute of Technology, Hertzstrasse, Karlsruhe, Germany; General Sir John Monash
Scholar, The General Sir John Monash Foundation, Melbourne, Victoria, Australia; SOS
Earthquakes, Earthquake-Report.com web service, Cederstraat, Mechelen, Belgium*

ABSTRACT

With around 250–300 damaging earthquakes each year reported worldwide, and 100
causing significant damage, the inhabitants of countries affected by earthquakes are
subjected to hardship and disaster. In the past few years, earthquakes in China, Haiti,
Chile, Japan, New Zealand, and Italy have had the far-reaching impact of earthquake
socioeconomic losses on the inhabitants and governments worldwide. In this chapter, an
increased global database of historical earthquake information combined with socioeconomic analysis presents the statistics, impacts, and historical oddities with respect to
deaths, homelessness, and both direct and indirect economic losses worldwide. The
chapter shows the increasing impact of earthquakes over time, which is synonymous
with increasing worldwide exposure. Data on over 13,000 damaging historical earthquakes causing a combined total of around 8.5 million deaths and over $4.3 trillion
(2013 adjusted) in economic losses have been collected through time.

9.1. INTRODUCTION

Earthquakes not only destroy buildings but also often the infrastructure, economy, and livelihoods of the individuals that inhabit the affected locations. As
yet, much effort has gone into collating the data from historical earthquakes in
databases (NGDC—National Geophysical Disaster Center—www.ngdc.noaa.gov,
PAGER—Prompt Assessment of Global Earthquakes for Response—
earthquake.usgs.gov/pager, Utsu Catalogue—iisee.kenken.go.jp/utsu/, EM-DAT—
Emergency Disasters Database—www.emdat.be, MunichRE NATCAT
Service—www.munichre.com); however, not as much validation and correcting of
errors in records have been undertaken. It is not only an earthquake shaking that
plays a major role toward losses, but also secondary effects such as tsunami, fire,

Earthquake Hazard, Risk, and Disasters. http://dx.doi.org/10.1016/B978-0-12-394848-9.00009-2

landslides, liquefaction, and fault rupture. The CATDAT Damaging Earthquakes Database (Daniell et al., 2011b) was developed to validate, remove discrepancies, and expand greatly upon existing global databases by examining the original sources—and to better understand the trends in vulnerability, exposure and possible future impacts of events similar to historical ones.

In this chapter, we discuss an increased database of historical earthquake information combined with socioeconomic analysis, including the creation and collection of indices not only spatially or for current conditions but also back through time. Data for over 13,000 damaging historical earthquakes have been collected and validated through time. These disasters have caused a total of around 8.5 million deaths and over $4.3 trillion (2013 HNDECI adjusted) in economic losses.

9.2. DEVELOPMENT OF A DATABASE TO ASSESS SOCIOECONOMIC IMPACTS OF EARTHQUAKES

Started in 2003, this database has been significantly improved during the past few years by increasing the number of earthquake records, increasing the quality of earthquake loss statistics, and adding many additional checks. Each validated earthquake includes many parameters, including seismological information (magnitude, hypocenter, location, intensity, and spectral values), building damage data (damage levels, important infrastructure, etc.), ranges of social losses to account for varying sources (deaths, injuries, homeless, and affected), and economic losses (direct, indirect, aid, and insured) as well as NaTech (Natural hazards triggering Technological accidents) data for disasters, and there is a comparison of the loss statistics with the socioeconomic status of the region.

Globally, depending on the source considered, a large range in death-toll estimates results; one example being the Xining earthquake, which affected China in 1927 and which reportedly caused between 40,000 and 200,000 deaths. It is difficult to quantify the exact number of deaths after an earthquake due to the often chaotic post-disaster situation such as quick burials, ad hoc and uncoordinated counting of bodies, inaccurate counting, and other reasons; however, with careful analysis of all sources detailing effects relating to an earthquake, an expert judgment can be made as to a reasonable range of fatalities. Detailed analysis is undertaken in CATDAT by going back to original sources to check. In this case, the EM-DAT value of 200,000 deaths is an aberration, and the 40,000 deaths estimate was much more reasonable. There had been confusion with the Haiyuan earthquake of 1920 as this earthquake caused 273,400 deaths (Zhang et al., 2010), whereas earlier values had a total of 200,000.

The 2010 Haiti earthquake is another example of the issue of reporting death-toll estimates, with counts ranging from 46,000 to 316,000 in the global literature one year after the disaster. An exact value of fatalities will never be known; however, estimates by Ambraseys and Bilham (2011) who prefer around 150,000, a detailed ground survey of Melissen (2010), which shows

82,000 and the SNGRD (Système National de Gestion des Risques et des Désastres—the Haiti government disaster authority) estimate of 212,000, all seem reasonable. Daniell et al. (2013) detailed a methodology demonstrating that around 110,000 to 170,000 can be a reasonable range taking into account the lack of consistency in reporting. In addition, a Michigan University study calculated a total value of 149,095 deaths for the event using a detailed ground survey sample of 1,900 households (Kolbe et al., 2010).

This detailed analysis of estimates is similarly undertaken for estimates of injured, homeless, affected, building damage, economic losses, and other socioeconomic consequences of earthquakes globally within the database.

The CATDAT Database was built using the following regime:

1. An initial database constructed by the author out of interest during an earthquake-related geophysics class at the University of Adelaide in 2003.
2. A review of the original sources contained in many estimates, including ReliefWeb archives (NLA—National Library of Australia archives), NGOs—nongovernmental organizations, insurance companies, papers such as Ambraseys and Melville (1982), books, seismological notes from BSSA (Bulletin of the Seismological Society of America) and many other nondigital databases.
3. Use and checking of databases such as the Mallet (1850), Milne (1911), Sieberg (1932), Montandon (1953), Kárník (1969) and many regional databases such as Gu et al. (1989), Kondorskaya and Shebalin (1982), and Coffman et al. (1982) as well as archived data from libraries.
4. A review of the newer existing major global earthquake socioeconomic effect databases (Ganse and Nelson (1981), Dunbar et al. (1992), EM-DAT, NGDC, Utsu, MRNATHAN, PAGER) in order to look for differences between them and existing issues.
5. Foreign language sources were also searched because it was seen that using only English-speaking references reduces the volume and accuracy of the earthquake record collection. Thus, by using foreign sources i.e., Silgado (1968, 1978) (Spanish), Rothe (1965) (French), Postpischl (1980) (Italian), Gu et al. (1989) (Chinese), KOERI (2010) (Turkish), as well as Portuguese, Russian, Dutch (old Indonesian records), etc., the number of discovered earthquakes, social losses, economic loss values, and building damage was significantly increased when compared to other databases.
6. The work of Nicolas Ambraseys was a key source in this database. Searches were made in both the language of colonization as well as the official current languages of the respective countries. In this way, many old records were discovered.
7. On a country-by-country basis (and province and district basis where available), levels of socioeconomic information for the human development index (life expectancy, literacy rate, GDP (gross domestic product) per capita, and gross education enrollment ratio), gross domestic product

(PPP (purchasing power parity) and Nominal), Consumer Price Index, wages, exchange rate, capital stock, urban and rural population, building inventory, codes and practice, as well as construction-based indices, were created. In addition, 120+ key social indicators and 90+ key economic indicators were selected, collated, and harmonized to compare historical quakes (Daniell, 2009−2013).

8. Since 2010, the database has been integrated into Earthquake-report.com to bring the latest earthquake data and news for every damaging earthquake worldwide. Data are collected in near-real time from many different agencies and then are entered into the database. The CATDAT rating system of historical earthquakes is also presented to show the impact of the most significant past world events, which will be explained in this chapter.

9. Another tool is the disaggregation of social losses of deaths from over 2000 fatal earthquakes since 1900 into deaths from various building typologies, nonstructural deaths related to shaking, and secondary effect and indirect effect deaths, as well as the economic losses from the earthquakes, in order to look at the various impacts of major earthquakes (Figure 9.1).

The type of expert validation procedure described above has been undertaken for all earthquake entries in CATDAT and hence a range of social and economic losses is gained with a higher confidence.

The criteria for an earthquake to be inserted into the database is as follows and forms the CATDAT Orange ranking (for full explanation see Daniell et al. (2012b)):

- Any earthquake causing death, direct injury, or homelessness (structural-related).
- Any earthquake causing structural damage or associated effects exceeding a loss of 105,000 international dollars, Hybrid Natural Disaster Economic Conversion Index (HNDECI), adjusted to 2012.

Since 1900, the reported number of damaging earthquakes has increased due to better recording of small losses, more infrastructure and population being exposed to earthquake losses, and better media coverage. The first human development index (life expectancy, education, and GDP per capita) from 1800 to 2010 worldwide was produced (Daniell, 2010), which documents this increase. This index gives a unique view of the number of damaging earthquakes occurring in highly developed nations versus developing nations. In 1900, most of the world was still developing and therefore the comparison of earthquakes with those of 2012 must be undertaken with caution (Daniell et al., 2012c); for example, an earthquake in the nation of Afghanistan in 2012 occurred in approximately the same socioeconomic climate in the USA in 1900 (Figure 9.2).

From 1900 to 2012, 3,000 casualty-bearing earthquakes have occurred (over 2,000 of which have caused fatalities), leading to over 2.38 million fatalities. There has been over $3 trillion loss in terms of direct and indirect

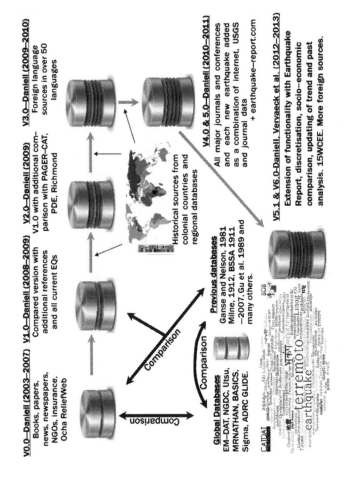

FIGURE 9.1   The process used to create the Damaging Earthquake database (v0.0–v6.0-2003–2013).

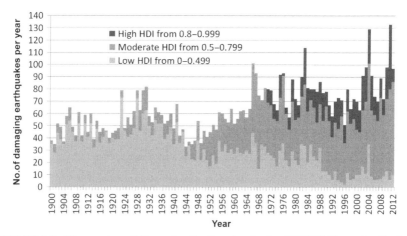

FIGURE 9.2   **The number of Damaging Earthquakes in the CATDAT Damaging Earthquakes database as compared to three levels of Human Development Index (HDI) of the nation at the time of the disaster.**

reconstruction costs needed as a result of earthquakes (2012-HNDECI adjusted).

A comparison of total fatalities and total monetary loss from earthquakes in each country is presented in Figure 9.3. An economic loss per fatality showing the relative economic vs social risk can be seen for these countries. The current

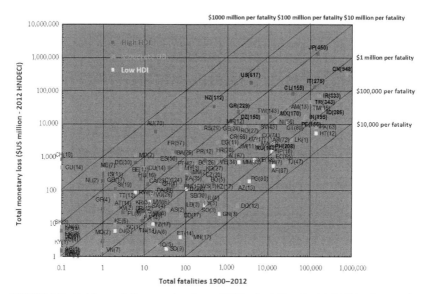

FIGURE 9.3   Total cumulative monetary losses (2012 adjusted) and total fatalities from earthquakes between 1900 and 2012 for each ISO-code country, including the number of damaging earthquakes.

HDI of these countries is also shown from 2012, as well as the total monetary loss in 2012 dollars; however, fatalities are unadjusted through time. Developed nations generally have greater economic losses per fatality and much of the scatter can be explained as the economic losses have been converted to 2012, whereas the fatalities have generally occurred in the now highly developed nations when they were still developing (i.e., Japan 1923, USA 1906), thus skewing the results.

9.3. SOCIAL LOSSES FROM EARTHQUAKES FROM 1900 TO 2012

9.3.1. Deaths from Earthquakes

Over eight million deaths have been recorded as having occurred due to earthquakes through time. Of these, around 2.3 million have occurred in the last 11 decades. Population has increased significantly over the past 11 decades, leading to part of the reason for this increase (Figure 9.4).

One hundred and forty-four earthquakes (and their associated effects) from 1900 to 2012 have caused over 1,000 fatalities, with the last earthquake being the Tohoku earthquake in March 2011. This has changed from 147 earthquakes in the past editions of the CATDAT database, with increased uncertainty on three earthquakes in the earlier half of the twentieth century being placed with a median estimate of just under 1,000 deaths. Through the death tolls researched, huge discrepancies exist in death toll estimates for the Xining 1927 and Haiti 2010 examples above. Additional uncertainties to those mentioned include inaccurate population counts at the time of event, people not registered in census, the general scale of disaster, double counting, copying errors, speed of death registers being produced, and cremation and burying of bodies never being recorded. In addition, problems in the past, such as corruption and government censoring or manufacturing of death tolls (both positively and negatively), have also played a role in which data are only now coming to light. However, key discrepancies can be avoided in the production of a database e.g., transfer errors from historical documents, lack of use of foreign languages, lack of research, lack of updating of sources, and general errors.

Another example of this problem of discrepancies is the Shemakha earthquake of 1902 in Azerbaijan in the NGDC, MunichRe, NATHAN, Utsu, EM-DAT and PAGER-CAT databases. EM-DAT does not include this earthquake in its database, having only the El Salvador, Guatemala, and Uzbekistan (Andizhan) earthquakes for 1902. Utsu includes 86 deaths and 60 injured as its main estimate but does have a note that it could have caused 10,000 or 20,000 deaths. NGDC also records a total of 86 deaths and 60 injured. PAGER-CAT uses the Utsu catalogue value of 86 deaths and 60 injured, due to the algorithm that they utilize to choose between databases.

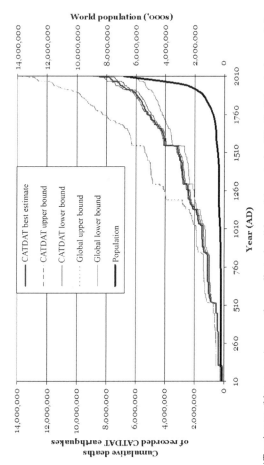

FIGURE 9.4   The CATDAT estimates (blue curves) versus the smallest plausible and largest plausible fatalities (red curves) from earthquakes from various literature sources. This is compared with the global population (black curve).

During the process of cross-validating, CATDAT uses a large number of different sources, including the initial source in the database (in this case that of Ganse and Nelson (1981) and Kondorskaya and Shebalin (1982), where the value of 86 deaths comes about by only including deaths from villages around Shemakha and not the city, Shemakha, itself). The 20,000 deaths mentioned in the notes section of Utsu is probably an exaggeration from newspapers combining the number of homeless (18,000) with deaths (2,000). An acceptable death toll range is anywhere from 1714 to 5000 deaths, which has been quoted by many sources (Kondorskaya and Shebalin, 1982; London Times 1902; New York Times 1902, Russian and Azerbaijani websites) and is allocated as the CATDAT accepted range for the death toll for this event. A count of 1,714 deaths was recorded in the first week after the earthquake by being entered into the Moslem Cemetery record, as stated by Kozák and Čermák (2010). This therefore correlates quite well with the median 2,000 death estimate. The literature bounds, however, for this event are between 86 and 20,000.

Through reanalysis of all events, a preferable difference in estimates is given by the CATDAT upper and lower values. The median value is the preferred value in the database and bars are shown with the range of the losses that are most plausible. The difference between this global upper (red diamonds) and lower (blue triangles) values from all literature sources as compared to the CATDAT median death toll (the straight line as denoted with a cross for a particular event) has shown that huge differences exist in some existing earthquakes such as Xining 1927, Turkmenistan 1948 and Haiti 2010 (Ambraseys and Bilham, 2011; Daniell et al., 2013) in global databases, as seen by the large number of earthquakes outside the error bounds (Figure 9.5).

Since 1900, six events have caused over 100,000 deaths each. Of these, two occurred in the 1920s and two in the 2000s. New research in the last few years has demonstrated that the Haiyuan earthquake in 1920 (Ml7.8–Ml8.6, 10 km depth) in China killed the most people, with a new research in 2010 by Zhang et al. (2010) showing 273,400 people died in this event, with a significant proportion of these deaths due to landslides. The Indian Ocean earthquake of 2004 (M_w9.1, 30 km depth) killed nearly all of the victims via the tsunami. Similarly, the Great Kanto earthquake in 1923 (M_w7.9, 15 km depth) killed 105,000–107,500 people, with 94,000 of these killed by fire. Tangshan 1976 (M_w7.6, 17 km depth), followed by Haiti 2010 (M_w7.0, 13 km depth) and Ashgabad in 1948 (M_w7.3, shallow), were events where infrastructure damage and collapse as a result of earthquake shaking were the main fatality cause. These 10 events have contributed to approximately 60 percent of the fatalities in earthquakes since 1900 (Figure 9.6, Table 9.1).

By normalizing population against the number of fatalities in an event, the relative risk in terms of the last 113 years of fatal earthquakes can be seen in Figure 9.7 as a percentage. Countries with a smaller population that have

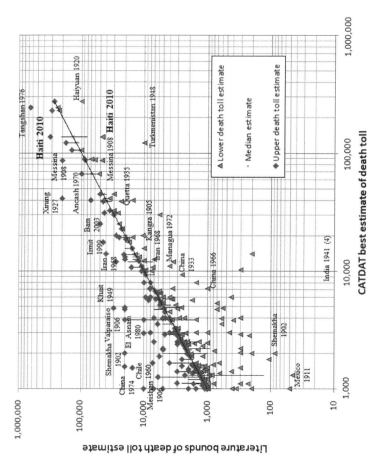

FIGURE 9.5 The upper- and lower-bound death toll estimates of earthquakes in global literature compared to the median CATDAT death toll for all earthquakes over 1000 deaths.

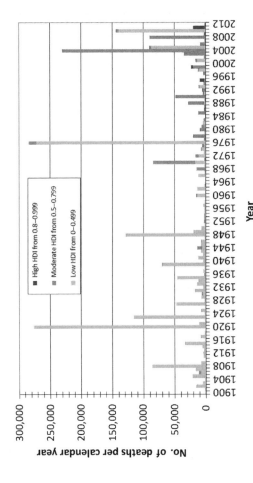

FIGURE 9.6 Fatalities per year based on the HDI of the country at time of event.

TABLE 9.1 The Parameters Included in the CATDAT Damaging Earthquakes Database for Each Earthquake, as of 2013

Rank	Earthquake	Main Country	Date	Median Fatalities	CATDAT Lower–Upper	Pref. Source
1	Haiyuan	China	December 16, 1920	273,465	258,707–283,407	Zhang et al. (2010)
2	Tangshan	China	July 27, 1976	242,419	240,000–255,000	Yong et al. (1988)
3	Indian Ocean	Indonesia, etc	December 26, 2004	228,194	227,640–230,210	Indiv. Country reports
4	Haiti[1]	Haiti	January 12, 2010	140,000	82,000–212,000[1]	Daniell et al. (2013), Ambraseys and Bilham (2011)
5	Ashgabad	Turkmenistan	October 5, 1948	122,000	110,000–176,000	CATDAT
6	Great Kanto	Japan	September 1, 1923	107,385	105,385–143,000	Moroi and Takemura (2004)
7	Sichuan	China	May 12, 2008	88,287	87,476–89,000	Govt.
8	Kashmir	Pakistan, etc.	October 8, 2005	87,364	73,338–87,364	ReliefWeb
9	Messina	Italy	December 28, 1908	85,926	80,000–90,000	CATDAT
10	Ancash	Peru	May 31, 1970	66,794	52,000–67,000	CATDAT

[1] The range is still being quantified and will be reduced or increased if more evidence of the jump of 100,000 deaths from Jan 29th to Feb 4th 2010 in SNGRD statistics is explained.

Source: de Ville de Goyet (2011), Ambraseys and Bilham (2011), Daniell et al. (2013).

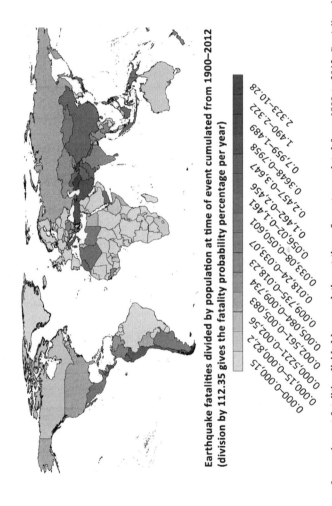

FIGURE 9.7   Number of annual event fatalities divided by population at time of event cumulated from 1900 to 2012. *Daniell and Vervaeck (2013).*

experienced a large quake, such as Turkmenistan, Armenia, Haiti, and Guatemala, can be seen to have a high relative loss. China has not only a lower relative risk, but also the highest absolute risk, given the large population.

The ratio of deaths occurring from earthquakes as compared to total worldwide population is reducing through time (as denoted by the black line in Figure 9.8) although there is a large periodic difference when using a 10-year average. This makes sense due to better engineering practices worldwide. However, the ratio of deaths occurring from earthquakes as compared to total worldwide deaths per year is increasing (red line in Figure 9.8). This is because the worldwide death rate itself is reducing as the average life expectancy increases worldwide. In 1900, approximately 49 million died worldwide. This increased in 1918 to 77.9 million, given the World War I and the influenza pandemic. Other peaks occurred in the World War II and in 1958−1961 with the Chinese famine. The minimum occurred in 1972 with 48.1 million deaths, and in 2013 about 58.6 million deaths are expected worldwide.

As the death percentage from earthquakes as a percentage of total worldwide deaths is increasing, governments need to do more to protect their citizens by choosing higher levels of seismic zonation than those based on 475-year hazard maps (10 percent in 50 years) (Stein et al., 2012; Wyss et al., 2012), controlling corruption (Ambraseys and Bilham, 2011), and educating people who build nonengineered structures.

In the top 10 events by number of deaths, many deaths were seen to have been caused by secondary effects such as tsunami, landslide, or fire. By disaggregating fatalities in the 2,020 fatal earthquakes from 1900 to 2012 into the different causes of fatality, whether from direct structural collapse or secondary effects, an in-depth view can be made of worldwide causes. Masonry buildings via roof collapse, falling debris, or structural members have contributed to over 57 percent of deaths. Approximately 8.5 percent have died in concrete buildings and 3 percent in timber buildings. When looking at the total number, 29 percent of deaths have occurred due to secondary effects and 71 percent of fatalities from damaged infrastructure as a result of direct earthquake shaking. It should be noted, however, that the database is a dynamic entity and further reanalysis of past events continues to take place, including separating heart attack deaths and nonstructural deaths from previous estimates.

A number of other authors have also attempted a split of secondary effects and damaged infrastructure through shaking effects for earthquakes, with this study agreeing well with their findings. Coburn and Spence (1992) show that for 1100 fatal earthquakes from 1900 to 1990 around 76 percent of fatalities were from damaged infrastructure through shaking and 24 percent from secondary effects. Bird and Bommer (2004) looked at 50 earthquakes from 1980 to 2003, concluding that 90 percent of earthquake deaths are due to damaged buildings. A study of Marano et al. (2010) of 749 fatal earthquakes from

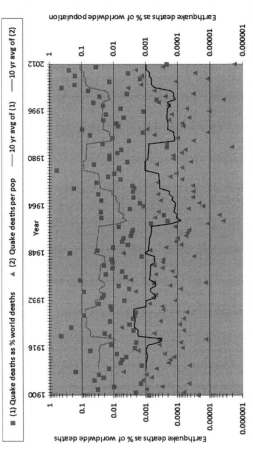

FIGURE 9.8   Yearly earthquake deaths as a proportion of total global deaths and total global population from 1900 to 2012.

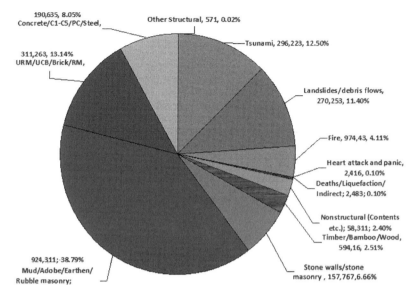

FIGURE 9.9   **The percentage of earthquake deaths due to various causes since 1900.**

September 1968 to June 2008 showed that 25 percent of fatalities from earthquakes were due to secondary effects. For the same period, 913 fatal earthquakes were recorded in the CATDAT database (Figure 9.9).

9.3.2. Homelessness from Earthquakes

Earthquakes since 1900 have caused a combined total of over 65 million people to become homeless. Although building collapse is the main cause of long-term homelessness post-earthquake, there are many reasons, such as serviceability of buildings, other building damage, weather conditions, utilities (power/water), government mandates/forced evacuation, and fear of after-shocks, which play a major role in influencing the total homeless number. With increasing populations, we can see that many of the highest homeless totals have occurred in earthquakes from the latter half of the twentieth century and the start of the twenty-first century (Table 9.2).

Approximately 450 earthquakes have validated homeless and building damage data and the relationship is shown in the following diagram. A log-arithmic trend can be fitted; however, it can be seen that there is a large difference in numbers, as homelessness is not only related to building damage, but is also related to building serviceability. In the 2011 Van earthquake (Daniell et al., 2011a), it was seen that the homeless number far outweighed the occupancy of the destroyed or severely damage buildings, given the cold weather conditions and other factors (Figure 9.10).

TABLE 9.2 The Top 10 Earthquakes in Terms of the Greatest Number of Homeless Since 1900

Date (UTC Time)	Location	Total Homeless
May 12, 2008	Sichuan, China	4,800,000 (Long term)—11,000,000 (short term)
October 8, 2005	Kashmir, Pakistan	3,500,000
September 1, 1923	Great Kanto, Japan	3,400,000
May 22, 1960	Chile tsunami	2,000,000
July 27, 1976	Tangshan, China	2,000,000
May 26, 2006	Yogyakarta, Indonesia	1,845,352
January 12, 2010	Port-au-Prince, Haiti	1,500,000—1,800,000
January 26, 2001	Gujarat, India	1,790,000
May 31, 1970	Ancash, Peru	1,700,000
December 26, 2004	Indian Ocean tsunami	1,690,000

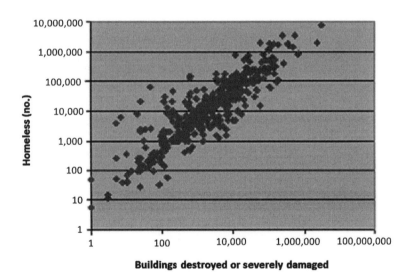

FIGURE 9.10   The ratio of heavily damaged buildings to homelessness in major historic earthquakes.

The ratio of fatalities to homelessness shows a large scatter. This is also due to the fact that smaller earthquakes can cause much utility damage that renders buildings unusable without causing fatalities. Foreshocks, historical experience of earthquakes, and many fatalities occurring in damage classes

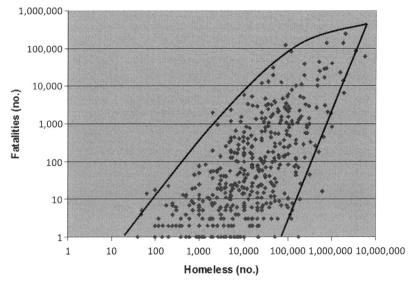

FIGURE 9.11   **Structural damage fatalities vs homeless (no.).**

other than collapse are just some reasons for why there is such a range (Figure 9.11).

The ratio of fatalities to homeless from 1900 to 2012 was also explored and showed a significant reduction through time. This finding comes about for two main reasons: (1) higher population in the shaking area means that more people will be in red-tagged buildings (buildings that are deemed unusable after an earthquake); and (2) better building practices reduce the fatality ratio as a total of all buildings containing many people (multistorey buildings), as shown from the Nicoya 2012 earthquake ($M_w7.6$) as compared to the Nicoya 1950 earthquake (M7.7) (Figure 9.12).

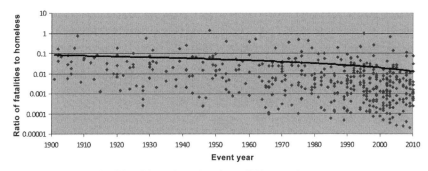

FIGURE 9.12   **Ratio of fatalities to homeless from 1900 onwards.**

9.4. ECONOMIC LOSSES FROM EARTHQUAKES FROM 1900 TO 2012

9.4.1. Historical Economic Losses from Earthquakes

Many nations have experienced damaging earthquakes with large economic consequences in the past few years (Sichuan 2008, Haiti 2010, Chile 2010, Tohoku 2011, Mirandola 2012).

To gain a reasonable comparison of the value of an earthquake in the past in the currency and value of today, it is important to look at correct conversion of parameters into the dollars of today. Most databases use a simple conversion of Consumer Price Index (CPI) based in the USA, as it is difficult to collect CPI data worldwide. CPI, however, is only a useful measure for a basket of goods, in many ways does not relate to an earthquake in the past, and is not a reasonable measure to convert historical losses to today. Thus, the concept of the HNDECI was produced (Daniell et al., 2010) to solve this problem (Table 9.3).

The HNDECI provides a method to combine wages, construction costs, CPI, GDP, worker's production and other indices to measure worth in order to take into account the change for each country through time. In much the same way as an engineering project escalation index taking into account an inflation adjustment, the direct and indirect losses at the time of the event are then brought to a comparable present day value. It is important to note that this is not a normalization method, but simply a more correct way of inflation adjusting the historical natural disaster losses. Using the HNDECI, more accurate comparison than ever before was produced for historical earthquakes in terms of today's dollar values (Table 9.4).

Applying the HNDECI to the damaging earthquake database brings the losses to a comparable USD amount via the HNDECI and exchange rate changes in 2012 dollars. Developed countries unsurprisingly dominate economic losses worldwide; however, as a percentage of proxies such as capital stock and GDP, smaller countries and less developed countries can be seen to be affected when a large city is hit by an earthquake. Major events occurring once every decade, such as Tohoku 2011, Sichuan 2008, Great Kanto 1923, dominate the total losses. The key earthquakes are shown on Figure 9.13 where over $75 billion in losses (2012-HNDECI adjusted) were caused.

In the last 113 years, over $3 trillion (in 2012 dollars) has been lost due to earthquakes. However, given that the global GDP produced during this time was around $2500 trillion, this covers about 0.12 percent of global GDP. In various countries, however, the relative percentage of losses has been much higher. A single incurred loss of infrastructure, assets, and livelihood in various regions can set back countries a number of years.

Japan has had the highest economic loss from earthquakes within the time period, with $1.421 trillion+ (2012 HNDECI Adjusted Dollars) economic losses compared to the nation with the second highest losses, China, with

TABLE 9.3 Details of Common Inflation Adjustment Measures for Economic Costs

Adjustment Parameter	Explanation
Consumer price index	Most common method. It is a cost comparison of consumer goods and services. Forms the basis for a country's inflation rate. Earthquake economic loss is generally not related to food and electrical goods costs so CPI is not reasonable.
GDP deflator	This is a measure of average prices, including not just consumer goods and services, but also comparison of the cost of housing, transportation, food, etc.
The consumer bundle	This is the average annual expenditure of a family or household—refers to how much goods are used as well as cost of the goods.
Unskilled wage	Wage of an unskilled worker. Unskilled work remains constant through time—so is a good measure of wage.
Worker's production index	An index based on the wage of a production worker in manufacturing (i.e., in a specific job)—earnings as well as the increase in added benefits through time.
The average wage	Average of all wages in the country. Influenced by changes in the composition of the workforce toward more skilled labor and also higher wages on the top end. Not as good a measure through time, as it is difficult to know what was included in 1900.
Project escalation index	A combination of construction materials, wages, inflation, and other measures attempting to account for the cost of an engineering project over a number of years.
Capital stock	The capital stock of a location is the amount of all tangible assets in terms of infrastructure, buildings, and exposure. It does not include GDP, but would include the products and equipment already built. There are two kinds of capital stock: gross (theoretical replacement value) and net (actual depreciated value).
GDP per capita	Gross domestic product produced per capita is also a good measure of the general output of a single person and has a good correlation with average income.
Gross domestic product	The gross domestic product is the market value of all goods and services produced in a country in a year. This overestimates greatly the cost of a natural disaster in current terms due to the large increase in population and associated infrastructure.

Source: Daniell (2013).

TABLE 9.4 The Assumptions for Adjustment within the Hybrid Natural Disaster Economic Index for Calculating Gross Capital Stock Replacement

Natural Disaster Parameter	Adjustment to Future Terms	Reason
Property loss (commercial, industrial and public buildings)	Country-based unskilled wage index, or net capital stock	Historical trends have been matched to property loss with good correlation
Reconstruction cost of residential buildings	Country-based construction cost indices, capital stock through time, individual studies.	Building costs analysis via historical components of houses gives closest value.
Crops, pastures, livestock	Using historical databases—if not, CPI.	CPI is most likely closest to the cost of crops and livestock.
Life insurance and intangible costs (deaths, injuries, disability)	Proxy on premiums. Country-based average wage or worker's production index or 1.5 times unskilled wage.	Bureau of Transport Economics (BTE) (2001) trended most of this cost to above an unskilled wage trend—with increasing GDP playing a role.
Indirect losses via business interruption	Consumer price index or GDP deflator	Economic values should be CPI adjusted (or interest rate)
Cleanup	A Combination of material costs (CPI) and demand surge wage. This is constant through time.	A 50–50 combination of CPI and unskilled wage.
Utilities and transport damage	Unskilled wage index.	Tied closer to construction materials and labor.

Source: Daniell et al. (2010).

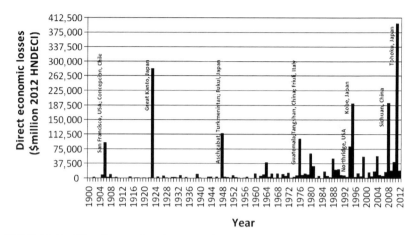

FIGURE 9.13   **Historical earthquake economic losses per year in 2012-HNDECI-adjusted dollars.**

$478 billion. It is important to note that these values change daily with changing exchange rates and economic data such as building prices.

When comparing the GDP (PPP) of a nation at the time of a particular earthquake and comparing the loss, an aggregated GDP view of 113 years of earthquakes is achieved (see Figure 9.14). Countries such as Macedonia, Armenia, Nicaragua, Chile, Haiti, and Turkmenistan have extremely high losses exceeding 1 year of GDP in 113 (or around 1 percent of GDP in the country). These losses take into account the repair and reconstruction cost and therefore a direct comparison with GDP is reasonable.

The highest absolute economic losses of all time have resulted from the earthquake/tsunami/powerplant disaster of Tohoku 2011 (Khazai et al., 2011). This showed the economic impact of cascading effects of earthquakes with the earthquake, then the tsunami, and then the powerplant disaster, compounding the economic losses. However, in terms of relative loss for a country, the Armenian earthquake of 1988, although technically occurring under the former USSR, occurred at a point when the effects were borne mostly by the country of Armenia. The central American countries also show a large relative exposure to earthquakes, with the Managua and Cartago quakes being just two examples (Table 9.5).

Historical earthquakes have generally had a minor impact in terms of insurance losses as compared to the total losses, as insurance takeout was historically low. In recent years, insurance takeout has increased, and with the increasing exposure in many countries it can be expected that insured losses will increase. There are some exceptions to the rule, however, as earthquakes such as San Francisco 1906 and Great Kanto 1923 had large insured loss totals, given the large fires started. When using the HNDECI, the Great Kanto earthquake in 1923 produced the biggest insurance loss of up to $55 billion as

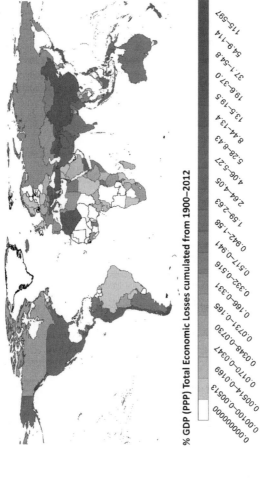

FIGURE 9.14 Economic loss due to earthquakes divided by GDP (PPP) at time of event, accumulated from 1900 to 2012.

TABLE 9.5 The Highest 10 Direct Economic Losses in Terms of Nominal GDP at the Time of Event Since 1900 and Also in Absolute Losses as per the 2012 HNDECI-Adjusted US Dollar Direct Economic Loss Value

Rank	Earthquake	Total Economic Loss ($billion 2012-HNDECI)	Earthquake	Relative % of Nominal GDP (Non-PPP)
1	Tohoku 2011	329.0	Armenia ([1]SSR) 1988	358.9
2	Great Kanto 1923	274.7	Managua 1972	67.1 to 105
3	Sichuan 2008	196.2	Cartago 1910	90.0
4	Great Hanshin 1995	190.6	Concepcion 1906	55.0 to 82.9
5	Fukui 1948	113.8	Haiti 2010	70.0
6	Northridge 1994	79.8	Wallis/Futuna 1993	54.0
7	Tangshan 1976	60.4	Great Kanto 1923	52.8
8	Irpinia 1980	58.0	Nicaragua 1931	51.0
9	San Francisco 1906	55.5	Guatemala 1976	50.3
=10	Niigata 2004	36.1	Maldives 2004	46.0
=10			Jamaica 1907	46.0

[1]Accounts for a partial Soviet Union response—doubling the 1990 Nominal GDP and GDP (PPP) of Armenia. Hyperinflation and devaluation made it very difficult to properly determine the GDP of the time; thus a range has been given incorporating different sources from 1988 to 1998 by using an average value through this period consistent with the reconstruction payout through time. Modeling leads to values as high as 594% of nominal GDP, of which approximately 25–30% were capital stock losses and 70–75% were GDP (nonconstruction) losses—see Daniell (2013) for more information.

compared to Tohoku 2011 with approximately $37 billion. However, using country-based CPI, the Tohoku event resulted in the highest loss. The insurance takeout within New Zealand, which was triggered by the Christchurch 2011 earthquake, was the highest insured loss as a proportion of GDP over $1 billion, with insured losses for the first time exceeding 10 percent of GDP from an event (NZ Reserve Bank, 2011) (Table 9.6).

9.4.2. Economic Losses from Secondary Effects

The relative influence of historical losses from infrastructure damage due to shaking, in comparison to tsunami, fire, landslides, liquefactions, fault rupture,

TABLE 9.6 List of Highest Insured Losses (1900–2012) in 2012 Country CPI Adjusted $ International (Year Average)

Rank	Earthquake	Country	Date	Insured Loss Range (in billion$)	Pref. Source for Loss
1	Tohoku	Japan	March 11, 2011	33.9–38.75	Industry estimates
2	Northridge	USA	January 17, 1994	23.53	Industry estimates
3	Christchurch	NZ	February 21, 2011	10.97–16.46	Industry estimates
4	Great Kanto	Japan	September 1, 1923	8.46–14.58	Daniell et al. (2010)
5	Maule	Chile	February 27, 2010	8.39–13.30	Industry estimates
6	Kobe	Japan	January 16, 1995	6.56	Industry estimates
7	San Francisco	USA	April 18, 1906	6.14	Daniell (2003–2013)
8	Izmit	Turkey	August 17, 1999	3.63–8.46	Industry estimates
9	Darfield	NZ	September 3, 2010	2.19–4.94	Industry estimates
10	Sumatra	Many	December 26, 2004	2.375–4.224	Daniell (2003–2013)
10	Loma Prieta	USA	October 18, 1989	2.58	Daniell (2003–2013)

and other type losses, is important if we are to understand the highest contributing factors to earthquake losses.

Disaggregation of secondary effect economic losses has been undertaken by many authors; however, these studies have not put dollar values to the losses in all cases e.g., Bird and Bommer (2004), Marano et al. (2010) for all types, Keefer (1984), Rodriguez et al. (1999) for landslide losses and NGDC/NOAA (2010) for tsunami losses. Although most historical losses have been infrastructure-damage related via earthquake shaking, the influence of the recent 2011 Tohoku earthquake has changed the historical percentages significantly for tsunami. Fire in San Francisco 1906 and Great Kanto 1923 caused a high proportion of losses (>50 percent), but since then important losses have also occurred in many earthquakes. Liquefaction has occurred in many earthquakes; its effects are also difficult to disaggregate for older historical earthquakes but were major effects in the Kobe 1995 and Christchurch 2011 earthquakes.

A detailed study of all 7208 damaging earthquakes from January 1, 1900 to December 31, 2012 has been undertaken by reviewing the original sources, descriptions, and expert opinion, where exact dollar amount losses with regard to disaggregation have been calculated. Thirty percent of direct economic losses have occurred due to secondary effects of earthquakes, whereas around 70 percent have come from direct earthquake effects. For total economic losses, taking into account the indirect losses, this percentage increases to 38 percent with the impacts of the Tohoku earthquake playing a major role. In terms of earthquake research, the focus on just infrastructure damage via shaking should be changed to one of holistic strategies for infrastructure damage as a result of shaking and secondary effects losses in the future. The following diagram shows direct losses and total economic losses from earthquakes (Figure 9.15).

Landslides can be seen to cause over 5 percent of economic losses currently and this would be higher were it not for the relatively low populations living worldwide in mountainous areas exposed to earthquakes since 1900. China has experienced major losses through the Haiyuan 1920 and Sichuan 2008 earthquakes. Khait 1949 and Ancash 1970 were also major landslide-bearing earthquakes generating major economic losses in their respective countries. The Tohoku 2011 and Indian Ocean 2004 earthquakes have both caused much of the economic losses due to tsunami in recent years; however, many tsunami-bearing earthquakes have caused losses, such as Chile 1960, Alaska 1964, Moro Gulf 1976 with over 10 percent of total losses generated by tsunami and additional NaTech losses via the powerplant disaster in Tohoku.

9.4.3. Losses in Economic Sectors from Historical Earthquake Events

Sectoral analysis of historical earthquake economic losses by disaggregating into component parts reveals that residential losses are not always the greatest

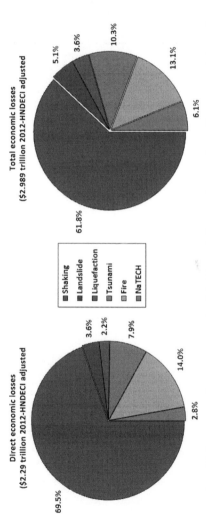

Direct economic losses
($2.29 trillion 2012-HNDECI adjusted)

69.5%
2.8%
14.0%
7.9%
2.2%
3.6%

- Shaking
- Landslide
- Liquefaction
- Tsunami
- Fire
- NaTECH

Total economic losses
($2.989 trillion 2012-HNDECI adjusted)

61.8%
6.1%
13.1%
10.3%
3.6%
5.1%

FIGURE 9.15   Disaggregation of infrastructure damage as a result of shaking and damage via secondary effects economic losses from 7208 earthquakes from 1900 to 2012. Left: Direct Economic Losses (losses occurring as a direct result of the earthquake); Right: Total Economic Losses (including indirect losses—all losses that are not directly as a result of the earthquake).

loss sector with respect to earthquakes. Sixty one major earthquakes between 1907 and 2012 around the world are shown in Figure 9.16, splitting direct earthquake losses into the various social (buildings-residential, and other buildings), infrastructure, production sectors, and cross-sectoral (banking, etc.) losses. In total, 47 classifications have been used for sectors for direct losses for earthquakes; however, five sectors are simply shown in the diagram. For all sectors, see Daniell et al. (2012a). Fifty percent of earthquake direct losses on average result from various types of buildings (around 45 percent being from residential buildings, and the other 55 percent consisting of commercial, industrial, governmental, educational, cultural, and health buildings and, in addition, contents). Nonstructural and structural losses in terms of these buildings have not been split at this time within the buildings. Approximately 30 percent of the direct losses generally come from infrastructure losses (transport/communications, pipelines, energy supply systems, etc.). About 15–20 percent of direct economic losses came from the production and cross-sectoral sectors where economic effects and goods and services are counted. Large indirect losses usually occur in the production sectors; however, these have not been included in this diagram as the focus is on direct losses, and so production sectors when taking total losses into account would have a much higher proportion of losses. One of the best examples of a direct production loss that has occurred due to an earthquake is from the Limon, Costa Rica, earthquake of 1991, where the entire banana stock ready for export was lost.

Unfortunately, there is no one value that can be applied exactly for capital stock and GDP loss discretization. Buildings and infrastructure are essentially the main losses to capital stock, whereas productive and cross-sectors are essentially GDP-based losses; however, separating these values will need additional study of past and future events, given the "value added" when rebuilding as compared to the pre-earthquake built value. This quantification of reconstruction GDP needs further analysis, and is dealt with in Daniell (2013).

9.4.4. Indirect Losses

Indirect losses following an earthquake can be modeled in many ways such as input–output (Cochrane, 1997; Toyoda, 2008; Okuyama, 2008 etc.), Social Accounting Matrices (Cole, 1995), econometric (Kuribayashi et al., 1987), Computable general Equilibrium models (Brookshire and McKee, 1992; Rose and Liao, 2005), surveys (Tierney, 1997) or indicator systems (Kundak, 2004; Hiete and Merz, 2009). Indirect losses are notoriously difficult to quantify; however, existing studies such as Toyoda (2008) for Kobe, CEPAL (1987) for South American earthquakes, and Okuyama (2008) have created estimates for individual earthquakes.

The trend of indirect to direct loss ratios in developing economies has shown that higher indirect losses generally occur with the size of economic

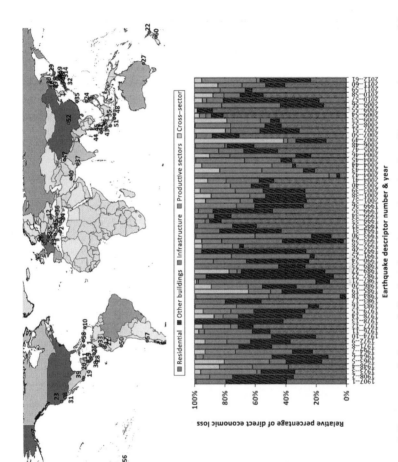

FIGURE 9.16   Disaggregation of direct economic losses of 61 selected major earthquakes into five economic sectors of loss. (Numbers correlate to earthquake locations in the above diagram.)

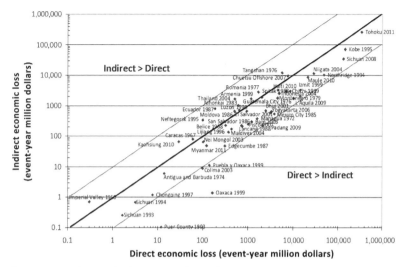

FIGURE 9.17   **Direct vs indirect economic loss from selected events in event-year mill.** US dollars within the CATDAT Damaging Earthquakes Database. *Daniell et al. (2012a)*.

loss and with decreasing development index (Daniell et al., 2012a). This is counterintuitive, with the size of business interruption in developed country economies exhibiting a much higher absolute value; however, because of the lack of redundancy in production sector practices in developing country economies, the relative influence of indirect losses is higher. This can be seen in Figure 9.17.

Indirect losses will continue to increase with production of various items being shared across countries, such as microchips being produced in different locations from where the end-electrical products are made, or cars being produced from car parts from two or more countries, with most cars having over 1500 components.

The 1987 earthquake in Ecuador is a good example of downtime of destroyed pipelines in oil-dependent nations. Thirty-three kilometer of pipelines were destroyed and although the pipeline damage was $122 million (1987), the associated losses with lack of supply due to a lack of redundancy totaled $766 million (1987). Other sectors such as tourism are also commonly greatly affected such as in the Maldives in the 2004 Tsunami.

9.5. CONCLUSION

A summary of the socioeconomic impact of earthquakes has been presented in this chapter. Earthquakes can be seen to impact economies in a crippling blow within seconds. However, earthquakes also allow for renewing of capital stock and may promote growth in certain regions. The CATDAT Damaging

Earthquakes Database collates over 20,000+ sources in more than 50 languages, combining the studies of authors around the world with socioeconomic index databases to produce a validated dataset that reduces erratic values of socioeconomic losses quoted wrongly through literature and gives a deeper analysis and range of earthquake losses and their trends than ever before.

It can be seen that the death rate from earthquakes as a proportion of all worldwide deaths is increasing, as was shown in Figure 9.8, and the economic losses can also be seen to be increasing in absolute numbers (but are decreasing as a percentage of total world capital stock and GDP). Thus, more work still needs to be undertaken in earthquake-resistant building, policy implementation, and reduction of corruption in building practices to ensure more safety of life during and after earthquakes.

REFERENCES

Ambraseys, N., Bilham, R., 2011. Corruption kills. Nature 469 (7329), 153−155.

Ambraseys, N.N., Melville, C.P., 1982. A History of Persian Earthquakes. Cambridge University Press, 219 pp.

Bird, J.F., Bommer, J.J., 2004. Earthquake losses due to ground failure. Eng. Geol. 75 (2), 147−179.

Brookshire, D.S., McKee, M., 1992. Other Indirect Costs and Losses from Earthquakes: Issues and Estimation. In Indirect Economic Consequences of a Catastrophic Earthquake. FEMA, National Earthquake Hazards Reduction Program, Washington, DC.

BSSA (Bulletin of the Seismological Society of America), 1911−2012. Seismological Notes. Various Years.

CEPAL, 1987. The natural disaster of March 1987 in Ecuador and the socio-economic repercussions. CEPAL, Report of April 22, 1987. (in Spanish).

CERESIS, 1985. Terremotos Destructivos en America del Sur, 1530−1894 (Destructive Earthquakes of South America 1530−1894). Lima, 328 pp.

Coburn, A., Spence, R., 1992. Earthquake protection John Wiley and Son Ltd., 1st Edn. West Sussex, England.

Cochrane, H., 1997. Indirect economic losses. In: Development of Standardized Earthquake Loss Estimation Methodology, vol. II. Risk Management Solutions, Inc., Menlo Park, CA.

Coffman, J.L., von Hake, C.A., Stover, C.W. (Eds.), 1982. Earthquake History of the United States. Revised edition (Through 1970), Reprinted 1982 with supplement (1971−80), NOAA and USGS Publication 41-1, 258 pp.

Cole, S., 1995. Lifelines and livelihood: a social accounting matrix approach to calamity preparedness. J. Conting. Crisis Manage. 3 (4), 228−246.

Daniell, J.E., 2003−2013. The CATDAT Damaging Earthquakes Database. searchable integrated historical global catastrophe database, Digital Database, updates v1.0 to latest update v6.01−nb. v0.0−v0.99 refers to all work and updates done 2003 to 2007.

Daniell, J.E., 2009−2013. CATDAT Global Socio-economic Databases. Digital Database and Reports, Karlsruhe, Germany.

Daniell, J.E., 2010. A Complete Country-based Temporal and Spatial Human Development Index − 1800−2010. Digital Database and Reports, Karlsruhe, Germany.

Daniell, J.E., 2013. The development of socio-economic fragility functions for use in worldwide rapid earthquake loss estimation procedures. (Ph.D. thesis) (unpublished), KIT, Karlsruhe.

Daniell, J.E., Vervaeck, A., 2013. The CATDAT Damaging Earthquakes Database—2012—Year in Review. CEDIM Research Report 2013-01, Karlsruhe, Germany.

Daniell, J.E., Wenzel, F., Khazai, B., 2010. The cost of historic earthquakes today—economic analysis since 1900 through the use of CATDAT. In: AEES 2010 Conference, Perth, Australia, vol. 21. Paper No. 7.

Daniell, J.E., Khazai, B., Kunz-Plapp, T., Wenzel, F., Vervaeck, A., Muehr, B., Markus, M., Erdik, M., 2011a. Comparing the current impact of the Van Earthquake to past earthquakes in Eastern Turkey. CEDIM Forensic Earthquake Analysis Group—Report #4, Karlsruhe, Germany.

Daniell, J.E., Khazai, B., Wenzel, F., Vervaeck, A., 2011b. The CATDAT damaging earthquakes database. Nat. Hazards Earth Syst. Sci. 11, 2235—2251.

Daniell, J.E., Khazai, B., Wenzel, F., Vervaeck, A., 2012a. The worldwide economic impact of earthquakes. In: 15th WCEE, Lisbon, Portugal. Paper No. 2038.

Daniell, J.E., Vervaeck, A., Khazai, B., Wenzel, F., 2012b. Worldwide CATDAT Damaging Earthquakes Database in conjunction withEarthquake-report.com—presenting past and present socio-economic earthquake data. In: 15th WCEE, Lisbon, Portugal. Paper No. 2025.

Daniell, J.E., Wenzel, F., Vervaeck, A., 2012c. The normalisation of socio-economic losses from historic worldwide earthquakes from 1900 to 2012. In: 15th WCEE, Lisbon, Portugal. Paper No. 2027.

Daniell, J.E., Khazai, B., Wenzel, F., 2013. Uncovering the mystery of the Haiti death toll. Nat. Hazards Earth Syst. Sci. Discuss. 1, 1913—1942.

de Ville de Goyet, C., 2011. Health Response to the Earthquake in Haiti. January 2010: Lessons to be Learned for the Next Massive Sudden-onset Disaster. Pan American Health Organization. http://reliefweb.int/sites/reliefweb.int/files/resources/Full_Report_3342.pdf (accessed 12.05.12.).

Dunbar, P.K., Lockridge, P.A., Whiteside, L.S., 1992. Catalog of Significant Earthquakes, 2150 B.C. − 1991 A.D. Including Quantitative Casualties and Damage, NOAA National Geophysical Data Center Report SE-49, Boulder, Colorado.

EM-DAT, 2008, 2010. Emergency Management Database. CRED, Catholic University of Louvain. Available from URL: http://www.emdat.be (last accessed 08.08 and 08.10.).

Ganse, R.A., Nelson J.B., 1981. Catalog of significant earthquakes 2000 B.C. to 1979, including quantitative casualties and damage. Report SE-27, World Data Center A for Solid Earth Geophysics, 145 pp.

Gu, G., Lin, T., Shi, Z., 1989. Catalogue of Chinese Earthquakes (1831 B.C.—1969 A.D.). Science Press, Beijing, China (English translation).

Gutenberg, B., Richter, C.F., 1948. Seismicity of the Earth and Associated Phenomena, second ed. Princeton Univ. Press, Princeton, NJ. 310 pp.

Hiete, M., Merz, M., May 2009. An indicator framework to assess the vulnerability of industrial sectors against indirect disaster losses. In: International ISCRAM Conference, Gothenburg, Sweden.

Kárník, V., 1969. Seismicity of the European Area − Part I. D. Reidel, Dordrecht, Holland.

Keefer, D.K., 1984. Landslides caused by earthquakes. BSSA 95, 406—421.

Khazai, B., Daniell, J.E., Wenzel, F., 2011. The March 2011 Japan earthquake—analysis of losses, impacts, and implications for the understanding of risks posed by extreme events. Technikfolgenabschätzung—Theorie Praxis 20 (3).

KOERI, 2010. Türkiye Deprem Katalogları. Kandilli Observatory and Earthquake Research Institute, Istanbul. http://www.koeri.boun.edu.tr/sismo.

Kolbe, A.R., Hutson, R.A., Shannon, H., Trzcinski, E., Miles, B., Levitz, N., Puccio, M., James, L., Noel, J.R., Muggah, R., 2010. Mortality, crime and access to basic needs before and after the Haiti earthquake: a random survey of Port-au-Prince households. Med. Conflict Survival 26 (4), 281—297.

Kondorskaya, N.V., Shebalin, N.V. (Eds.), 1982. New Catalog of Strong Earthquakes in the USSR from Ancient Times through 1977, World Data Center A for Solid Earth Geophysics, Report SE-31, US Department of Commerce, NOAA.

Kozák, J., Čermák, V., 2010. The Illustrated History of Natural Disasters. Springer.

Kundak, S., August 2004. Economic loss estimation for earthquake hazard in Istanbul. In: 44th European Congress of the European Regional Science Association Regions and Fiscal Federalism, pp. 25−29.

Kuribayashi, E., Ueda, O., Tazaki, T., 1987. An econometric model of long-term effects of earthquake losses. In: 13th Joint Meeting of US-Japan Panel on Wind and Seismic Effects, Tsukuba.

Mallet, R., 1852. Third report on the facts of earthquake phenomena-catalogue of recorded earthquakes from 1606 B.C. to A.D. 1850, Rep. British Association for 1852, 1-176, ibid for 1853, 118-212, ibid for 1854, 1−326.

Marano, K.D., Wald, D.J., Allen, T.I., 2010. Global earthquake casualties due to secondary effects: a quantitative analysis for improving rapid loss analyses. Nat. Hazards 52 (2), 319−328.

Melissen, H.J., February 23, 2010. Haiti Quake Death toll well under 100,000. Radio Netherlands Worldwide. http://www.rnw.nl/english/article/haitiquakedeathtollwellunder-100000.

Milne, J., 1911. Catalogue of Destructive Earthquakes A.D. 7−1899. Report 81st Meeting. British Association for the Advancement of Science, Portsmouth, 1912.

Montandon, F., 1953. Les tremblements de terre destructeurs en Europe. UNESCO, Geneva.

Moroi, T., Takemura, M., 2004. Mortality estimation by causes of death due to the 1923 Kanto earthquake. J. Jpn. Assoc. Earthquake Eng. 4, 21−45 (in Japanese).

MunichRe (MRNATHAN), 2009. Globe of Natural Disasters. MRNATHAN DVD, Munich Reinsurance Company.

New York Times, February 18, 1902. 2000 Dead at Shemakha.

NGDC/NOAA, 2010. Significant Earthquakes Database and Significant Tsunami Database, 2010 online searchable catalogue at: http://ngdc.noaa.gov (last accessed 08.10.).

NLA, 2010. National Newspaper Archives (1802−1954), Online Searchable Database of Newspapers from Around Australia including The Advertiser, The Argus, Sydney Morning Herald, Brisbane Courier, Hobart Mercury etc., http://newspapers.nla.gov.au.

NZ Reserve Bank, 2011. Financial Stability Report May 2011. http://www.rbnz.govt.nz/finstab/fsreport/fsr_may2011.pdf.

Okuyama, Y., 2008. Critical review of methodologies on disaster impacts estimation. Background Paper for EDRR Report.

PAGER-CAT., 2008. PAGER-CAT Earthquake Catalog as described in Allen, T.I., Marano, K., Earle, P.S., Wald, D.J., 2009. PAGER-CAT. A composite earthquake catalog for calibrating global fatality models. Seism. Res. Lett. 80 (1), 50−56.

Papazachos, B.C., Papazachou, C., 1997. The Earthquakes of Greece. Ziti Editions, Thessaloniki, 304 pp.

Postpischl, D. (Ed.), 1980. Catalogo dei Terremoti Italiani dall'Anno 1000 al 1980 (Catalog of Italian Earthquakes from 1000 up to 1980). Bologna, 239 pp and subsequent updates.

ReliefWeb, 2010. OCHA ReliefWeb. Available from: http://www.reliefweb.int.

Rodriguez, C.E., Bommer, J.J., Chandler, R.J., 1999. Earthquake-induced landslides:1980−1997. Soil Dyn. Earthquake Eng. 18, 325−346.

Rose, A., Liao, S.-Y., 2005. Modeling regional economic resilience to disasters: a computable general equilibrium analysis of water service disruptions. J. Reg. Sci. 45 (1), 75−112.

Rothe, J.P., 1965. The Seismicity of the Earth 1953−1965. UNESCO, Paris, 335 pp − and associated references.

Sieberg, A., 1932. Erdbebengeographie. In: Gutenberg (1929–1932), pp. 688–1005.

Silgado, E., 1968. Historia de los Sismos más notables ocurridos en el Perú (1513–1960). Bol. Bibliográfico Geofísica y Oceanografía Americana 4. México.

Silgado, E., 1978. Historia de los Sismos mas notables del Peru (1913–1974). Instituto de Geologia y Mineria, Journal No. 3, Series C, Lima.

Stein, S., Wysession, M., 2003. An Introduction to Seismology, Earthquakes, and Earth Structure. Blackwell Publishing, Oxford, UK, 498 pp.

Stein, S., Geller, R.J., Liu, M., 2012. Why earthquake hazard maps often fail and what to do about it. Tectonophysics.

Swiss Re.: Sigma, Economic Research and Consulting, Swiss Reinsurance Company Ltd., 1999–2009.

Tierney, K., 1997. Impacts of recent disasters on businesses: the 1993 midwest floods and the 1994 Northridge earthquake. In: Jones, B. (Ed.), Economic Consequences of Earthquakes: Preparing for the Unexpected. NCEER, Buffalo, NY.

London Times, New York Times, February 21, 1902. Seismic Disturbances Recommence – Number of Killed Now Placed at 5,000.

Toyoda, T., May 2008. Economic impacts of Kobe earthquake: a quantitative evaluation after 13 years. In: Proceedings of the 5th International ISCRAM Conference. Washington, DC, USA, pp. 606–617.

University of Richmond, 2010. The Disaster Database Project (database on the internet). University of Richmond, Richmond, VA. http://learning.richmond.edu/disaster/index.cfm.

USGS, 2010. Historic World Earthquakes. http://earthquake.usgs.gov/earthquakes/. – links to PDE.

Utsu, T., 2002. A list of deadly earthquakes in the world: 1500–2000. In: Lee, W.K., Kanamori, H., Jennings, P.C., Kisslinger, C. (Eds.), International Handbook of Earthquake Engineering and Seismology. Academic Press, Amsterdam, pp. 691–717.

Wyss, M., Nekrasova, A., Kossobokov, V., 2012. Errors in expected human losses due to incorrect seismic hazard estimates. Nat. Hazards 62 (3), 927–935.

Yong, C., Tsoi, K.-L., Feibi, C., Zhenhuan, G., Qijia, Z., Zhangli, C. (Eds.), 1988. The Great Tangshan Earthquake of 1976: An Anatomy of Disaster. Pergamon Press, Oxford, p. 153.

Siyuan, Z., et al., 2010. Chinese seismologists increase Haiyuan earthquake death toll to 273000. In: Haiyuan County Earthquake Conference (90th Anniversary) (in Chinese).

Chapter 10

The Contribution of Paleoseismology to Earthquake Hazard Evaluations

Mustapha Meghraoui[1] and Kuvvet Atakan[2]

[1] Institut de Physique du Globe, UMR 7516, University of Strasbourg, France, [2] Department of Earth Science, University of Bergen, Bergen, Norway

ABSTRACT

Paleoseismology is a relatively young method of earthquake studies at the interface between geology and seismology. Paleoseismic investigations have enriched the fault rupture database in some active zones (e.g., California, Turkey, Italy) and contribute to a significant progress in the concept of earthquake cycle. The increasing number of paleoseismic studies in the last decades was dedicated primarily to the major continental faults that experienced large earthquakes (Mw > 7). The validity of paleoseismic results was tested in regions with long historical catalogs and well-documented past earthquake ruptures as along the North Anatolian Fault and Dead Sea Fault. Major advances in the calculation of realistic earthquake hazard assessment are achieved when earthquake geology and paleoseismic data are available. Among case studies, the probabilistic seismic hazard assessment of the Istanbul (Turkey) region, based on earthquake scenarios revealed the influence of paleoseismic data on the ground motion equations. The integration of paleoseismology in earthquake hazard projects is, however, not yet systematic and consequently many earthquake hazard maps include large uncertainties and approximations.

10.1 INTRODUCTION

Geological investigations applied to the study of past earthquakes are now at the forefront of seismological and seismic hazard research. Also known as paleoseismology, these investigations include the study of active faulting, seismites (sedimentary structures produced by shaking), liquefaction features, and other effects of earthquake shaking (e.g., rockfalls and landslides) that can be recorded in superficial geological units (Soloneko, 1973; McCalpin, 1996). The record of permanent earthquake deformation at the surface is essential in

Earthquake Hazard, Risk, and Disasters. http://dx.doi.org/10.1016/B978-0-12-394848-9.00010-9

paleoseismology as it infers that many major shallow earthquakes may leave a recognizable and measurable trace. The occurrence of past earthquakes belongs necessarily to a tectonic process and a related seismic cycle that implies cumulative surface slip along the same fault system (Yeats et al., 1997). Difficulties in the identification of paleoseismic features may arise, however, when the recurrence time for past earthquakes exceeds a few tens of thousands years, particularly in active zones where surface deformation due to fault slip competes with erosional processes.

Earthquake geology and its "alter ego" paleoseismology have been applied to a large number of seismically active zones worldwide. Paleoseismology is nowadays a fully accepted field in Earth Sciences by academics and industrial institutions; however, its initiation in the 1970s as a key analysis of the coseismic slip (i.e., total slip associated with one seismic event) in past earthquakes was limited to some sections of the San Andreas Fault (SAF) (California; Sieh, 1978; Wallace, 1981). Since then, the limited time window of the historical seismicity catalog (average 500 years for very large events) contributed to the increase of paleoseismological projects in active zones worldwide. With the field identification and characterization of seismogenic fault sources that document the timing and size of past large seismic events, paleoseismic studies (that may also include archeoseismic investigations) became a challenging objective in several active tectonic zones and they are now a necessity for a realistic seismic hazard assessment (SHA).

Initially, the integration of paleoseismic results into the seismic hazard and risk analysis was made necessary due to the incompleteness of earthquake catalogs. In fact, early studies of segmented active faults and their long-term seismicity distribution revealed that in most active zones instrumental seismicity catalogs do not cover the seismic cycle (Schwartz and Coppersmith, 1984). Even in active zones where the historical seismicity covers more than 1,000 years (e.g., the Dead Sea Fault), paleoseismology provides significant results on the recurrence time of large earthquakes with fault parameters and related size of past seismic events that complement old manuscripts (Deng and Liao, 1996; Meghraoui et al., 2003; Galli et al., 2008). Results on the timing of past large earthquakes along faults (with Mw > 7) revealed that interseismic periods may range from ~100 years to 100 ka, depending on the seismotectonic framework and interplate/intraplate domains (Yeats et al., 1997; Meghraoui and Crone, 2001). Studies on the long-term behavior of seismogenic faults and related paleoseismic data are nowadays often required in seismic risk specifications of proposals for building facilities and as input to national and international regulations for seismic safety (e.g., International Atomic Energy Agency Safety Standard Series, 2002). More importantly, recent works question the validity of seismic hazard maps pointing out the poorly constrained earthquake parameters resulting from ignoring paleoseismic data in the hazard analysis (Stein et al., 2012).

The purpose of this chapter is to present the recent progress in paleoseismological investigations and its impact on the seismic hazard evaluation in regions with long historical earthquake records. Combined with earthquake geology, paleoseismology appears to be a powerful approach to primarily constrain the return period of large earthquakes along rupture zones including blind faults. Modern paleoseismology consists in the application of new techniques in order to perform a comprehensive analysis of the timing of past earthquake activity. Selected case studies of field investigations in diverse active zones offer the possibility to measure the efficiency of paleoseismic data in seismic hazard estimates. Although the integration of paleoseismic data in SHA has been widely applied for the SAF activity by the Working Group on California Earthquake Probabilities (2003), here we focus on the SHA calculations in regions where paleoseismic results can be compared to historical seismic catalogs. The determination of past earthquake fault parameters and timing that should cover several seismic cycles may be strengthened by a comparison with a historical seismicity and archeoseismic database. Whether deterministic or probabilistic, a completed analysis of the SHA requires the constraint of earthquake faulting parameters and related paleoseismic database.

10.2 MODERN TECHNIQUES FOR PALEOEARTHQUAKE STUDIES

The reading of successive coseismic displacements at the surface and subsequent collection of paleoseismic data require detailed geological and geophysical investigations in active zones. Surface faulting associated with large to moderate earthquakes in the last decades were tremendously helpful in the application and calibration of paleoseismic methods of investigations (Yeats et al., 1997). Using the basic trenching method across fault scarps, numerous studies provided noteworthy results with paleoseismic data (McCalpin, 1996). Other significant paleoseismic studies are those conducted in regions with subtle, hidden, or hitherto not visible surface deformation. However, in any active region without recent or historical coseismic surface ruptures, the primary difficulty is the access to the geological earthquake record. The identification of late Pleistocene and/or Holocene tectonic features and active fault characterization are major field investigations that precede trenching operations. The goal of any paleoseismic study is to find evidence of a recent faulting event or coseismic deformation, and construct a paleoearthquake database with slip/event, slip rate, location and size of successive past events, recurrence time and estimate a probability for the occurrence of a large event in a given active zone.

10.2.1 Field Investigations in Paleoseismology

Field investigations have been in continuous development in the last decades as they were applied to different fault zones and interplate or intraplate tectonic domains. However, in the absence of a clearly identified seismogenic fault zone, other methods focusing on secondary earthquake-induced features have been used in order to have access to the paleoearthquake archive. Besides the standard exploratory trenching, tools for paleoseismic studies are developed to include:

1. Coring and excavating seismite features and seiche deposits in lakes and seas.
2. Coastal coseismic uplift or subsidence and their cumulative signature.
3. Excavation of earthquake-induced liquefaction features.
4. Broken or deformed stalagmites and stalactites in caves.
5. Earthquake-induced massive rockfalls and landslides.
6. Exposed fault planes in hard rock with successive vertical slip.
7. Exposed minor faulting events, and/or earthquake-induced sedimentary deposits (e.g., seismites or tsunami deposits) in cores or "Geoslicer."

Trench-excavation methods across fault zones were often adapted to the seismotectonic framework, fault geometry, and kinematics associated with the characteristics of sedimentary deposits. Although basic observations as faulting event horizons and major vs secondary tectonic features can be considered as constant features in paleoseismology, the record of successive coseismic slip in sedimentary units at the surface closely depends on the faulting mechanism. Therefore, trenching normal, thrust, or strike-slip faults requires primarily field investigations in tectonic geomorphology, quaternary geology, and includes a careful trench-site selection and specific trench-excavation procedures. While two-dimension (2D) trenching are more appropriate to the investigation of vertical surface slip or deformation on dip-slip faults (normal or reverse-thrust faulting), lateral slip on strike-slip faults entails both two and three-dimension (3D) trenching (Figure 10.1). In some cases, investigating a relatively wide fault zone with distributed surface deformation needs several tens of meters-long excavations with 2D or 3D trenching, or several trenches across major and secondary surface ruptures.

Among the techniques applied in paleoseismology, the use of Lidar or total station measurements of geomorphic features and shallow geophysical prospecting (i.e., mainly Ground Penetrating Radar (GPR), electrical resistivity, high-resolution seismics) largely contribute to the paleoseismic site selection before trenching operations. Tectonic geomorphology and shallow geophysics are powerful field investigations that expose the signature of repeated coseismic rupture and related cumulative slip at the surface and subsurface (Ferry et al., 2004; Haddad et al., 2012). Total station and differential GPS equipments have been recently extensively used to measure lateral and or vertical offsets along faults (Figure 10.2; Kondo et al., 2005; Meghraoui et al., 2012).

With the advent of numerical cameras, the development of automatically corrected photomosaic for trench-wall logging significantly improved the accuracy of paleoseismic data. Although, unfortunately rarely used, a powerful technique that allows an accurate trench-log is the portable hyperspectral

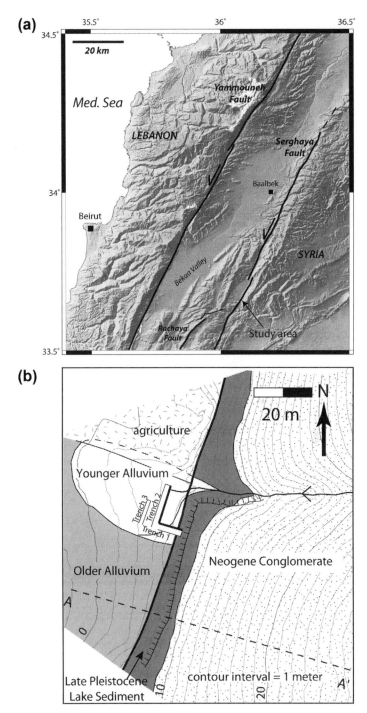

FIGURE 10.1 Trench site selection and 2D–3D configuration (Gomez et al., 2003): (a) Simplified morphotectonic map of the Lebanese–Syrian restraining bend of the Dead Sea fault (DSF) system (red box for study area); (b) Trench site across the Serghaya fault branch of the DSF, c1, c2 and c3 indicate channel trend and long black lines are fault zones; (c) Trench 1, north wall across the Serghaya Fault with the exposed fault zone and related colluvial wedges of past earthquakes; (d) Trench 2, parallel to the Serghaya Fault with displaced channels C1, C2, and C3 the younger slip probably corresponding to the Mw 7.3, 1757 earthquake.

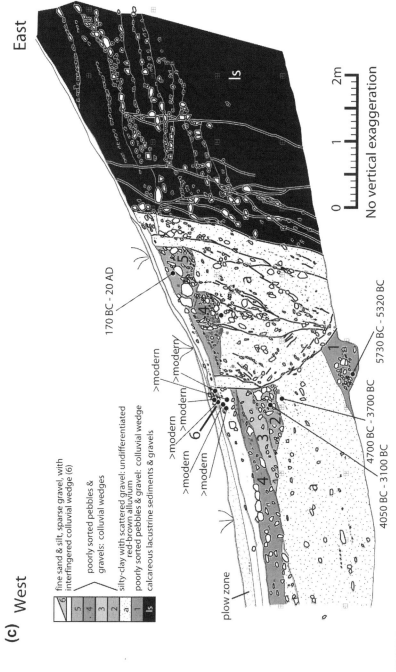

(c)

FIGURE 10.1   cont'd

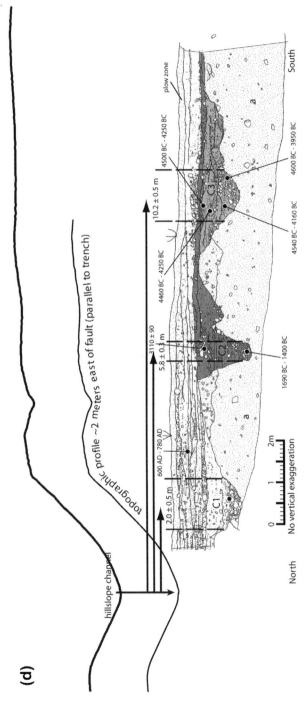

FIGURE 10.1 cont'd

(a)

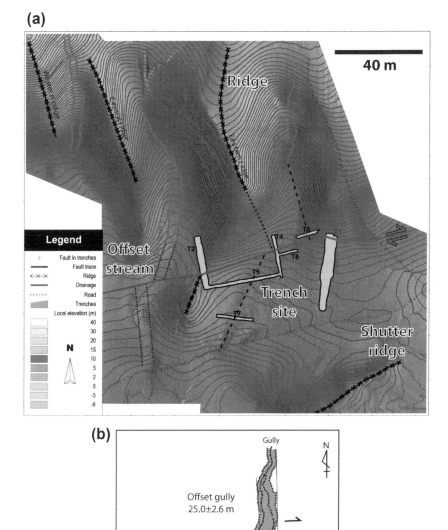

FIGURE 10.2 (a) Combined DGPS and total station microtopographic survey (thick contour lines are 5 m) of the 1912 earthquake segment of the North Anatolian Fault (NAF) at Guzelkoy, showing the location of seven trench-excavations and their relation to the fault. Two cumulative

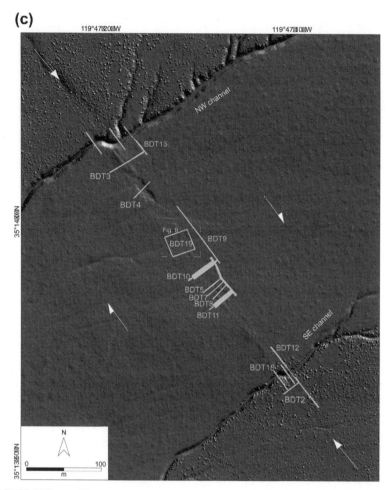

FIGURE 10.2   cont'd

offset streams are measured from the channel configurations. Using the surface morphology only, the measured stream offsets indicate 26.6 ± 1 m and 11 ± 0.5 m (for eastern and western streams, respectively) and 29 ± 1 m ridge offset (Meghraoui et al., 2012). (b) Total station survey map of the cumulative offset of a gully along the NAF at the 1944 Bolu-Gerede earthquake Rupture. The fault zone forms a linear depression on an alluvial fan surface. A gully incising the fan surface crosses the fault zone perpendicularly with a slight bend at the fault. This gully exhibits a right-lateral offset of 25 ± 2.6 m (Kondo et al., 2005). (c) Airborne Lidar-derived topography (0.25-m resolution) along a portion of the San Andreas Fault (SAF) in the Carrizo Plain. The solid white arrows delineate surface ruptures related with the 9 January 1857 Fort Tejon earthquake (Mw 7.9) affect and deflect two stream channels. BDT 19 box is location of the terrestrial laser scanning. Yellow boxes indicate paleoseismic excavations (BDT) and white half arrows indicate dextral motion along the SAF. By scanning a low-relief channel, the Lidar survey indicates subtle geomorphic markers relevant for paleoseismic site selection along the fault (Haddad et al., 2012).

camera to acquire field-based visible near-infrared and shortwave infrared high spatial/spectral resolution images (Ragona et al., 2006). Another method of data collection is taking a peel (i.e., glue a wall section on a gunny piece) from a trench wall stratigraphic section that allows a detailed analysis of the stratigraphy and identification of faulting events. This latter technique provides a stratigraphic section comparable to the "Geoslicer" that consists in a drill of two steel frames in a muddy (marsh or lake deposits) that may reveal faulting of earthquake-induced structures (Sawai et al., 2007). These techniques can be considered as among the modern methods in paleoseismology that greatly improved the description, interpretation, and archiving of paleoseismic data of faulted exposures and other paleoearthquake features. Either through a trench photomosaic or trench-wall imaging spectroscopy, the numerical paleoseismic data storage is becoming a widely used method of field investigations.

10.2.2 Dating Methods

The dating of past earthquakes is a critical step in paleoseismic studies. In trenches, the relative chronology of sedimentary units is often complemented by isotopic dating and other methods (Walker, 2005). The commonly used isotopic methods are:

1. The radiocarbon dating using the Accelerator Mass Spectrometry (AMS).
2. The Uranium series for samples with carbonate calcium ($CaCo_3$) content.
3. The Potassium—Argon (K—Ar) for travertine and tufa units.
4. The Thermo-Luminescence and Optically Stimulated Luminescence for sandy units (quartz and feldspar content).
5. The dating of geomorphic surfaces by Cosmogenic Al-Be and Chlorine.
6. The Lead-210 isotopes for very young sediments (<100 years).

Nowadays, the dating practice in paleoseismology consists of the collection of datable material (e.g., charcoal fragments, bones, tests of marine organisms, organic deposits, and peat) from sedimentary layers that allow the bracket of faulting events or deformation level using radiocarbon dating by AMS (see also p.141 in McCalpin, 1996). If collected samples present no contamination and a good amount of carbon content (>1 mg) after preliminary treatments, uncertainties of dating may be as low as ±10 to ±30 years. The standard laboratory dating is then calibrated (using radiocarbon calibration curves; Reimer et al., 2004) to provide the real age of dated units and thus, an inferred age-range for the occurrence of a given paleoearthquake. The calibration curve is accessible through different programs (e.g., http://calib.qub.ac.uk/calib/, Stuiver et al., 2005; http://c14.arch.ox.ac.uk/oxcalhelp/hlp_contents.html, Bronk Ramsey, 2009). At this point, the bracketed dates for a faulting event with a low level of uncertainties of isotopic dating can bear a direct correlation with the record of an earthquake as reported in a historical catalog. The stratigraphic analysis and estimated uncertainties of dating are crucial in paleoseismology. When a stratigraphic sequence with faulting events and consistent dating is determined, the age of past earthquakes can be calculated using a Bayesian analysis and a related probability density function for the occurrence of events within a time range (Bronk Ramsey, 2009).

A useful nondestructive method is the morphologic dating of fault scarps inferred from the erosional diffusivity, K, which is the amount of sedimentary material removed from a surface ΔS vs the time t of erosion, and the diffusion equation of elevation z as a function of distance x in the form $\frac{\delta z}{\delta t} = K \frac{\partial^2 z}{\partial^2 x}$ (Nash, 1980; Hanks, 2000). The age of cumulative vertical slip obtained from the modeling of fault scarps may be quite comparable to the isotopic dating if the constant mass diffusivity K is well estimated. Other indirect methods of dating include the dendrochronology, lichenometry, pollen analysis, and the soil development with related horizons (A, B, and C in general) over the Holocene in regions with wet to semiarid climate (Noller et al., 2000).

10.3 PALEOSEISMOLOGY AND SEISMIC SOURCE CHARACTERIZATION

The significance of paleoseismic investigations is recognized under the assumption that an earthquake rupture and its recurrence are preserved at the surface. In any tectonic domain (interplate or intraplate), the geological process of earthquake faulting would require a threshold magnitude Mw > 5.5 to see a coseismic rupture and its record within the superficial deposits (see page 16 in McCalpin, 1996). The understanding of faulting parameters and pattern of surface deformation is a critical aspect for the interpretation of paleoseismic data. Using the knowledge of the regional seismogenic layer W and rigidity modulus μ of a given tectonic zone, earthquake geology may access the seismic moment $Mo = \mu L W \Delta U$ by means of the average coseismic slip ΔU and fault length L. The interpretation of slip per event, slip rate, rupture dimensions, elapsed time, and recurrence time of large seismic events as fault parameters largely depends on the level of uncertainty of these parameters (Pantosti and Yeats, 1993). Trenching across recent surface faulting compared to the coseismic slip distribution along strike shows the complexity of the recurrent nature of surface deformation and ultimately, the importance of the determined uncertainties.

Models of earthquake faulting were first based on the fault segmentation with persistent structural boundaries and characteristic earthquake behavior that suggests the repetition of same size events with similar coseismic slip along strike (Figure 10.3(a); Schwartz and Coppersmith, 1984; Wesnousky, 2006). The long-term variation of slip distribution observed along numerous coseismic faults indicates that moderate-size earthquakes (Mw < 6.5) with coseismic slip may occur on rupture patches, and they often complement the total slip in larger event (Figure 10.3(a); Sieh, 1996). The coseismic slip

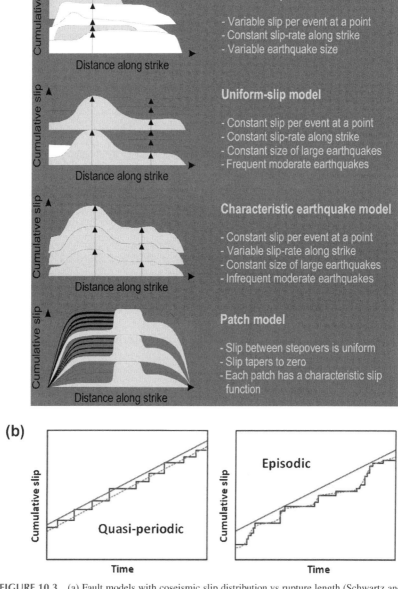

FIGURE 10.3 (a) Fault models with coseismic slip distribution vs rupture length (Schwartz and Coppersmith, 1984; Sieh, 1996). (b) Faulting behavior with constant strain rates and quasiperiodic earthquake recurrence (Shimazaki and Nakata, 1980), and variable slip with episodic occurrence of large or moderate earthquakes also called Wallace-type faulting behavior (Weldon et al., 2004).

variation along strike and hypothesis of long-term characteristic behavior have major implications on field investigations in paleoseismology. Ideally, the significance of a paleoseismic study relies on the spatial distribution of trench excavations along a fault zone and long-term record of past earthquakes (Weldon et al., 2004). In most case studies, only the increase of paleoseismic investigating sites (including dated cumulative slip from geomorphology) along a fault is able to constrain the slip rate during a period that comprises several seismic cycles (Figure 10.3(b)). The identification of a paleoseismic site with a maximum coseismic slip along surface faulting, related stress drop $\Delta\sigma = C\mu\frac{\Delta U}{L}$, and cumulative slip recorded by the geomorphology (e.g., from offset streams and terrace deposits) would also contribute to assessing the size of past earthquakes. Regardless of the geometrical complexities (step over, releasing-bend, and restraining-bend) along a fault segment, the study of historical coseismic slip distribution is decisive for site selection in paleoseismology and the determination of the size of past earthquakes (Yeats et al., 1997).

10.4 CASE STUDIES WITH LONGEST EARTHQUAKE RECORDS

The detailed studies of recent large and giant earthquake faulting and related surface deformation have contributed to an improvement of paleoseismic field investigations. The visible surface ruptures that were obvious targets for paleoseismic trenching allowed a better perception of fault geometries and related pattern of repeated earthquake faulting. Furthermore, fault parameters collected from earthquake geology and paleoseismic studies were consistent with results obtained from earthquake seismology and earthquake geodesy. In this context, major continental faults with large recent earthquakes appeared to be real, open access natural laboratories for paleoseismology.

The increase of paleoseismic studies in the last two decades points out the need of qualitative and quantitative data collection and their calibration with historical earthquakes. Besides the SAF, active zones with long historical record (>1,000 years) such as the Dead Sea Fault (DSF), the North Anatolian Fault (NAF), the East Anatolian Fault (EAF), and diverse late Quaternary fault ruptures in Turkey, Italy, Greece, Spain, Belgium, Algeria, Argentina, Venezuela, Iran, Japan, and China have been the locus of numerous research projects in paleoseismology. Among striking examples of correlation between paleoseismic data and historical earthquakes, the 88-km-long surface rupture with 1.45-m right-lateral (and 0.95-m vertical) coseismic slip across the Great Wall in china were associated to the M8 Yinchuan-Pingluo earthquake of 1739; 15 trenches dug across the fault revealed the occurrence of four large earthquakes in 8400 BP, 4600–6300 (or 5700) BP, 2600 BP, and in 1739 (Deng and Liau, 1996). The giant earthquakes in Sumatra and Japan subduction zones have unearthed previous paleoseismic studies of repeated coral reef uplift and subsidence and paleotsunami records, respectively, also showing the value of historical accounts of past earthquakes (Meltzner et al., 2010; Sawai et al., 2012). In all these active zones, the analysis of faulting events were confronted to the local or regional earthquake catalogs with a comparison between the surface deformation and size of past seismic events.

10.4.1 Paleoseismology of the NAF and the Marmara Seismic Gap

The occurrence of the 1999-Izmit earthquake (Turkey, Mw 7.4) with approximately 120-km-long left-lateral strike-slip surface faulting, correlated with the 1939−1967 westward earthquake sequence and ruptures along the NAF (Barka and Kadinsky-Cade, 1988), and the potential for a large forthcoming earthquake near Istanbul (Stein et al., 1997), prompted a keen interest in the paleoseismic history of the fault segments.

The well identified 1939−1999 North Anatolian earthquake fault segments and related slip distribution (Barka and Kadinsky-Cade, 1988; Barka, 1999) prepared the path toward a spatiotemporal analysis of past earthquake distribution compared to the historical catalog (Barka, 1996). This previous work in earthquake geology and the occurrence of the 1999-Izmit event are a unique example of an active tectonic framework that allows the unraveling of the paleoseismic history of a major continental fault. Eastern and central fault segments of the NAF were also the site of numerous paleoseismic investigations. However, the proximity of the Istanbul seismic gap located between the 1999 and 1912 fault segments in the Mamara Sea (Barka, 1999; Ambraseys, 2009), with fault interaction, stress change, and earthquake forecast in the next decades (Hubert-Ferrari et al., 2000), called for urgent paleoseismic investigations on the western NAF segments.

Among the first paleoseismic trenches across the 1999-surface faulting, onshore field investigations combined six excavations with GPR to resolve right-lateral slip events and constrain the earthquake history on a channel−fan complex (Ferry et al., 2004; Rockwell et al., 2009). The paleoseismic study identifies two moderate faulting events and two major surface rupturing events during the last 400 years that can be correlated with historical events in 1754, 1878 (with Mw ∼ 6) and in 1719, 1894 (with Mw ∼ 7), respectively (Figure 10.4).

West Marmara case study: Earthquake geology and paleoseismic onshore investigations along the 1912-earthquake fault segment of the NAF west of the Marmara Sea document the coseismic slip distribution and establish the occurrence of faulting events corresponding to the historical events of 1063, 1343−1354, 1716, and 1912 (Rockwell et al., 2009; Aksoy et al., 2010; Meghraoui et al., 2012). Figure 10.5 summarizes the main results of paleoseismic investigations that confirm the seismic gap location over the last 2,000 years record and suggest periodic occurrence of large earthquakes with

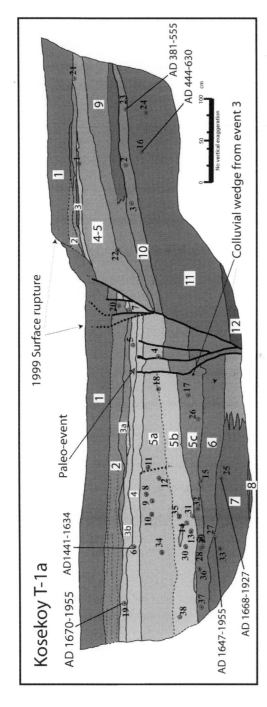

FIGURE 10.4 Log of the Köseköy trench across the 1999 surface ruptures of the NAF. The 1999 rupture in this fault branch is purely dip-slip, based on the lack of lateral offset of an adjacent flume and berm (Ferry et al., 2004; Rockwell et al., 2009). Note the older fractures that did not rupture in 1999. Black dots with numbers are detrital charcoal sample locations for dates. Larger numbers in white boxes are unit designations. 1. Topsoil (A) horizon; 2. Bedded gravel; 3. Stratified fine to coarse sand, 4. Clayey silt sand, 5. (a) Clayey silt with fine sand, iron staining in pores grading downward to (5b/6a) clayey sand silt with lenses of fine sand; 6b. Sandy clayey with silt pebbles interpreted as colluviums; 6c. Massive plastic clay, grading coarser away from fault; 7. Gravelly clayey silt with gravel content increasing toward fault; 8. Coarse sand capped by 1-cm-thick clay layer; 11. (a) Coarse sandy clayey gravel grading downward ad laterally to clayey sandy gravel; 12. Silty fine sand with scattered pebbles; 13. Coarse sandy gravel with common fragments of fired bricks; 14. Coarse sand.

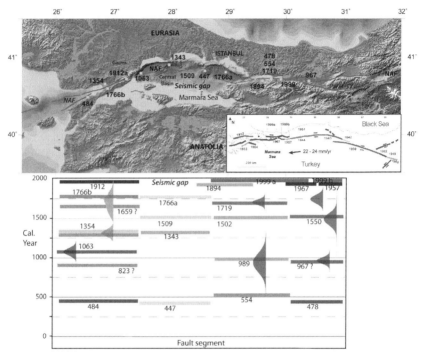

FIGURE 10.5 Earthquake sequences on each fault segment of the NAF in the Marmara region of Turkey (Meghraoui et al., 2012). The analysis of spatial and temporal distribution of earthquakes consists in the combination of the historical seismicity (Ambraseys, 2009), the fault geometrical complexities, and the paleoseismic timing of past earthquakes (*1912 segment*: This study, Rockwell et al., 2009; Aksoy et al., 2010, *1999 Izmit segment*: Klinger et al., 2003; Pavlides et al., 2006; Dikbaş and Akyüz, 2011, *1999 Duzce segment*: Pantosti et al., 2008, *1967 segment*: Palyvos et al., 2007). The probability density function indicates the error bar in the timing of past events. Each fault segment shows a periodic succession of five large events except for the Marmara segment.

an average 300 years recurrence interval, and 17 mm/year average slip rate for the past two millenniums (Klinger et al., 2003; Pavlides et al., 2006; Palyvos et al., 2007; Pantosti et al., 2008; Aksoy, 2009; Rockwell et al., 2009; Dikbaş and Akyüz, 2011; Meghraoui et al., 2012). The geodetic (GPS) rate of deformation being 22−26 mm/year (Reilinger et al., 2006), the ~17 mm/year tectonic rate of active faulting in the Marmara Sea region suggests a slip deficit necessarily accommodated by more forthcoming displacements along the NAF.

The integration of the results of earthquake geology and paleoseismic data in the SHA and risk assessment of the Grand Istanbul region and central Marmara Sea fault zones became a challenge after the 1999-earthquake sequence (Atakan et al., 2002). The recent works that determine the fault segment boundaries, coseismic slip distribution of the 1999-Izmit and 1912-Ganos fault ruptures, paleoseismic events and slip rates of the NAF east and west of the Marmara Sea (Barka, 1999; Armijo et al., 2005; Aksoy et al., 2010) provide unprecedented data on earthquake fault characterizations. They also resolve the development of scenarios of earthquake rupture propagations necessary for the SHA models in the Marmara Sea (see the SHA sections therein).

10.4.2 Paleo- and Archeo-seismology of the DSF and Slip Deficit

Among paradoxes in the seismic activity along large continental faults, the apparent quiescence of the DSF illustrates the variability of earthquake faulting. The lack of major seismic events with Mw > 6.0 in the last centuries along the continental DSF is in contradiction with the historical catalog and related report of faulting events over the last 3,000 years or so (Figure 10.6(a); Guidoboni et al., 1994; Ambraseys and Jackson, 1998; Sbeinati et al., 2005). Previously, the low-level instrumental seismicity and temporal seismic quiescence erroneously suggested inactive faulting implying serious under-estimation of the SHA in the Middle East. First accounts of seismic structures in geological deposits reported in the Lisan lacustrine deposits of the Dead Sea describe continuous large earthquake sequences in the last 50 ka (El-Isa and Mustafa, 1986; Marco et al., 1996). Recent studies describing obvious faulting outcrops and clear accounts of historical earthquakes with surface ruptures generated a strong interest in the development of paleoseismic excavations along strike.

The richness of archeological sites along the DSF constitutes an excep-tional advantage for dating prehistorical earthquakes. Field investigations revealed left-lateral faulting of the following archeological sites: a total 2.1-m offset walls of a Crusader castle during the 20 May 1202 and 30 October 1759 earthquakes (Ellenblum et al., 1998); the cumulative 13.6-m offset of the Al Harif aqueduct since the Roman time, the most recent event being the 29 June 1170 earthquake (Figure 10.6(b); Meghraoui et al., 2003; Sbeinati, 2010); the 2.2-m offset wall of Qasr Tilah can be related to an earthquake event in AD 608−826 (Klinger et al., 2000a); the offset Galei-Kinneret city walls during the 18 January 749 earthquake (Marco et al., 2003); the minimum 1-m vertical slip of a Qanat during the 1068 earthquake on the Southern DSF near Eilat (Zilberman et al., 2005); cumulative offset of the Roman-Byzantine-Islamic Qasr Tilah site of the northern Araba valley during the 634 or 659/660, 873, 1068, and 1546 earthquakes (Haynes et al., 2006); a total ∼42 m of the c.5000 BC Tell Sicantarla, Hittite road, and Roman walls, with ∼9-m offset during the 1408 and 1872 earthquakes (Altunel et al., 2009), and Tell Saidiyeh in c.759 BC, 1150 BC, 2300 BC, and 2900 BC. These archeo-seismological studies of the DSF are complemented with tectonic geomorphology and

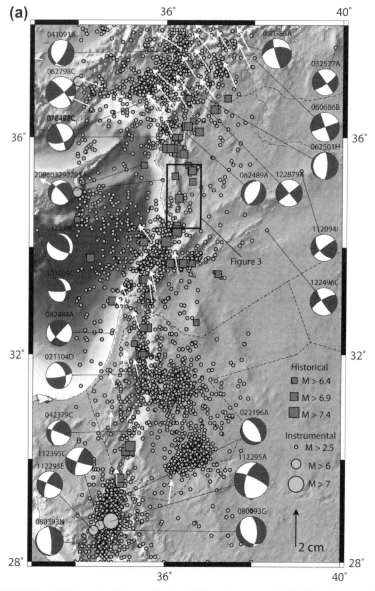

FIGURE 10.6 (a) Seismicity (historical before 1900 and instrumental till 2004) along the Dead Sea fault (data from merged ISC, EMSC and the APAME Project catalog). Except for the 1995 Gulf of Aqaba large seismic event, major earthquakes (with M > 7) date from the Middle age (Ambraseys, 2009). Focal mechanism solutions are from Harvard CMT. (b) Satellite view from Google Earth of the 13.6 m left-lateral offset Al Harif aqueduct (black arrow) along the Missyaf segment of the Dead Sea fault in Syria (Sbeinati et al., 2010).

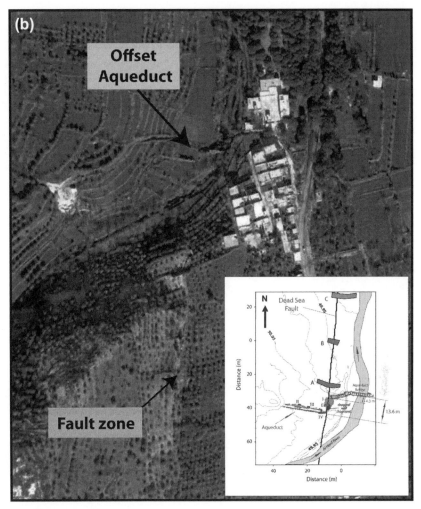

FIGURE 10.6   cont'd

paleoseismic investigations that constrain paleoearthquake faulting episodes and slip rates. Left-lateral offset archeological features, streams, and alluvial fans provide slip rates ranging from 3.5 to 6.5 mm/year (Klinger et al., 2000b; Meghraoui et al., 2003; Marco et al., 2005; Ferry et al., 2007; Altunel et al., 2009; Sbeinati et al., 2010; Karabacak et al., 2010; Le Béon et al., 2010; Ferry et al., 2011).

In trenches, radiocarbon dating bracket the 1759 earthquake on the northern Jordan Valley fault segment and the Serghaya fault branch through 3D trenching with 0.5- and 2.5-m left-lateral slip, respectively (Gomez et al., 2003; Marco et al., 2005). The major 1202 earthquake (Mw 7.6, Ambraseys and Jackson, 1998) is bracketed in the northern Jordan Valley at Beyt Zayda with 1.6-m left-lateral slip (Marco et al., 2005) and in the Lebanese segment of the DSF at the Yammouneh pull-apart basin (Figure 10.7; Daëron et al., 2007; Nemer et al., 2008). Faulting of the 749 earthquake of the Jordan Valley segment is resolved at Tell Saidiyeh site and Beyt Zayda (Marco et al., 2005; Ferry et al., 2011). Paleoseismic trenches near the Al Harif aqueduct resolve three faulting events in AD 160−510, AD 625−690, and

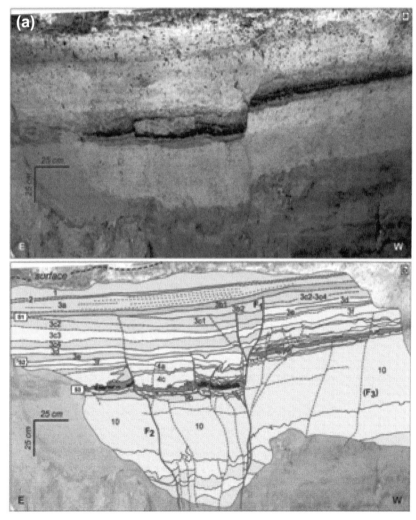

FIGURE 10.7   Trenches across the Yammouneh Basin site of Lebanese segment of the DSF. The uppermost event horizons correspond to the 20 May 1202 earthquake ((a) Daëron et al., 2007; (b) Nemer et al., 2008).

(b)

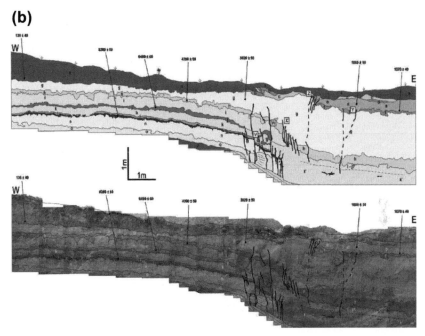

FIGURE 10.7   cont'd

AD 1010—1210, the latter being correlated with the 1170-large earthquake (Sbeinati et al., 2010). At the northern end of the DSF, faulting in trenches provide evidence for three historical earthquakes in 859 AD, 1408, and 1872 (Akyüz et al., 2006). The longest paleoseismic records on the DSF are determined in the Jordan Valley with a sequence of 12 coseismic surface ruptures over 14 ka (Ferry et al., 2011) and 10—13 faulting events over the last 12 ka on the Yammouneh fault of the DSF restraining bend in Lebanon (Daëron et al., 2007).

The average 5 mm/year left-lateral slip obtained from the geological or geodetic rate of deformation, compared to the seismic quiescence for large earthquakes in the last 1,000 years or so (Meghraoui et al., 2003; Marco et al., 2005; Ferry et al., 2007; Alchalbi et al., 2010; Al Tarazi et al., 2011) imply 4—5 m slip deficit along the fault. Taking into account the DSF segmentation, the coseismic slip distribution (from paleoseismology) and maximum size of historical earthquakes, the estimated slip deficit corresponds to the occurrence of Mw > 7 events. An attempt of seismic hazard evaluations including earthquake geology and paleoseismic data is provided for the Northern DSF in Syria and Lebanon (Nemer, 2005; Sbeinati, 2010) and in Northern Araba Valley and the Jordan Valley segment (Al Tarazi and Sandvol, 2007).

10.5 PALEOEARTHQUAKE STUDIES AND THEIR INTEGRATION IN SHA

It is now commonly accepted that paleoearthquake studies provide important constraints in a seismic hazard analysis. Since the late 1980s the importance of paleoseismic data in probabilistic seismic hazard assessment (PSHA) was recognized and applied. The most common way of integrating paleo-earthquake data into the seismic hazard calculations today is to include paleoearthquakes in the seismicity catalogs to expand the limited time frame of the instrumental and historical earthquake records. Usually the instrumental seismicity catalogs extend over roughly 100 years, whereas the historical data may go further back in time, depending upon the region of interest. In areas where long historical earthquake records are kept such as China, Middle East and parts of Europe, the catalog time span for earthquakes with Mw > 5 may be as large as 1,000−2,000 years in most cases, and to 4,000 years in extreme cases. Other active zones in North and South America, Africa, Australia, and New Zealand that cover a significant part of the world, only have earthquake records for a few centuries. In cases where the deformation rates are high (>1 cm/year), such as the active plate boundaries, a complete earthquake cycle may be embedded within the instrumental seismicity catalogs. In slowly deforming areas such as plate interiors, on the other hand, the earthquake cycle is far beyond any catalog time span even if these may extend more than 1,000 years. Therefore, statistical treatment of earthquake recurrence in time, based only on the historic seismicity catalog, usually carries large uncertainties in their completeness for the earthquake record (Meghraoui et al., 2001). Paleoearthquake studies provide therefore an important contribution in this regard (Figure 10.8).

10.5.1 Paleoseismic Data and the Probabilistic Seismic Hazard Evaluation

Apart from the simple improvement by including more earthquakes from the geological past, there are other important constraints that paleoearthquake studies may bring to seismic hazard analysis. Paleoseismic data favor the time-dependent earthquake occurrence, which supports the deterministic method of SHA rather than the probabilistic approach. In classical PSHA studies, the seismic sources are defined based on either area or line sources representing individual faults. For the latter, paleoearthquake studies provide important information on fault geometry, segmentation, and the size of the possible earthquakes. Various aspects of paleoearthquake studies that can bring useful information to PSHA are shown in Figure 10.9 (Schwartz and Coppersmith, 1986).

Typical parameters that can be obtained by paleoearthquake studies are fault characteristics and kinematics (i.e., strike, dip, rake, and mechanism),

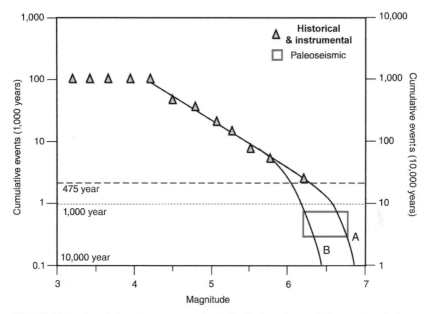

FIGURE 10.8   Cumulative frequency-magnitude distribution of the 1,000 years historical and instrumental seismicity in the greater Basel region, the site of the large and damaging earthquake in 1356 ($\sim$M 6.5). The activity rates constrained by the instrumental and historical records are similar at low magnitudes but different at higher magnitudes due to paleoseismic data. The difference is also modest for the 475-year return period commonly assumed in hazard mapping but becomes dominant for longer return periods and low occurrence probabilities. *(Modified after Meghraoui et al. (2001).)*

displacement per event and slip rate, identification of individual paleo-earthquakes and their recurrence time. Regarding the latter, the constraints provided by paleoseismology in assessing the maximum magnitude on a given fault, are especially relevant. In cases where several events are identified, additional parameters can be deduced such as slip-rate, recurrence interval, and elapsed time since last earthquake. Through detailed studies of the fault segmentation, rupture length can be estimated, which then can lead to the calculation of maximum paleoearthquake size using empirical relations such as those presented in Wells and Coppersmith (1994).

In PSHA methods, the long-term earthquake potential along a given fault is dependent upon a recurrence model. Primarily there are two basic categories. The first and most commonly used recurrence model is based on a Poissonian distribution, where the occurrence of a future earthquake is assumed to be independent of the occurrence of the previous events. Here, the activity rate and the b-value (showing the relations between the large and the small earthquakes), are computed using the Gutenberg–Richter relations, where the frequency of occurrence of the various earthquake magnitude levels are deduced based on a cumulative distribution over the catalog time span. The probability

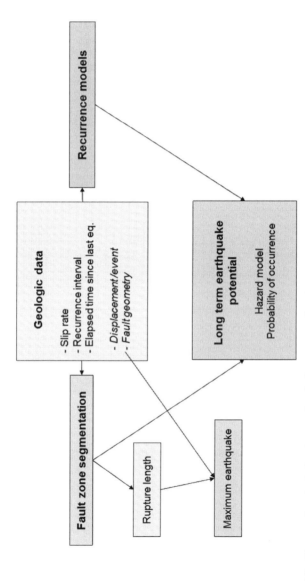

FIGURE 10.9   The use of geologic data in assessing the long-term earthquake potential. *(Redrawn from Schwartz and Coppersmith (1986).)*

of exceeding a chosen level of ground motion (GM) within a given time is then calculated based on these parameters, for a given geographical location.

The following relation presents the standard PSHA calculation for several seismic sources (Cornell, 1968; McGuire, 1993):

| Annual probability (P) of exceeding a given ground motion (a^*) in a given time (t) | Probability density functions of m and r integrated over all magnitudes (m) and distances (r) |

$$P[A > a^* \ in \ time \ t]/t = \sum_i v_i \iint G_{A|m,r}(a^*) f_M(m) f_R(r|m) \, dm \, dr$$

| Annual rate of earthquakes for all seismic sources (v_i) | Probability of exceeding a ground motion level given an earthquake with m and r (based on a ground motion prediction model) |

The second category of recurrence models assumes that earthquake occurrence is not random and is dependent on both time and slip. The simplest time-dependent recurrence model (also called "renewal model"; Cornell, 1968), assumes periodic occurrence of earthquakes along a given fault, where earthquakes occur in fixed time intervals. This time interval is controlled by the time needed for the level of crustal stresses to reach a stage where the frictional capacity is overridden and rupture takes place. In such a simplistic model of periodic earthquake occurrence, the time elapsed since the occurrence of the last earthquake becomes crucial, due to the conditional probability calculations. Paleo-earthquake studies can provide information about this within the limitations of the dating methods available. Other more complex recurrence models exist, such as exponential, Weibull, Brownian, gamma, and lognormal distributions, where the large earthquakes occur in clusters in time (Mathews et al., 2002).

10.5.2 Paleoseismology Promotes Earthquake Scenarios

In cases where the rate of deformation is low and recurrence of earthquakes is beyond the limits of instrumental and historical seismic catalogs, or there is a controlling earthquake at a close distance from the site of interest, the use of probabilistic seismic hazard methods based on statistical treatment of earthquake catalogs is probably not sufficient. Scenario-based hybrid GM simulation methodologies seem to provide a promising alternative for estimating the seismic hazard in these cases. In recent years, due to the large amount of data gathered from the large earthquakes, deterministic SHA methods based on GM simulations of individual earthquake rupture scenarios are being used. These changing trends in SHA are partly due to the challenges in the built environment and engineering applications, which require more sophisticated seismic hazard analysis incorporating the state-of-the-art knowledge on the fault behavior. The physics based deterministic methods have two main types: kinematic and dynamic. Among these, the hybrid broadband GM simulations using kinematic assumptions for the fault rupture complexity may benefit substantially from the paleoearthquake studies. This is especially true for faults with no large recent earthquakes to constrain the rupture details. Paleoseismic data can provide constraints on input earthquake rupture scenarios for hybrid GM simulations. These include; fault geometry and kinematics, fault segmentation, location of the asperities (areas of largest slip on repeated ruptures), rupture size (earthquake magnitude/seismic moment), and rupture initiation point (directivity effects). Paleoearthquake studies have a great potential on providing valuable information on most of these parameters.

10.5.3 Integration of Paleoearthquake Studies in SHA: The Example Case of Istanbul

There are several examples of kinematic GM simulations based on rupture scenarios (e.g., Pulido et al., 2004; Sørensen et al., 2007a; Cultrurera et al., 2010; Imperatori and Mai, 2012). Most of these are for retrospect studies following the large recent earthquakes. Although there is a lot to be learned from these retrospect studies, as they refine the methodologies used, the application of kinematic GM simulations in predictive studies may have important implications for the mitigation efforts in regions where seismic risk is high. One such example is the case of Istanbul. It is highly relevant to perform predictive studies of GM simulations for this case, due to a number of studies that have all pointed out the high hazard and risk in the city of Istanbul with a significant population density (e.g., Parsons et al., 2000; Hubert-Ferrari et al., 2000; Atakan et al., 2002; Erdik et al., 2004; Parsons, 2004). In the following, we discuss the importance of paleoearthquake data in developing realistic input scenarios for a possible future earthquake rupture along the Marmara segment of the NAF.

In a predictive kinematic GM-simulation study, Pulido et al. (2004) have computed the expected level of GM based on various rupture scenarios. In their study the details of the fault segmentation and location of asperities are based on the previous studies in the region where the paleoearthquake data from the 1912 Ganos and 1999 Izmit and Düzce earthquake rupture segments were used to develop an input model for the fault behavior in the Marmara Sea. The results indicate significant differences depending upon the chosen scenario. Geographical variation of GM is mainly controlled by the rupture initiation point, which affects the results significantly (Figure 10.10). The frequency dependent variation in the GM (i.e., spectral acceleration and spectral velocity values) in the metropolitan area of Istanbul is the most important aspect for engineering applications.

In a similar study on the sensitivity of GM simulations to input parameters for the case of Istanbul, Sørensen et al. (2007b), have found that the location of the rupture initiation, stress drop, rise time, rupture velocity, and the anelastic attenuation for the studied region are the most important parameters for the ground-motion modeling, in terms of ground-shaking levels. However, the impact of these parameters in frequency bands of engineering interest varies

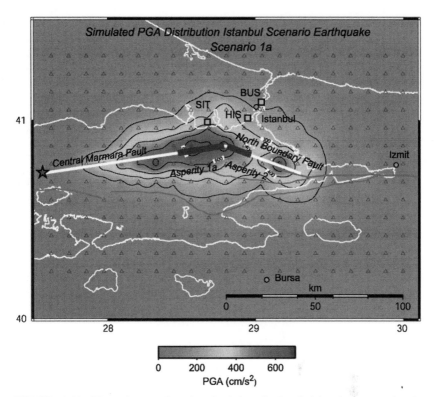

FIGURE 10.10   Kinematic ground motion simulations for Istanbul based on an earthquake rupture scenario along the Marmara segment of the North Anatolian Fault (NAF) *(from Pulido et al., 2004)*. Blue lines are line sources assumed to be asperities along the NAF. The star represents the earthquake rupture initiation point that determines the subsequent rupture propagation. PGA: Peak Ground Acceleration; BUS: Business district; HIS: Historical center; SIT: Area close to the airport where local site effects are significant.

(Figure 10.11). From an engineering perspective, the most important parameters are the location of rupture initiation point and the stress drop.

In addition to the above, there is a need to constrain the dynamic rupture parameters such as, rupture velocity, rise time, and stress drop. In dynamic GM simulation methods (e.g., Olsen et al., 2008; Oglesby and Mai, 2012; Pulido and Dalguer, 2009), these parameters are dynamically varied based on constituent relations. The contribution of paleoearthquake data on these parameters, however, does not seem obvious. In most of the dynamic GM-simulation studies the use of paleoearthquake data is restricted more on the kinematic parameters that define the earthquake source.

In recent years there has also been a focus on stochastic simulation of earthquake occurrence, so-called earthquake forecasting studies. However, using only statistical analysis of earthquake occurrence is limited in

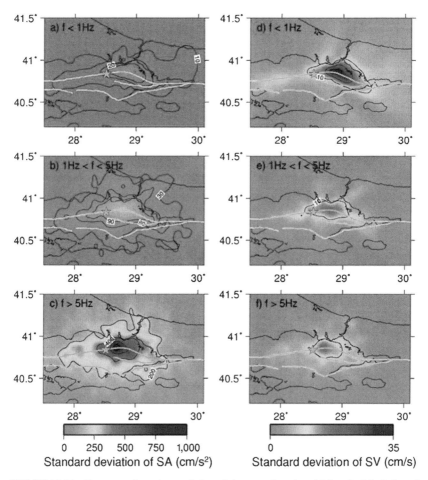

FIGURE 10.11 Frequency dependent variation of the ground motion. (a) Standard deviation of Spectral Accelerations (SA) for the frequencies f < 1 Hz (b) Standard deviation of Spectral Accelerations (SA) for the frequencies 1 Hz < f < 5 Hz (c) Standard deviation of Spectral Accelerations (SA) for the frequencies f > 5 Hz (d) Standard deviation of Spectral Velocities (SV) for the frequencies f < 1 Hz (e) Standard deviation of Spectral Velocities (SV) for the frequencies 1 Hz < f < 5 Hz (f) Standard deviation of Spectral Velocities (SV) for the frequencies f > 5 Hz. *(From Sørensen et al. (2007b).)*

reproducing the characteristics of strongly coupled faults when compared to the physics based deterministic models (e.g., Hainzl et al., 2013).

10.6 DISCUSSION—CONCLUSION

The visibility of earthquake ruptures and their induced effects at the surface determines the consistency of paleoseismic data. The detailed studies of coseismic surface ruptures associated with large earthquakes in the last decades were fundamental for the understanding of repeated tectonic movement along a fault zone and subsequent application of the paleoseismological approach (McCalpin, 1996). Hence, the practice of paleoseismology has also developed successfully for other faults that do not expose fresh surface ruptures. Indeed, the identification and study of seismogenic faulting, tectonic geomorphology, coseismic slip distribution, and trench site selection necessary for any paleoearthquake investigation reached a standard practice. The field investigation technology in paleoseismology acknowledges significant progresses with the use of new methods (Differential GPS, Lidar, optical and digital—hyperspectral cameras—for the photomosaic of trench walls, Geoslicer, and dating methods). Hence, paleoseismic data and results constitute a significant constraint in models of earthquake faulting behavior and the seismic hazard analysis.

Large continental and seismogenic faults such as the NAF, clearly expose outcrops of coseismic surface ruptures in regions that experienced large historical earthquakes (Ambraseys and Jackson, 1998). Field investigations of coseismic rupture and paleoearthquakes provide access to fault parameters with physical characteristics often consistent with results of seismic source and geodetic studies. The experience in earthquake geology gained following the study of coseismic surface ruptures in diverse earthquake areas (with different fault mechanisms) and its application to active faults without recent coseismic faulting is essential in paleoseismology (Yeats et al., 1997). The 1939—1999 NAF earthquake sequence and related segmentation constitute an open laboratory for the study of serial ruptures and long-term behavior of seismic sources. The well-documented earthquake ruptures, historical seismic events and related fault segments, including the 1668 unusually large earthquake involving several fault segments (Barka, 1996) served as a remarkable guideline for paleoseismic studies (e.g., Aksoy, 2009). The paleoseismic trenching in the Marmara Sea region of the NAF assessed the value of the size of past rupture events, slip rate, and earthquake recurrence period for the seismic hazard and risk assessment in the Istanbul region (Atakan et al., 2002; Pulido et al., 2004).

Although earthquake hazard projects involve paleoseismology with mainly trench investigations, most of PSHA and strong motion calculations still rely on seismicity catalogs with scarce database and poorly constrained long—term faulting behavior. Approximations and errors in seismic hazard and risk maps of many parts of the active zones are often due to short seismicity catalogs that do not include earthquake geology and paleoseismic data (Stein et al., 2012).

REFERENCES

Aksoy, M.E., Meghraoui, M., Vallee, M., Cakir, Z., 2010. Rupture characteristics of the A.D. 1912 Murefte (Ganos) earthquake segment of the North Anatolian Fault (western Turkey). Geology 38 (11), 991—994. http://dx.doi.org/10.1130/G31447.1.

Aksoy, M.E., 2009. Active Tectonics and Paleoseismology of the Ganos Fault Segment and Seismic Characteristics of the 9 August 1912 Mürefte Earthquake of the North Anatolian Fault (Western Turkey) (Ph.D. dissertation). İstanbulTeknik Univ., Istanbul, Turkey, 304 pp.

Akyüz, H., Altunel, E., Karabacak, V., Yalciner, C., 2006. Historical earthquake activity of the northern part of the Dead Sea Fault zone, southern Turkey. Tectonophysics 426, 281−293.

Alchalbi, A., et al., 2010. Crustal deformation in northwestern Arabia from GPS measurements in Syria: slow slip rate along the northern Dead Sea Fault. Geophys. J. Int. 180, 125−135 http://dx.doi.org/10.1111/j.1365-246X.2009.04431.x.

Al-Tarazi, E., Sandvol, E., 2007. Alternative models of seismic hazard evaluation along the Jordan Dead Sea transform. Earthquake Spectra 23 (1), 1−19.

Al Tarazi, E., Abu Rajab, J., Gomez, F., Cochran, W., Jaafar, R., Ferry, M., 2011. GPS measurements of near-field deformation along the southern Dead Sea Fault System. G-Cubed 12 (12), Q12021. http://dx.doi.org/10.1029/2011GC003736.

Altunel, E., Meghraoui, M., Karabacak, V., Akyüz, S., Ferry, M., Yalçiner, Ç., Munschy, M., 2009. Archaeological sites (Tell and Road) offset by the Dead Sea Fault in the Amik Basin, Southern Turkey. Geophys. J. Int. 179, 1313−1329.

Ambraseys, N.N., Jackson, J.A., 1998. Faulting associated with historical and recent earthquakes in the Eastern Mediterranean region. Geophys. J. Int. 133 (2), 390−406.

Ambraseys, N.N., 2009. Earthquakes in the Mediterranean and Middle East: A Multidisciplinary Study of Seismicity up to 1900. Cambridge University Press, Cambridge, UK, 947 pp.

Armijo, R., et al., 2005. Submarine fault scarps in the Sea of Marmara pull-apart (North Anatolian Fault): implications for seismic hazard in Istanbul. Geochem. Geophys. Geosyst. 6, Q06009. http://dx.doi.org/10.1029/2004GC000896.

Atakan, K., Ojeda, A., Meghraoui, M., Barka, A., Erdik, M., Bodare, A., 2002. Seismic Hazard in Istanbul following the 17 August 1999 Izmit and 12 November 1999 Duzce earthquakes. Bull. Seismol. Soc. Am. 92 (1), 466−482.

Barka, A.A., Kadinsky-Cade, K., 1988. Strike-slip fault geometry in Turkey and its influence on earthquake activity. Tectonics 7, 663−684.

Barka, A., 1996. Slip distribution along the North Anatolian Fault associated with the large earthquakes of the period 1939 to 1967. Bull. Seismol. Soc. Am. 86, 1238−1254.

Barka, A., 1999. The 17 August 1999 Izmit earthquake. Science 285, 1858−1859.

Bronk Ramsey, C., 2009. Bayesian analysis of radiocarbon dates. Radiocarbon 51 (1), 337−360.

Cornell, C.A., 1968. Engineering seismic risk analysis. Bull. Seismol. Soc. Am. 58, 1583−1906.

Cultrurera, G., Cirella, A., Spagnuolo, E., Herrero, A., Tinti, E., Pacor, F., 2010. Variability of kinematic source parameters and its implication on the choice of the design scenario. Bull. Seismol. Soc. Am. 100 (3), 941−953. http://dx.doi.org/10.1785/0120090044.

Dikbaş, A., Akyüz, H.S., 2011. Palaeoseismological investigations on the Karadere Segment, North Anatolian Fault zone, Turkey. Turk. J. Earth Sci. 20 (4), 395−410. http://dx.doi.org/10.3906/yer-0911-50.

Daëron, M., Klinger, Y., Tapponnier, P., Elias, A., Jacques, E., Sursock, A., 2007. 12,000-year-long record of 10 to 13 paleo-earthquakes on the Yammoûneh Fault (Levant Fault System, Lebanon). Bull. Seismol. Soc. Am. 97 (3), 749−771.

Deng, Q., Liao, Y., 1996. Paleoseismology along the range-front fault of Helan Mountains, north central China. J. Geophys. Res. 101 (B3), 5873−5893.

El-Isa, Z.H., Mustafa, H., 1986. Earthquake deformations in the Lisan deposits and seismotectonic implications. Geophys. J. R. Astr. Soc. 86, 413−424.

Ellenblum, R., Marco, S., Agnon, A., Rockwell, T., Boas, A., 1998. Crusader castle torn apart by earthquake at dawn, 20 May 1202. Geology 26, 303–306.

Erdik, M., Demircioglu, M., Sesetyan, K., Durukal, E., Siyahi, B., 2004. Earthquake hazard in Marmara region, Turkey. Soil Dyn. Earthquake Eng. 24, 605–631.

Ferry, M., Meghraoui, M., Giraud, J.-F., Rockwell, T.K., Kozaci, A., Akyuz, S., Barka, A., 2004. Ground-penetrating radar investigations along the North Anatolian Fault near Izmit, Turkey: constraints on the right-lateral movement and slip history. Geology 32 (1), 85–88.

Ferry, M., Meghraoui, M., AbouKaraki, N., Al-Taj, M., Amoush, H., Al-Dhaisat, S., Barjous, M., 2007. A 48-kyr-long slip rate history for the Jordan Valley segment of the Dead Sea Fault. Earth Planet. Sci. Lett. 260, 394–406.

Ferry, M., Meghraoui, M., Abou Karaki, N., Al-Taj, M., Khalil, L., February 2011. Episodic behavior of the Jordan Valley section of the Dead Sea Fault from a 14-kyr-long integrated catalogue of large earthquakes. Bull. Seismol. Soc. Am. 101 (1), 39–67. http://dx.doi.org/10. 1785/0120100097.

Galli, P., Galadini, F., Pantosti, D., 2008. Twenty years of paleoseismology in Italy. Earth-Sci. Rev. 88 (1–2), 89–117. http://dx.doi.org/10.1016/j.earscirev.2008.01.001.

Gomez, F., Meghraoui, M., Darkal, A.N., Hijazi, F., Mouty, M., Suleiman, Y., Sbeinati, R., Darawcheh, R., Al-Ghazzi, R., Barazangi, M., 2003. Holocene faulting and earthquake recurrence along the Serghaya branch of the Dead Sea Fault System in Syria and Lebanon. Geophys. J. Int. 153, 658–674.

Guidoboni, E., Comastri, A., Traina, G., 1994. Catalogue of Ancient Earthquakes in the Mediterranean Area up to the 10th Century, ING. Roma-SGA, Bologna, 504 pp.

Haddad, D.E., Akçiz, S.O., Arrowsmith, J.R., Rhodes, D.D., Oldow, J.S., Zielke, O., Toké, N.A., Haddad, A.G., Mauer, J., Shilpakar, P., 2012. Applications of airborne and terrestrial laser scanning to paleoseismology. Geosphere 8 (4), 771–786. http://dx.doi.org/10.1130/ GES00701.1.

Hainzl, S., Zöller, G., Brietzke, G.B., Hinzen, K.-G., 2013. Comparison of deterministic and stochastic earthquake simulators for fault interactions in the Lower Rhine Embayment, Germany. Geophys. J. Int. 195 (1), 684–694. http://dx.doi.org/10.1093/gji/ggt271.

Hanks, T.C., 2000. The age of scarplike landforms from diffusion-equation analysis. In: Nollet, J.S., Sower, J.M., Lettis, W.R. (Eds.), Quaternary Geochronology: Methods and Applications. AGU, Washington, DC, pp. 313–338.

Haynes, J., Niemi, T., Atallah, M., 2006. Evidence for ground-rupturing earthquakes on the Northern Wadi Araba fault at the archaeological site of Qasr Tilah, Dead Sea Transform Fault System, Jordan. J. Seismol. 10, 415–430.

Hubert-Ferrari, A., Barka, A., Jacques, E., Nalbant, S., Meyer, B., Armijo, R., Tapponnier, P., King, G.C.P., 2000. Seismic hazard in the Marmara Sea region following the 17 August 1999 Izmit earthquake. Nature 404, 269–273.

Imperatori, W., Mai, M., 2012. Sensitivity of broad-band ground-motion simulations to earthquake source and earth structure variations: an application to the Messina Straits (Italy). Geophys. J. Int. 188 (3), 1103–1116. http://dx.doi.org/10.1111/j.1365-246X.2011.05296.x.

International Atomic Energy Agency, Safety Standard Series, 2002. Evaluation of Seismic Hazards for Nuclear Power Plants: Safety Guide No NS-G-3.3. IAEA, Vienna, 31 pp.

Karabacak, V., Altunel, E., Meghraoui, M., Akyüz, H.S., 2010. Field evidences from northern Dead Sea Fault zone (South Turkey): new findings for the initiation age and slip rate. Tectonophysics 480 (1–4), 172–182.

Klinger, Y., Avouac, J.P., Dorbath, L., AbouKaraki, N., Tisnerat, N., 2000a. Seismic behaviour of the Dead Sea Fault along the Araba valley, Jordan. Geophys. J. Int. 142, 769–782.

Klinger, Y., Avouac, J.-P., AbouKaraki, N., Dorbath, L., Bourles, D., Reyss, J.L., 2000b. Slip rate on the Dead Sea transform in northern Araba valley (Jordan). Geophys. J. Int. 142, 755−768.

Klinger, Y., et al., 2003. Paleoseismic evidence of characteristic slip on the western segment of the North Anatolian Fault, Turkey, Bull. Seismol. Soc. Am. 93, 2317−2332.

Kondo, H., et al., 2005. Slip distribution, fault geometry, and fault segmentation of the 1944 Bolu-Gerede earthquake rupture, North Anatolian Fault, Turkey. Bull. Seismol. Soc. Am. 95 (4), 1234−1249. http://dx.doi.org/10.1785/0120040194.

Le Béon, M., et al., 2010. Early Holocene and Late Pleistocene slip rates of the southern Dead Sea Fault determined from 10Be cosmogenic dating of offset, alluvial deposits. J. Geophys. Res. 115, B11414. http://dx.doi.org/10.1029/2009JB007198.

McGuire, R., 1993. Computations of seismic hazard. Ann. Geofis. XXXVI (3−4), 181−200.

Marco, S., Stein, M., Agnon, A., Ron, H., 1996. Long-term earthquake clustering: a 50,000-year paleoseismic record in the Dead Sea Graben. J. Geophys. Res. 101, 6179−6191.

Marco, S., Hartal, M., Hazan, N., Lev, L., Stein, M., 2003. Archaeology, history, and geology of the A.D. 749 earthquake, Dead Sea transform. Geology 31, 665−668.

Marco, S., Rockwell, T., Heimann, A., Frieslander, U., 2005. Late Holocene activity of the Dead Sea transform revealed in 3D paleoseismic trenches on the Jordan Gorge segment. Earth Planet. Sci. Lett. 234, 189−205.

Mathews, M.V., Ellsworth, W.L., Reasenberg, P.A., 2002. A Brownian model for recurrent earthquakes. Bull. Seismol. Soc. Am. 92 (6), 2233−2250. http://dx.doi.org/10.1785/0120010267.

McCalpin, J.P. (Ed.), 1996. Paleoseismology. Academic Press, San Diego, California, 588 pp.

Meghraoui, M., Crone, A., 2001. Earthquakes and their preservation in the geological records. J. Seismol. 5, 281−285.

Meghraoui, M., Delouis, B., Ferry, M., Giardini, D., Huggenberger, P., Spottke, I., Granet, M., 2001. Active normal faulting in the upper Rhine graben and paleoseismic identification of the 1356 Basel earthquake. Science 293, 2070−2073.

Meghraoui, M., et al., 2003. Evidence for 830 years of seismic quiescence from palaeoseismology, archaeoseismology and historical seismicity along the Dead Sea Fault in Syria. Earth Planet. Sci. Lett. 210, 35−52.

Meghraoui, M., Aksoy, M.E., Akyuz, H.S., Ferry, M.A., Dikbas, A., Altunel, E., 2012. Paleoseismology of the North Anatolian Fault at Güzelköy (Ganos segment, Turkey): size and recurrence time of earthquake ruptures west of the Sea of Marmara. Geochem. Geophys. Geosyst. 13. http://dx.doi.org/10.1029/2011GC003960.

Meltzner, A.J., Sieh, K., Chiang, H.-W., Shen, C.-C., Suwargadi, B.W., Natawidjaja, D.H., Philibosian, B.E., Briggs, R.W., Galetzka, J., 2010. Coral evidence for earthquake recurrence and an A.D. 1390-1455 cluster at the south end of the 2004 Aceh-Andaman rupture. J. Geophys. Res. 115, B104022. http://dx.doi.org/10.1029/2010JB007499.

Nash, D.B., 1980. Morphologic dating of degraded normal fault scarps. J. Geol. 88, 353−360.

Nemer, T., 2005. Sismotectonique et comportement sismique du relais transpressif de la faille du Levant: roles et effets des branches de failles sur l'alea sismique au Liban (Ph.D. thesis). Universite Louis Pasteur, Strasbourg, 206 pp.

Nemer, T., Gomez, F., Al Haddad, S., Tabet, C., 2008. Coseismic growth of sedimentary basins along the Yammouneh strike-slip fault (Lebanon). Geophys. J. Int. 175, 1023−1039.

Noller, J.S., Sowers, J.M., Lettis, W.R. (Eds.), 2000. Quaternary Geochronology: Methods and Applications. AGU publications, Washington, DC, 581 pp.

Oglesby, D.D., Mai, M., 2012. Fault geometry, rupture dynamics and ground motion from potential earthquakes on the North Anatolian Fault under the Sea of Marmara. Geophys. J. Int. 188, 1071−1087. http://dx.doi.org/10.1111/j.1365-246X.2011.05289.x.

Olsen, K.B., Day, S.M., Minster, J.B., Cui, Y., Chourasia, A., Okaya, D., Maechling, P., Jordan, T., 2008. TeraShake2: spontaneous rupture simulations of Mw 7.7 earthquakes on the southern San Andreas Fault. Bull. Seismol. Soc. Am. 98, 1162−1185.

Palyvos, N., Pantosti, D., Zabcı, C., D'Addezio, G., 2007. Paleoseismological evidence of recent earthquakes on the1967 Mudurnu valley earthquake segment of the North Anatolian Fault zone. Bull. Seismol. Soc. Am. 97 (5), 1646−1661. http://dx.doi.org/10.1785/0120060049.

Pantosti, D., Pucci, S., Palyvos, N., DeMartini, P.M., D'Addezio, G., Collins, P.E.F., Zabcı, C., 2008. Paleoearthquakes of the Düzce Fault (North Anatolian Fault zone): insights for large surface faulting earthquake recurrence. J. Geophys. Res. 113, B01309. http://dx.doi.org/10. 1029/2006JB004679.

Pantosti, D., Yeats, R., 1993. Paleoseismology of great earthquakes of the Late Holocene. Ann. Geofis. 36, 237−257.

Parsons, T., Toda, S., Stein, R.S., Barka, A., Dieterich, J.H., 2000. Heightened odds of large earthquakes near Istanbul: an interaction based probability calculation. Science 288, 661−665.

Parsons, T., 2004. Recalculated probability of M ≥ 7 earthquakes beneath the Sea of Marmara, Turkey. J. Geophys. Res. 109, B05304. http://dx.doi.org/10.1029/02003JB002667.

Pavlides, S., Chatzipetros, A., Tutkun, S., Özaksoy, V., Doğan, B., 2006. Evidence for Late Holocene activity along the seismogenic fault of the 1999 İzmit earthquake, NW Turkey. In: Robertson, A.H.F., Mountrakis, D. (Eds.), Tectonics Development of the Eastern Mediterranean Region, vol. 260. Geol. Soc. Spec. Publ., pp. 635−647.

Pulido, N., Dalguer, L.A., 2009. Estimation of the high-frequency radiation of the 2000 Tottori (Japan) earthquake based on a dynamic model of fault rupture: application to the Strong ground motion simulation. Bull. Seismol. Soc. Am. 99 (4), 2305−2322. http://dx.doi.org/10. 1785/0120080165.

Pulido, N., Ojeda, A., Atakan, K., Kubo, T., 2004. Strong ground motion estimation in the Sea of Marmara region (Turkey) based on a scenario earthquake. Tectonophysics 391, 357−374.

Ragona, D., Minster, B., Rockwell, T., Jussila, J., 2006. Field imaging spectroscopy: a new methodology to assist the description, interpretation, and archiving of paleoseismological information from faulted exposures. J. Geophys. Res. 111 (B10), B10309. http://dx.doi.org/10. 1029/2006JB004267.

Reilinger, R., et al., 2006. GPS constraints on continental deformation in the Africa-Arabia-Eurasia continental collision zone and implications for the dynamics of plate interactions. J. Geophys. Res. 111, B05411. http://dx.doi.org/10.1029/2005JB004051.

Reimer, P.J., et al., 2004. IntCal04 terrestrial radiocarbon age calibration, 26-0 ka BP. Radiocarbon 46, 1029−1058.

Rockwell, T., et al., 2009. Paleoseismology of the North Anatolia Fault near the Marmara Sea: implications for fault segmentation and seismic hazard. Geol. Soc. Lon. Spec. Publ. 316, 31−54. http://dx.doi.org/10.1144/SP316.3.

Sawai, Y., Shishikura, M., Okamura, Y., Takada, K., Matsu'ura, T., Aung, T., Komatsubara, J., Fujii, Y., Fujiwara, O., Satake, K., Kamataki, T., Sato, N., 2007. A Study on Paleotsunami

Using Handy Geoslicer in Sendai Plain (Sendai, Natori, Iwanuma, Watari, and Yamamoto), Miyagi, Japan, Tech. Report No. 7, Geological Survey of Japan. AIST, National Institute of Advanced Industrial Science and Technology, Tsukuba, Japan, pp. 31—46.

Sawai, Y., Namegaya, Y., Okamura, Y., Satake, K., Shishikura, M., 2012. Challenges of anticipating the 2011 Tohoku earthquake and tsunami using coastal geology. Geophys. Res. Lett. 39 (L21309). http://dx.doi.org/10.1029/2012GL053692.

Sbeinati, M.R., Darawcheh, R., Mouty, M., 2005. The historical earthquakes of Syria: an analysis of large seismic events from 1365 B. C. to1900 A. D. Ann. Geophys. 48, 347—435.

Sbeinati, M.R., 2010. Historical Seismology, Paleo-archeoseismology and Seismic Hazard along the Dead Sea Fault in Syria (Ph.D. thesis dissertation). Université de Strasbourg, 378 pp.

Sbeinati, M.R., Meghraoui, M., Suleyman, G., Gomez, F., Grootes, P., Nadeau, M., Al Najjar, H., Al-Ghazzi, R., 2010. Timing of earthquake ruptures at the Al Harif Roman Aqueduct (Dead Sea Fault, Syria) from archeoseismology and paleoseismology. Special volume "Archaeoseismology and paleoseismology". In: Sintubin, M., Stewart, I.S., Niemi, T.M., Altunel, E. (Eds.), Ancient Earthquakes: Geological Society of America Special Paper 471. http://dx.doi.org/10.1130/2010.2471(20).

Schwartz, D.P., Coppersmith, K.J., 1984. Fault behavior and characteristic earthquakes: examples from the Wasatch and San Andreas Fault zones. J. Geophys. Res. 89, 5681—5698.

Shimazaki, K., Nakata, T., 1980. Time-predictable recurrence model for large earthquakes, GeoPhys. Res. Lett. 7, 279—282.

Schwartz, D.P., Coppersmith, K.J., 1986. Seismic hazards: new trends in analysis using geologic data. In: Active Tectonics. National Academy Press, Washington, DC., pp. 215—230.

Sieh, K.E., 1978. Prehistoric large earthquakes produced by slip on the San Andreas Fault at Pallet Creek, California. J. Geophys. Res. 83, 3907—3939.

Sieh, K., 1996. The repetition of large-earthquake ruptures. Proc. Natl. Acad. Sci. 93, 3764—3771.

Soloneko, V.P., 1973. Paleoseismogeology. Izv. Acad. Sci. USSR, Phys. Solid Earth 9, 3—16 (in Russian).

Stein, R.S., Barka, A.A., Dieterich, J.H., 1997. Progressive failure on the North Anatolian Fault since 1939 by earthquake stress triggering. Geophys. J. Int. (Oxford University Press) 128, 594—604.

Stein, S., Geller, R.J., Liu, M., 2012. Why earthquake hazard maps often fail and what to do about it. Tectonophysics 562—563, 1—25.

Stuiver, M., Reimer, P.J., Reimer, R.W., 2005. CALIB 5.0 (WWW program and documentation).

Sørensen, M.B., Atakan, K., Pulido, N., 2007a. Simulated strong ground motions for the great M 9.3 Sumatra—Andaman earthquake of 26 December 2004. Bull. Seismol. Soc. Am. 97 (1A), S139—S151. http://dx.doi.org/10.1785/0120050608.

Sørensen, M.B., Pulido, N., Atakan, K., 2007b. Sensitivity of ground-motion simulations to earthquake source parameters: a case study for Istanbul, Turkey. Bull. Seismol. Soc. Am. 97, 881—900.

Walker, M.J.C., 2005. Quaternary Dating Methods. Wiley, UK, 286 pp.

Wallace, R.E., 1981. Active faults, paleoseismology, and earthquake hazards in the western United States. In: Simpson, D.W., Richards, P.G. (Eds.), Earthquake Prediction: An International Review, Maurice Ewing Ser, vol. 4. Am. Geophys.Union, Washington, DC, pp. 209—216.

Weldon, R., Fumal, T., Biasi, G., 2004. Wrightwood and the earthquake cycle: what a long recurrence record tells us about how faults work. GSA Today 14 (9), 4—10.

Wells, D.L., Coppersmith, K.J., 1994. New empirical relationships among magnitude, rupture length, rupture width, rupture area, and surface displacement. Bull. Seismol. Soc. Am. 75, 939—964.

Wesnousky, S.G., 2006. Predicting the endpoints of earthquake ruptures. Nature 444 (7117), 358–360.

Working Group on California Earthquake Probabilities, 2003. Earthquake Probabilities in the San Francisco Bay region: 2002 to 2031. USGS Open-File Report 03–214.

Yeats, R.S., Sieh, K., Allen, C.R., 1997. The Geology of Earthquakes. Oxford Univ. Press, 568 pp.

Zilberman, E., Amit, R., Porat, N., Enzel, Y., Avner, U., 2005. Surface ruptures induced by the devastating 1068 AD earthquake in the southern Arava valley, Dead Sea rift, Israel. Tectonophysics 408, 79–99.

Chapter 11

The Role of Microzonation in Estimating Earthquake Risk

Imtiyaz A. Parvez [1] and Philippe Rosset [2]

[1] *CSIR Centre for Mathematical Modelling and Computer Simulation, Bangalore, India,*
[2] *WAPMERR, Geneva, Switzerland*

ABSTRACT

This chapter is dedicated to understanding the role of seismic zonation and micro-zonation, as well as understanding seismic risk analysis and mitigation strategy. The merits and demerits of various approaches to estimating earthquake hazard are discussed in terms of whether it is probabilistic, deterministic, or neodeterministic. The importance of geotechnical, geomorphological, and geological databases for seismic microzonation has been highlighted along with various techniques available to characterize site conditions. A variety of tools currently in use illustrate the basic principles of microzonation mapping at different scales. The main parameters involved in earthquake loss assessments and evaluating the influence of soil conditions on these estimates are discussed using QLARM, an advanced seismic risk estimation tool, for a few case histories.

11.1 INTRODUCTION

Earthquakes constitute the most feared of natural hazards, and they occur with no warning and can result in great destruction and loss of life. As one of the most devastating natural events, earthquakes impose economic challenges on society and governments. Megacities and urban areas are being developed all over the world, resulting in a considerable growth of earthquake risks to the number of lives and economic assets. One way to mitigate the destructive impact of earthquakes is to conduct a seismic hazard assessment and take remedial measures.

Earthquake hazard zonation for urban areas, mostly referred to as seismic microzonation, is the first step toward a seismic risk analysis and mitigation strategy in densely populated urban regions. One would like to quantify the spatial variation of the subsurface geological response due to a typical earthquake that can be expected in the area. In order to do so a preventive tool

Earthquake Hazard, Risk, and Disasters. http://dx.doi.org/10.1016/B978-0-12-394848-9.00011-0

that computes ground motion associated with selected earthquake scenarios can be based upon realistic modeling of ground motion developed from the knowledge of the seismic source, the propagation of seismic waves, and local site effects. The obtained synthetic seismograms contain information on an earthquake as an energy source and information on the medium through which the radiated energy propagates. Due to the complex nature of the earthquake source and the structure of the earth as the propagation medium for elastic waves, a seismogram is usually very complicated in its appearance. With proper selection of the parts of an earthquake record and with a proper choice of the theoretical model at the base of the experimental design, it is possible to extract information on selected parameters of earthquake source or earth structure relevant to seismic hazard. Usually, the areas with low seismic velocity sediments overlying bedrock with high seismic velocity give large amplification to the seismic signals.

It is therefore essential to establish a good database of the local subsurface geological conditions. However, it is challenging to obtain such a database in the urban areas, particularly in the developing countries because borehole data, geophysical surveys, and laboratory tests are often very limited. The other way to tackle such situations is to divide the area of interest into large subareas where the subsurface geological model and ground conditions are similar and the response curves may be assumed to be the same for the subarea and can be extended to the entire zone.

In this chapter, we will try to discuss the role of seismic zonation, microzonation, building typology, and population in the assessment of seismic risk and loss.

11.2 GROUND MOTION ESTIMATE AT THE REGIONAL SCALE

A quantitative ground motion prediction is a key for assessing the seismic hazard and mitigating the earthquake risk. Usually, such a quantification of ground motion at the bedrock level is often carried out at the regional/national or global scale, but the complexity starts when one looks at the effect of the subsurface geological sediments, as has been observed from many earthquake scenarios, the major damage to the buildings and man-made structures is mostly found in the area of soft sediments. The main factors that control the level of ground motion are source, path, and site effects. Among them, site effects have many times played a principal role on damage to buildings, as seen from the Mexico, Kobe, Loma Prieta, Izmit, Bhuj, and many other earthquakes. For the purpose of quantifying the ground motion at the surface, it is keenly required to develop the method for characterizing site effects and to understand soil behaviors during strong shaking. The constructive interference of incoming waves due to the effect of 2D or 3D geological structure can induce strong amplification. One of the basic problems associated with the study of microzonation is to determine the seismic ground motion, at a given site, due to an earthquake with a given magnitude (or moment) and epicentral distance. Now, with available computational powers, it is possible to simulate the ground motion using the available knowledge of seismic waves, their propagation, and excitation due to soft sedimentary layers. Alternatively, one has to choose some analytical, empirical approach and ambient noise at the particular site to quantify the site-specific ground motion and respective amplification.

Seismic zonation at the bedrock level depends on the regional seismicity, attenuation of ground motion intensity, and definition of potential seismic source zones. Regional and local seismicities can be investigated using seismological, geological, and paleoseismological data. Seismological data are collected from catalogs of historical and instrumentally located earthquakes. Geological data are collected from active fault maps, which are available for most areas. Paleoseismology looks at geologic sediments and rocks, for signs of ancient earthquakes. It is used to supplement seismic monitoring, for the calculation of seismic hazard.

There are two approaches to evaluate the earthquake hazard at bedrock: deterministic (Deterministic Seismic Hazard Analysis) and probabilistic (PSHA—Probabilistic Seismic Hazard Analysis, e.g., Stirling, 2014). The deterministic approach is based on selected scenario earthquakes and specified ground motion probability level. The probabilistic approach encompasses all possible earthquake scenarios and all ground motion probabilities, then computes the probability of earthquake occurrence during a certain time period. Probabilistic methods can be viewed as inclusive for all deterministic events with a finite probability of occurrence. In this context, proper deterministic methods that focus on a single earthquake ensure that each event is realistic, i.e., has a finite probability of occurrence (McGuire, 2001).

11.2.1 Seismicity and Attenuation

The basic assumption of hazard assessments is that the earthquake activity will recur where it was observed in the past or where it can be expected in the future. This information can be estimated from the seismicity of the region of interest. It can also include areas, which are tectonically prone to earthquakes, which were not recorded, in the historical past. Seismicity of the region, where one wants to study the earthquake hazard, are among the most important products as they provide a comprehensive database useful for numerous studies, and an outstanding one is investigating the seismicity of an area. There are many global earthquake catalogs, for example, the United States Geological Survey (USGS)/National Earthquake Information Center catalog, International Seismological Center catalog, the National Oceanic and Atmospheric Administration catalog, and many more who provide a fairly complete list of earthquakes from approximately 1960 with a magnitude of approximately 5. Here, it is necessary to integrate the global earthquake catalog with local and historical events in the region of interest. The clustering of these events along with their focal mechanisms, geology, and tectonics in including active faults and lineaments are used to prepare seismogenic zones. For the probabilistic earthquake hazard assessment, the completeness of the earthquake catalog and the attenuation law of the area of interest play a very important role.

The next important step is to determine an appropriate attenuation law or Ground motion prediction equations, giving ground motion-intensity measures such as peak ground motions or response spectra as a function of earthquake magnitude and distance. They are important tools in the analysis of probabilistic seismic hazard. These equations are typically developed empirically and semiempirically by a regression of recorded strong-motion amplitude data versus magnitude, distance, and possibly other predictive variables (Parvez et al., 2001; Atkinson and Boore, 2006; Boore and Atkinson, 2008; Pezeshk et al., 2011 and many more). The amount of data used in regression analysis to prepare the attenuation law for a region is an important issue as it bears heavily on the reliability of the results, particularly the magnitude and distance ranges. Parvez et al., 2001 have shown systematic regional effects (Figure 11.1) in estimating the attenuation law for Western Himalayas and Eastern Himalayas (Shillong Plateau). They may be considered as a basis for future regionalized seismic hazard assessment in the Himalayan region.

11.2.2 Ground Motion Modeling

A viable numerical and analytical alternative to model and simulate the ground motion is possible now using current computational resources and physical knowledge of the seismic wave generation and propagation processes, along with the improving the quantity and quality of geophysical data. A set of scenarios of expected ground shaking due to a wide set of potential earthquakes can be defined by means of full waveform modeling, based on the possibility to efficiently compute synthetic seismograms in complex laterally heterogeneous anelastic media. In this way, a set of scenarios of ground motion can be defined, either at a national or local scale, the latter considering the 2D and 3D heterogeneities of the medium traveled by the seismic waves.

For microzonation or site-specific earthquake hazard assessment, the ground motion modeling is a more appropriate tool than the others based on some empirical approach. It has been proven that the soft sediments play a very important role in amplification of ground motion and their response to various structures. A recently published Neodeterministic Seismic Hazard Assessment (NDSHA) by Peresan et al., 2011 permits one to integrate the available information provided by the most updated seismological, geological, geophysical, and geotechnical databases for the site(s) of interest, as well as

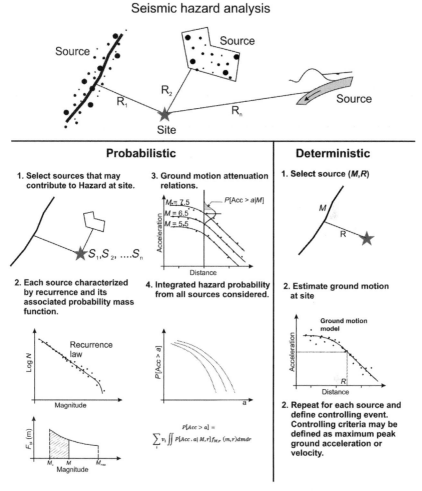

FIGURE 11.1   Schematic diagram for probabilistic and deterministic seismic hazard mapping.

advanced physical modeling techniques, to provide a reliable and robust basis for the development of a deterministic design basis for cultural heritage and civil infrastructures in general (Field, 2000; Panza et al., 2001). The neo-deterministic approach has been different from the deterministic approach because it means scenario-based methods for seismic hazard analysis. The attenuation relations and other assumptions about local site responses based on linear convolution questionable on mathematical and physical ground are not included here. The NDSHA procedure provides strong ground motion parameters based on the physical modeling of seismic wave propagation at different scales—regional, national and metropolitan—accounting for a wide set of possible seismic sources and for the available information about the mechanical properties of the propagation media.

11.2.3 Probabilistic Seismic Hazard Assessment

The main goal of earthquake engineers is to ensure that a designed structure can resist a given level of ground motion while maintaining a desired level of performance (e.g., Tolis, 2014). But how can one quantify the level of ground shaking used to perform this analysis? There are uncertainties about the location, size, and resulting ground motion of future earthquakes. PSHA aims to quantify these uncertainties and combine them to produce an explicit description of the distribution of future shaking that may occur at a site. This integrates over all possible earthquake ground motions at a site to develop a composite representation of the spectral amplitudes and hazards (annual frequencies of exceedance) at that site. The analysis has a strong basis in earth sciences and earthquake engineering, and allows decisions on seismic design levels for a facility to be made in the context of the earthquake magnitudes, locations, and ground motions (including the effects of local site conditions on amplitudes of strong shaking) that may occur. The use of PSHA is common throughout the world for determining seismic design levels.

The classical PSHA (Cornell, 1968) determines the probability of exceedance, over a specified period of time for various levels of ground motion. The main elements of a PSHA are the seismic sources/seismogenic zones, within which the seismogenic process is frequently assumed to be rather uniform; the characteristics of the earthquake recurrence within the seismogenic zones, which are assumed to be poissonian; and the attenuation relations, which provide estimates of ground motion parameters at different distances from the sources. The hazard at a site is given in terms of the probability of exceeding different levels of ground motion during a specified period of time. This is achieved through the calculation of the frequency of earthquakes with some damaging potential and the calculation of the conditional probability of exceedance of a given ground motion level, for each of these contributing earthquakes (summed over all potentially contributing sources). Thus, the PSHA aims at the statistical characterization of ground motion at a site.

The basic formulation of the PSHA was derived in the 1970s using the "Total Probability Theorem":

$$P(Y > y) \simeq \sum v_i \iint P[Y \cdot y | M, R] f_{M,R}(m, r) \mathrm{d}m \mathrm{d}r,$$

where Y and y are earthquake ground motions, M and m are magnitudes, R and r are distances, v_i is the rate of earthquakes for source i, and the summation is over all sources.

The development of quantitative ground motion equations has helped in the advancement of PSHA. Many researchers have made remarkable developments to the basic formulation of PSHA as it is used today. First, earthquakes were recognized to rupture a finite segment of the causative fault, thus becoming a source of energy with finite dimensions rather than a point source as assumed by Cornell (1968). Der Kiureghian and Ang (1977) described this effect and recommended that the distance from the closest point of rupture to the site of interest was the best distance measure to use for ground motion estimation. This effect is important only for large magnitude earthquakes, of course, but these are often the events that dominate the seismic hazard in plate margin areas. The final outcome was that the seismic hazard results were more accurate at near-fault sites affected by large magnitude earthquakes, and that seismic hazard maps for regions with major faults were more realistic. A development that had a large impact on seismic hazard calculations was the recognition that ground motion equations and seismic hazard curves could be developed directly on spectral response (McGuire, 1974), leading to the concept of a uniform hazard spectrum. Prior to that, seismic hazard curves were developed for Peak Ground Acceleration (PGA), Peak Ground Velocity (PGV), and perhaps Peak Ground Displacement (PGD), and the spectrum was constructed by amplifying these peak motions measures (e.g., Newmark and Hall, 1982).

11.2.4 Deterministic Seismic Hazard Assessment

The deterministic seismic hazard involves four basic steps: identification of sources; determination of the controlling earthquake for each source; selection of a ground motion relationship; and the computation of the design ground motion parameter(s). When identifying the sources, it is important to review all the possible seismic sources that could produce damaging ground motion at the site of interest, mainly on the basis upon the interpretation of geological, geophysical, and seismological data. The sources can be represented by points, lines, or areas; however, most are defined as a capable finite fault with appropriate focal mechanism because the preciseness of a point or line may not properly depict the knowledge of the earthquake mechanisms.

The next step, the definition of "controlling earthquake" depends on the maximum earthquake that can be generated in the source zones. This can be the maximum credible earthquake, the expected earthquake, or any other type of earthquake. The decision of the choice of controlling earthquakes is critical, and this constitutes the most vulnerable part in the deterministic analysis. For a facility whose failure only presents a relatively low hazard, the controlling earthquake might be defined as one that would be reasonably expected to occur during the operating life of the facility. Earthquake magnitude or intensity is commonly used to define the size of the earthquake and in addition to these sizes, there is an appropriate distance, which is the distance between the source and the site. For critical facilities, facilities whose failures present high hazards (e.g., nuclear power plants and chemical plants)—the controlling earthquake is usually defined as the maximum magnitude earthquake, *Mmax* that a given source is believed to be capable of generating.

Ground motion parameters (PGA, PGV, or PGD and response spectra) are the important hazard parameters required in assessing deterministic seismic hazard, based on the attenuation properties that vary from region to region. These can be established in the form of equations that are typically developed empirically/semiempirically by a regression of recorded strong-motion amplitude data versus magnitude, distance, and possibly other predictive variables. An attenuation expression needs to be selected appropriately based on the type of source mechanism (e.g., shallow crustal and subduction) and the path. There are many statistical studies that provide estimates of PGA, PGV as well as pseudo-absolute-acceleration spectra (PSA) and pseudo-relative-velocity spectra. The detailed schematic diagram for both the probabilistic and deterministic seismic hazard mapping is shown in detail in Figure 11.1.

Finally, the site design ground motion is determined by using the attenuation law/expression selected above to compute the ground motions (PGAs or PGVs, the parameter used to describe the site ground motion) corresponding to the controlling earthquakes associated with each source zone identified. The earthquake associated with the largest of these site PGAs is typically used to define the site's design ground motion. The design PGA can be used to scale fixed-shape spectra, or its associated magnitude—distance pair can be used in conjunction with a spectral attenuation relationship to directly compute the site's design spectra.

11.2.5 Neodeterministic Seismic Hazard Assessment

Seismic hazard assessment can also be performed following a neodeterministic approach (NDSHA), which allows giving a realistic description of the seismic ground motion due to an earthquake of a given distance and magnitude (Costa et al., 1993; Panza et al., 1996; Parvez et al., 2003). The approach is based on modeling techniques that have been developed from a detailed knowledge of both the seismic source process and the propagation of seismic waves. This permits us to define a set of earthquake scenarios and to simulate the associated synthetic signals without having to wait for a strong event to occur. The NDSHA can be applied at the regional scale, computing seismograms at the nodes of a grid with the desired spacing, or at the local scale, taking into account the source characteristics, the path, and local geological and geotechnical conditions. Synthetic signals can be used as seismic input in subsequent engineering analyses aimed at the computation of the full nonlinear seismic response of the structure or simply the earthquake damaging potential. Massive parametric tests to explore the influence not only of deterministic source parameters and structural models but also of random properties of the same source model enable a realistic estimate of seismic hazard and their uncertainty.

In the NDSHA approach, the definition of the space distribution of seismicity accounts only for the largest events reported in the earthquake catalog grouped into 0.2° × 0.2° cells, assigning to each cell the maximum magnitude recorded within it. A smoothing procedure is then applied to account for spatial uncertainty and for source dimensions (Panza et al., 2001). Only cells located within the seismogenic zones are retained. This procedure for the definition of earthquake locations and magnitudes for NDSHA makes the method robust against uncertainties in the earthquake catalog, which is not required to be complete for magnitudes <5. A double-couple point source is placed at the center of each cell, with a focal mechanism consistent with the properties of the corresponding seismogenic zone and a depth, which is a function of magnitude. To define the physical properties of the source–site paths, the territory is divided into 1.0° × 1.0° cells, each characterized by a structural model composed of flat, parallel anelastic layers that represent the average lithosphere properties at the regional scale (Brandmayr et al., 2010). Synthetic seismograms are then computed by the modal summation technique for sites placed at the nodes of a grid with a step of 0.2° × 0.2° that covers the national territory, considering the average structural model associated with the regional polygon that includes the site. Seismograms are computed for an upper frequency content of 1 Hz, which is consistent with the level of detail of the regional structural models, and the point sources are scaled for their dimensions using the spectral scaling laws proposed by (Gusev, 1983), as reported in Aki (1987). An example of such a neodeterministic hazard for the Indian subcontinent prepared by Parvez et al. (2003) is shown in Figure 11.2.

At the local scale, the neodeterministic method is based on a hybrid technique that combines two methods: the analytical modal summation (Panza, 1985; Florsch et al., 1991; Panza et al., 2001) and the numerical finite-difference methods (Virieux, 1984, 1986; Levander 1988) taking advantage of the characteristics of both. Each of the two methods is applied in that part of the structural model where it works most efficiently. The modal summation is applied to simulate wave propagation from the source to the sedimentary basin of interest. Being a purely analytical technique, there is no penalty applied in this part of the modeling associated with the model size. The finite difference method is applied to propagate the incoming wave field in the laterally heterogeneous part of the structural model that contains the sedimentary basin.

11.3 LOCAL SITE RESPONSE AND MICROZONATION

Mapping of seismic hazards at local scales to incorporate the local site response is called seismic microzonation (Finn et al., 2004). The study of various strong earthquakes, the world over, has demonstrated that the degree of damage to structures is influenced by the ground conditions, which include the geotechnical properties of rocks and unconsolidated deposits, the geomorphic features, and the tectonic fabric. Long back, Milne (1898), on the basis of his

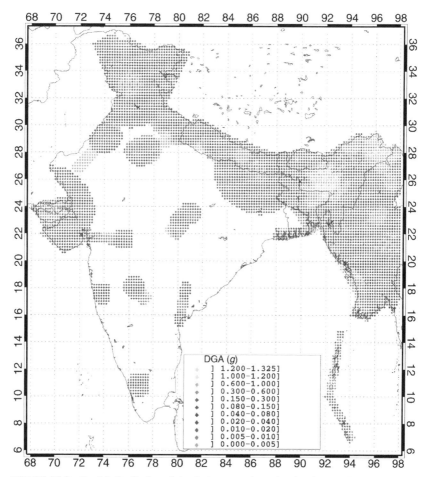

FIGURE 11.2 Spatial distribution of the design ground acceleration in *g*. *(After Parvez et al. (2003).)*

study of the 1819 Japan earthquake, had observed that ground motions were higher at sediment sites than on harder rocks. This site response factor was best exemplified in the case of the September 19, 1985, Mexico earthquake originating along the west coast of Mexico. The effect of the seismic event was seen to be a maximum 300 km away in Mexico City, where a large number of 6−10 story buildings were destroyed, in which 20,000 people were killed (Tobin, 1997). It was interpreted that the heightened damage in the city was a result of amplification of certain frequencies of seismic waves passing through the sedimentary layers of the lake bed zone, over which the buildings were founded.

Three major factors that control the level of strong ground motion are source, path, and site effects (Figure 11.3). Among them, site effects have

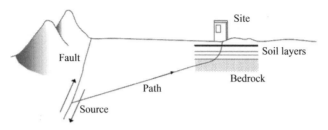

FIGURE 11.3   Ground motion propagation from source to site (*Kramer, 1996*).

sometimes played a principal role in generating damage to buildings, as seen from the Mexico, Kobe, Loma Prieta, Izmit, Bhuj, and several other earthquakes. In order to have a better understanding on site effects, it is keenly required to develop the method for characterizing site effects and to understand soil behaviors during strong shaking. In the present scenario, the H/V spectral ratio method (Nakamura, 1989) is probably one of the easiest and most commonly adopted approaches to study the site response. Despite the apparent appeal of Nakamura's method, the physical basis and actual relevance for site effect estimates of this method have never reached a scientific consensus. The constructive interference of incoming waves due to the effect of a 2D or 3D geological structure produce very strong ground motions. One of the basic problems associated with the study of seismic zonation/microzonation is to determine the seismic ground motion, at a given site, due to an earthquake with a given magnitude (or moment) and epicentral distance. Actually, the number of recorded signals is relatively low and the installation of local arrays in each zone with a high level of seismicity is too expensive an operation that requires a long time interval to gather statistically significant data sets. Alternatively, one has to choose some analytical, empirical, or numerical solutions based on the theoretical knowledge of seismic waves, their propagation and excitation due to soft sedimentary layers (Parvez et al., 2002, 2004; Parvez and Madhukar 2006).

11.3.1 Local Site Effects

The local site effects are the influence of the soil response on the seismic motion at the ground surface due to an earthquake. Once the probable earthquake characteristics are determined, the most important step is evaluating the ground motion characteristics on the surface while accounting for local geological and geotechnical site conditions. Local site effects are considered the most significant factors in the microzonation of ground motions.

It is inevitable to validate the effects of local site for estimating strong motions and mitigating earthquake risk and losses. As it has been observed from many earthquakes, the major damage to the buildings and man-made structures is mostly found in the area of soft sediments that induced amplification of seismic waves. The ideal solution for such a problem could be to use a wide database of recorded strong motions and to group those accelerograms that have similar source, path, and site effects. In practice, however, such a database is not available particularly in a country like India. Actually, the number of recorded signals is relatively low and the installation of local arrays in each zone with a high level of seismicity is too expensive and requires a long time interval to gather statistically significant data sets. Alternatively, one has to choose some analytical, empirical, or numerical solutions based on theoretical knowledge of seismic waves, their propagation, and excitation due to soft sedimentary layers.

There are several methods proposed to investigate the behavior of a soft sedimentary structure in the excitation of seismic waves. The most common procedure, introduced by Borcherdt (1970), and applied by numerous researchers, is to compare the spectra of seismograms of earthquakes with the ones obtained at a nearby reference station located on competent bedrock. The factors of epicentral distance and source radiation, therefore, are practically the same for both neighboring sites, and the differences in the response can be ascribed to the local geological or topographical characteristics of the site. This technique needs the occurrence of earthquakes and the assumption that the radiation pattern and epicentral distances for both the sites are similar. Besides, one needs to deploy instruments at all the sites of interest.

Nakamura (1989) developed a simple technique based on the ratio of the spectra of the horizontal to the vertical components of ground motion generated by microtremors or ambient noise. Lermo and Chavez-Garcia (1994) indicate that the method assumes that the surface layers do not amplify the "vertical tremor". Besides, it is assumed that, for a wide range of frequencies, the ratio of the horizontal to the vertical spectrum at the base of the system has a value near unity. According to them, this latter assumption was experimentally verified by Nakamura using microtremors recorded in a borehole. Thus, Nakamura concluded that the spectral ratio between the horizontal and vertical components of motion at the same site can be used as an estimate of site effects for internal waves.

Field and Jacob (1993) worked on a 3D model in a simpler layer on a half space with microtremors under a random distribution in space and time of forces applied in selected points on top of the layer. By using Green functions, the horizontal and vertical amplitudes were evaluated and compared with the response spectrum at the surface for the incident SH waves to a vertical plane. The peak frequency in both the cases was coincident with the natural resonance frequency of the layer for shear waves vertically incident. Lachet and Bard (1995) concluded that Nakamura's technique may be used to determine the natural resonance frequency of a soft layer, but it fails to predict the amplification of surface waves.

Moreover, they showed that the natural frequency of the layer obtained with Nakamura's technique and ambient noise simulations is independent of the excitation source, dependent on Poisson's ratio, and controlled by the polarization curve of the Rayleigh waves. However, the method based on the assumption, not always fulfilled, that the propagation of the vertical component of motion is not perturbed by the uppermost surface layers, and can therefore be used to remove source and path effects from the horizontal components.

Instead of waiting for data accumulation, either based on the computation of the spectral ratio between the signal recorded at soft soil and nearby bedrock site, or generation of microtremors of ambient noise to be used for the H/V ratio technique, it is wiser to apply the preventive tool given by realistic modeling, based on computer codes developed from the knowledge of the seismic source and of the propagation of seismic waves associated with the given earthquake scenario. With such an approach, source, path, and site effects are all taken into account, and a detailed study of the wave field that propagates at large distances from the epicenter is possible. Actually, the realistic modeling of ground motion requires a simultaneous knowledge of the geotechnical, lithological, geophysical parameters and topography of the medium, on one side, and tectonic, historical, paleo-seismological (e.g., Meghraoui and Atakan, 2014), seismotectonic models, on the other, for the best possible definition of the probable seismic source. The initial stage of the realistic modeling is thus devoted to the collection of all available data concerning the shallow geology, and the construction of a three-dimensional structural model to be used in the numerical simulation of ground motion.

A powerful hybrid technique has been developed by Fäh et al., 1993a and 1993b, which combines the modal summation (Panza, 1985; Panza and Suhadolc, 1987; Florsch et al., 1991; Panza et al., 2001) and Finite-difference scheme (Virieux, 1984, 1986), exploiting both the methods to their best. However, the most fundamental data, such as geotechnical information, S-wave velocity structure at a site where a prediction of ground motion is required, are generally insufficient; therefore, the reliability of the modeling of strong ground motion due to a 2D/3D structure is highly dependent on the structure of shallow geology over the bedrock. There are several exploration techniques used to obtain the S-velocity structure, but the conventional seismic methods are difficult or impossible to implement in urban areas or environmentally sensitive areas. To overcome this difficulty, recently a popular technique of "Microtremor Array Observation" is being applied, which makes use of microtremors (ambient noise) found in abundance anywhere on the surface of the earth. The Array measurement of microtremors to obtain the S-velocity structure of surface geology, H/V ratio technique, and numerical simulation technique to ground motion simulation will be discussed in the next part.

11.3.2 Fundamental Frequency and Site Amplification

As discussed, the fundamental frequency and respective amplification in the ground motion amplitudes are the most important factors for the site-specific microzonation. Assessments of these two parameters are executed in different ways; some of the tools currently used are discussed below.

11.3.2.1 Array Measurement of Microtremors

Array observations of the vertical component of microtremors are frequently conducted to estimate subsurface structures as a shear-wave velocity distribution in urban or environmentally sensitive areas. A key technique is the "microtremor survey method" (Okada, 1998; Okada et al., 1990) that uses microtremors found in abundance anywhere on the surface of the earth. Analysis of the records are done using the frequency—wave number power spectral (f—k) method (Asten and Henstridge, 1984; Horike, 1985; Matsushima and Okada, 1990; Yamanaka et al., 1999) or the spatial autocorrelation (SPAC) method (Aki, 1957; Henstridge, 1979; Okada, 1998). In recent years, both the methods have become of major interest as a tool that could yield more quantitative information such as shear-wave velocity and thickness of sediments over basement. Both the f—k and SPAC methods are based on the assumptions that microtremors are a spatiotemporally stationary stochastic process. The f—k method is an application of the technique developed to detect nuclear explosion using a seismic network with a diameter as large as 200 km. The statistical parameter called the frequency—wave number power spectral density (f—k spectrum) played a central role in the detection of nuclear explosions. Its principle is to detect relatively powerful seismic signals from noise, and it can separate multimode surface waves as well as body waves, but it requires a seismometers array with sets of stations that are distributed uniformly in azimuth, with a variety of distances between stations to ensure high-resolution estimates for the f—k power spectrum. This technique requires a large number of stations in the array at least seven for reliable results (Kudo et al., 2002).

On the other hand, the SPAC method requires fewer stations (practically three or four stations as a minimum requirement) and is based on the theory developed by Aki (1957). He gave a theoretical basis of the SPAC coefficient defined for microtremor data and developed a method to estimate the phase velocity dispersion of surface waves contained in the signal using a specially designed circular array. Henstridge (1979) also introduced a licit expression of the relationship between the SPAC coefficient and the phase velocity of fundamental-mode Rayleigh waves. Okada (1998) extended it to an exploration method, and it is currently called the SPAC method. The flow of the observation and analysis in applying the method is shown in Figure 11.4.

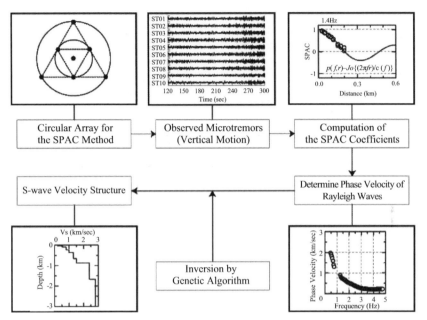

FIGURE 11.4   A flow of observation and analysis in the SPAC method for estimating S-wave velocity structures using array observation of microtremors. *(After Kudo et al. (2002).)*

For a circular array of stations for microtremor observation, let us represent harmonic waves of frequency ω of microtremors by the velocity waveforms $u(0, 0, \omega, t)$ and $u(r, \theta, \omega, t)$ observed at the center of the array $C(0, 0)$ and at the point $X(r, \theta)$ of the array. The SPAC function is defined as

$$\phi(r, \theta, \omega) = \overline{u(0, 0, \omega, t) \cdot u(r, \theta, \omega, t)}, \qquad (11.1)$$

where $\overline{u(t)}$ is the average velocity of the waveform in the time domain. The SPAC coefficient ρ is defined as the average of the autocorrelation function ϕ in all directions over the circular array:

$$\rho(r, \omega) = \frac{1}{2\pi \cdot \phi(0, \omega)} \int_0^{2\pi} \phi(r, \theta, \omega) d\theta, \qquad (11.2)$$

where $\phi(0, \omega)$ is the SPAC function at the center $C(0, 0)$ of the circular array. By integration of Eqn (11.2), one can find

$$\rho(r, \omega) = J_0 \left(\frac{\omega r}{c(\omega)} \right), \qquad (11.3)$$

where $J_0(x)$ is the zero-order Bessel function of the first kind of x and $c(\omega)$ is the phase velocity at frequency ω. The SPAC coefficient $\rho(r, \omega)$ may be obtained in the frequency domain using the Fourier Transform of the observed microtremors:

$$\rho(r, \omega) = \frac{1}{2\pi} \int_0^{2\pi} \frac{R_e[S_{CX}(\omega, r, \theta)]}{\sqrt{S_C(\omega) \cdot S_X(\omega, r, \theta)}} d\theta, \tag{11.4}$$

where $S_C(\omega)$ and $S_X(\omega, r, \theta)$ are the power spectral densities of microtremors at sites C and X, respectively, and $S_{CX}(\omega, r, \theta)$ is the cross-spectrum between ground motions at these two sites. Thus, the SPAC coefficients may be obtained from averaging the normalized coherence function defined as the cospectrum between points C and X in the direction. From the SPAC coefficients $\rho(r, \omega)$, the phase velocity is obtained for every frequency from the Bessel function argument of Eqn (11.3), and the velocity model can be inverted.

By numerical simulations using an array of seven stations, as shown in Figure 11.4, Miyakoshi et al., 1996 concluded that the observable maximum wavelength is approximately 10 times the array radius by the SPAC method, independent of the directions of the waves and their numbers, and roughly five times or less of the array radius by the $f-k$ method in the case of plural wave propagations. The observable minimum wavelength, which is essentially limited by the spatial aliasing or by the minimum distance between stations, has no significant difference between the SPAC and $f-k$ methods (Okada, 1998).

11.3.2.2 Nakamura H/V Ratio Technique for Resonance Frequency

The other most popular technique of Nakamura (1989) is widely used to obtain a quick microzonation map of any large urban area. It uses microtremor measurements for the estimation of resonance frequency and site effects. However, the site amplification obtained by this technique is questionable, as reported in many papers (Lermo et al., 1988; Kobayishi et al., 1991; Morales et al., 1991; Field et al., 1990; Akamatsu et al., 1991; Dravinski et al., 1991). Here, the technique has been described very well and is based on the assumptions for the fundamental characteristics of microtremors.

Usually, it is assumed that the transfer functions of surface layers can be given by the ratio

$$S_T = \frac{H_S}{H_B}.$$

However, considering the great contribution of Rayleigh wave propagation for the ambient noise, it will be necessary to convert the ratio H_S/H_B, in order to estimate a transfer function for microtremor measurements. Assuming that the vertical tremor is not amplified by the surface layers (Figure 11.5), the ratio E_R defined below should represent the effect of the Rayleigh wave on the vertical motion.

$$E_R = \frac{V_S}{V_B}.$$

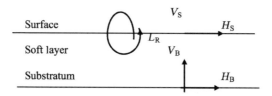

FIGURE 11.5 **Illustration of the simple model assumed for the interpretation of the microtremor H/V ratio as defined by** *Nakamura (1989).*

Assuming that the effect of the Rayleigh wave is equal for vertical and horizontal components, it is possible to define a corrected modified spectral ratio,

$$S_M = \frac{S_T}{E_R} = \frac{H_S/H_B}{V_S/V_B}.$$

As a final condition, it is assumed that for all frequencies of interest

$$\frac{H_B}{V_B} = 1.$$

Thus, an estimate of the transfer function is given by the spectral ratio between the horizontal and the vertical components of the motion at the surface

$$S_M = \frac{H_S}{V_S}.$$

Some of the above conditions were already tested, experimentally and theoretically by different authors (Jensen, 2000; Bour et al., 1998; Teves-Costa et al., 1996; Lermo and Chavez-Garcia, 1994; Nakamura, 1989).

11.3.2.3 Numerical Simulation of Strong Ground Motion

Fäh et al. (1993a,b) developed a hybrid method that combines the modal summation technique, valid for laterally homogeneous anelastic media, with a finite difference that includes the lateral heterogeneity of the 2D subsurface geological structure and optimizes the use of the advantage of both methods. The modal summation technique is applied to simulate propagation from the source position to the sedimentary basin or the local irregular feature of interest and the finite-difference method is used in the laterally heterogeneous part of the structural model, which contains the sedimentary basin (Figure 11.6).

This hybrid approach allows us to calculate the local wave field from a seismic event, for both small (a few kilometers) and large (a few hundreds of kilometers) epicentral distances. The use of the mode summation method helps to include an extended source, which can be modeled by a sum of point sources appropriately distributed in time and space. This allows the simulation

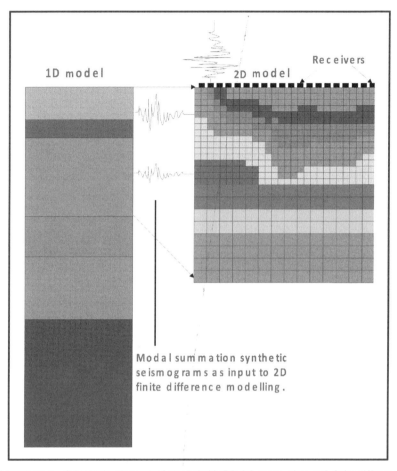

FIGURE 11.6   **Schematic diagram of the hybrid (Modal summation and finite-difference method).**

of a realistic rupture process of the fault. The path from the source position to the sedimentary basin or the local heterogeneity can be approximated by a structure composed of flat 1D homogeneous layers. The finite-difference method applied to treat wave propagation in the sedimentary basin permits the modeling of wave propagation in complicated and rapidly varying 2D velocity structures. The coupling of the two methods is carried out by introducing the resulting time series obtained with the mode summation method into the finite-difference computations. The ground motion time series computed for the 1D modal contain all possible body waves and surface waves consistent with the preassigned phase velocity and frequency interval. To excite the finite-difference grid, ground motion time series are computed at adjacent points lying on one side of the 2D part of the model and sampling different depths along two vertical lines that belong to the regular grid used for discretization of the medium (Fäh et al., 1990). In this way, the seismic wave field generated and propagated in the 1D medium is used to excite the 2D part of the structural model.

The anelasticity is treated in the finite-difference computations by introducing into the equation of motion a convolution term. This additional term is represented by a system of differential equations that define a low-order rational function. This function approximates the viscoelastic modules of the generalized Maxwell body and is introduced into the stress—strain relation. The solution of this equation developed as a system of n ordinary differential equations of the first order is possible using a numerical algorithm if the quality factor is constant within a defined frequency band. For the SH wave propagation, the computation scheme follows Emmerich and Korn (1987), for P-SV case by Emmerich, (1992). Furthermore, the ground agreement of the 1D analytical and hybrid method results allows establishing to which depth the grid has to be extracted to guarantee the completeness of all the signals introduced into the 2D model. In this way, it is possible to study the wave propagation in 2D heterogeneous media, as sedimentary basins, with a significant accuracy in computations (Panza et al., 2001).

The initial stage of this work requires collection of all available data concerning the shallow geology, and the construction of cross-sections along which to model the ground motion. It is a multidisciplinary activity by nature, since the required information is obtained from different disciplines, such as seismology, history, archeology, geology, and geophysics. The final product is a map of expected ground motions, which in turn constitutes the basis for realistic modeling of ground motion. Thus, a complete database in terms of hazard parameters can be constructed immediately, until more experimental evidence and recordings become available. This database would then naturally be updated continuously by comparison with incoming new experimental data.

11.4 LIQUEFACTION

Liquefaction occurs in saturated soils when vibrations or water pressure within a mass of soil cause the soil particles to lose contact with one another. This water pressure influences how tightly the particles themselves are pressed together. Prior to an earthquake, the water pressure is relatively low. However, earthquake shaking can cause the water pressure to increase to the point where the soil particles can readily move with respect to each other. Earthquake shaking often triggers this increase in water pressure; as a result, the soil behaves like a liquid, has an inability to support weight, and can flow down very gentle slopes. This condition is usually temporary and is most often caused by an earthquake vibrating water-saturated fill or unconsolidated soil.

The upward propagation of shear waves through the ground during an earthquake generates shear stresses and strains that are cyclic in nature (Seed and Idriss, 1982). If cohesionless soil is saturated, excess pore pressure may accumulate during seismic shearing and lead to liquefaction. Unsaturated soils are not subject to liquefaction because volume compression does not generate excess pore pressure. Large deformations and liquefactions are more prone with contractive soils while dilative soils may associate with cyclic softening and limited deformations. Since liquefaction is associated with the tendency for soil grains to rearrange when sheared, anything that impedes the movement of soil grains will increase the liquefaction resistance of a soil deposit. Particle cementation, soil fabric, and aging—all related to the geologic formation of a deposit—are important factors that can hinder particle rearrangement (Seed, 1979). Soil deposits prior to the Holocene epoch (>10,000 years old) are usually not prone to liquefaction (Youd and Perkins, 1978), perhaps due to weak cementation at the grain contacts. However, conventional sampling techniques inevitably disturb the structure of cohesionless soils such that laboratory test specimens are usually less resistant to liquefaction than the in situ soil. Even with reconstituted laboratory samples, the soil fabric and resistance to liquefaction are affected by the method of preparation such as dry pluviation, moist tamping, and water sedimentation. After liquefaction has occurred, the initial soil fabric and cementation have very little influence on the shear strength beyond about 20 percent strain (Ishihara, 1993).

Characteristics of the soil grains (distribution of sizes, shape and composition.) influence the susceptibility of a soil to liquefy (Seed, 1979). While liquefaction is usually associated with sands or silts, gravelly soils have also been known to liquefy. Rounded soil particles of a uniform size are generally the most susceptible to liquefaction (Poulos et al., 1985). Well-graded sands with angular grain shapes are generally less prone to liquefy because of a more stable interlocking of the soil grains. On the other hand, natural silty sand sediments tend to be deposited in looser state, and thus are more likely to

TABLE 11.1 Classification of Soil Liquefaction Consequences

In Situ Stress Condition	Soil Behavior	Typical Field Observation
No driving shear stress	• Volume decrease • Pore pressure increase	• Ground settlement • Sand boils and ejection from surface fissures
Driving shear greater than residual strength	• Loss of stability • Liquefaction	• Flow slides • Sinking of heavy buildings • Floating of light structures
Driving shear less than residual strength	• Limited shear distortion • Soil mass remains stable	• Slumping of slopes • Settlement of buildings • Lateral spreading

After Castro (1987).

exhibit contractive shear behaviors than clean sands. The nature and severity of liquefaction damage is a function of the reduced shear strength and the magnitude of the static shear loads supported by the soil deposit (Ishihara et al., 1991). On the basis of the relative magnitude of static driving shear stresses that may be present due to a surface slope or a foundation bearing load, Castro (1987) classifies the possible consequences of liquefaction as shown in Table 11.1. Mapping of liquefaction hazard potential is one of the key factors in microzonation. There are several studies available of liquefaction hazard maps. For example, Broughton et al. (2001) mapped the liquefaction potential hazard, based on detailed field investigations of the geologic units in Memphis, Tennessee. For those soils that have moderate to high potential for liquefaction, a further evaluation can be performed based on in situ tests such as the standard penetration test, cone penetration test, and shear-wave velocity (Kramer, 1996).

11.5 CASE HISTORIES OF SOME INDIAN MEGACITIES

All over the world, the seismic hazard and risk assessment studies of the urban complexes at a microlevel are now several decades old. Here, we present a couple of examples of Indian megacities.

Delhi represents a typical example of a megacity, which is under severe seismic threats not only from the local earthquakes but also from the Himalayan earthquakes, located 200−250 km from the city. The city has already suffered serious damage in the past because of the degraded conditions of the historical built environment, and because of severe local site amplification (Iyengar, 2000). In the present scenario, the high density of the population and the kind of built environment increase the vulnerability of many parts of this megacity. Sound antiseismic construction requires the knowledge of seismic site response, in terms of both peak ground acceleration and response spectral ratio.

Parvez et al. (2002 and 2004) have worked on site-specific microzonation of Delhi city using the hybrid technique discussed above for the realistic modeling of ground motion along 2D structures. They generated synthetic seismograms along two representative geological cross-sections in Delhi City; the NS cross-section runs from Inter State Bus Terminus to Sewanagar and another EW cross-section from Tilak Bridge to Punjabibagh. These profiles, initially available up to 30−35 m of depth, have been further extended down, to approximate the bedrock depth level, using the Iyengar (2000) data. The details of the material properties of these cross-sections are given in Parvez et al., 2002 and 2004.

The synthetic seismograms (SH- and P-SV waves) have been computed with the hybrid method for an array of 100 sites regularly spaced, every 100 m, along the NS cross-section for a source of July 15, 1720 ($M = 7.4$). Figure 11.7 shows the three-component synthetic strong-motion accelerograms computed, which clearly defines the trend of the amplification effects that very well

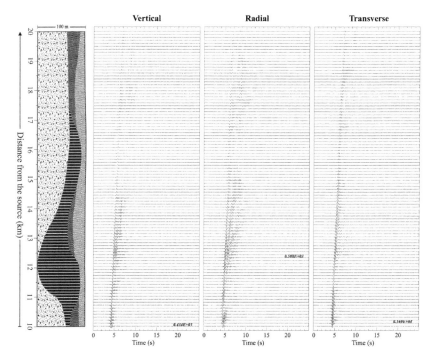

FIGURE 11.7 The North-South cross-section and corresponding synthetic strong-motion records computed for the 1720 earthquake. The maximum amplitude value in centimeters per squared second is also indicated. *(Modified after Parvez, et al. (2002).)*

reflect the geometry of the cross-section models. Peak acceleration (AMAX) of 1.6 g is estimated in the transverse component at 10.2 km of the epicentral distance from the source, which is a quite large value and represents a severe seismic hazard, as can be expected in the epicentral area of an event of magnitude 7.4. We believe that the peak value within 10 km of the epicentral distance is saturated for a large event in terms of damage/ground motion like that observed at the epicenter. Such high values of AMAX are in agreement with the reports of the damage caused by the 1720 earthquake (Iyengar, 2000). The radial components of ground motion exhibit peak values in the range from 0.41 to 0.59 g, and the vertical components reaches similar peak amplitudes, in the range from 0.42 to 0.43 g.

The RSR, that is, the response spectra computed from the signals synthesized along the local model normalized by the response spectra computed from the corresponding (same epicentral distance) signals synthesized for the regional bedrock model is another parameter known as site amplification, relevant for earthquake engineering purposes. The distribution of RSR as a function of frequency and epicentral distance along the profile, up to a maximum frequency of 5 Hz, is shown for the three components in Figure 11.8. For each component of motion, the numbers in parentheses

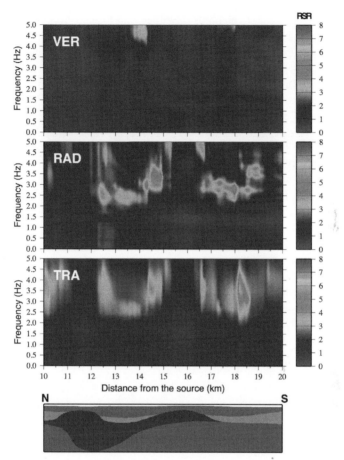

FIGURE 11.8   The NS cross-section and the corresponding plot of the response spectra ratio (RSR) versus frequency. The numbers in brackets represent in order the distance in kilometers, frequency in hertz, and value of peak RSR, where maximum amplification is found. *(Modified after Parvez et al. (2002).)*

identify the maximum amplification. In order, the distance from the source in kilometers, the frequency in hertz, and the value of RSR are given. A 5 percent damping of the response spectra is considered since reinforced concrete buildings are already or will be built in the area. There are sites, where the amplifications are relevant in all the three components, even if the maximum amplifications are always found in the horizontal components. The RSR is 5–10 in the frequency range of 2.8–3.7 Hz, for the radial and transverse components of motion. The amplification of the vertical component is large at high frequencies (>4 Hz) whereas it is negligible in a lower frequency range.

The site response study in Ahmedabad city using the H/V spectral ratio and 1D shear velocity of subsurface soil using microtremor array observations applying $f-k$ and SPAC methods have been performed by Parvez and Mad-hukar (2006). The microtremor arrays around 150 different sites in the city have been used to record the ambient noise. On most of the sites, more than one array was deployed to capture the lower and upper geological properties. To obtain the first-order response of site characteristics, the most commonly used Nakamura (H/V ratio) technique has been adopted using the ambient noise recorded by an array of seven Lennartz 5-s seismometers. Most of the sites have shown a fundamental resonance frequency at 0.6 Hz. Very few sites have a peak frequency between 2 and 6 Hz; however, the first peak at 0.6 Hz is also explicit in these sites. This indicates that the thickness of the upper soft soil is very deep (350–400 m), which corresponds to the frequency of 0.6 Hz. Geologically, Ahmedabad city is sitting on thick Quaternary sediments, and there is no direct evidence of basement rock. The H/V spectral ratio confirms that most of the sites have a fundamental frequency of around 0.6 Hz without any sharp peak that means there is no high impedance contrast in geological structure beneath the top layer and it changes gradually as semiconsolidated material overlying the beds below 400 m. Figure 11.9(A) shows the spatial distribution of the fundamental frequencies in Ahmedabad city. The sites along the river and other water bodies (e.g., lakes) show the fundamental frequency above 2 Hz, which means that the new soft sediments are brought to the sites by rivers. However, the 0.6-Hz fundamental frequencies still dominate these sites. There is no one-to-one correlation with the damage that occurred in Ahmedabad consequent to the Bhuj earthquake in 2001 with the site response results obtained in this study. Most of the newly constructed apartment buildings with eight to nine floors were damaged mainly in areas to the west of the Sabarmati River. However, fundamental frequencies of such buildings do not fall in the 0.6-Hz ranges. There were clusters of damage in the Maninagar area with relatively low-rise apartments (four to five floors) that lie between

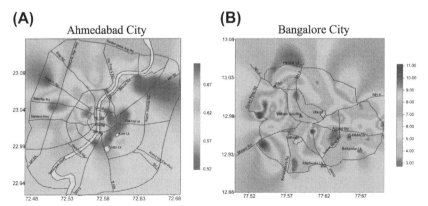

FIGURE 11.9 **Spatial distribution of fundamental frequencies in (A) Ahmedabad city and (B) Bangalore city.**

the Chandola and Kakaria Lake. This can be treated as damage due to site effects as in this region a higher resonance frequency has also been obtained.

A similar study has also been performed in Bangalore city as shown in Figure 11.9(B). Bangalore is a highly developed and dense city and is a center of attraction for youth and many industries due to its expansion in industry, trade, and commercial value leading to a rapid growth of the city and large-scale urbanization. More than 90 noise measurements were performed covering the Bangalore Metropolis area of about 220 sq km. The fundamental frequencies (f_0) from the ambient noise H/V spectral ratio for each site were calculated, and they are found in the range of 1.2−14 Hz. It has also been found that the frequency in the south-western part is the highest compared with that in the south-eastern part (Figure 11.9(B)). Comparing them with the borehole data validated the results. The soil depth obtained from the borehole data varies from 6 to 34 m. The corresponding contour maps were produced so that the predominant frequency could be compared one to one with the overburden thickness inferred from various borehole locations. The resonant frequencies and soil thickness compare very well.

11.6 INFLUENCE OF MICROZONATION DATA ON RISK ASSESSMENT

As explained earlier, classic examples of the role of surface geology on seismic waves were first provided by the 1906 San Francisco (Borcherdt and Gibbs, 1976) and the 1985 Mexico City earthquakes and later illustrated during several other destructive earthquakes (e.g.,, Kobe (1995) or Bam (2003)). Nowadays, it is generally recognized that so Northridge (1994)ft soils have a major influence on seismic ground motion and are then a major consideration in seismic risk estimate. Typical cases are cities built partially on solid rock and partially on rivers or more generally on soft deposits. In such a case, some neighborhoods experience no amplification, while others do. Among all data needed for risk assessment, the implementation of micro-zonation data is one of the challenging tasks to achieve in loss estimation.

11.6.1 Involved Data in Loss Estimation

Seismic loss modeling needs specific data at the different steps of the chain of calculation (e.g., Jaiswal et al., 2011; Trendafiloski et al., 2011). The different involved data are briefly detailed in the next few paragraphs.

Strong ground motions damage buildings, sometimes bringing about collapse and killing people. Shaking of the ground decreases with distance from the release of the energy, the hypocenter, or, more accurately expressed, from the entire area of rupture. To calculate the intensity of shaking at a given settlement, the computer looks up the attenuation (decrease in amplitude) for seismic waves that travel the distance to the settlement in question. Locations

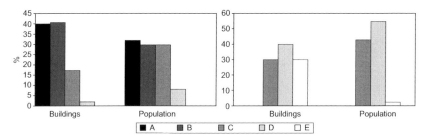

FIGURE 11.10   Examples of distributions for buildings and population in these buildings in developing (left) and industrialized countries (right). The building classes are those of the EMS98 scale with A being the weakest and F the strongest. The weak constructions of classes A and B are largely absent in the industrialized world. *(Source: Wikipedia, 2013).*

of the seismic source as well as the energy release are then the basics for loss estimation (Wyss, 2014). Wyss et al. (2011) discussed the influence of the epicenter errors on loss calculations.

For estimating losses to the built environment, one needs to calculate the damage expected for each type of building present in a given settlement. For each settlement, one needs to know the distribution of buildings into classes with different resistances to strong shaking. There are many scales for classifying building types, some considering the construction types as the Hazus scale and others grouping different construction types (buildings) into vulnerability classes as the European Macroseismic Scale (EMS98). It is obvious that the distribution of building types is different in industrialized and developing countries (Figure 11.10) and also in villages compared to those in cities in the same country. A uniform classification scheme is under review in the frame of the Global Earthquake Model initiative in order to create a unique description for a building or a building typology worldwide.

After one knows the distribution of buildings into classes, one needs to estimate how the population is distributed into these building types. These distributions are not identical because the higher quality houses tend to shelter more people per building. Consequently, the availability and reliability of the population data are the main factors that need to be complemented by information on the occupancy rate during a journey to integrate the occurrence time of the earthquake in the loss estimate.

The probability that a building of a given type may collapse if subjected to a certain intensity of shaking is the next important parameter for calculating expected human losses. The weak buildings that are present in developing countries are the ones that are likely to collapse at moderate intensities. The numbers of fatalities and injured people (casualties are the sum of these two parameters) are estimated, using a casualty matrix. Depending on the building type and damage degree, these tables contain the probabilities that a person in a particular building ends up dead, injured, or unscathed after damage to the building by an earthquake. The data in casualty matrixes are so poorly known that we cannot give uncertainties here.

To some extent, data related to critical facilities are also to be considered in the loss estimation because they could induce secondary effects such as fire, pollution, and flooding.

11.6.2 Case Histories at Regional and Local Scales

The influence of site amplification on losses is illustrated by two case studies using QLARM at regional and local scales. QLARM is an expert software tool to estimate losses (building damage, injured, and fatalities) due to earthquakes with the purposes to trigger rapid humanitarian response and to analyze the risk in the scenario or probabilistic mode. It has a global scope, with focus on developing countries. Each settlement of the QLARM database is populated with its name, coordinates, population, a distribution of building stock, and population by EMS98 vulnerability class. The soil amplification factor for different ground motion parameters is also included when available. Settlements (or city models) in the database are mostly defined as a single point or divided into districts when data regarding city subdivisions are available. In general, building and population distributions are defined for three different sizes of settlements; rural, midurban, and urban. QLARM calculates the mean damage for the different building classes and the range of casualties for each settlement around the epicenter (Trendafiloski et al., 2011). Wyss discusses the alert service provided with QLARM in this book.

Ground shaking intensity is calculated using IPE for the settlements of the database around the epicenter. A soil amplification factor is then applied when available to intensity values that are directly obtained from intensity-related mapping or indirectly derived from PGA, PSA(T), or Vs_{30}-related mapping.

At the regional scale, a database was built to estimate human losses in Algeria for seismic scenarios replicating three major historical earthquakes (Rosset et al., 2009). Six major and recent earthquakes with available intensity and fatalities data (Mw ranges from 5.4 to 5.9) are used to calibrate IPEs and building distribution (Rosset et al., 2010). We then applied the proxy method based on the relationships between topographic slope and Vs_{30} (Allen and Wald, 2007) to estimate soil amplification since no local data existed. For each settlement of the database, Vs_{30} is averaged using the values of the regular grid calculated for the country within a circled area of search. The latter is proportional to the population of the settlement considering an average density of 22,000 in/km^2 estimated for a sample of cities on satellite images.

The map in Figure 11.11 shows the soil class estimated for each city in Algeria following the NEHRP (1994) classification. More than 60 percent of the settlements are classified as soil type D (stiff soil), and others range from soft to hard rock. Soil classes were converted to additional intensity values using the amplification factor given by the NEHRP (1994) and the ground

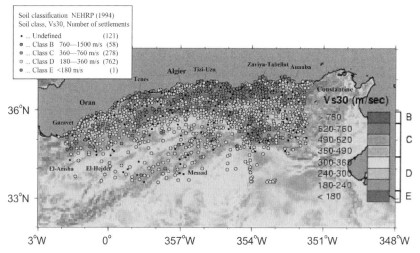

FIGURE 11.11 Estimated soil classes using Vs_{30} derived from topographic slope. Each colored dot corresponds to a settlement of the database and is ranked following the soil classification of the NEHRP (1994). Undefined class corresponds to class B and is given to settlements where no values of Vs_{30} exist within the search area. Background map and Vs_{30} data are downloaded from the Global Vs30 Map Server (http://earthquake.usgs.gov/hazards/apps/vs30/).

motion-intensity conversion equations of Wald et al. (1999). The added intensity values are 0, 0.16, and 0.32 for soil types B, C, and D, respectively.

Three past major earthquakes were selected as deterministic scenarios. Table 11.2 shows the calculated intensities and the ranges of killed and injured people for the three scenarios during the day and the night (Rosset et al., 2010). If soil amplification based on average Vs_{30} is included in the calculation, the increase of fatalities is 40 percent, 10 percent, and 20 percent for Algiers, Oran, and Djidjelli scenarios, respectively. Values for injuries increasing are 40 percent, 15 percent, and 20 percent.

The map in Figure 11.12 shows the mean building damage to each settlement in the case of a repeat of the M7.5 earthquake of Oran (1790) during the day (Rosset et al., 2010).

At the regional scale, it is sufficient to consider the entire population concentrated in a single point of latitude and longitude in order to get stable results for the damage estimate and the human losses. Some buildings may collapse because they are built on poor soils, while others of the same type remain standing, and we do not have the detailed information that would be necessary to calculate losses for individual houses.

For large cities located in seismogenic areas, a world database of city models with subdivisions is under construction in QLARM because, at this scale, the distribution of soft soils could play a major role in the final estimates of losses. To date, it concerns 48 cities around the world. The influence of site effects has been investigated in Bucharest (Trendafiloski et al., 2009) and Lima. The city model for

TABLE 11.2 QLARM Loss Estimate for three Scenarios Repeating Historical Earthquakes. Estimated Fatalities and Injuries are Given With and Without Site Amplification

Scenario	Algier		Oran		Djidjelli	
Latitude/longitude depth/Mw	36.67N/2.95E 15 km/7.0		35.7N/0.64W 15 km/7.5		36.70N/6.08E 10 km/7.3	
Time of the day	Night	Day	Night	Day	Night	Day
Intensity	IX		X		X	
Fatalities (in thousands)	8.4–16.3	5.3–10.2	17.1–26.3	10.7–16.4	5.9–9.7	3.70–6.0
Injuries (in thousands)	37.3–69.3	23.3–43.3	45.2–64.4	25.0–41.8	22.8–36.9	14.2–23.0
Fatality with site amplification	+40 percent		+10 percent		+20 percent	
Injury with site amplification	+40 percent		+15 percent		+20 percent	

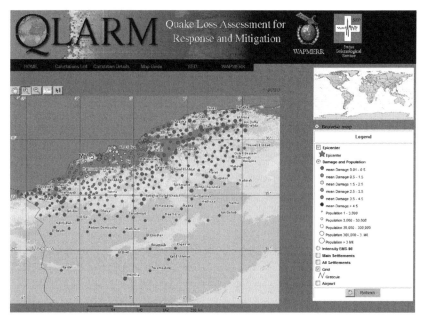

FIGURE 11.12   QLARM mapping tool in the case of a repeating Oran, 1790 earthquake (Mw = 7.5) during the day. Each dot is a settlement. The size of the dots is proportional to its population. The colors represent the mean damage degree to the different classes of buildings. The red star locates the epicenter.

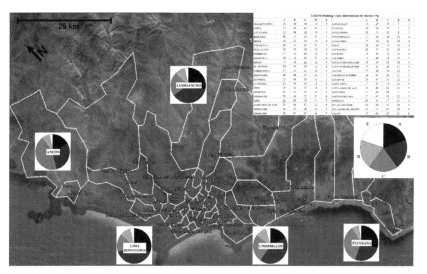

FIGURE 11.13   Building distribution for the city model of Lima metropolitana. A building distribution using the EMS98 classes (A—E) is proposed for each of the 42 districts as shown in the table and for a few of them in the pie.

FIGURE 11.14   Microzonation of Lima metropolitana. The urban area is divided into colored zones corresponding to four types of soil; S1 for rock, S2 intermediate soils, S3 flexible soils, and S4 for exceptional conditions (map from CISMID, 2004). The table in the top-right side lists out the proposed additional intensity value ΔI based on the preponderant soil type in each district.

Lima Metropolitana is divided into 43 administrative districts where a specific building and population distribution is provided (Figure 11.13).

The amplification factor is derived from a microzonation map proposed by CISMID (2004), which identifies four types of soils (S1–S4). For each district, a preponderant soil type is considered based on the relative extension of its urbanized area. The attributed soil factor to PGA given by the seismic code of Peru (NTE-E30, 2003) is then converted to an additional intensity value. It is 0 for rock and soil type S1, 0.3 for S2 (intermediate soils), and 0.5 for S3 (flexible soils). An intermediate value 0.15 is given when soils S1 and S2 equally cover a district. Figure 11.14 shows the distribution of the soil types and the attributed increase in intensity by districts.

Several scenarios based on this city model were proposed by Wyss et al. (2009) and more specifically an M8, Lima offshore scenario at 20 km of the coast (77.2W, 12.18S, 33 km). On average, fatalities are around 10,000 and injuries 2.8 times more. Those values increase by 15 percent if soil amplification is considered. In this case, 58 percent of the population is affected by the increase of the intensity value due to soft soil.

The purpose of these two examples is not to estimate the influence of microzonation data on the loss estimate but more to illustrate the approach applied in QLARM to include it in the city models. Nevertheless, the results indicate that site conditions should not be neglected in urban areas of important seismic zones worldwide since it could increase the casualties by ≥25 percent.

11.7 CONCLUSIONS

The purpose of seismic zonation and microzonation is to identify zones where the last meters of soil could affect the amplitude of ground shaking. This information is crucial since it contributes locally to the seismic hazard.

Various numerical and experimental methods express the local soil response to ground shaking. Discussions among scientists on the best parameter or parameters to define site conditions are not closed. Vs_{30} values that tend to be commonly used in national building codes and recent GPMEs are one of them, but recent studies indicate that it could misestimate site amplification (e.g., Luzi et al., 2011; Castellero et al., 2008). We presented few other parameters that help in defining site effect.

For example, the advanced study aimed at NDSHA allow for the development of a set of regional and local scale maps of the expected ground motion at bedrock and surface level, based on the physical modeling of ground shaking from a wide set of possible scenario earthquakes. These maps can be defined in terms of peak ground displacement, velocity, acceleration, or any other ground motion parameter that can be extracted from the set of complete synthetic seismograms describing ground shaking at the different sites. Recently, displacement-based and energy approaches have made significant advances. The former approach is based on the assumption that structural and nonstructural damage can be reduced by limiting displacement demands; the displacement response spectrum, as the input, is of great significance to displacement-based design just as the acceleration response spectrum is to the traditional force-based design. The latter approach, beyond the potentiality of designing earthquake-resistant structures by balancing energy demands and supplies, allows us to properly characterize the different types of time histories (impulsive, periodic or with long duration pulses) that may correspond to an earthquake ground shaking, simultaneously considering the dynamic response of a structure.

Earthquake risk assessment requires various types of data that are often disconnected from microzonation ones because their implementation remains tricky. Indeed, it is not straightforward to use existing microzonation data in seismic scenarios because relationships to derive amplification factor from parameters used to define site conditions are missing. Research in this direction should improve the estimate in the future.

REFERENCES

Akamatsu, J., Fujita, M., Kameda, H., 1991. Long-period (1−10 s) microtremor measurement in the areas affected by the 1989 Loma Prieta earthquake. In: Proc. 4th International Conference on Seismic Zonation, vol. 3. Earthquake Eng. Res. Inst, Stanford, California, pp. 393−399.

Aki, K., 1957. Space and time spectra of stationary stochastic waves, with special reference to microtremors. Bull. Earthquake Res. Inst. 35, 415−456.

Aki, K., 1987. Strong motion seismology. In: Erdik, M., Toksoz, M. (Eds.), Strong Ground Motion Seismology, NATO ASI Series C, Mathematical and Physical Science, D, vol. 204. Reidel Publishing Company, Dordrecht, pp. 3−39.

Allen, T.I., Wald, D.J., 2007. Topographic Slope as a Proxy for Global Seismic Site Conditions (VS30) and Amplification around the Globe, U.S. Geological Survey, Open File Report 2007-1357, 69 pp.

Asten, M.W., Henstridge, J.D., 1984. Array estimators and the use of microseisms for reconnaissance of sedimentary basins. Geophysics 49, 1828−1837.

Atkinson, G.M., Boore, D.M., 2006. Earthquake ground motion prediction equations for eastern America. Bull. Seism. Soc. Am. 96, 2181−2205.

Boore, D.M., Atkinson, G.M., 2008. Ground-motion prediction equations for the average horizontal component of PGA, PGV, and 5%-damped PSA at spectral periods between 0.01 s and 10.0 s. Earthq. Spectra 24 (1), 99−138.

Borcherdt, R.D., Gibbs, J.F., 1976. Effects of local geological conditions in the San Francisco Bay region on ground motions and the intensities of the 1906 earthquake. Bull. Seism. Soc. Am. 66, 467−500.

Borcherdt, R.D., 1970. Effects of local geology on ground motion near San Francisco Bay. Bull. Seism. Soc. Am. 60, 29−61.

Brandmayr, et al., 2010. The lithosphere in Italy: structure and seismicity. In: Beltrando, Marco, Peccerillo, Angelo, Mattei, Massimo, Conticelli, Sandro, Doglioni, Carlo (Eds.), Journal of the Virtual Explorer, vol. 36. paper 1, doi: 10.3809/jvirtex.2010.00224.

Broughton, A.T., Van Arsdale, R.B., Broughton, J.H., 2001. Liquefaction susceptibility mapping in the city of Memphis and Shelby County, Tennessee. Eng. Geo. 62, 201−222.

Bour, M., Fouissac, D., Dominique, P., Martin, C., 1998. On the use of microtremor recordings in seismic Microzonation, Soil Dyn. Earthquake Eng. 17, 465−474.

Castellaro, S., Mulargia, F., Rossi, P.L., 2008. V_{S30}: proxy for seismic amplification? Seismol. Res. Lett. 79, 540−543.

Castro, G., 1987. On the behaviour of soils during earthquakes - liquefaction, Soil dynamics and liquefaction. In: Cakmak, A.S. (Ed.). Elsevier Science Pub., Amsterdam, pp. 169−204.

CISMID, 2004. Microzonificación Sísmica de Lima. In: Aguilar Bardales, Z., Alva Hurtado, J.E. (Eds.) http://www.cismid.uni.edu.pe/descargas/redacis/redacis32_p.pdf, p. 55.

Cornell, C.A., 1968. Engineering seismic risk analysis. Bull. Seism. Soc. Am. 58 (5), 1583−1606.

Costa, G., Panza, G.F., Suhadolc, P., Vaccari, F., 1993. Zoning of the Italian territory in terms of expected peak ground acceleration derived from complete synthetic seismograms. In: Cassinis, R., Helbig, K., Panza, G.F. (Eds.), Geophysical Exploration in Areas of Complex Geology, II, J. Appl. Geophys, 30, pp. 149−160.

Kiureghian, Der, Ang, A., 1977. A fault rupture model for seismic risk analysis. Bull. Seismol. Soc. Am. 67 (4), 1173−1194.

Dravinski, M., Yamanaka, H., Nakajima, Y., Kagami, H., et al., 1991. Observation of long period microtremors in San Francisco metropolitan area. In: Proc. 4th International Conference on Seismic Zonation, vol. 3. Earthquake Eng. Res. Inst., Stanford, California, 401M07.

Emmerich, H., 1992. PSV-wave propagation in a medium with local heterogeneities: A hybrid formulation and its application, Geophys.AuthorAnonymous. J. Int. 109, 54−64.

Emmerich, H., Korn, M., 1987. Incorporation of attenuation into time-domain computations of seismic wave fields. Geophysics 52, 1252−1264.

Fäh, D., Suhadolc, P., Panza, G.F., 1990. Estimation of strong ground motion in laterally heterogeneous media: modal summation—finite differences. In: Proc. of the 9th European Conference of Earthquake Engineering, Moscow, USSR, 11−16 September 1990, Vol. 4A, pp. 100−109.

Fäh, D., Suhadolc, P., Panza, G.F., 1993a. Variability of seismic ground motion in complex media: the Friuli area (Italy). In: Cassinis, R., Helbig, K., Panza, G.F. (Eds.), Geophysical Exploration in Areas of Complex Geology, II, J. Appl. Geophys., vol. 30, pp. 131−148.

Fäh, D., Iodice, C., Suadolc, P., Panza, G.F., 1993b. A new method for the realistic estimation of seismic ground motion in megacities: the case of Rome. Earthq. Spectra 9, 643−668.

Field, E., Jacob, K., 1993. The theoretical response of sedimentary layers to ambient seismic noise. Geophys. Res. Lett. 20, 2925−2928.

Field, E.H., Hough, S.E., Jacob, K.H., 1990. Using microtremors to assess potential earthquake site response: a case study in Flushing Meadows, New York City. Bull. Seism. Soc. Am. 80, 1456−1480.

Field, E.H., and the SCEC Phase III Working Group, 2000. Accounting for site effects in probabilistic seismic hazard analyses of Southern California: overview of the SCEC Phase III Report. Bull. Seism. Soc. Am. 90, S1−S31.

Finn, W.D.L., Onur, T., Ventura, C.E., 2004. Microzonation: developments and applications. In: Ansal, A. (Ed.), Recent advances in earthquake geotechnical engineering and microzonation. Kluwer Academic Publishers, Dordrecht, the Netherlands, pp. 3−26.

Florsch, N., Fäh, D., Suhadolc, P., Panza, G.F., 1991. Complete synthetic seismograms for high-frequency multimode SH-waves. Pure Appl. Geoph. 136, 529−560.

Gusev, A.A., 1983. Descriptive statistical model of earthquake source radiation and its application to an estimation of short period strong motion. Geophys. J.R. Astr. Soc. 74, 787−808.

Henstridge, J.D., 1979. A signal processing method for circular arrays: Geophysics 44, 179−184.

Ishihara, K., 1993. Liquefaction and flow failure during earthquakes. Geotechnique 43 (3), 351−451.

Ishihara, K., Verdugo, R., Acacio, A.A., 1991. Characterization of cyclic behaviour of sand and post-seismic stability analyses. Proc., 9th Asian Reg. Conf. Soil Mech. Found. Engrg. vol. 2., 45−67. Bangkok, Thailand, December.

Iyengar, R.N., 2000. Seismic status of Delhi megacity. Curr. Sci. 78 (5), 568−574.

Jasiwal, K.S., Wald, D.J., Earle, P.S., Porter, K.A., Hearne, M., 2011. Earthquake casualty models within the USGS prompt assessment of global earthquakes for response (PAGER) system. In: Spence, R., So, E., Scawthornp, C. (Eds.), Human Casualties in Natural Disasters: Progress in Modeling and Mitigation. Springer, Dordrecht, Heidelberg, London, New York, pp. 83−94.

Jensen, V.H., 2000. Seismic microzonation in Australia. J. Asian Earth Sci. 18, 3−15.

Kobayashi, H., Set, K., Midorikawa, S., Samano, T., Yamazaki, Y., 1991. Seismic microzoning study of Mexico City by means of microtremor measurements. In: Proc. 4th International Conference on Seismic Zonation, vol. 3. Earthquake Eng. Res. Inst, Stanford, California, pp. 557−564.

Kramer, S.L., 1996. Geotechnical Earthquake Engineering. Prentice Hall, New Jersey.

Kudo, K., Kanno, T., et al., 2002. Site-specific issues for strong motions during the Kocaeli, Turkey earthquake of 17 August 1999, as inferred from Array observations of microotremors and aftershocks. Bull. Seism. Soc. Am. 92, 448−465.

Lachet, C., Bard, P., 2−7 April 1995. Theoretical Investigations on the Nakamura's technique. In: Proceedings of the Third International Conference on Recent Advances in Geotechnical Earthquake Engineering and Soil Dynamics, vol. II. St. Louis Missouri.

Lermo, J., Chavez-Garcia, F.J., 1994. Are microtremors useful in site response evaluation? Bull. Seism. Soc. Am. 84, 1350−1364.

Lermo, J., Rodriguez, M., Singh, S.K., 1988. The Mexico earthquake of September 19, 1985−natural period of sites in the Valley of Mexico from microtremor measurements and strong motion data. Earthq. Spectra 4, 805−814.

Levander, A.R., 1988. Fourth-order finite difference P-SV seismograms. Geophysics 53 (11), 1425−1436.

Luzi, L., Puglia, R., Pacor, F., Gallipoli, M., et al., 2011. Proposal for a soil classification based on parameters alternative or complementary to V_{S30}. Bull. Earthq. Eng. 9 (6), 1877−1898.

Matsushima, T., Okada, H., 1990. Determination of deep geological structures under urban areas. Butsuri Tansa 43, 21−33.

McGuire, R.K., 1974. Seismic structural response risk analysis, incorporating peak response regressions on earthquake magnitude and distance. Res. Report R74-51, MIT, Department of Civil Engineering.

McGuire, R.K., 2001. Deterministic vs. probabilistic earthquake hazards and risks. Soil Dyn. Earthq. Eng. 21, 377−384.

Meghraoui, M., Atakan, K., 2014. The Contribution of Paleoseismology to Earthquake Hazard Evaluations. In: Shroder, J., Wyss, M. (Eds.), Earthquake Hazard, Risk, and Disasters. Elsevier, London, pp. 237−271.

Milne, J., 1898. Seismology, first ed. Kegan Paul, Trench, Trube, London.

Miyakoshi, K., Okada, H., Ling, S., 1996. Maximum wavelength possible to estimate phase velocities of surface waves in microtremors. In: Proc. 94th SEGJ Conf. Society of Exploration Geophysicists, Japan, pp. 178−182.

Morales, J., Vidal, F., Pena, J.A., et al., 1991. Microtremor study in the sediment-filled basin of Zafarraya, Granada (Southern Spain). Bull. Seism. Soc. Am. 81, 687−693.

Nakamura, Y., 1989. A method for dynamic characteristics estimation of subsurface using microtremor on the ground surface. Q. Rep. Railway Tech. Res. Inst. 30 (1), 25−33.

Newmark, N.M., Hall, W.J., 1982. Earthquake spectra and design. Earthquake Engineering Research Institute, Berkeley, Calif.

Okada, H., 1998. Microtremors as an exploration method, geo-exploration handbook. Soc. Explor. Geophys. Jpn. 2, 203−211.

Okada, H., Matsushima, T., Moriya, T., Sasatani, T., 1990. An exploration technique using long-period microtremors for determination of deep geological structures under urbanized areas. Butsuri Tansa 43, 402−417.

Panza, G.F., Suhadolc, P., 1987. Complete strong motion synthetics. In: Bolt, B.A. (Ed.), Seismic Strong Motion Synthetics, Computational Techniques 4. Academic Press, Orlando, pp. 153−204.

Panza, G.F., 1985. Synthetic seismograms: the Rayleigh waves model summation. J. Geophys. 58, 125−145.

Panza, G.F., Romanelli, F., Vaccari, F., 2001. Seismic wave propagation in laterally heterogeneous anelastic media: theory and application to seismic zonation. In: Dmowska, R., Saltzman, B. (Eds.), Advances in Geophysics, vol. 43. Academic Press, San Diego, USA, pp. 1−95.

Panza, G.F., Vaccari, F., Costa, G., Suhadolc, P., Faeh, D., 1996. Seismic input modelling for zoning and microzoning. Earthq. Spectra 12, 529−566.

Parvez, I.A., Vaccari, F., Panza, G.F., 2003. A deterministic seismic hazard map of India and adjacent areas. Geophys. J. Int. 155, 489−508.

Parvez, I.A., Vaccari, F., Panza, G.F., 2004. Site-specific microzonation study in Delhi metropolitan city by 2-D modelling of SH and P-SV waves. Pure Appl. Geoph. 161, 1165−1184.

Parvez, I.A., Gusev, A.A., Panza, G.F., Petukhin, A.G., 2001. Preliminary determination of the interdependence among strong motion amplitude, earthquake magnitude and hypocentral distance for the Himalayan region. Geophys. J. Int. 144, 577−596.

Parvez, I.A., Panza, G.F., Gusev, A.A., Vaccari, F., 2002. Strong-motion amplitudes in Himalayas and a pilot study for the deterministic first-order microzonation in a part of Delhi city. Curr. Sci. 82, 158−166.

Parvez, I.A., Madhukar, K., 2006. Site response in Ahmedabad city using microtremor array observation: a preliminary report, CSIR C-MMACS Report PD CM 602.

Peresan, A., Zucolo, E., Vaccari, F., Gorshkov, A., Panza, G.F., 2011. Neo-Deterministic seismic hazard and pattern recognition techniques: time dependent scenarios for north-eastern Italy. Pure Appl. Geoph. 168, 583−607.

Pezeshk, S., Zandieh, A., Tavakoli, B., 2011. Hybrid empirical ground motion prediction equations for eastern north America using NGA models and updated seismological parameters. Bull. Seism. Soc. Am. 101 (4), 1859−1870.

Poulos, S.J., Castro, G., France, W., 1985. Liquefaction evaluation procedure. Journal of Geotechnical Engineering Division. ASCE 1ll (6), 772–792.

Rosset, P., Trendafiloski, G., Wyss, M., 2010. Project ALGER: Capacity Building in Earthquake Surveillance and Rapid Information for Algeria. Internal Final Report, WAPMERR, 22 pp.

Rosset, P., Trendafiloski, G., Yelles, K., Semmane, F., Wyss, M., 2009. Application of the loss estimation tool QLARM in Algeria. Geophys. Res. Abstr. 11 (EGU2009-9167, 2009, EGU General Assembly 2009).

Seed, H.B., Idriss, I.M., 1982. Ground Motion and Soil Liquefaction during Earthquakes. Earthquake Engineering Research Institute Monograph, Oakland, Calif.

Seed, H.B., 1979. Soil liquefaction and cyclic mobility evaluation for level ground during earthquakes. J. Geotech. Eng. Div. ASCE 105 (2), 201–255.

Stirling, M.W, 2014. The Continued Utility of Probabilistic Seismic Hazard Assessment. In: Shroder, J., Wyss, M. (Eds.), Earthquake Hazard, Risk, and Disasters. Elsevier, London, pp. 359–376.

Teves-Costa, P., Matias, L., Bard, P.Y., 1996. Seismic behaviour estimation of thin alluvium layers using microtremor recordings. Soil Dyn. Earthq. Eng. 15, 201–209.

Tobin, G.A., 1997. Natural Hazards: Explanation and Integration. Guilford Press.

Tolis, S.V., 2014. To what extent can engineering reduce seismic risk? In: Shroder, J., Wyss, M. (Eds.), Earthquake Hazard, Risk, and Disasters. Elsevier, London, pp. 531–541.

Trendafiloski, G., Wyss, M., Rosset, P., Marmureanu, G., 2009. Constructing city models to estimate losses due to earthquakes worldwide: application to Bucharest, Romania. Earthq. Spectra 25 (3), 665–685.

Trendafiloski, G., Wyss, M., Rosset, P., 2011. Loss estimation module in the second generation software QLARM. In: Spence, R., So, E., Scawthornp, C. (Eds.), Human Casualties in Natural Disasters: Progress in Modeling and Mitigation. Springer, Dordrecht, Heidelberg, London, New York, pp. 95–104.

Virieux, J., 1984. SH-wave propagation in heterogeneous media: velocity–stress finite-difference method. Geophysics 49, 1933–1957.

Virieux, J., 1986. P-SV wave propagation in heterogeneous media: velocity–stress finite-difference method. Geophysics 51, 889–901.

Wald, D.J., Quitoriano, V., Heaton, T.H., Kanamori, H., 1999. Relationship between peak ground acceleration, peak ground velocity, and modified Mercalli intensity in California. Earthq. Spectra 15 (3), 557–564.

Wikipedia, 2013. Loss Estimates in Real Time for Earthquakes Worldwide. http://en.wikipedia. org/wiki/Loss_estimates_in_real_time_for_earthquakes_worldwide.

Wyss, M., Elashvili, M., Jorjiashvili, N., Javakhishvili, Z., 2011. Uncertainties in teleseismic epicenter estimates: implications for real-time loss estimates. Bull. Seism. Soc. Am. 101 (3), 1152–1161.

Wyss, M., Trendafiloski, G., Rosset, P., Wyss, B., 2009. Preliminary Loss Estimates for Possible Future Earthquake Near Lima. WAPMERR Report, 65 pp.

Wyss, M., 2014. Ten years of real-time earthquake loss alerts. In: Shroder, J., Wyss, M. (Eds.), Earthquake Hazard, Risk, and Disasters. Elsevier, London, pp. 143–165.

Yamanaka, H., Sato, H., Kurita, K., Seo, K., 1999. Array measurements of long-period microtremors in southwestern Kanto plain, Japan, − Estimation of S-wave profiles in Kawasaki and Yokohama cities. J. Seism. Soc. Jpn. 51, 355–365.

Youd, T.L., Perkins, D.M., 1978. Mapping of liquefaction induced ground failure potential. J. Geotech. Eng. Div. 104, 433–446.

Chapter 12

Why are the Standard Probabilistic Methods of Estimating Seismic Hazard and Risks Too Often Wrong

Giuliano Panza [1,2,3,6], Vladimir G. Kossobokov [2,4,5,6], Antonella Peresan [1,2,6] and Anastasia Nekrasova [2,4]

[1] Department of Geosciences, University of Trieste, Trieste, Italy, [2] The Abdus Salam International Centre for Theoretical Physics — SAND Group, Trieste, Italy, [3] China Earthquake Administration, Institute of Geophysics, Beijing, China, [4] Institute of Earthquake Prediction Theory and Mathematical Geophysics, Russian Academy of Sciences, Moscow, Russian Federation, [5] Institut de Physique du Globe de Paris, France, [6] International Seismic Safety Organization, ISSO

Ne quid falsi dicere audeat, ne quid veri non audeat

De oratore II, 15, 62 (Cic)

ABSTRACT

According to the probabilistic seismic hazard analysis (PSHA) approach, the deterministically evaluated or historically defined largest credible earthquakes (often referred to as Maximum Credible Earthquakes, MCEs) are "an unconvincing possibility" and are treated as "likely impossibilities" within individual seismic zones. However, globally over the last decade such events keep occurring where PSHA predicted seismic hazard to be low. Systematic comparison of the observed ground shaking with the expected one reported by the Global Seismic Hazard Assessment Program (GSHAP) maps discloses gross underestimation worldwide. Several inconsistencies with available observation are found also for national scale PSHA maps (including Italy), developed using updated data sets. As a result, the expected numbers of fatalities in recent disastrous earthquakes have been underestimated by these maps by approximately two to three orders of magnitude. The total death toll in 2000—2011 (which exceeds 700,000 people, including tsunami victims) calls for a critical reappraisal of GSHAP results, as well as of the underlying methods.

In this chapter, we discuss the limits in the formulation and use of PSHA, addressing some theoretical and practical issues of seismic hazard assessment, which range from the overly simplified assumption that one could reduce the tensor problem of seismic-wave generation and propagation into a scalar problem (as implied by ground motion

Earthquake Hazard, Risk, and Disasters. http://dx.doi.org/10.1016/B978-0-12-394848-9.00012-2

309

prediction equations), to the insufficient size and quality of earthquake catalogs for a reliable probability modeling at the local scale. Specific case studies are discussed, which may help to better understand the practical relevance of the mentioned issues. The aim is to present a critical overview of different approaches, analyses, and observations in order to provide the readers with some general considerations and constructive ideas toward improved seismic hazard and effective risk assessment. Specifically, we show that seismic hazard analysis based on credible scenarios for real earthquakes, defined as neo-deterministic seismic hazard analysis, provides a robust alternative approach for seismic hazard and risk assessment. Therefore, it should be extensively tested as a suitable method for formulating scientifically sound and realistic public policy and building code practices.

12.1 INTRODUCTION

A simple answer exists to the question in the title of this chapter: most, if not all, the standard probabilistic methods to assess seismic hazard, namely PSHA, and associated risks are based on subjective, commonly unrealistic, and even erroneous assumptions about seismic recurrence. The explosion of hazard papers, particularly in the last 25–30 years, can be correlated to unrestrained promotion of the PSHA approach with its required time-dependent recurrent earthquake inputs. This requirement generated various research projects for justified funding to study faults, local and regional seismicity, crustal deformation, and uncertainties of these parameters.

After years with many publications, we know that recurrent earthquake hazard results have failed us. Acceleration estimates based on PSHA were exceeded in several recent earthquakes with disastrous consequences. Seismic hazard assessment (SHA) can be done better without overcomplication and using economic and realistically robust assumptions by deterministic seismic hazard analysis (DSHA) or in its modern enhanced variant, the neo-deterministic seismic hazard analysis (NDSHA). Regretfully, most of the state-of-the-art theoretical approaches to assess probability of seismic events are based on too simplistic (e.g., Poisson or periodic occurrence) or, conversely, delicately designed (e.g., STEP and ETAS) models of earthquake sequences. Some of these models are inappropriate in our view, some can be rejected by the existing statistics, and others are hardly testable in our lifetime. In the early history of instrumental seismology some subjective assumptions were necessary due to setup of a too few seismographic stations delivered sparse observations of seismic activity. Some of these assumptions are no longer necessary nowadays, when archives of comprehensive seismic monitoring, from the global to regional scales and lasting for decades, are available for data mining, which exposes the complexity of earthquake recurrence.

Nevertheless, probabilistic analyses, including SHA and earthquake forecasting, based on assumptions that are unproven, keep being widely advertised and used in practical applications. These analyses can lead to inaccurate advice to decision makers and the population at risk, who may take inappropriate seismic safety decisions. As a result, the population of seismic regions continues facing unexpected risks and losses. It is enough to mention (Kossobokov and Nekrasova, 2012) that all 12 of the deadliest earthquakes that happened during 2000−2011 (total number of deaths exceeding 700,000, including tsunami victims) were unexpected "surprises" considering the maps of the Global Seismic Hazard Assessment Program (GSHAP) (Giardini et al., 1999) endorsed in the framework of the United Nations International Decade for Natural Disaster Reduction (UN/IDNDR). This result shows that the GSHAP map is not suitable for critical risk assessments intended to reduce disasters caused by earthquakes. Moreover it points to the limitations in the underlying methods that have been analyzed in the literature during the last two decades and are discussed in this chapter.

Comparison of observed numbers of fatalities with those calculated by the QLARM program (Trendafiloski, et al., 2011) for the reported magnitude and the magnitude implied by GSHAP (Wyss et al., 2012), indicates that the numbers of fatalities in recent disastrous earthquakes were underestimated by the world seismic hazard maps by approximately two to three orders of magnitude. Thus, seismic hazard maps based on the standard probabilistic method: (1) cannot be used to estimate the risk to which the population is exposed due to large earthquakes in many regions; and (2) earthquake mitigation measures in areas where large earthquakes are possible should not be based on GSHAP maps. It follows that the international project Global Earthquake Model (http://www.globalquakemodel.org/) is on the wrong track, if it continues to base seismic risk estimates on the standard probabilistic method to assess seismic hazard.

Some supporters of probabilistic SHA put forward some rather weak justifications for the PSHA approach, which are exemplified by the following quotation from Marzocchi (2013):

A final consideration derives from Italian experience. Some criticisms of PSHA are often based on the fallacy that earthquakes caused casualties because the seismic building code was wrong. This is not necessarily true. In Italy, to my knowledge, the last two earthquakes (L'Aquila, 2009, and Emilia, 2012) did not cause any casualties among people living in structures constructed according to the present building code.

No physics and no evidence occur in this paper that supports PSHA capability in anticipating ground shaking, but just generic considerations and opinions, shifting the focus from hazard mapping to risk mitigation issues. Suffice it to say that a number of buildings in the city of L'Aquila "constructed according to the present building code" did collapse on April 06, 2009. According to a comprehensive analysis of data on mortality in the 2009 L'Aquila earthquake (Alexander and Magni, 2013): "Sixteen deaths (7.9 percent) occurred in dispersed locations outside the city center, whereas the other 186 (92.1 percent, or 60.4 percent of the total mortality in the earthquake) were highly concentrated in central locations. Despite all that has been written about the relative safety of reinforced concrete frame (RC) buildings in comparison with unreinforced masonry buildings (URMB), 39 deaths (21.0 percent) occurred in the area of URMB and 147 (79 percent) in areas of newer RC buildings.

To understand this situation it needs to be borne in mind that in L'Aquila city center four areas had different seismic performance, and thus different patterns of mortality (Alexander, 2011). A similar observation applies in the case of May 20, 2012, M5.9 Emilia earthquake (Wyss, 2012) "The dismal performance of the many new industrial plants in Emilia Romagna, which collapsed, lost their roofs, or their walls, shows that regional building practices can remain hidden from the world community trying to estimate earthquake risk and lead to surprises and unnecessary fatalities." On the other hand, the relatively small number of victims in Emilia Romagna is connected with the occurrence time—very early morning—of the earthquake: on account of the many churches collapsed or severely damaged, an origin time in the late morning of the same Sunday might have caused hundreds of victims participating in religious ceremonies.

When advocating PSHA methods, Marzocchi (2013) also seems not acquainted with the significant underestimation, by the PSHA maps, of the ground shaking during the 2012 Emilia earthquake. According to Peresan et al. (2013) the PSHA map, forming the basis for the Italian building code, predicts peak ground acceleration (PGA) to be less than 0.175 times the acceleration of gravity (g), whereas the observed ground motion on May 20, 2012 at Emilia exceeded 0.25g. An in-depth engineering analysis carried out for specific structures in the Emilia region (Artioli et al., 2013) shows that the actual ground acceleration spectra was much higher than the one defined for a 475-year return period (i.e., having 10 percent probability of being exceeded in 50 years), and rather close to the spectra estimated for a 2475-year return period. The same authors conclude that the 2012 earthquakes ground shaking was seriously underestimated by design criteria indicated in the engineering norms for this geographical region.[1]

---

1. Chapman et al. (2014) described, an instructive case history of reevaluation of the hazard assessments at Diablo Canyon Nuclear Power Plant in central coastal California after the Hosgri fault zone was discovered 5 km offshore from the plant. It is notable that despite the U.S. Nuclear Regulatory Commission was motivated to move toward more probabilistic approaches, the U.S. Geological Survey recommended to follow a deterministic one by postulating that a magnitude 7.5 earthquake may occur anywhere along the Hosgri fault zone, including the point closest to the plant. In agreement with this recommendation the Diablo Canyon NPP was strengthened for the postulated Hosgri event. Evidently, a probabilistic fault-displacement hazard analysis suggested, "to explore 'what-if' scenarios. Incorporating expert, evidence-based judgments on the likelihood of movement and possible magnitudes of displacement" is similar to NDSHA, which can supply information also about recurrences without biasing hazard estimation (Peresan et al., 2013).

This kind of PSHA "surprises" is common worldwide. For example, the recent July 21, 2013 M6.0 Gansu (China) earthquake produced ground acceleration of 0.3g while the GSHAP map of PGA predicts just 0.11g, with an underestimation factor of about 3, a similar degree of underestimation as that for six additional earthquakes for which accelerations were measured instrumentally (Zuccolo et al., 2011). The same can be stated about its M5.6 aftershock on July 22, 2013 for which the underestimate was a factor of about 2.3.

In this chapter, we consider: (1) theoretical and practical limitations of PSHA; (2) provide short descriptions of alternative methods of SHA and simple tests of validation by comparison with real seismic events; (3) demonstrate how errors in seismic hazard estimates propagate into assessment of expected risks; and (4) exemplify a multitude of seismic risks for the same hazard and population due to variable vulnerability.

12.2 THEORETICAL LIMITS OF PSHA

The typical seismic hazard problem lies in the estimate of the ground motion characteristics associated with future earthquakes, both at regional and local scale. The first scientific and technical methods developed for SHA were deterministic and based on the observation that damage distribution is commonly correlated with the spatial distribution of sismicity and the physical properties of the underlying soil. The 1970s saw the beginning of the development of probabilistic seismic hazard maps on a national, regional, and urban (microzoning) scale. In the 1990s these instruments for the mitigation of seismic hazard came to prevail over deterministic cartography.

The classical PSHA (Cornell, 1968) determines the probability of exceeding, over a specified period of time, various levels of ground motion. The main elements of a PSHA are: (1) the seismic sources (e.g. the seismogenic zones) for which the seismogenic process is frequently assumed to be rather uniform; (2) the characteristics of the earthquakes recurrence associated with seismic sources, which is assumed to be Poissonian; and (3) the attenuation relations, which provide estimates of ground motion parameters (usually PGA) at different distances from the sources. The estimate of PGA is achieved by calculating the probability of occurrence of earthquakes with some damaging potential and the calculation of the conditional probability of exceedance of a given ground motion level, for each of these contributing earthquakes (summed over all potentially contributing sources, as shown in Figure13.1 of Stirling, 2014). Thus, PSHA aims at the statistical characterization of ground motion at a site, although, at most of the sites the available data are not sufficient to verify the assumptions nor to adequately constrain the parameters of the statistical model. Since the 1970s many variants of the original algorithm by Cornell (1968) have been proposed (e.g., McGuire, 1995; Bazzurro and Cornell, 1999), but they all basically follow the same underlying scheme, as reported in Anderson et al. (2010). Also, there are differences between standard PSHA maps at regional scale and PSH analysis carried out for specific sites (e.g., SSHAC, 1997; Stepp et al., 2001; Renault et al., 2010); the latter can be more detailed, though not necessarily more reliable (e.g. Klügel, 2005).

Although the seismic zonation adopted in many countries, either on a national or a regional scale, has been defined according to the classical probabilistic approach (Bommer and Abrahamson, 2006, and references therein), PSHA is controversial and the subject of long-standing debate (e.g., Castaños and Lomnitz, 2002; Panza et al., 2011a) that points to errors in the mathematics and physical assumptions of PSHA (Wang, 2011; Panza et al., 2013). Specifically, probabilistic, seismic hazard maps are: (1) strongly dependent on the available observations, unavoidably incomplete due to the long timescales involved in geological processes leading to the occurrence of a strong earthquake; (2) do not adequately consider the source and site effects, since they resort to linear convolutive techniques, e.g., empirical ground motion prediction equation (GMPE) (e.g., Boore and Atkinson, 2008), which cannot be applied when dealing with complex geological structures, because the ground motion generated by an earthquake can be formally described as the tensor product of the earthquake source tensor with the Green function of the medium (Aki and Richards, 2002); (3) do not properly consider the temporal properties of seismicity, being based on the assumption of memoryless random occurrence of earthquakes (Bilham, 2009).

In spite of the numerous theoretical shortcomings (see Klügel (2011) and Wang and Cobb (2012) for a detailed discussion) and of its poor performance, PSHA is still widely applied at regional and global scales (e.g., SHARE and GEM large-scale projects). Although a number of specific variants and applications of PSHA have been proposed (e.g. Stirling, 2014), most of the attempts to improve seismic hazard mapping do not include substantial methodological advances and basically rely on the collection and revision of input data, which are commendable, but not sufficient to address the abovementioned problems that are generally overlooked. Therefore, in this section we discuss some of the theoretical limitations of PSHA and the way they may significantly impede the reliability of the resulting maps.

The PSHA claims to incorporate the information on seismic source zones, magnitude-recurrence relations, and ground motion attenuation relations to estimate the seismic hazard in terms of the annual probability of exceedance (or return period) of a given threshold of the ground motion at a site. Under the assumption of a homogeneous Poisson distribution of earthquakes—thus under the assumption of a time-independent process without memory—the probability of exceeding the chosen threshold Y of the ground motion in a time interval of duration T is:

$$P(Y, T) = 1 - e^{\lambda_C(Y)T}$$

where:

- $P(Y,T)$ is the ground motion exceedance probability curve;
- $\lambda_C(Y)$ is the ground motion exceedance rate curve and estimates the number of times per year that an earthquake ground motion exceeds the amplitude Y. The homogeneous Poisson approach is commonly used in practice, where $\lambda_C(Y)$ does not depend on time.

The general expression of $\lambda_C(Y)$ for a single source is:

$$\lambda_C(Y) = \iiint n(M,R,X)\Phi\left(y \geq Y|\widehat{\mathbf{Y}}(M,R,X),\sigma_T\right)\mathrm{d}M\mathrm{d}X\mathrm{d}R$$

The quantity R is the distance of the site from the source, the quantity X is an expression of additional explanatory variables depending on fault style and site-specific features. The integral is characterized by two basic terms: a seismicity model n and a probability Φ, which derives from an attenuation relation or ground motion prediction equation (GMPE), whose characterization is debated and affected by large uncertainties. In practical applications, the quantity $\lambda_C(Y)$ commonly becomes a summation of the hazard integrals from several individual sources; through this procedure PSHA claims to consider all possible scenarios, because it considers all the possible sources.

12.2.1 Limits Related to Earthquake Probability Models

The seismicity model is described through $n(M,R,X)$ and gives the number of earthquakes per year in a specific magnitude and area range. This model is mostly determined by the definition of the source zones (i.e. seismogenic zones), whose boundaries are uncertain, and by the characterization of earthquakes recurrence, which is another debated topic. Standard zone-based PSHA requires the definition of the geometry of seismic source zones, where seismicity is typically assumed to be rather uniform, and a maximum expected magnitude; currently other definitions of sources are used as well, including line sources and zoneless approaches (e.g., Frankel, 1995; Woo, 1996). Extended sources and other variants can be naturally handled by NDSHA, as illustrated in Section 12.4.2. Assessing the probability that a given ground shaking would actually occur within a specified timespan strongly depends on assumptions about the recurrence of large earthquakes that have large uncertainties and often turn out to be incorrect, as discussed by Kagan et al. (2012). Seismological observations about large earthquakes, in fact, are too limited to constrain the probabilities associated with their occurrence and related ground shaking; therefore the maps, resulting from extrapolations without data support, may turn out to be poor approximations of reality.

An essential step in SHA consists of the definition of the area considered to estimate earthquake recurrence (typically the seismogenic zone), which is strictly interrelated with the size of the events that may affect it. According to the Multiscale Seismicity (MS) model (Molchan et al., 1997), the

Gutenberg—Richter (GR) law describes adequately only the ensemble of earthquakes that are geometrically small with respect to the dimensions of the analyzed region. In the original global formulation by Gutenberg and Richter (1956) all the earthquakes could be approximated by points, since even the dimensions of the sources of the largest events are negligible with respect to the size of the Earth, and the linearity of the GR is holding up to the largest magnitudes. On the contrary, when focusing on a limited region, the point approximation may no longer be valid, for example, an event with $M = 7$, whose surface rupture length can be estimated around 50 km (Wells and Coppersmith, 1994) can be considered a point such as only within a region of linear dimensions larger than 500 km. In agreement with the MS model, where seismicity is analyzed over relatively small regions, the frequency—magnitude relation is linear (self-organized criticality (SOC)) only up to a certain magnitude, while for the larger events it usually exhibits an upward bend (often defined as Characteristic Earthquake (CE)). Specifically, within the seismo-genic zones delimited for SHA, the number of strong earthquakes (whose source size becomes comparable with the region size) generally exceeds the estimation based on the extrapolation of the GR law (SOC) and hence they can be considered abnormally strong (CE) within the given region. The dependence of the linearity of the GR relation on the dimensions of the investigated area is illustrated considering different regions of the World by Kossobokov and Mazhkenov (1994), who proposed to generalize the GR law in the form

$$\log_{10}N = A + B \cdot (5 - M) + C \cdot \log_{10}L$$

where $N = N(M,L)$ is the expected annual number of earthquakes with magnitude larger than M in an area of linear dimension L; A and B are similar to the coefficients of the GR relation, while C estimates the fractal dimension of the earthquake-prone faults. Such a Unified Scaling Law for Earthquakes (Kossobokov and Nekrasova, 2007) states that the distribution of rates or waiting times between earthquakes depends only on the local value of the control parameter $10^{-BM} \cdot L^C$, which represents the average number of earth-quakes per unit time, with magnitude greater than M occurring in the area of size $L \times L$. With increasing magnitude the linearity truncates with decreasing dimension of the area, as soon as the recurrence of such earthquakes in it compares to the timespan of the catalog used.

Even in the simplest (and idealized) case of CEs considered by Bizzarri and Crupi (2013), in which the limiting cycle is reached by the system, and even in the framework of a simplified fault model, the possibility to a priori predict an impending earthquake based on the interevent time estimated through a universal analytical relation, still remains not feasible. On the other hand, a numerical estimate of T-cycle would require the exact knowledge of the rheological model (and its parameters at all times over the entire life of the fault) and of the actual state of the fault, which parameters are often unknown. Furthermore, the competing processes potentially occurring during faulting, even in the simplest and idealized condition of an isolated fault, can significantly complicate the regular cyclicity of earthquakes predicted by the analog fault system (Bizzarri, 2012).

Besides this basic difficulty in the identification of the appropriate model of earthquake recurrence, in many instances, the overlooking of the early definitions by Cornell led to the confusing use of the terms probability and occurrence frequency (or return period) (Wang, 2011). These may cause PSHA to become a purely numerical artifact; for example, by considering very low probabilities of exceedance, PHSA could in principle extrapolate ground motion with a return period of million years from a few hundred years of the available earthquake catalogs. The heuristic limitations are, indeed, a major limit of PHSA—the available short earthquake catalogs worldwide do not allow the statistics inference theory to project the probabilistic estimation for long period of time as 1000—10,000 years. Clearly, the map tries to predict events that do not have a significant statistics within the input data, and most probably many of them did not even appear yet. This calls for the need to integrate the information provided by earthquake catalogs with additional geological and geophysical evidences, independent from observed seismicity. However, with PSHA it is quite difficult, if not is quite difficult to take into account the information given by paleoseismicity (e.g., Michetti et al., 2012), morphostructural analysis (Gorshkov et al., 2004), or similar studies (e.g., Brune and Whitney, 2000). In fact, though geological and geophysical evidences may contribute to identification of areas where large, yet unobserved earthquakes may occur, defining the associated occurrence probability is by far a more difficult problem, particularly in view of the large uncertainties and incompleteness of such information. NDSHA is naturally free of this limitation.

12.2.2 Limits of Empirical Attenuation Relations

The term Φ gives the probability that an earthquake, with magnitude M at a distance R from the site, will cause a ground motion that equals or exceeds the chosen threshold Y. It derives from an attenuation relation that provides estimates of ground motion parameters at different distances from the source. In principle, the definition of attenuation relation may account for different types of data, eventually including information from waveform modeling; in most cases, however, empirical GMPE from pooled ergodic data are used. The available observations for a single site typically do not cover a period long enough to allow a statistical analysis, therefore records from many different sites and even different regions are used to represent the variability of the ground motion. However, trading of spatial information for temporal information under the ergodic assumption often results in mixing of seismologically incompatible data. Even though the GMPE represents a mean regression equation, $\widehat{Y}(M, R, X)$ from the GMPE is treated as the median value of Y, which is not necessarily true unless Φ is chosen as a distribution in which mean and median coincide. Φ is usually approximated with a normal distribution of log y; in this case log $\widehat{Y}(M, R, X)$ is really the median value of log y, as mean and median are the same for the normal distribution. Although the lognormal distribution can be convenient in the framework of GMPE development (see Anderson et al., 2010) and is therefore frequently adopted, it supports an unbounded upper tail, which is nonphysical.

Assuming that the attenuation relation is a monotonically decreasing function of distance is suggested by the mathematics, but is not universally true in the physical reality. In fact, the amplitude of the peak of the ground motion tends to diminish as long as the seismic waves essentially propagate in the crust. Once the waves reach either the Moho (i.e., the boundary between the Earth's crust and the mantle) or a strong vertical discontinuity within the crust, the S-waves can be postcritically reflected and interfere with the surface waves, the result being an increase in the amplitude of the latter. This can lead to an increase of the ground motion with distance. Theoretical examples can be found in Fäh and Panza (1994) for the city of Rome (Italy) and in Panza and Suhadolc (1989) for the Friuli Venezia Giulia region (Italy), whereas empirical examples for eastern North America are given by Burger et al. (1987).

Anderson and Brune (1999) used precarious rocks for testing the predictions of PSHA analysis, showing that many structures are too old and fragile to be consistent with recent estimates of $\lambda_C(Y)$ which, therefore, appear exceedingly large over the long timespan considered. Similar results are provided by Brune (1999) for Southern California and by Stirling and Anooshehpoor (2006) for New Zealand. The cause of the overestimation of $\lambda_C(Y)$ for rare and strong events (which have a low probability of exceedance) can probably be ascribed to the nonphysical unbounded tail of the distribution. PSHA assigns to an earthquake of a given magnitude increasing values of Y with decreasing probability. This means that to a low-energy seismic event can be assigned infinitely high values of Y (with a small probability), which reality will not support. As small magnitude events are more frequent than larger ones, the product $n\Phi$ for a small earthquake (which has a high n and a small Φ) can become comparable to the same product for a strong event (with a small n but a higher Φ). This affects the seismic hazard estimation diluting the hazard background in strongly seismic areas and enhances the effects of small earthquakes in areas of low seismicity, i.e., the artificial high values of Y provided by small events would lead to an underestimation of the energy content (which is linked to the magnitude) in strongly seismic areas and to an overestimation of the ground motion that can possibly be felt in low seismic areas, in agreement with independent observations by Zuccolo et al. (2011). Imposing an upper bound to Y is not a solution, but a mitigation strategy. As a matter of fact, the bound just provides a value, which the seismic hazard asymptotically converges to, regardless of the physical reality (i.e., the magnitude).

Besides the demonstrated shortcomings, the nonlinearity of the seismic process questions the pertinence to PSHA estimates of the attenuation relation itself. The amplitude spectrum Ω of the jth event recorded at the ith site is usually represented as (e.g., Paskaleva et al., 2007):

$$\Omega_{ij}(t) = E_j(t) * P_{ij}(t) * S_i(t)$$

where $E_j(t)$ describes the source, $P_{ij}(t)$ the path, and $S_i(t)$ the site effects and $*$ indicates the convolution operator. The inverse problem, i.e., the separation of source, path, and site effects terms in this equation, is nonlinear, even if just the simplest constraints (e.g., far-field approximation and constant focal mechanism) are requested. The attenuation relation aims to describe the changes in the amplitude of the ground motion as the distance from the epicenter increases, thus it should represent the path term. But the path term cannot be separated from the source and the site effects terms: the result is a biased description of the effects of propagation as it is meant in the framework of the PSHA analysis.

The choice of the most appropriate model for earthquakes recurrence and seismic-wave attenuation is a hard task. In general PSHA allows for the use of source- and site-specific additional variables X, GMPE, and n, but the problem is usually simplified adopting the same regional models for all sources and sites. Another strategy used in order to bypass this difficulty is the adoption of a logic tree with multiple input models. Each branch j of the tree supplies an estimate of the hazard curve, $\lambda_{Cj}(Y)$ according to a specific combination jth of the seismicity and attenuation relation models and the ground motion exceedance rate curve, $\lambda_C(Y)$ is given by their weighted sum.

The use of the logic tree is another feature of PSHA that may lead to an inappropriate estimation of the total uncertainty, due to an unclear understanding or separation between aleatoric and epistemic components (e.g., Klügel, 2007a,b, 2011). The calculation of the total uncertainty should be adjusted so that the contribution of these quantities is not counted twice. Let us consider an example of this problem. The attenuation relations are calculated using all of the data available for the region under study, usually without discriminating records based on the specific focal mechanisms. This procedure contributes to increase the data dispersion. When the attenuation relations are finally defined, their standard deviation accounts for the differences in focal mechanisms, which are aleatoric variables of the model. In some other branch of the logic tree, however, the focal mechanisms can be epistemic variables of a model that accounts for the features of the sources, and epistemically contribute to the uncertainty. As a result, if the uncertainty of the attenuation relation and the uncertainty of the sources description are considered independent, the focal mechanisms will contribute twice to the total uncertainty.

12.3 PRACTICAL LIMITS OF PSHA

Seismic hazard maps seek to predict the shaking that would occur, therefore to test the available models for SHAs, including earthquake forecasts and predictions (Jordan et al., 2011; Peresan et al., 2012), against the real seismic activity is the necessary precondition of any responsible seismic hazard and risk estimation. Otherwise, the use of untested maps may be misleading, which may be considered similar to negligence as in medical malpractice, although at much higher level of resulting losses (Wyss et al., 2012).

The usefulness of any seismic hazard map can be determined in a process of deterministic comparison to reality, namely by the comparison between the map's estimates and deterministically observed earthquake effects (i.e., macroseismic intensity, PGA, and peak velocity values (PGV)). Let us consider the crucial role of a sample size required to test probabilistic model estimates of ground motion effects, given on a seismic hazard map, against the actual effect of the occurrence of many earthquakes. It does not matter to estimate the intrinsic parameters of a given model, but simply to find out how often the value on the map was exceeded in reality before and after its publication. This type of count evaluates the map's performance, which is its capability to anticipate the observed ground shaking, although with a bias related to testing mainly on the learning sample. Nevertheless, according to Kossobokov and Nekrasova (2012) in the case of the GSHAP map testing on data in the time interval 1990−1999 before and 2000−2009 after its publication, the bias is negligibly small when compared to the observed misfit. As shown in Albarello and D'Amico (2008) and Mucciarelli et al. (2008), such a counting procedure allows us to check some internal parameters of the model used in the compilation of the map, e.g., the expected repeat time of exceedance, and, in principle, to reject the model if this number contradicts expectations.

Albarello and D'Amico (2008) have demonstrated that in Italy "a significant mismatch exists between PGA values characterized by an exceedance probability of 10 percent in 30 years and what has actually been observed at 68 accelerometric stations located on stiff soil, where continuous seismicity monitoring has been performed in the last 30 years." In this study we do not go that far, but just compare the performances of different maps in-between themselves.

The expected repeat time of exceedance is an internal parameter of a special class of models and it should not be confused with the time period during which the model map is applicable. In fact, 10 percent probability of exceedance in 50 years corresponds to the return period of 475 years given a certain (i.e., Poisson distribution) probabilistic assumption and this number is a control parameter of the model PGA values on the map. The validity of the basic probabilistic model could be tested by the achieved statistics of model expectations in agreement or in contradiction with the rate of occurrence over a period of time long enough to make a statistically reliable judgment (Albarello and D'Amico, 2008; Kossobokov and Nekrasova, 2012).

We presume any hazard map is applicable to and fits the past evidence and, hopefully, predicts hazardous events in the future. Probabilistic estimates depend on an accepted probability model, which might be inadequate for practical applications. The dependence of ground shaking on the selected choice of control parameters is one of the basic disadvantages of PSHA. The seismic hazard outcomes, specifically, of the strongest earthquakes may be stochastic but they happen with certainty (i.e., 100 percent) despite being infrequent happenings.

12.4 POSSIBLE ALTERNATIVES TO PSHA: THE NEO-DETERMINISTIC APPROACH (NDSHA)

The persistence in resorting to the PSHA probabilistic approach, in spite of its poor performances and basic shortcomings (Stein et al., 2012), is often explained by the need to deal with uncertainties related to ground shaking and earthquake recurrence, because there were no possible alternatives to this scheme of analysis. However, the arguments supporting the continued use of PSHA (e.g., Stirling, 2014) do not appear convincing (see Frankel, 2013; related reply by Stein et al. (2013)). Alternative methods do exist and should be explored, so that the resulting maps can be cross-checked and verified against real seismicity, and their anticipatory capability can be assessed.

NDSHA is an innovative, but already well-consolidated, scenario-based procedure (Panza et al., 2001, 2013, 2012) that supplies realistic time histories of strong ground motions from which it is natural to retrieve peak values for ground displacement, velocity and design acceleration based on earthquake scenarios (e.g., Parvez et al., 2011; Paskaleva et al., 2011). The procedure is particularly suitable for the optimum definition of the characteristics of modern antiseismic devices, when the accelerometer data available are not representative of the possible scenario earthquakes—as it is often the case—and when nonlinear dynamic analysis is necessary. By sensitivity analysis, knowledge gaps, related to endemic lack of data, can be addressed, due to the limited amount of scenarios to be investigated, in comparison with the computational resources needed.

NDSHA addresses some issues largely neglected in traditional hazard analysis, namely how crustal properties affect attenuation: ground motion parameters are not derived from overly simplified attenuation relations, but rather from synthetic time histories. Starting from the available information on the Earth's structure (mechanical properties), seismic sources, and the level of seismicity of the investigated area, it is possible to estimate PGA, PGV, and peak ground displacement (PGD) or any other parameter relevant to seismic engineering, which can be extracted from the computed theoretical signals. Synthetic seismograms can be efficiently constructed with the modal summation technique (e.g., Panza et al., 2001; La Mura et al., 2011) to model ground motion at sites of interest, using knowledge of the physical process of earthquake generation and wave propagation in realistic three-dimensional media.

Where the numerical modeling is successfully compared with records, the synthetic seismograms permit the microzoning, based upon a set of possible scenario earthquakes. Where no records are available the synthetic signals can be used to estimate the ground motion without having to wait for a strong earthquake to occur (predisaster microzoning). In both cases the use of modeling is necessary since the so-called local site effects can be strongly dependent on the properties of the seismic source and can be properly defined only by means of envelopes. In fact, several techniques that have been proposed to empirically estimate the site effects using observations (records) convolved with theoretically computed signals corresponding to simplified models, supply reliable information about the site response to noninterfering seismic phases, but they are not adequate in most of the real cases, when the seismic sequel is formed by several interfering waves.

One of the most difficult tasks in earthquake scenario modeling is the treatment of uncertainties, because each of the key parameters has its own uncertainty and intrinsic variability, parameters that commonly are not quantified explicitly. A possible way to handle this problem is to vary systematically (within the range of related uncertainties) the modeling parameters associated with seismic sources and structural models, i.e., to perform a parametric study to assess the effects of the parameters describing the mechanical properties of the propagation medium and of the earthquake focal mechanism (i.e., strike, dip, rake, depth). The parametric studies allow us to generate advanced ground-shaking scenarios for the proper evaluation of the site-specific seismic hazard, with the necessary and complementary check based on both probabilistic and empirical procedures.

Once the gross features of the seismic hazard are defined by means of the parametric analyses, a more detailed modeling of the ground motion can be carried out for sites of specific interest. Such a detailed analysis duly takes into account the earthquake source characteristics, the mechanical properties of the path and of the local geology, and it can be easily performed using widely available computational tools, such as modern laptops or, for very complex situations, worldwide grid-and-cloud advanced e-infrastructures (e.g., Magrin et al., 2012; Peresan et al., 2014).

12.4.1 Basic Differences between PSHA, DSHA, and NDSHA Approaches

The neo-deterministic approach, NDSHA (Panza et al., 2012 and references therein) means scenario-based methods for seismic hazard analysis, where attenuation relations and other similarly questionable assumptions about local site responses, all implying some form of physically unsound linear convolution, are not allowed in. Instead, realistic synthetic time series are used to construct earthquake scenarios. The NDSHA procedure provides strong ground motion parameters based on the seismic-wave propagation modeling at different scales—regional, national, and metropolitan—accounting for a wide set of possible seismic sources and for the available information about structural models. NDSHA is a statistical approach since it can account for a large number of possible earthquakes and the related uncertainties are considered with some detail. This scenario-based method relies on observable data being complemented by physical—mathematical modeling techniques that can be submitted to a formalized validation process. NDSHA can use, in addition, the information about the location of potential earthquakes above a given magnitude, which did not occur in historical time (e.g., seismogenic nodes, as shown in Peresan et al. (2011)).

The basic differences between the NDSHA and the classical DSHA approach, which is a statistical approach as well, although it is called deterministic, because it relies on the use of empirical attenuation relations and on a small set of scenario earthquakes, are illustrated in Figure 12.1. In DSHA, seismic hazard is defined as the median or certain percentile (e.g., 84 percent) ground motion from a single earthquake or set of earthquakes, and it is calculated from simple statistics of earthquakes and ground motion (Krinitzsky, 2002). The major drawbacks of DSHA are: (1) to rely on GMPEs or attenuation relations; and (2) to obscure occurrence interval (frequency). Local site effects can be strongly dependent on the characteristics of the

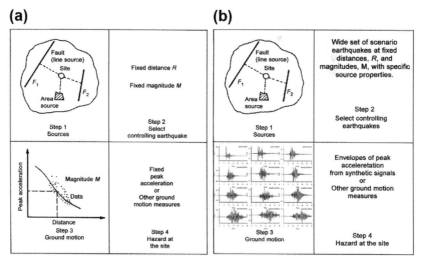

FIGURE 12.1   Basic steps of (a) classical deterministic seismic hazard analysis and (b) neodeterministic seismic hazard analysis (NDSHA) analysis. In the NDSHA the expected ground motion is modeled at any site of interest (e.g., nodes of a regular grid), considering a wide set of scenario events, starting from the available information about seismic sources and regional structural models. NDSHA computes earthquake ground motions as a tensor product between the earthquake source tensor with the Greens function for the medium (realistic synthetic seismograms at different source-site distance—Step (3)) and it avoids the use of an incorrect scalar quantity implied in the ground motion prediction equations or attenuation relationships. *Modified after Reiter (1990).*

seismic source (Field, 2000; Panza et al., 2001). NDSHA derives earthquake ground motions as a tensor product (of earthquake source tensor with the Greens function for the medium) and it avoids using an approximate scalar quantity implied in the GMPEs and/or attenuation relationships. Convolutive methods (e.g., using GMPEs) can be severely in error, and the use of synthetic seismograms is fundamental, even when relevant observational data are available, in order to explore the local responses that may correspond to sources that are different from the known ones, that rarely are representative of all possible events.

Being based on a large set of possible scenario earthquakes, the use of appropriate frequency–magnitude relations eventually permits association with NDSHA peak value maps the corresponding recurrence estimates, as well as the generation of neo-deterministic ground-shaking maps for specified return period (Peresan et al., 2013; Magrin et al., 2014). Moreover, the related uncertainties can be explicitly defined by massive parametric tests.

Although a statistical analysis is essential in assessment of seismic hazard, because of the nature of Earth processes and because the results from any approach have to be validated against observations, the classical PSHA is not a valid probabilistic approach (Wang, 2011) because: (1) it is based on a oversimplified earthquake source model (single point source which is no longer valid for earthquakes having damage potential); (2) mathematical pitfalls (i.e., incorrect use of statistics and probability theory such as GR curve and ground motion attenuation); and (3) misinterpretation and use of the annual probability of exceedance (i.e., the probability of exceedance in 1 year—a dimensionless quantity) as a frequency (per year—a dimensional quantity). For more shortcomings of traditional PSHA see Sections 12.2 and 12.3.

As discussed in Section 12.2, the use of attenuation relations is really one of the weak points of any hazard estimation based on convolutive concepts. Therefore it appears necessary to resort to the computation of realistic time series, taking advantage of the already existing databases about global Earth models (e.g., http://igppweb.ucsd.edu/~gabi/crust1.html; http://earthquake.usgs.gov/research/structure/crust/database.php). For a given area, in a first approximation, computations can be made using the existing data, which are already provided at a sufficient resolution; these estimates, if necessary, as in the case of special objects, can be naturally refined using ad hoc information from local studies. The NDSHA method has been already successfully applied in many areas worldwide (e.g., Panza et al., 2012 and references therein), to strategic buildings, lifelines, and cultural heritage sites (e.g., Vaccari et al., 2009), and for the purpose of seismic microzoning, to several urban areas (e.g., Indirli et al., 2011), also in the framework of UNESCO/IUGS/IGCP projects "Realistic Modeling of Seismic Input for Megacities and Large Urban Areas" (Panza et al., 2002), "Seismic microzoning of Latin

America cities", and "Seismic Hazard and Risk Assessment in North Africa" (Mourabit, 2014).

12.4.2 National Scale NDSHA Hazard Maps

In the NDSHA approach the definition of the space distribution of seismicity accounts essentially for the damaging events ($M \geq 5$) reported in the national earthquake catalogs. Italy is the only country where an earthquake catalog is available, sufficiently complete for $M \geq 5$ for about a millennium (e.g., Gasperini et al., 2004 and its updates). The flexibility of NDSHA permits incorporatation of the additional information about the possible location of strong earthquakes provided by morphostructural analysis, thus significantly reducing gaps in known seismicity, as reported in parametric catalogs. Specifically, the areas prone to strong earthquakes are identified based on the morphostructural nodes, which represent specific structures formed around the intersections of lineaments. Lineaments are identified by the morphostructural zonation (MZ) method (Alekseevskaya et al., 1977) that, independently from information about earthquakes, delineates a hierarchical block structure of the study region, using tectonic and geological data, with special care to topography. The MZ method differs from the standard morphostructural analysis where the term "lineament" (Hobbs, 1904, 1912) is used to define the complex of alignments detectable on topographic maps or on satellite images. According to that definition the lineament is locally defined and the existence of the lineament does not depend on the surrounding areas. In MZ, the primary element is the block—a relatively homogeneous area—whereas the lineament is a secondary element of the morphostructure. The boundary zones between blocks are the lineaments and the nodes are formed at the intersections or junctions of two or more lineaments. Among the defined nodes, those prone to strong earthquakes are then identified by pattern recognition on the basis of the parameters characterizing indirectly the amount of neo-tectonic movements and fragmentation of the crust at the nodes (e.g., elevation and its variations in mountain belts and watershed areas; orientation and density of linear topographic features; type and density of drainage pattern). For this purpose, the nodes are defined as circles of radius $R = 25$ km surrounding each point of intersection of lineaments. The MZ of Italy and surrounding regions, as well as the identification of the sites where strong events can nucleate, has been performed by Gorshkov et al. (2002, 2004) considering two magnitude thresholds: $M \geq 6.0$ and $M \geq 6.5$.

The identified seismogenic nodes are used, along with the seismogenic zones (Meletti and Valensise, 2004), to characterize the earthquake sources used in the seismic ground motion modeling, as described by (Peresan et al., 2011). The earthquake epicenters reported in the catalog are grouped into $0.2° \times 0.2°$ cells, assigning to each cell the maximum magnitude recorded within it. A smoothing procedure is then applied, to account for spatial uncertainty and source dimensions. Only the sources located within the seismogenic zones, as well as the sources located within the earthquake-prone nodes, are considered; moreover, if the smoothed magnitude M of a source inside a node is lower than the magnitude threshold, M_0, identified for that node, in the computation of the synthetic seismograms M_0 is used.

NDSHA has been extended to account for the source process in some detail (rupture process at the source and the consequent directivity effect). The preliminary results provided by this ongoing research (i.e., the regression relations between the strong motion parameters and the macroseismic intensities) confirm the results obtained with a 1 Hz cutoff frequency in the point-source approximation (Panza et al., 1997, 2013).

Considering specific faults included within alerted nodes, with this second variant of NDSHA it is possible to perform parametric studies, which permit to single out the relevance of source-related effects, like directivity. In Figure 12.2 we supply an example of scenario corresponding to the fault ITIS038 from the database DISS3 (Basili et al., 2008), which falls within the node I26 (Gorshkov et al., 2002). The rupture process at the source and the consequent directivity effect (i.e., radiation at a site depends on its azimuth with respect to rupture propagation direction) is modeled by means of the algorithm developed by Gusev and Pavlov (2006) and Gusev (2011), that simulates the radiation from a fault of finite dimensions, named PULSYN (PULse-based wide band SYNthesis).

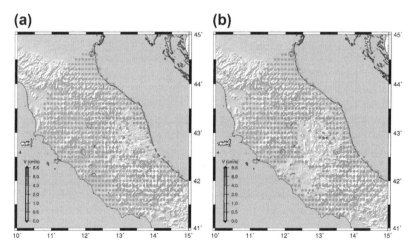

(a)   **(b)**

FIGURE 12.2 Ground-shaking scenarios at bedrock (peak velocity values) (a) for source directivity south-east; (b) for source directivity north-west. The fault ITIS038 from the database DISS3 (Basili et al., 2008) is considered. Cellular structural models of the lithosphere are considered to model wave propagation. *After Panza et al. (2013).*

An updated and detailed review of NDSHA is given by Panza et al. (2011b, 2013).

12.5 PERFORMANCES OF PSHA: THE VALIDATION PROBLEM

One is well advised, when traveling to a new territory, to take a good map and then to check the map with the actual territory during the journey.

Wasserburg (2010)

The GSHAP project was launched in 1992 by the International Lithosphere Program with the support of the International Council of Scientific Unions, and endorsed as a demonstration program in the framework of UN/IDNDR. The GSHAP Project terminated in 1999 when the probabilistic SHA maps (Giardini et al., 1999, 2003; Shedlock et al., 2000) and digital data got published. The majority of recent disastrous earthquakes, like March 11, 2011 Tohoku (Japan), April 13, 2010 Southern Qinghai (China), January 12, 2010 Port-au-Prince (Haiti), September 30, 2009 Padang (Southern Sumatra, Indonesia), May 12, 2008 Wenchuan (Sichuan, China), October 8, 2005 Kashmir (northern border region of India and Pakistan), March 28, 2003 Nias Island (Indonesia), December 26, 2004 Sumatra–Andaman Islands ("Indian Ocean Disaster"), May 21, 2003 Boumerdes (Algiers), January 26, 2001 Bhuj (Gujarat, India), and many other, prove that the maps produced by GSHAP are evidently misleading.

The M6.9 earthquake that struck China (ESE of Hotan) on February 18, 2014 is just a recent reminder; it caused a PGA >42%g, which is more than 1.75 times higher than the PGA value assigned by GSHAP at the epicentral location. Thus this is another underestimation of about one unit in macroseismic intensity.

The extreme catastrophic nature of earthquakes is known for centuries due to the devastation caused by many of them. Earthquakes are naturally characterized by their effect on the Earth's surface. There are several traditional scales (e.g., MCS-Mercalli, Cancani, Sieberg; MSK-Medvedev, Sponheuer, Karnik, MMI-Modified Mercalli Intensity) that measure damage that result from earthquakes, of which the MMI scale is the most commonly used. According to Reiter (1990) and Decanini et al. (1995) the following gross empirical relations hold: $I_{MM} \sim I_{MSK}$ and $I_{MM} \sim (5/6) I_{MCS}$, respectively. The contemporary seismic databases provide information on the size of an earthquake in terms of some physically different magnitude scales. The classical empirical relation linking the magnitude M and MMI intensity at the epicenter I_0 for shallow depth earthquakes was suggested by Gutenberg and Richter (1956):

$$M = 2/3 \, I_0 + 1 \qquad (12.1)$$

Seismic engineering requires different measures that characterize earthquake ground motion. One of those is PGA. Linking magnitude of an earthquake to intensity of the ground shaking it produces, is a complex and difficult problem (Sauter and Shah, 1978; Wald et al., 1999). However, Figure 12.3 shows some empirical relations between PGA in m/s^2 and MMI at the epicenter, I_0, suggested by different authors, i.e., Aptikaev et al. (2008) Eqn (12.2), Shteinberg et al. (1993) Eqn (12.3), Sauter and Shah (1978) Eqn (12.4), and Murphy and O'Brien (1977) Eqn (12.5):

$$I_0 = 2.50 \log_{10}(\text{PGA}) + 6.89 \tag{12.2}$$

$$I_0 = 2.97 \log_{10}(\text{PGA}) + 6.71 \tag{12.3}$$

$$I_0 = 3.62 \log_{10}(\text{PGA}) + 6.34 \tag{12.4}$$

$$I_0 = 2.86 \log_{10}(\text{PGA}) + 6.96 \tag{12.5}$$

Formula (12.2) is the inverse of the original $\log_{10}(\text{PGA}) = 0.4\text{MMI} - 0.755 \pm 0.08$ (Aptikaev et al., 2008), while Formula (12.3) is the best exponential fit (with $R^2 = 0.9867$) of the data from the original table (Shteinberg et al., 1993) plotted as squares in Figure 12.3.

The MMI scale discrete values (12 integer classes from I to XII) are hardly in a logical agreement with the estimates with accuracy of two decimals of the empirical relations (12.2−12.5), which formulas and their plots in Figure 12.3 illustrate clearly and confirm a robust, essentially, consensus fit of PGA ranges suggested by different authors (including those not mentioned in this publication) to characterize the qualitatively described classes of the MMI scale. The log-linear regression between maximum observed macroseismic intensity, I(MCS), and computed peak values of ground motion (A), considering historical events, has a slope of about 0.3. Cancani (1904) modified the Mercalli scale with the declared intent to get such relationship between I(MCS) and ground motion A.

The combination of the classical relation (12.1) with any of the empirical relations (12.2−12.5) provides all necessary links for a comparison of the GSHAP final PGA estimates with the actual seismic activity.

FIGURE 12.3 Some empirical relations for PGA in m/s^2 as a function of Modified Mercalli Intensity at epicenter, I_0.

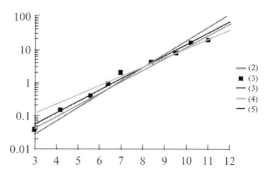

The results of the comparison by Kossobokov and Nekrasova (2012) can be summarized as follows. Some 1181 out of 1347 and 1021 out of 1201 strong normal depth earthquakes were reported in the global hypocenters database (GHDB, 1989), for the years 2000−2009 and 1990−1999, respectively, which are associated maximum GSHAP PGA values. These earthquakes sample $I_0(M)$ and $I_0(\text{mPGA})$ along with their difference $\Delta I_0 = I_0(M) - I_0(\text{mPGA})$ over the two decades considered. Figure 12.4 displays all 1181 pairs of $I_0(M)$ and $I_0(\text{mPGA})$ related to strong earthquakes in 2000−2009 for each of the four Formulas (12.2−12.5). Each of the four plots discloses the evident underestimation by GSHAP of the potential ground shaking: all the data points below the diagonal represent the "surprises" to PGA prediction given by GSHAP in a decade following its release. By inspection with bare eyes it is evident that the percentage of "surprises" when the GSHAP predicted ground shaking was exceeded in seismic events is by far larger than "a 10 percent chance", as stated in the definition of PGA plotted on the GSHAP maps.

The distribution of ΔI_0 is summed up in Table 12.1. The poor performance of the GSHAP product evident from the table could have been found at the time of its publication. The performance evaluation should have been done by the contributors to the GSHAP as the first order validation test. The statistics that one would get from a similar analysis of ΔI_0 on a learning sample, e.g., in 1990−1999 are not that different from those in 2000−2009 (Table 12.1) and do not justify using the GSHAP map as an endorsed demonstration of scientific achievement. The claim of a 10 percent chance of exceedance in

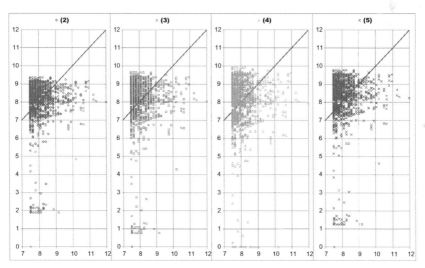

FIGURE 12.4   Pairs of $I_0(\text{mPGA})$ (ordinate) and $I_0(M)$ (abscissa) for strong surface earthquakes recorded in 2000−2009 for the empirical relations (12.2−12.5), respectively. Points below the diagonal indicate "surprises" of the Global Seismic Hazard Assessment Program project in the first decade after its completion.

TABLE 12.1 Number of Shallow Earthquakes (in One Decade) That Violate GSHAP PGA Prediction, Using the Relationships 5.2–5.5 Defined Above

	Total	$\Delta I_0 = I_0(M) - I_0(mPGA)$											
		(2)			(3)			(4)			(5)		
		>0	>1	>2	>0	>1	>2	>0	>1	>2	>0	>1	>2
2000–2009 (Test on Control Sample, After Publication)													
M = 6 or more	1,181	529	191	83	530	204	89	559	232	105	426	164	78
M = 7 or more	113	113	79	28	112	76	30	106	74	35	105	65	25
1990–1999 (Test on Learning Sample)													
M = 6 or more	1,021	471	182	66	463	185	69	487	203	91	385	137	61
M = 7 or more	129	124	74	15	120	65	16	117	63	22	115	46	11

50 years appears to be violated in 1990—1999 for more than one-third of the strong events and for about 90 percent or more of the earthquakes above magnitude 7.0 (Table 12.1, last two rows).

Table 12.2 discloses clearly the inconsistency of GSHAP PGA for each of the major earthquakes with magnitude 7.5 or more in 2000—2010: The absolute minimum of ΔI_0 over all estimates by Formulas (12.2—12.5) equals 0.6, whereas the average and median of ΔI_0 are about 2. This means that each of the shallow earthquakes with magnitude 7.5 or larger that happened in 10 years after publication of the GSHAP results was a "surprise" leading to unexpected damage and loss. Obviously, such a level of contradiction cannot be attributed to imperfections of the GSHAP models in some regions as shown by Figure 12.5. In the figure more than 7,000 epicenters of strong shallow earthquakes in 1900—2009 are shown color-coded with respect to the difference ΔI_0 computed by using Formulas (12.1) and (12.3). This is a map of GSHAP past "surprises" where dark yellow and red colors of "big surprises" cover widespread seismic regions of the entire globe.

Finally, measurements of peak accelerations and field observations of intensity in the epicentral zones of catastrophic earthquakes are consistent with and confirm our results obtained by robust calculations. Examples include the following events: January 12, 2010 Port-au-Prince (Haiti), May 12, 2008 Wenchuan (Sichuan, China), May 21, 2003 Boumerdes (Algiers), December 26, 2003 Bam (Iran), January 26, 2001 Bhuj (Gujarat, India), and January 17, 1995 Kobe (Japan). For megathrusts off the Pacific coast of the Tohoku Region, Japan, such measurements are not available because the rupture occurred beneath the sea.

However, numerous records of PGA observed on March 11, 2011 on Honshu Island testify to the multiple exceedence of the GSHAP PGA expectations (2.4—3.2 m/s^2). For instance, according to the report of Japan's National Research Institute for Earth Science and Disaster Prevention, accelerations along the coast of the Miagi Prefecture at about 75 km from the epicenter attained 29.33 m/s^2, which is almost six times higher than the GSHAP PGA values in the epicentral zone and 10 times higher in the measurement point. Moreover, peak accelerations recorded by NEID on March 11, 2011 on the surface exceeded 4 m/s^2 on 10—15 percent of the Honshu Island, whereas the GSHAP values do not exceed 3.2 m/s^2 in the same region. More than a quarter of the population living in the largest Japanese Island (more than 30 million people) had experienced unexpected ground shaking of MMI intensity 7 or more.

Regretfully, the list of GSHAP "surprises" keeps growing. The Port-au-Prince, M7.3 earthquake on January 12, 2010, which became the final motivation to perform a systematic check of the GSHAP results for such kind of "surprises", is included in Table 12.2 in spite its lower magnitude. This disastrous event brings attention to another statistics associated with the GSHAP performance: Table 12.3 summarizes the number of fatalities along

TABLE 12.2 Shallow Magnitude 7.5 or More Earthquakes Associated with GSHAP PGA after 1999

Date	φ	λ	M	mPGA	n	$I_0(M)$	I_0(mPGA)				ΔI_0			
							1	2	3	4	1	2	3	4
2000/06/04	−4.72	102.09	8.0	3.874	9	10.5	8.4	8.5	8.5	8.6	2.1	2.0	2.0	1.9
2000/06/18	−13.80	97.45	7.8	0.153	6	10.2	4.8	4.3	3.4	4.6	5.4	5.9	6.8	5.6
2000/11/16	−3.98	152.17	8.2	4.876	6	10.8	8.6	8.8	8.8	8.9	2.2	2.0	2.0	1.9
2000/11/16	−5.23	153.10	7.8	7.341	4	10.2	9.1	9.3	9.5	9.4	1.1	0.9	0.7	0.8
2000/11/17	−5.50	151.78	8.2	7.364	9	10.8	9.1	9.3	9.5	9.4	1.7	1.5	1.3	1.4
2000/12/06	39.57	54.80	7.5	3.817	6	9.8	8.3	8.4	8.4	8.6	1.4	1.3	1.3	1.1
2001/01/13	13.05	−88.66	7.8	3.488	4	10.2	8.2	8.3	8.3	8.5	2.0	1.9	1.9	1.7
2001/01/26	23.42	70.23	8.0	2.050	9	10.5	7.7	7.6	7.5	7.9	2.8	2.9	3.0	2.6
2001/06/23	−16.26	−73.64	8.2	6.059	4	10.8	8.8	9.0	9.2	9.2	2.0	1.8	1.6	1.6
2001/07/07	−17.54	−72.08	7.6	5.443	6	9.9	8.7	8.9	9.0	9.1	1.2	1.0	0.9	0.8
2001/10/19	−4.10	123.91	7.5	0.847	9	9.8	6.7	6.5	6.1	6.8	3.0	3.3	3.7	3.0
2001/11/14	35.95	90.54	8.0	0.913	6	10.5	6.8	6.6	6.2	6.8	3.7	3.9	4.3	3.7
2002/01/02	−17.60	167.86	7.5	5.602	6	9.8	8.8	8.9	9.0	9.1	1.0	0.8	0.7	0.6
2002/03/05	6.03	124.25	7.5	2.901	4	9.8	8.0	8.1	8.0	8.3	1.7	1.7	1.7	1.5
2002/09/08	−3.30	142.95	7.8	2.695	6	10.2	8.0	8.0	7.9	8.2	2.2	2.2	2.3	2.0

2002/10/10	−1.76	134.30	7.7	3.388	6	10.1	8.2	8.3	8.3	8.5	1.8	1.8	1.8	1.8	1.6
2002/11/02	2.82	96.08	7.6	2.864	9	9.9	8.0	8.1	8.0	8.3	1.9	1.8	1.8	1.9	1.6
2002/11/03	63.52	−147.44	8.5	4.023	9	11.3	8.4	8.5	8.5	8.7	2.8	2.7	2.7	2.7	2.6
2003/01/20	−10.49	160.77	7.8	7.822	6	10.2	9.1	9.4	9.6	9.5	1.1	0.8	0.8	0.6	0.7
2003/01/22	18.77	−104.10	7.6	6.409	6	9.9	8.9	9.1	9.3	9.3	1.0	0.8	0.8	0.6	0.6
2003/07/15	−2.60	68.38	7.6	4.224	4	9.9	8.5	8.6	8.6	8.7	1.4	1.3	1.3	1.3	1.2
2003/08/04	−60.53	−43.41	7.6	1.844	10	9.9	7.6	7.5	7.3	7.7	2.3	2.4	2.4	2.6	2.2
2003/08/21	−45.10	167.14	7.5	4.364	12	9.8	8.5	8.6	8.7	8.8	1.3	1.1	1.1	1.1	1.0
2003/09/25	41.81	143.91	8.3	5.869	9	11.0	8.8	9.0	9.1	9.2	2.1	2.0	2.0	1.8	1.8
2003/09/27	50.04	87.81	7.5	4.303	8	9.8	8.5	8.6	8.6	8.8	1.3	1.2	1.2	1.1	1.0
2003/11/17	51.15	178.65	7.8	6.464	8	10.2	8.9	9.1	9.3	9.3	1.3	1.1	1.1	0.9	0.9
2004/02/07	−4.00	135.02	7.5	3.103	9	9.8	8.1	8.2	8.1	8.4	1.6	1.6	1.6	1.6	1.4
2004/11/11	−8.15	124.87	7.5	5.864	8	9.8	8.8	9.0	9.1	9.2	0.9	0.8	0.8	0.6	0.6
2004/12/23	−49.31	161.35	8.1	0.890	12	10.7	6.8	6.6	6.2	6.8	3.9	4.1	4.1	4.5	3.8
2004/12/26	3.30	95.98	9.0	2.784	9	12.0	8.0	8.0	7.9	8.2	4.0	4.0	4.0	4.1	3.8
2004/12/26	6.91	92.96	7.5	2.705	4	9.8	8.0	8.0	7.9	8.2	1.8	1.8	1.8	1.8	1.6
2005/03/28	2.09	97.11	8.6	2.956	9	11.4	8.1	8.1	8.0	8.3	3.3	3.3	3.3	3.4	3.1
2005/07/24	7.92	92.19	7.5	0.690	9	9.8	6.5	6.2	5.8	6.5	3.3	3.5	3.5	4.0	3.3
2005/10/08	34.54	73.59	7.7	2.235	6	10.1	7.8	7.7	7.6	8.0	2.3	2.3	2.3	2.4	2.1

Continued

TABLE 12.2 Shallow Magnitude 7.5 or More Earthquakes Associated with GSHAP PGA after 1999—cont'd

Date	φ	λ	M	mPGA	n	$I_0(M)$	I_0(mPGA)				ΔI_0			
							1	2	3	4	1	2	3	4
2006/02/22	−21.32	33.58	7.5	0.762	9	9.8	6.6	6.4	5.9	6.6	3.2	3.4	3.8	3.1
2006/04/20	60.95	167.09	7.6	1.011	10	9.9	6.9	6.7	6.4	7.0	3.0	3.2	3.5	2.9
2006/05/03	−20.19	−174.12	8.0	1.838	9	10.5	7.6	7.5	7.3	7.7	2.9	3.0	3.2	2.8
2006/07/17	−9.28	107.42	7.7	1.848	9	10.1	7.6	7.5	7.3	7.7	2.5	2.5	2.7	2.3
2006/11/15	46.59	153.27	8.3	8.022	12	11.0	9.2	9.4	9.6	9.5	1.8	1.6	1.3	1.4
2007/01/13	46.24	154.52	8.2	7.815	8	10.8	9.1	9.4	9.6	9.5	1.7	1.4	1.2	1.3
2007/01/21	1.07	126.28	7.5	3.251	6	9.8	8.2	8.2	8.2	8.4	1.6	1.5	1.6	1.3
2007/04/01	−8.47	157.04	8.1	3.075	4	10.7	8.1	8.2	8.1	8.4	2.5	2.5	2.5	2.3
2007/08/15	−13.39	−76.60	8.0	4.154	12	10.5	8.4	8.5	8.6	8.7	2.1	2.0	1.9	1.8
2007/09/12	−4.44	101.37	8.5	3.151	4	11.3	8.1	8.2	8.1	8.4	3.1	3.1	3.1	2.9
2007/09/12	−2.63	100.84	8.1	2.610	4	10.7	7.9	7.9	7.8	8.2	2.7	2.7	2.8	2.5
2007/11/14	−22.25	−69.89	7.7	3.429	6	10.1	8.2	8.3	8.3	8.5	1.8	1.8	1.8	1.6
2008/02/20	2.77	95.96	7.5	2.857	4	9.8	8.0	8.1	8.0	8.3	1.7	1.7	1.8	1.5
2008/05/12	31.00	103.32	8.1	1.751	9	10.7	7.5	7.4	7.2	7.7	3.2	3.2	3.4	3.0

2009/01/03	−0.41	132.89	7.7	3.154	9	10.1	8.1	8.2	8.1	8.4	1.9	1.9	1.9	1.7
2009/01/15	46.86	155.15	7.5	2.341	4	9.8	7.8	7.8	7.7	8.0	1.9	1.9	2.1	1.7
2009/03/19	−23.04	−174.66	7.6	1.719	4	9.9	7.5	7.4	7.2	7.6	2.4	2.5	2.7	2.3
2009/07/15	−45.76	166.56	7.8	4.218	8	10.2	8.5	8.6	8.6	8.7	1.7	1.6	1.6	1.5
2009/08/10	14.10	92.89	7.6	3.050	9	9.9	8.1	8.1	8.1	8.3	1.8	1.8	1.8	1.6
2009/09/29	−15.49	−172.10	8.1	2.028	9	10.7	7.7	7.6	7.5	7.8	3.0	3.0	3.2	2.8
2009/10/07	−12.52	166.38	7.9	5.870	9	10.4	8.8	9.0	9.1	9.2	1.5	1.4	1.2	1.2
2009/10/07	−13.01	166.51	7.7	5.048	9	10.1	8.6	8.8	8.9	9.0	1.4	1.3	1.2	1.1
2010/01/12	18.44	−72.57	7.3[1]	1.511	6	9.5	7.3	7.2	7.0	7.5	2.1	2.2	2.5	2.0
2010/06/12	7.85	91.92	7.5	0.409	6	9.8	5.9	5.6	4.9	5.9	3.8	4.2	4.8	3.9
2010/04/06	2.38	97.05	7.9	2.876	6	10.4	8.0	8.1	8.0	8.3	2.3	2.3	2.3	2.1
2010/02/27	−36.12	−72.90	8.8	2.857	9	11.7	8.0	8.1	8.0	8.3	3.7	3.6	3.7	3.4
2011/03/11	38.32	−142.37	9.0	4.974	9	12	8.6	8.8	8.9	9.0	3.4	3.2	3.1	3.1

[1] Big "Surprises" are Marked With Shades of Grey: Δl_0 from 1 to 2 (Light Grey) and $\Delta l_0 = 2$ or more (Darker Grey).

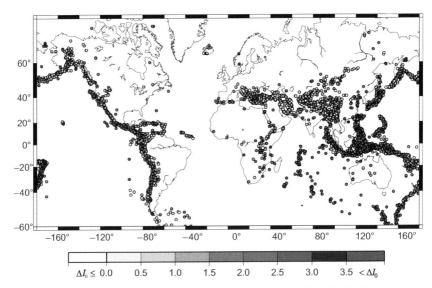

FIGURE 12.5   Spatial distribution of ΔI_0, the difference between the factual effect (macroseismic intensity) and that predicted by GSHAP for strong surface earthquakes in 1900–2009. "Surprises" ($\Delta I_0 > 0$) are ubiquitous, whereas "big surprises" ($\Delta I_0 > 1$) are widespread almost in all seismic regions of the world.

with the discrepancy of GSHAP predictions for the top 12 deadliest earthquakes since 2000. In total, the number of fatalities comes to about 700,000 people, including tsunami victims. The average and the median of the GSHAP discrepancy, measured by underestimation difference ΔI_0, are both about two units of intensity.

Thus, a systematic quantitative comparison of the GSHAP PGA estimates with those related to actual strong earthquakes disclose inadequacy of this "probabilistic" product, which appears inappropriate for responsible seismic risk evaluation and knowledgeable disaster prevention. Making quantitative probabilistic claims in the frameworks of the most popular objectivists' viewpoint on probability requires a long series of recurrences, which cannot be obtained at local scale from the existing catalogs of earthquakes. The evident failure of GSHAP appeals for an urgent revision, starting from first principles of the global seismic hazard maps, including of the basic methods involved and hazard assessment justification on learning and control.

12.6 PERFORMANCE OF NDSHA

A possible alternative to the conventional probabilistic approach represented by GSHAP, on a global scale, and by the official seismic hazard maps for Italy, on a regional scale, is provided by NDSHA (Panza et al., 2001, 2012; Peresan et al., 2011). In the case of the recent May 20, 2012, Emilia, M5.9 earthquake

TABLE 12.3 Earthquakes Caused the Greatest Population Losses Since 2000

Region	Date	M	Number of Fatalities Including Tsunami Victims	Difference ΔI_0
Sumatra Island—Andaman Islands ("Indian Ocean disaster")	26/12/2004	9.0	227,898	4.0
Port-au-Prince (Haiti)	12/01/2010	7.3	222,570	2.2
Wenchuan (Sichuan, China)	12/05/2008	8.1	87,587	3.2
Kashmir (northern border region of India and Pakistan)	08/10/2005	7.7	86,000	2.3
Bam (Iran)	26/12/2003	6.6	31,000	0.2
Bhuj (Gujarat, India)	26/01/2001	8.0	20,085	2.9
Tohoku region (Honshu Island, Japan)	11/03/2011	9.0	15,870 (2,814 missing)[1]	3.2
Yogyakarta (Java Island, Indonesia)	26/05/2006	6.3	5,749	0.3
Southern Qinghai (China)	13/04/2010	7.0	2,698	2.1
Boumerdes (Algiers)	21/05/2003	6.8	2,266	2.1
Nias Island (Indonesia)	28/03/2005	8.6	1,313	3.3
Padang (Southern Sumatra, Indonesia)	30/09/2009	7.5	1,117	1.8

[1] The data on fatalities in Tohoku mega-earthquake and tsunami are given as reported on September 19, 2012 by the Japanese National Police Agency.

in Northern Italy (Brandmayr et al., 2013) the NDSHA map, published in 2001 (Panza et al., 2001), is closer to reality than the official ones, based on PSHA, which provide the basis for the Italian building code (Peresan and Panza, 2012). In order to assess whether this is just a sporadic case or a paradigmatic example, a systematic analysis was performed, to better understand the performances and possible limits of the different methods. The assessment is made investigating their performances with respect to past earthquakes (Nekrasova et al., 2014): the seismic hazard maps for the territory of Italy, obtained by the PSHA (Gruppo di Lavoro, 2004; Meletti and Montaldo, 2007) and NDSHA (Peresan et al., 2013) approaches are compared with the observed seismicity.

Table 12.4 gives the relations between the intensity on the MCS scale, and the ground acceleration values derived from the comparison between the map

TABLE 12.4 Relation between I_{MCS} and Model Ground Motion, mGA, Corresponding Either to PGA(g) or DGA(g) for the Territory of Italy

I_{MCS}	VI	VII	VIII	IX	X	XI
mGA(g)	0.01−0.02	0.02−0.04	0.04−0.08	0.08−0.15	0.15−0.3	0.3−0.6

Source: After Indirli et al. (2011).

of maximum felt intensities in Italy (Boschi et al., 1995) and the maximum design ground acceleration (DGA), obtained from modeling the ground motion generated by past seismicity (Panza et al., 2001). These relations are used for converting the ground motion data from PGA10%, PGA2%, DGA, DGA10%, DGA2% into the MCS scale values. The five pairs of the linked MCS values on a ground motion model map I_{mGA} (i.e., $I_{PGA10\%}$, $I_{PGA2\%}$, I_{DGA}, $I_{DGA10\%}$, $I_{DGA2\%}$) and the actual ones I_{obs}—are analyzed to find inconsistencies. This counting procedure is rather general and it has been already used to find a significant mismatch "between PGA values characterized by an exceedance probability of 10 percent in 30 years and what has actually been observed at 68 accelerometric stations located on stiff soil" (Albarello and D'Amico, 2008).

According to Indirli et al. (2011) the link between PGA and MCS over the Italian territory, summarized in Table 12.4, survived the rigid test by the empirical seismological evidence since its publication by Panza et al. (1997). Moreover, these conversion rules are found applicable not only in Italy but in other seismic regions worldwide. The robust relationship between PGA and MCS, as given in Table 12.4, allows for an uncertainty of a factor of 2 in the ranges of PGA attributed to MCS integers (specifically, the upper limit of a range is by a factor of 2 larger than its lower limit) and it is consistent with the purpose of Cancani (1904). Naturally, this kind of uncertainty is intrinsic for sizing earthquakes either in terms of energy or macroseismic effect (Bormann, 2012).

12.6.1 Validation of Estimated against Observed Epicentral Intensity

Figure 12.6 shows the six intensity maps obtained: (1) from the real seismicity I_{obs} (Figure 12.6(a)), as well as (2) from the ground motion estimates $I_{PGA10\%}$, $I_{PGA2\%}$, I_{DGA}, $I_{DGA10\%}$, $I_{DGA2\%}$ (Figure 12.6(b)−(f), respectively). The percentage of the points with intensity $I \geq$ VIII for each of these maps is summarized in Table 12.5. One can see that PGA10%, PGA2%, DGA, and DGA2% model maps assign intensity $I \geq$ VIII to more than 90 percent of the study area, whereas in the I_{obs} map (observed intensities in about 2000 years) $I \geq$ VIII occupies just 38 percent.

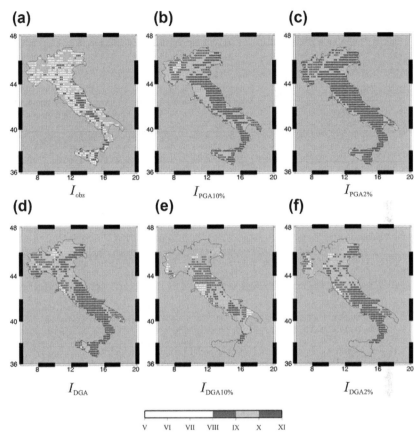

FIGURE 12.6 Intensity maps in comparison: (a) I_{obs}—obtained from the real seismicity, CPTI04 catalog (Gruppo di Lavoro, 2004); PSHA method maps: (b) 475 return period map—$I_{PGA10\%}$; (c) 2475 return period map—$I_{PGA2\%}$; NDSHA method maps; (d) standard—I_{DGA}; (e) 475 return period map—$I_{DGA10\%}$; (f) 2475 return period map—$I_{DGA2\%}$. PGA, peak ground acceleration; DGA, design ground acceleration (after Nekrasova et al., 2014).

TABLE 12.5 The Percentage of I_{MCS} from Different Ranges in the Six Intensity Maps

I_{MCS}	$P(I_{MCS})$, %					
	Observation	PGA10%	PGA2%	DGA	DGA10%	DGA2%
$\geq$XI	3.01	—	33.75	15.75	0.67	12.37
$\geq$X	11.70	43.88	73.50	45.88	13.84	42.88
$\geq$IX	23.05	76.25	90.63	78.38	45.09	73.73
$\geq$VIII	37.77	97.88	100	96.88	87.05	95.42

The empirical density distributions of the difference, $\Delta I = I_{mGA} - I_{obs}$, derived from the comparison between the model and the observed intensity maps, for each of the five methods under consideration, are shown in Figure 12.7. The density distributions of ΔI are shown as functions of I_{obs}. Color indicates one of the three intervals of the observed intensity: $VI \leq I_{obs} \leq VII$ (blue), $VIII \leq I_{obs} \leq IX$ (green), and $I_{obs} \geq X$ (red).

The statistical significance of the observed differences and their interpretation should be considered in further investigations that will require special analysis of the basic complexity and dependencies of the observed ground shaking and earthquake location, besides a straightforward comparison of the distribution functions. However it can be observed that both PGA10% and PGA2% maps provide a higher rate of overestimations ($\Delta I > 0$) and a lower rate of presumably correct estimates ($\Delta I \leq 0$), with respect to DGA maps. Specifically, the ratio between the two rates equals 7.06, 21.56, 6.83, 1.66, and 4.09 for PGA10%, PGA2%, DGA, DGA10%, and DGA2%, correspondingly.

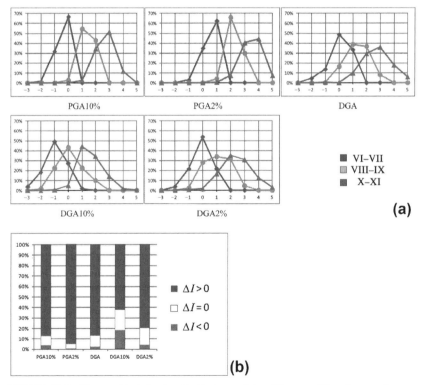

FIGURE 12.7 The empirical density distributions of the difference, $DI = I_{mGA} - I_{obs}$, as function of I_{obs} (a), where I_{mGA} is the intensity from a model map mGA and I_{obs} is the observed macroseismic intensity; the balance, in percent, between under- and over-estimation of ground shaking by each model (b).

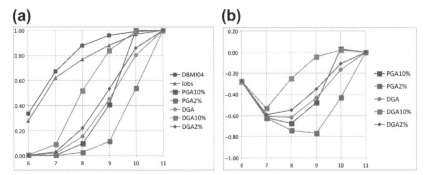

FIGURE 12.8   (a) The empirical probability functions of the macroseismic intensity and (b) the difference between a model and the real intensities $F_i(I) - F_0(I)$. DGA, design ground acceleration; PGA, peak ground acceleration.

Figure 12.8(a) shows the empirical cumulative distributions $F_i(I)$ of embedment of intensity VI or more from the CPTI04 catalog (Gruppo di Lavoro, 2004), I_{obs}, and from the five model estimates $I_{PGA10\%}$, $I_{PGA2\%}$, I_{DGA}, $I_{DGA10\%}$, $I_{DGA2\%}$. The fit of F_is related to DGA10% and to the CPTI04 real catalog data is remarkable at intensities IX and X, when compared with the other four models. Figure 12.8(b) shows the five curves of the difference $F_i(I) - F_0(I)$ between each of the models and the real data. The maximum absolute difference of the empirical distributions is commonly used in the Kolmogorov–Smirnov (K–S) two-sample criterion to distinguish whether or not the values from the two samples are drawn from the same statistical distribution of independent variables. The two sample K–S statistic λ_{K-S} applied to a model and the real catalog is defined as $\lambda_{K-S}(D, n, m) = [nm/(n+m)]^{1/2}D$, where $D = \max|F_i(I) - F_0(I)|$ is the maximum value of the absolute difference between the empirical distributions $F_i(I)$ and $F_0(I)$, $I =$ VI, VII, VIII, IX, X, XI, whose sizes are n and m, respectively. Table 12.6 summarizes the results of comparison for each of the five model maps in terms of D and λ_{K-S}.

TABLE 12.6 The Kolmogorov–Smirnov Two-Sample Statistic λ_{K-S} Applied to a Model versus the Real Seismic Intensity Maps

Statistic	Model Seismic Intensity Map				
	PGA10%	PGA2%	DGA	DGA10%	DGA2%
Sample size	564	564	564	372	458
D	0.67	0.77	0.62	0.53	0.59
λ_{K-S}	11.314	12.892	10.332	7.949	9.408

The K–S test results confirm quantitatively what can be seen from Figure 12.8: The values of seismic intensity attributed by any model considered and reported by CPTI04 are hardly from the same distribution (the significance level is by far less than 1 percent, i.e., confidence more than 99 percent). On the other hand the DGA10% map appears to be "the best fit" among the five models available. When looking at Figure 12.8(a) one may think that the distribution based on PGA10% also fits well the observed one. This is true only for the quite small sample of intensity $I \geq$ X (which is just 3 percent of the total). This fact is better understood when looking at Figure 12.8(b), where the quantified value of the fit (which maximum deviation from 0 determines the Kolmogorov–Smirnov statistic) is plotted for each model.

Finally, the hazard maps of $I_{PGA10\%}$, $I_{PGA2\%}$, I_{DGA}, $I_{DGA10\%}$, and $I_{DGA2\%}$ are compared with the location of seismic events that produced intensity $I \geq$ VIII at the epicenter. The results are summarized in Table 12.7. As a reference observational data set, the comparative analysis is extended also to the macroseismic database DBMI04, used in the compilation of the Italian catalog CPTI04, (Stucchi et al., 2007). Table 12.7(a) lists the empirical counts required for the calculation of the statistical significance P given in Table 12.7(b), estimated as:

$$P = 1 - B(N_{s+} - 1, N_s, N_{I+}/N_{all}),$$

where $B(m,n,p)$ is the standard binomial distribution function that provides the probability of m or less successes at random in n trials, with probability p of success in a single trial; N_{s+} and N_s are the numbers of the strong seismic events in agreement with the intensity map and the total for the region under investigation; N_{all} is the total number of grid nodes of an intensity map; and N_{I+} is the number of the nodes with intensity I or more.

According to the binomial probability test P, for the intensity range VIII the correspondence between reported intensities and all hazard maps can be attributed to a random coincidence. In the other three higher ranges of intensity the test P might be indicative of the nonrandomness for each of the five model maps.

12.6.2 Results of Comparison with Improved Seismic and Macroseismic Data

For the three basic maps PGA10%, PGA2%, and DGA an additional comparison with an updated, presumably improved, compilation of seismic data is performed. The catalog of the $M \geq 5$, earthquakes with hypocentral depth $h \leq 70$ km, as reported in the up-to-date version of UCI2001 (for reference see Peresan and Panza (2002)) for the Italian territory is used for this purpose. In addition to a comparison at epicenters, a comparison with direct macroseismic observations reported in DBMI04 (Stucchi et al., 2007) and rounded to 0.5 units of intensity is performed. The corresponding control maps are shown in Figure 12.9.

TABLE 12.7 The Binomial Test of the Five Hazard Maps and Macroseismic Observations (Last Row) against Earthquakes from the Four Intensity Ranges[2]

(a) Empirical Counts

Map	N_{all}	N_s VIII	IX	X	XI	N_{I+} VIII	IX	X	XI	N_{s+} VIII	IX	X	XI
PGA10%	800	204	103	35	8	783	610	351	—	193	92	24	—
PGA2%	800	204	103	35	8	800	725	588	270	203	102	34	8
DGA	800	204	103	35	8	775	627	367	126	198	97	29	6
DGA10%	448	153	81	25	7	390	202	62	3	110	47	11	—
DGA2%	590	177	90	31	8	563	435	253	73	171	86	23	4
DBMI04	48,126	246	112	36	8	4459	1575	536	65	207	97	31	7

(b) Binomial Probability $P = 1 - B(N_{s+} - 1, N_s, N_{I+}/N_{all})$

Map	VIII	IX	X	XI
PGA10%	99.84%	0.06%	0.28%	Not Applicable
PGA2%	100.00%	0.05%	0.03%	0.02%
DGA	54.54%	0.00%	0.00%	0.03%
DGA10%	100.00%	1.31%	0.02%	NA
DGA2%	29.54%	0.00%	0.04%	1.08%
DBMI04	0.00%	0.00%	0.00%	0.00%

(c) Sum of Errors According to Empirical Counts

Map	VIII	IX	X	XI
PGA10%	103.27%	86.93%	75.30%	100.00%
PGA2%	100.49%	91.60%	76.36%	33.75%
DGA	99.82%	84.20%	63.02%	40.75%
DGA10%	115.16%	87.06%	69.84%	100.67%
DGA2%	98.81%	78.17%	68.69%	62.37%
DBMI04	25.12%	16.67%	15.00%	12.64%

[2]This is an application of the binomial test to the earthquake data that were available and presumably used for the seismic hazard estimates (i.e., not independent data); therefore, the obtained probabilities are not unbiased values, although they might be used as a quantitative indicator of the model fit to the real data.

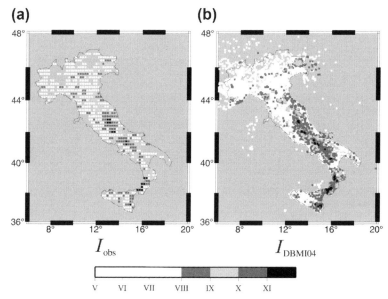

FIGURE 12.9 The intensity maps in comparison: (a) I_{obs}—same as Figure 12.1(a); (b) I_{DBMI04}—database of reported macroseismic data.

Thus, in terms of efficiency, measured by the sum of the error percentages (smaller value corresponds to better efficiency), i.e., $102.86\% = 33.63\% + 69.23\%$ for PGA10%, $99.65\% = 68.88\% + 30.77\%$ for PGA2%, and $81.81\% = 39.50\% + 42.31\%$ for DGA, the neo-deterministic model appears to outscore the probabilistic ones. Similarly, the efficiency measured accounting only for "large errors" in expectation ($\Delta M > 0.5$), is estimated with the $64.40\% = 33.63\% + 30.77\%$ for PGA10%, $72.73\% = 68.88\% + 3.85\%$ for PGA2%, and $51.04\% = 39.50\% + 11.54\%$ for DGA.

Table 12.8 lists the number of intensity records that violate mGA-maps prediction, i.e., those with the positive difference $\Delta I = I_{DBMI04} - I_{mGA}$, where I_{mGA} is the value attributed to the nearest SHA-map grid point within 20-km distance from the I_{DBMI04} data point location (Figure 12.9(b)). The degree of violation is qualitatively the same as in Tables 12.7(c) and 12.9 obtained when using CPTI04 and UCI2001 catalogs.

Thus, once again, in terms of the sum of the error percentages (Tables 12.7(c), 12.9, and 12.10), the neo-deterministic model appears to outscore the probabilistic ones. It could be argued that the comparison of macroseismic observations with model macroseismic parameters presented here could be sufficient and the only one to be considered. However, a single earthquake may invalidate more than a single value on a given map, so that testing the model hypothesis would be biased by dependencies in a control sample. That is why, in the main part of our study, each seismic event is attributed with only one

TABLE 12.8 Number of DBMI04 Intensity That Violate Model Ground Motion-Maps Prediction

	DBMI04	PGA10		PGA2		DGA	
I	N_{I+}	N_{I+}	$\Delta I > 0$	N_{I+}	$\Delta I > 0$	N_{I+}	$\Delta I > 0$
XI	65	—	65	270	0	126	8
X	535	351	201	588	2	367	27
IX	1572	610	206	725	3	627	50
VIII	4421	783	208	800	3	775	56

TABLE 12.9 Number of Shallow Earthquakes That Violate Model Ground Motion-Map Predictions

	UCI2001	PGA10			PGA2			DGA		
		%	ΔM		%	ΔM		%	ΔM	
M_{max}	N_{max+}	M_{max+}	>0	>0.5	M_{max+}	>0	>0.5	M_{max+}	>0	>0.5
7	4	—	4	4	0.75	4	1	1.13	4	1
6.5	10	—	10	6	33.50	6	1	15.75	6	2
6	26	33.63	18	8	68.88	8	1	39.50	11	3
5.5	75	71.38	21	9	86.50	9	1	72.50	19	5
5	233	92.13	25	9	99.75	10	1	92.50	30	6

TABLE 12.10 Sum of Errors for the Three Model Maps Compared to DBMI04

I	PGA10%	PGA2%	DGA
XI	100.00	33.75	28.06
X	81.45	73.87	50.92
IX	89.35	90.82	81.56
VIII	102.58	100.07	98.14

value of macroseismic intensity, which is the empirical MCS intensity at the epicenter, I_0, provided in the CPTI04 catalog.

The comparison of the model intensity maps against the real seismic activity in Italy reveals many discrepancies regarding several aspects of models of seismic ground-shaking distribution in space and size. In terms of efficiency in anticipating ground shaking, measured by the sum of the error percentages, the NDSHA models appear to outscore the PSHA ones and might be a better fit to the real seismicity.

Therefore seismic hazard maps, which seek to predict the shaking that would actually occur, must be tested against the real seismic activity. Otherwise, the use of untested maps may mislead to crimes of negligence, similar to medical malpractice, although at much higher levels of consequent losses (Wyss et al., 2012).

12.7 ESTIMATES OF SEISMIC RISKS

Risk estimates are the result of the convolution of the hazard with the exposed object under consideration along with its vulnerability:

$$R(c) = H(c) \otimes O(c) \otimes V(O(c))$$

where $H(c)$ is a hazard estimate at location c, $O(c)$ is the exposure of objects at risk at location c, and $V(O)$ is the vulnerability of objects at risk. In specific applications, c could be a point, or a line, or some area on or under the Earth surface, the distribution of potential hazards, as well as objects of concern and their vulnerability, could be time-dependent.

Thus, there exist many different risk estimates even if the same object of risk and the same hazard are involved. Specifically, the estimate may result from the different laws of convolution, as well as from different kinds of vulnerability of an object at risk under specific environments and conditions. Both conceptual issues must be resolved in a multidisciplinary, problem-oriented research performed by specialists in the fields of hazard, objects at risk, and object vulnerability. To illustrate this general concept, Parvez et al. (2014) have used the following simplified four convolutions of SHA map $H(c)$ with the population density distribution P.

All the four risk estimates consider the same population data (Gridded Population of the World, GPWv3 (2005), model estimates of the year 2010) as an object of risk and simply use the mathematical product as the convolution law. The first estimate in a cell c of the uniform grid is based on the constant equal vulnerability of an individual, $R_i(c) = H(c) \cdot \int_c P$, where $\int_c P$ is the integral of the population density over a given cell c, i.e., the number of individuals within the area of c. The second risk estimate differentiates individual vulnerability in proportion to the population density at a given site, $R_{ii}(c) = H(c) \cdot \int_c P \cdot P$. This and the other two assumptions—$R_{iii}(c) =$

$H(c) \cdot \int_c P \cdot P^2$ and $R_{iv}(c) = H(c) \cdot \int_c P \cdot P^3$ —appear to be rather natural due to specifics of artificial environment and infrastructure in the areas of high concentration of individuals, e.g., the number of floors in a typical residential building changes with population density.

The resulting maps of the four risks in arbitrary units are given in Figure 12.10. For illustration purposes, each of the four risk scales is covering just the top seven decimal orders of the risk values, so that the cells in red are 1,000,000 times more risky than those in blue. The collapse of the risky areas to the region of the densest population appears to demonstrate how the nonlinearity of conditions changes expectation of seismic risk. As expected, the megacities and their agglomerations are at the top of the risk distributions.

To avoid misleading and counterproductive interpretations, it should be emphasized that *the risk estimates presented for the region under study are given here for academic methodological purposes only.* They do not use procedures that might be more adequate for estimating hazard, objects at risk and their vulnerability, and are used here simply to illustrate the general

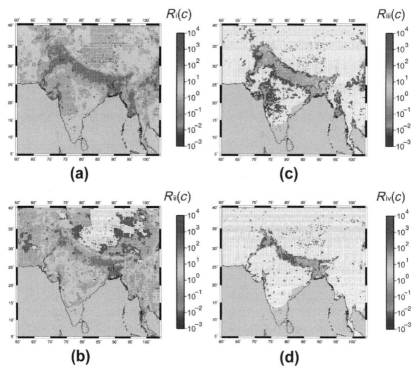

FIGURE 12.10 Simplified convolutions of seismic hazard map $H(g)$ with the population density distribution P for the Himalayas and surrounding regions: (a) $R_i(c) = H(c) \cdot \int_c P$, (b) $R_{ii}(c) = H(c) \cdot \int_c P \cdot P$, (c) $R_{iii}(c) = H(c) \cdot \int_c P \cdot P^2$, and (d) $R_{iv}(c) = H(c) \cdot \int_c P \cdot P^3$.

problem-oriented possibilities. Obviously, the estimations addressing more realistic and practical kinds of seismic risks should involve experts in distribution of objects at risk of different vulnerability, i.e., specialists in earthquake engineering, social sciences, and economics.

12.8 SUMMARY AND CONCLUSIONS

We discussed in this chapter some of the theoretical and practical limits of the widely applied PSHA approach to SHA. So far, the arguments supporting the continued use of PSHA, in spite of its poor anticipatory performances and basic shortcomings, do not appear convincing. Such arguments are briefly summarized in the conclusions of the paper by Stirling (2014), claiming that PSHA: (1) is simple; (2) provides useful information for engineering design; (3) it should be used until another alternative method is available that is shown to do better than PSHA.

As we pointed out in this chapter (e.g., Section 12.2), the simple formulation of PSHA is affected by several theoretical biases (e.g., use of convolutive techniques; inadequate treatment of uncertainties) and assumptions not supported by data, which may hamper the resulting hazard estimates, particularly in areas where historical information is scarce. Moreover, nowadays it is widely recognized that the ground motion description provided by PSHA as a single parameter (PGA) is not sufficient for the appropriate analysis of the seismic response of buildings and infrastructures; on the other hand, alternative methods exist, which have proved effective for engineering design (e.g., Vaccari et al., 2009; Panza et al., 2011b). Interestingly, the PSHA maps have never been supported by a formal validation process (Stein et al., 2011). The first systematic comparison of those to the real seismic activity on the global scale (Kossobokov and Nekrasova, 2012) disclosed clearly the statistics contradictory to 10 percent of exceedence in 50 years, i.e., 50 percent of surprises for strong earthquakes about magnitude 6 and 90 percent—for major events above magnitude 7. Evidently, on a local, or even on a regional scale, probabilistic estimates might be untestable in realistic time (e.g., Beauval et al., 2008). Therefore the application of alternative methods, which are already available (like NDHSA), will have the twofold effect to: (1) extensively test them, and eventually; (2) show that they perform more reliably than PSHA.

Timing and locations of earthquakes are not predictable with a precision of a certain date and rupture zone. Although the long-lasting global experiment testing intermediate-term and middle-range earthquake predictions supports the hypothesis that the M8 algorithm can be effective to globally reduce the impact of great earthquakes (magnitude M8.0+) in advance (Kossobokov, 2012; Davis et al., 2012), there is no link to implement measures and improve earthquake preparedness in response to them. This is not done yet, in part due to limited distribution of these predictions and the lack of applying existing methods for using intermediate-term predictions to make decisions and taking action. Operational aspects of predictions of earthquakes and related ground shaking are discussed in Peresan et al. (2012), with special emphasis on the Italian territory, where a rigorous prospective testing of two different prediction algorithms has been ongoing for more than a decade,[3] thus allowing the assessment of their statistical significance and the establishment of operational practice of definition of time-dependent scenarios for the territory of Italy, which has been carried on routinely in the framework of the Agreement with the Civil Defence of the Friuli Venezia Giulia Region (NE Italy).

On the other hand, the general locations of earthquakes are well known globally by earthquake monitoring and geologic studies. Strong disastrous earthquakes are low-probability events that happen with certainty (i.e., 100 percent probability). Long recurrence or low-probability earthquake information provided by any group as the basis for SHA leads to false comfort of ignoring such rare eventualities, so that they will happen again as unpredictable disastrous surprises.

Losses from natural hazards have continued to increase because population increases and corruption prevent the construction of buildings resistant to strong shaking (Bilham, 2014). As population increases, the obligation of experts to provide realistic estimates of the seismic hazard increases. Scientists, with their special education, knowledge, and skills, owe to society, which does not possess the same special knowledge and skills, to work for their protection. Errors in SHA propagate nonlinearly into unbearable errors in the expected human and economic losses (Wyss et al., 2012).

The largest expected earthquake size for earthquake-prone locations can be realistically estimated with high confidence. Preparing for such hazardous events ensures reduction or even avoidance of potential disasters. Earthquakes do not kill people but their consequences, i.e., tsunamis, landslides, collapsed buildings, bridges, and other constructions, do. The hazard scenarios for such an event estimated by deterministic computations, like NDSHA, provide the basis for necessary preparations, from land-use planning and building code regulation to emergency management. Many risks are generated by earthquakes. These should be not ignored in any realistic and responsible seismic risk evaluation and knowledgeable disaster prevention.

Contemporary science can do a better job in estimating natural hazards, assessing risks, and delivering such information before catastrophic events occur. "A return period of 475 years" implies using observations accumulated over geological timescales of about several thousand years. In lieu of local

---

3. The experiment started in Italy in 2002 (Peresan et al., 2005), aimed at a real-time testing of M8S and CN predictions for earthquakes with magnitude larger than a given threshold (namely 5.4 and 5.6 for CN algorithm, and 5.5 for M8S algorithm). Predictions are regularly updated every 2 months and a complete archive of predictions is made available online (http://www.ictp. trieste.it/www_users/sand/prediction/prediction.htm); the obtained results already allowed the assessment of the statistical significance of the issued predictions.

seismic observations covering such long periods, one may try using pattern recognition of earthquake-prone areas based on the appropriate geophysical and geological data sets that are available. The geomorphological pattern-recognition approach (Gelfand et al., 1972) is especially useful as a preparedness and mitigation tool in seismic regions that have no historical record, but are potentially prone to strong earthquakes. The validity of this method has been supported by statistics of strong earthquake occurrences after numerous publications of pattern recognition results for many seismic regions and magnitude ranges (Gorshkov et al., 2003; Soloviev et al., 2014). Also, the current knowledge of the physical process of earthquake generation and wave propagation in anelastic media help providing assessment of seismic hazard based on computation of synthetic seismograms in terms of realistic modeling of ground motion (Panza et al., 2001). These advances are contained in NDSHA, whereas PSHA can hardly benefit from such knowledge (it is notable that in California the deterministic approach is used for practical evaluations of seismic risk (Mualchin, 2011)).

The code for application of NDSHA is available (see www.xeris.it for the code version recently enabled on a cloud platform). The comparison of the model intensity mapped against the real seismic activity in Italy reveals that, in terms of efficiency in anticipating ground shaking, measured by the sum of the error percentages, the NDSHA models appear to outscore the PSHA ones and might be a better fit to the real seismicity. This property makes NDSHA application still worthwhile on account of its sensitivity to the quality of model input data. Also, the existing algorithms for diagnosis of times of increased probability, proved reliable in the ongoing earthquake prediction experiment (Kossobokov and Shebalin, 2003; Kossobokov, 2014) to deliver an intermediate-term, middle-range component to the NDSHA approach in the evaluation of time-dependent seismic hazard (Peresan et al., 2011).

Seismic hazard maps, which seek to predict the shaking expected to occur in the future, must be tested against the real seismic history and future earthquakes, in order to reliably assess their anticipatory capability, which is the key toward effective risk mitigation.

REFERENCES

Aki, K., Richards, P.G., 2002. Quantitative Seismology. University Science Books, Sausalito, CA, USA. ISBN 978-0935702965.

Albarello, D., D'Amico, V., 2008. Testing probabilistic seismic hazard estimates by comparison with observations: an example in Italy. Geophys. J. Int. 175, 1088−1094.

Alekseevskaya, M.A., Gabrielov, A.M., Gvishiani, A.D., Gelfand, I.M., Ranzman, E.Ya., 1977. Formal morphostructural zoningof mountain territories. J. Geophys. 43, 227−233.

Alexander, D.E., 2011. Mortality and morbidity risk in the L'Aquila, Italy, earthquake of 6 April 2009 and lessons to be learned. In: Spence, R., Ho, E., Scawthorn, C. (Eds.), Human Casualties in Earthquakes: Progress in Modelling and Mitigation. Advances in Natural and Technological Hazards Research No. 29. Springer, Berlin, pp. 185−197.

Alexander, D., Magni, M., January 7, 2013. Mortality in the L'Aquila (Central Italy) earthquake of 6 April 2009. PLoS Curr. 5 http://dx.doi.org/10.1371/50585b8e6efd1 e50585b8e6efd1. http://www.ncbi.nlm.nih.gov/pmc/articles/PMC3541886/.

Anderson, J.G., Brune, J.N., 1999. Methodology for using precarious rocks in Nevada to test seismic hazard models. Bull. Seismol. Soc. Am. 89, 456–467.

Anderson, J.G., Brune, J.N., Purvance, M., Biasi, G., Anooshehpoor, R., 2010. Benefits of the Use of Precariously Balanced Rocks and Other Fragile Geological Features for Testing the Predictions of Probabilistic Seismic Hazard Analysis. Draft position paper.

Aptikaev, F.F., Mokrushina, N.G., Erteleva, O.O., June 2008. The Mercalli family of seismic intensity scales. J. Volcanol. Seismol. 2 (3), 210–213 (Original Russian Text © Аптикаев Ф.Ф., Мокрушина Н.Г., Эртелева О.О., 2008. Категория сейсмических шкал семейства Меркалли. Вулканология и сейсмология, No 3, Май-Июнь 2008, С. 74–78).

Artioli, E., Battaglia, R., Tralli, A., 2013. Effects of May 2012 Emilia earthquake on industrial buildings of early '900 on the Po river line. Eng. Struct. 56, 1220–1233.

Basili, R., Valensise, G., Vannoli, P., Burrato, P., Fracassi, U., Mariano, S., Tiberti, M.M., Boschi, E., 2008. The Database of Individual Seismogenic Sources (DISS), version 3: summarizing 20 years of research on Italy's earthquake geology. Tectonophysics. http://dx.doi.org/10.1016/j.tecto.2007.04.014.

Bazzurro, P., Cornell, C.A., 1999. Disaggregation of seismic hazard. Bull. Seismol. Soc. Am. 89, 501–520.

Beauval, C., Bard, P.-Y., Hainzl, S., Guéguen, P., 2008. Can strong-motion observations be used to constrain probabilistic seismic-hazard estimates? Bull. Seismol. Soc. Am. 98 (2), 509–520. http://dx.doi.org/10.1785/0120070006.

Bilham, R., 2009. The seismic future of cities. Bull. Earthquake Eng. 7 (4), 839–887. http://dx.doi.org/10.1007/s10518-009-9147-0.

Bilham, R., 2014. Aggravated earthquake risk in South Asia: engineering vs. human nature. In: Shroder, J., Wyss, M. (Eds.), Earthquake Hazard, Risk, and Disasters. Elsevier, London, pp. 103–141.

Bizzarri, A., 2012. What can physical source models tell us about the recurrence time of earthquakes? Earth-Sci. Rev. 115, 304–318. http://dx.doi.org/10.1016/j.earscirev.2012.10.004.

Bizzarri, A., Crupi, P., 2013. Linking the recurrence time of earthquakes to source parameters: a dream or a real possibility? Pure Appl. Geophys. http://dx.doi.org/10.1007/s00024-013-0743-1.

Bommer, J.J., Abrahamson, N.A., 2006. Why do modern probabilistic seismic hazard analyses often lead to increased hazard estimates? Bull. Seismol. Soc. Am. 96, 1967–1977.

Boore, D.M., Atkinson, G.M., 2008. Ground-motion prediction equations for the average horizontal component of PGA, PGV, and 5%-damped PSA at spectral periods between 0.01 s and 10.0 s. Earthquake Spectra 24 (1), 99–138.

Bormann, P. (Ed.), 2012. New Manual of Seismological Observatory Practice (NMSOP-2). IASPEI, GFZ German Research Centre for Geosciences, Potsdam. http://dx.doi.org/10.2312/GFZ.NMSOP-2 urn:nbn:de:kobv:b103-NMSOP-2. http://nmsop.gfz-potsdam.de.

Boschi, E., Favalli, P., Frugoni, F., Scalera, G., Smriglio, G., 1995. Mappa Massima Intensità Macrosismica Risentita in Italia. Istituto Nazionale di Geofisica, Roma.

Brandmayr, E., Romanelli, F., Panza, G.F., 2013. Stability of fault plane solutions for the major N-Italy seismic events in 2012. Tectonophysics 608, 525–529. http://dx.doi.org/10.1016/j.tecto.2013.08.034.

Brune, J.N., 1999. Precariously rocks along the Mojave section of the San Andreas fault, California: constraints on ground motion from great earthquakes. Seism. Res. Lett. 70, 29–33.

Brune, J.N., Whitney, J.W., 2000. Precarious rocks and seismic shaking at Yucca Mountain, Nevada. In: Whitney, J.W., Keefer, W.R. (Eds.), Geologic and Geophysical Characterization Studies of Yucca Mountain, Nevada, A Potential High-Level Radioactive-Waste Repository. U.S. Geological Survey Digital Data Series 058 (Chapter M).

Burger, R.W., Sommerville, P.G., Barker, J.S., Herrmann, R.B., Helmberger, D.V., 1987. The effect of crustal structure on strong ground motion attenuation relations in eastern North America. Bull. Seismol. Soc. Am. 77, 1274—1294.

Cancani, A., 1904. Sur l'emploi d'une double echelle seismique des intesites, empirique et absolue. Gerlands Beitr. Geophys. 2, 281—283.

Castaños, H., Lomnitz, C., 2002. PSHA: is it science? Eng. Geol. 66 (3—4), 315—318.

Chapman, N., Berryman, K., Villamor, P., Epstein, W., Cluff, L., Kawamura, H., 2014. Active faults and nuclear power plants. EOS Trans. AGU 95 (4), 33—40.

Cornell, C.A., 1968. Engineering seismic risk analysis. Bull. Seismol. Soc. Am. 58, 1583—1606.

Davis, C., Keilis-Borok, V., Kossobokov, V., Soloviev, A., 2012. Advance prediction of the March 11, 2011 Great East Japan Earthquake: a missed opportunity for disaster preparedness. Int. J. Disaster Risk Reduct. 1, 17—32. http://dx.doi.org/10.1016/j.ijdrr.2012.03.001.

Decanini, L., Gavarini, C., Mollaioli, F., 1995. Proposta di definizione delle relazioni tra intensita' macrosismica e parametri del moto del suolo. Atti 70 Convegno L'ingegnaria Sismica in Italia, 1, pp. 63—72.

Fäh, D., Panza, G.F., 1994. Realistic modelling of observed seismic motion in complex sedimentary basins. Ann. Geofis. 37, 1771—1797.

Field, E.H., The SCEC Phase III Working Group, 2000. Accounting for site effects in probabilistic seismic hazard analyses of Southern California: overview of the SCEC Phase III Report. Bull. Seismol. Soc. Am. 90, S1—S31.

Frankel, A., 1995. Mapping seismic hazard in the Central and Eastern United States. Seism. Res. Lett. 66 (4), 8—21.

Frankel, A., 2013. "Why earthquake hazard maps often fail and what to do about it", by S. Stein, R.J. Geller, and M. Liu. Tectonophysics 592, 200—206.

Gasperini, P., Camassi, R., Mirto, C., Stucchi, M., 2004. Catalogo Parametrico dei Terremoti Italiani, versione 2004 (CPTI04). INGV, Bologna. http://emidius.mi.ingv.it/CPTI04/.

Gelfand, I., Guberman, Sh., Izvekova, M., Keilis-Borok, V., Rantsman, E., 1972. Criteria of high seismicity, determined by pattern recognition. Tectonophysics 13, 415—422.

GHDB, 1989. Global Hypocenters Data Base CD-ROM NEIC/USGS. Denver, CO and its updates through June 2010.

Giardini, D., Grünthal, G., Shedlock, K.M., Zhang, P., 1999. The GSHAP global seismic hazard map. Ann. Geofis. 42 (6), 1225—1228.

Giardini, D., Grünthal, G., Shedlock, K.M., Zhang, P., 2003. The GSHAP global seismic hazard map. In: Lee, W., Kanamori, H., Jennings, P., Kisslinger, C. (Eds.), International Handbook of Earthquake & Engineering Seismology, International Geophysics Series 81 B. Academic Press, Amsterdam, pp. 1233—1239.

Gorshkov, A., Kossobokov, V., Soloviev, A., 2003. 6. Recognition of earthquake-prone areas. In: Keilis-Borok, V.I., Soloviev, A.A. (Eds.), Nonlinear Dynamics of the Lithosphere and Earthquake Prediction. Springer, Heidelberg, pp. 239—310.

Gorshkov, A.I., Panza, G.F., Soloviev, A.A., Aoudia, A., 2002. Morphostructural zonation and preliminary recognition of seismogenic nodes around the Adria margin in peninsular Italy and Sicily. JSEE. Spring 2002 4 (1), 1—24.

Gorshkov, A.I., Panza, G.F., Soloviev, A.A., Aoudia, A., 2004. Identification of seismogenic nodes in the Alps and Dinarides. Boll. Soc. Geol. Ital. 123, 3—18.

Gridded Population of the World, Version 3 (GPWv3), 2005. Palisades, NY: Socioeconomic Data and Applications Center (SEDAC), Columbia University. Available at. http://sedac.ciesin. columbia.edu/gpw (2012.05.29).

Gruppo di Lavoro, 2004. Catalogo parametrico dei terremoti italiani, versione 2004 (CPTI04). INGV, Bologna. http://emidius.mi.ingv.it/CPTI04/.

Gusev, A.A., 2011. Broadband Kinematic Stochastic Simulation of an Earthquake Source: a Refined Procedure for Application in Seismic Hazard Studies. Pure Appl. Geophys. 168 (1—2), 155—200.

Gusev, A.A., Pavlov, V., 2006. Wideband simulation of earthquake ground motion by a spectrum-matching, multiple-pulse technique. First European Conference on Earthquake Engineering and Seismology, Geneva, Switzerland, 3—8 September 2006. Abstract Book, 408. 2006.

Gutenberg, B., Richter, C.F., 1956. Earthquake magnitude, intensity, energy, and acceleration. Bull. Seismol. Soc. Am. 46, 105—145.

Hobbs, W.H., 1904. Lineaments of the Atlantic border region. Geol. Soc. Am. Bull. 15, 483—506.

Hobbs, W.H., 1912. Earth Features and Their Meaning: An Introduction to Geology for the Student and General Reader. Macmillan Co., New York, 347 pp.

Indirli, M., Razafindrakoto, H., Romanelli, F., Puglisi, C., Lanzoni, L., Milani, E., Munari, M., Apablaza, S., 2011. Hazard evaluation in Valparaiso: the MAR VASTO Project. Pure Appl. Geophys. 168 (3—4), 543—582.

Jordan, T., Chen, Y., Gasparini, P., Madariaga, R., Main, I., Marzocchi, W., Papadopoulos, G., Sobolev, G., Yamaoka, K., Zschau, J., 2011. ICEF Report. Operational earthquake forecasting: state of knowledge and guidelines for utilization. Ann. Geophys. 54 (4), 315—391.

Kagan, Y.Y., Jackson, D.D., Geller, R.J., 2012. Characteristic earthquake model, 1884—2011, R. I. P. Seismol. Res. Lett. 83, 951—953.

Klügel, J.U., 2005. Problems in the application of the SSHAC probability method for assessing earthquake hazards at Swiss nuclear power plants. Eng. Geol. 78, 285—307.

Klügel, J.-U., 2007a. Error inflation in probabilistic seismic hazard analysis. Eng. Geol. 90, 186—192.

Klügel, J.-U., 2007b. Comment on "Why do modern probabilistic seismic-hazard analyses often lead to increased hazard estimates" by Julian J. Bommer and Norman A. Abrahamson. Bull. Seismol. Soc. Am. 97, 2198—2207.

Klügel, J.-U., 2011. Uncertainty analysis and expert judgment in seismic hazard analysis. Pure Appl. Geophys. 168 (1—2), 27—53.

Kossobokov, V.G., 2014. Times of Increased Probabilities for Occurrence of Catastrophic Earthquakes: 25 Years of Hypothesis Testing in Real Time. In: Shroder, J., Wyss, M. (Eds.), Earthquake Hazard, Risk, and Disasters. Elsevier, London, pp. 477—504.

Kossobokov, V.G., Mazhkenov, S.A., 1994. On similarity in the Spatial Distribution of Seismicity. In: Computational Seismology and Geodynamics, vol. 1. AGU, Washington, DC pp. 6—15.

Kossobokov, V., Shebalin, P., 2003. 4. Earthquake prediction. In: Keilis-Borok, V.I., Soloviev, A.A. (Eds.), Nonlinear Dynamics of the Lithosphere and Earthquake Prediction. Springer, Heidelberg, pp. 141—207.

Kossobokov, V.G., Nekrasova, A., 2007. Unified scaling law for earthquakes: implications for seismic hazard and risk assessment. In: IUGG2007, July 2—13, 2007, Perugia, Italy. Abstracts, SS002—65.

Kossobokov, V.G., 2012. Earthquake prediction: 20 years of global experiment. Nat. Hazards 69 (2), 1155—1177. http://dx.doi.org/10.1007/s11069-012-0198-1.

Kossobokov, V.G., Nekrasova, A.K., 2012. Global seismic hazard assessment program maps are erroneous. Seismic Instrum. 48 (2). http://dx.doi.org/10.3103/S0747923912020065. Allerton Press, Inc., 2012162-170.

Krinitzsky, E.L., 2002. How to obtain earthquake ground motions for engineering design. Eng. Geol. 65, 1−16. http://dx.doi.org/10.1016/S0013-7952(01)00098-9.

La Mura, C., Yanovskaya, T.B., Romanelli, F., Panza, G.F., 2011. Three-dimensional seismic wave propagation by modal summation: method and validation. Pure App. Geophy. 168, 201−216.

Magrin, A., Peresan, A., Vaccari, F., Cozzini, S., Rastogi, B.K., Parvez, I.A., Panza, G.F., 2012. Definition of seismic and tsunami hazard scenarios by exploiting EU-India Grid e-infrastructures, 30. Proceedings of the International Symposium on Grids and Clouds - ISGC 2012 (February 26−March 2, 201). Proceedings of Science, PoS (ISGC 2012) (ISSN: 1824-8039).

Magrin, A., Peresan, A., Vaccari, F., Panza, G.F., 2014. Neo-Deterministic Seismic Hazard Assessment and Earthquake Recurrence, in preparation.

McGuire, R.K., 1995. Probabilistic seismic hazard analysis and design earthquakes: closing the loop. Bull. Seismol. Soc. Am. 85, 1275−1284.

Marzocchi, W., 2013. Seismic hazard and public safety. Eos 94 (27), 240−241.

Meletti, C., Montaldo, V., 2007. Stime di pericolosità sismica per diverse probabilità di super-amento in 50 anni: valori di ag. http://esse1.mi.ingv.it/d2.html. Deliverable D2.

Meletti, C., Valensise, G., 2004. Zonazione sismogenetica ZS9 − App.2 al Rapporto Conclusivo. In: Gruppo di Lavoro MPS, 2004. Redazione della mappa di pericolosità sismica prevista dall'Ordinanza PCM 3274 del 20 marzo 2003. Rapporto Conclusivo per il Dipartimento della Protezione Civile. In italian.

Michetti, A.M., Giardina, F., Livio, F., Mueller, K., Serva, L., Sileo, G., Vittori, E., Devoti, R., Riguzzi, F., Carcano, C., Rogledi, S., Bonadeo, L., Brunamonte, F., Fioraso, G., 2012. Active compressional tectonics, quaternary capable faults, and the seismic landscape of the Po Plain (N Italy). Ann. Geophys. 55 (5), 969−1001. http://dx.doi.org/10.4401/ag-5462.

Molchan, G., Kronrod, T., Panza, G.F., 1997. Multi-scale seismicity model for seismic risk. Bull. Seismol. Soc. Am. 87, 1220−1229.

Mourabit, T., Abou Elenean, K.M., Ayadi, A., Benouar, D., Ben Suleman, A., Bezzeghoud, M., Cheddadi, A., Chourak, M., ElGabry, M.N., Harbi, A., Hfaiedh, M., Hussein, H.M., Kacem, J., Ksentini, A., Jabour, N., Magrin, A., Maouche, S., Meghraoui, M., Ousadou, F., Panza, G.F., Peresan, A., Romdhane, N., Vaccari, F., Zuccolo, E., 2014. Neo-deterministic seismic hazard assessment in North Africa. J. Seismol 18 (2), 301−318. http://dx.doi.org/10.1007/s10950-013-9375-2.

Mualchin, L., 2011. History of modern earthquake hazard mapping and assessment in California using a deterministic or scenario approach. Pure Appl. Geophys. 168, 383−407. http://dx.doi.org/10.1007/s00024-010-0121-1.

Mucciarelli, M., Albarello, D., D'Amico, V., 2008. Comparison of probabilistic seismic hazard estimates in Italy. Bull. Seismol. Soc. Am. 98 (6), 2652−2664.

Murphy, J.R., O'Brien, L.J., 1977. The correlation of peak ground acceleration amplitude with seismic intensity and other physical parameters. Bull. Seismol. Soc. Am. 67, 877−915.

Nekrasova, A., Kossobokov, V., Peresan, A., Magrin, A., 2014. The comparison of the NDSHA, PSHA seismic hazard maps and real seismicity for the Italian territory. Nat. Hazards 70 (1), 629−641. ISSN: 0921-030X. http://dx.doi.org/10.1007/s11069-013-0832-6.

Panza, G.F., Suhadolc, P., 1989. Realistic Simulation and Prediction of Strong Ground Motion. In: Computers and Experiments in Stress Analysis, vol. 82, pp. 77−98, Southampton Boston. Fourth International Conference on computational methods and experimental measurements. Computational Mechanics Publications and Springer-Verlag.

Panza, G.F., Vaccari, F., Cazzaro, R., 1997. Correlation between macroseismic intensities and seismic ground motion parameters. Ann. Geophys. 15, 1371—1382.

Panza, G.F., Romanelli, F., Vaccari, F., 2001. Seismic wave propagation in laterally heterogeneous anelastic media: theory and applications to seismic zonation. Adv. Geophys. 43, 1—95.

Panza, G.F., Alvarez, L., Aoudia, A., Ayadi, A., Benhallou, H., Benouar, D., et al., 2002. Realistic modeling of seismic input for megacities and large urban areas (the UNESCO/IUGS/IGCP project 414). Episodes 25, 160—184. ISSN: 0705-3797.

Panza, G., Irikura, K., Kouteva-Guentcheva, M., Peresan, A., Wang, Z., Saragoni, R. (Eds.), 2011a. Advanced Seismic Hazard Assessment, Pure Appl. Geophys. 168 (1—4), 752 pp.

Panza, G.F., Peresan, A., Vaccari, F., Romanelli, F., Martelli, A., 2011b. Scenario-based time-dependent definition of seismic input: an effective tool for engineering analysis and seismic isolation design. In: Proceedings del congresso SEWC2011-Structural Engineers World Congress (Cernobbio, Como, 4—6 April 2011).

Panza, G.F., La Mura, C., Peresan, A., Romanelli, F., Vaccari, F., 2012. Seismic hazard scenarios as preventive tools for a disaster resilient society. In: Dmowska, R. (Ed.), Adv Geophys. Elsevier, London, pp. 93—165.

Panza, G.F., Peresan, A., La Mura, C., 2013. Seismic hazard and strong ground motion: an operational neo-deterministic approach from national to local scale. In: UNESCO-EOLSS Joint Commitee (Ed.), Encyclopedia of Life Support Systems (EOLSS), Geophysics and Geochemistry. Developed under the Auspices of the UNESCO, Eolss Publishers, Oxford ,UK, pp. 1—49.

Parvez, I.A., Nekrasova, A., Kossobokov, V., 2014. Estimation of seismic hazard and risks for the Himalayas and surrounding regions based on Unified Scaling Law for Earthquakes. Nat. Hazards 71 (1), 549—562.

Parvez, I.A., Romanelli, F., Panza, G.F., 2011. Long period ground motion at bedrock level in Delhi city from Himalayan earthquake scenarios. Pure Appl. Geophy. 168 (3—4), 409—477. http://dx.doi.org/10.1007/s00024-010-0162-5.

Paskaleva, I., Dimova, S., Panza, G.F., Vaccari, F., 2007. An earthquake scenario for the micro-zonation of Sofia and the vulnerability of structures designed by use of the Eurocodes. Soil Dyn. Earthquake Eng. 27, 1028—1041.

Paskaleva, I., Kouteva-Guencheva, M., Vaccari, F., Panza, G.F., 2011. Some contributions of the neo-deterministic seismic hazard assessment approach to earthquake risk assessment for the city of Sofia. Pure Appl. Geophys. 168, 521—541.

Peresan, A., Panza, G.F., 2002. UCI2001: The Updated Catalogue of Italy. Internal report IC/IR/2002/3. ICTP, Trieste, Italy.

Peresan, A., Kossobokov, V., Romashkova, L., Panza, G.F., 2005. Intermediate-term middle-range earthquake predictions in Italy: a review. Earth Sci. Rev. 69 (1—2), 97—132.

Peresan, A., Kossobokov, V., Panza, G.F., 2012. Operational earthquake forecast/prediction. Rend. Fis. Acc. Lincei 23, 131—138.

Peresan, A., Magrin, A., Nekrasova, A., Kossobokov, V.G., Panza, G.F., 2013. Earthquake recurrence and seismic hazard assessment: a comparative analysis over the Italian territory. In: Proceedings of the ERES 2013 Conference. WIT Trans. Built Environ. vol. 132, 23—34. http://dx.doi.org/10.2495/ERES130031.

Peresan, A., Magrin, A., Vaccari, F., Panza, G.F., 2014. Is PSHA the only way to deal with earthquake recurrence and uncertainties? Proceedings of the 50th Annual Convention of the Indian Geophysical Union (Hyderabad, 8—12 January 2014), 96—99.

Peresan, A., Panza, G.F., 2012. Improving earthquake hazard assessment in Italy: an alternative to "Texas sharpshooting". EOS Trans. Am. Geophys. Union 93 (51), 538—539.

Peresan, A., Zuccolo, E., Vaccari, F., Gorshkov, A., Panza, G.F., 2011. Neo-deterministic seismic hazard and pattern recognition techniques: time-dependent scenarios for North-Eastern Italy. Pure Appl. Geophys. 168 (3−4). http://dx.doi.org/10.1007/s00024-010-0166-1.

Reiter, L., 1990. Earthquake Hazard Analysis. Columbia University Press, New York, p. 254.

Renault, P., Heuberger, S., Abrahamson, N.A., 2010. PEGASOS Refinement Project: An improved PSHA for Swiss nuclear power plants. Proceedings of 14ECEE—European Conference of Earthquake Engineering.

Sauter, F., Shah, H.C., 1978. Estudio de seguro contra terremoto. Franz Sauter y Asociados Ltda, San Jose, Costa Rica, 250 pp.

Shedlock, K.M., Giardini, D., Grünthal, G., Zhang, P., 2000. The GSHAP global seismic hazard map. Seismol. Res. Lett. 71 (6), 679−686.

Shteinberg, V.V., Saks, M.V., Aptikaev, F.F., Alkaz, V.G., Gusev, A.A., Yerokhin, L.Yu., Zahradník, I., Kendzera, A.V., Kogan, L.A., Lutikov, A.I., Popova, E.V., Rautian, T.G., Chernov, Yu.K., 1993. Methods of assessment of seismic effects, 1993. Probl. Eng. Seismol. 34, 5−94 (in Russian). (Original Russian Text ©Штейнберг В. В., Сакс М. В., Аптикаев Ф. Ф., Алказ В. Г., Гусев А. А., Ерохин Л. Ю., Заградник И., Кендзера А. В., Коган Л. А., Лутиков А. И., Попова Е. В., Раутиан Т. Г., Чернов Ю. К., Методы оценки сейсмических воздействий.//Вопросы инженерной сейсмологии. 1993. Вып. 34. С. 5−94.).

Soloviev, A.A., Gvishiani, A.D., Gorshkov, A.I., Dobrovolsky, M.N., Novikova, O.V., 2014. Recognition of Earthquake-Prone Areas: Methodology and Analysis of the Results. Izvestiya, Phys. Solid Earth vol. 50. (No. 2), 151−168. http://dx.doi.org/10.1134/S1069351314020116.

SSHAC, 1997. Review of Recommendations for Probabilistic Seismic Hazard Analysis: Guidance on Uncertainty and Use of Experts. Senior Seismic Hazard Analysis Committee (SSHAC), Panel on Seismic Hazard Evaluation, Committee on Seismology, Commission on Geosciences, Environment, and Resources. The National Academies Press, National Research Council.Washington, DC.

Stein, S., Geller, R., Liu, M., 2011. Bad assumptions or bad luck: why earthquake hazard maps need objective testing. Seismol. Res. Lett. 82, 5.

Stein, S., Geller, R., Liu, M., 2012. Why earthquake hazard maps often fail and what to do about it. Tectonophysics, 562−563, 1−25.

Stein, S., Geller, R., Liu, M., 2013. Reply to comment by Arthur Frankel on "Why Earthquake Hazard Maps Often Fail and What to do About It". Tectonophysics 592, 207−209.

Stepp, J., Carl, et al., 2001. Probabilistic seismic hazard analyses for ground motions and fault displacement at Yucca Mountain, Nevada. Earthquake Spectra 17.1, 113−151.

Stirling, M.W., Anooshehpoor, R., 2006. Constraints on probabilistic seismic hazard models from unstable landform features in New Zealand. Bull. Seismol. Soc. Am. 96, 404−414.

Stirling, M.W., 2014. The continued utility of probabilistic seismic hazard assessment, in earthquake hazard, risk and disasters. In: Shroder, J., Wyss, M. (Eds.), Earthquake Hazard, Risk, and Disasters. Elsevier, London, pp. 359−376.

Stucchi, M., Camassi, R., Rovida, A., Locati, M., Ercolani, E., Meletti, C., Migliavacca, P., Bernardini, F., Azzaro, R., 2007. DBMI04, il database delle osservazioni macrosismiche dei terremoti italiani utilizzate per la compilazione del catalogo parametrico CPTI04. Quad. Geofis. 49, 38. Available at. http://emidius.mi.ingv.it/DBMI04/.

Trendafiloski, G., Wyss, M., Rosset, P., 2011. Loss estimation module in the second generation software QLARM. In: Spence, R., So, E., Scawthorn, C. (Eds.), Human Casualties in Earthquakes: Progress in Modeling and Mitigation. Springer, pp. 381−391.

Vaccari, F., Peresan, A., Zuccolo, E., Romanelli, F., Marson, C., Fiorotto, V., Panza, G.F., 2009. Neo-deterministic seismic hazard scenarios: application to the engineering analysis of historical buildings. In: Proceedings of PROHITECH 2009-Protection of Historical Buildings Mazzolani. Taylor & Francis Group, London, pp. 1559−1564. ISBN 978-0-415-55803-7.

Wald, D.J., Quitoriano, V., Heaton, T.H., Kanamori, H., 1999. Relationships between peak ground acceleration, peak ground velocity and modified Mercalli intensity in California. Earthquake Spectra 15, 557−564.

Wang, Z.M., 2011. Seismic hazard assessment: issues and alternatives. Pure Appl. Geophys. 168, 11−25.

Wang, Z., Cobb, J.C., 2012. A critique of probabilistic versus deterministic seismic hazard analysis with special reference to the New Madrid seismic zone. Geological Soc. Am. Spec. Pap. 493, 259−275.

Wasserburg, G.J., 2010. Comment on "AGU Statement: Investigation of Scientists and Officials in L'Aquila, Italy, Is Unfounded". Eos 91 (42), 384.

Wells, D.L., Coppersmith, K.J., 1994. New empirical relationships among magnitude, rupture length, rupture width, rupture area, and surface displacement. Bull. Seismol. Soc. Am. 84, 974−1002.

Woo, G., 1996. Kernel Estimation Methods for Seismic Hazard Area Source Modeling. Bull. Seism. Soc. Am. 86 (2), 353−362.

Wyss, M., 2012. Origin of Human Losses Due to the Emilia Romagna, Italy, M5.9 Earthquake of 20 May 2012 and Their Estimate in Real Time. Abstract NH43A-1619 Presented at 2012 Fall Meeting, AGU, San Francisco, Calif., 3−7 Dec. http://fallmeeting.agu.org/2012/eposters/eposter/nh43a-1619/.

Wyss, M., Nekrasova, A., Kossobokov, V., 2012. Errors in expected human losses due to incorrect seismic hazard estimates. Nat. Hazards 62 (3), 927−935. http://dx.doi.org/10.1007/s11069-012-0125-5.

Zuccolo, E., Vaccari, F., Peresan, A., Panza, G.F., 2011. Neo-deterministic and probabilistic seismic hazard assessments: a comparison over the Italian territory. Pure Appl. Geophys. 168, 69−83.

Chapter 13

The Continued Utility of Probabilistic Seismic-Hazard Assessment

Mark W. Stirling

GNS Science, Lower Hutt, New Zealand

ABSTRACT

Probabilistic seismic-hazard assessment (PSHA) has been a standard input to the engineering, planning, and insurance industries for over four decades. The purpose of this chapter is to provide an overview of how PSHA is performing in the modern world. PSHA has been, after all, the focus of considerable criticism in the literature in recent years, particularly after the occurrence of major devastating earthquakes in many parts of the world. In this chapter, I discuss the advantages and limitations of PSHA in light of the recent criticisms, and then discuss the new developments that are contributing to PSHA, or are expected to do so in the future.

13.1 INTRODUCTION

The method of probabilistic seismic-hazard assessment (PSHA) has now been a standard input to engineering, planning, and insurance applications for several decades. The PSHA method defined by Allin Cornell in the late 1960s (Cornell, 1968) uses the location, recurrence behavior, and predicted ground motions of earthquake sources to estimate the frequency or probability of exceedance for a suite of ground-motion levels (Figure 13.1). Between 2002 and 2014, a great deal of debate has occurred in the literature regarding the validity of the PSHA (e.g., Castanos and Lomnitz, 2002; Stein et al., 2011; Stirling, 2012; Wyss et al., 2012; Kossobokov and Nekrasova, 2012; Panza et al., 2014). Many of the more recent criticisms claim that PSH models have been inadequate in anticipating the accelerations due to recent devastating earthquakes (e.g., M_w 9, March 11, 2011, Tohoku, Japan, and M_w, February 22, Christchurch, New Zealand earthquakes). The purpose of this chapter is to review the present utility of PSHA from the perspective of

Earthquake Hazard, Risk, and Disasters. http://dx.doi.org/10.1016/B978-0-12-394848-9.00013-4

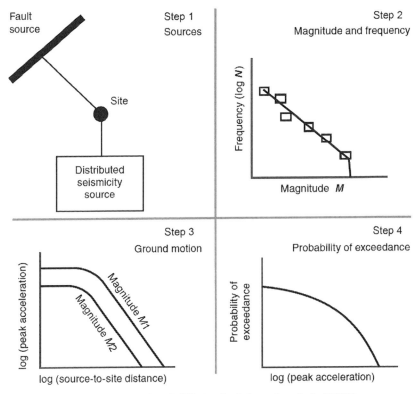

FIGURE 13.1   The four steps of probabilistic seismic hazard analysis (PSHA).

someone who has led the development of PSH models at national, regional, and site-specific scales over much of the last 20 years, and who regularly provides timely PSH solutions to end users.

13.2 THE LOGIC OF PSHA

PSHA is based on Cornell's fundamental logic that hazard at a site is based on the location, recurrence behavior, and predicted ground motions of earthquakes surrounding the site (Figure 13.1). PSH models use estimates of earthquake recurrence derived from seismicity catalogs, active fault data, and from geodetic strain rates derived from global positioning system (GPS) data to develop source models (Steps 1 and 2 in Figure 13.1). Ground-motion prediction equations (GMPEs; Step 3) are developed from strong motion data sets. The fundamental output of PSHA is the hazard curve (Step 4), which gives the expected frequency or probability of exceedance for a suite of earthquake shaking levels (e.g., peak ground acceleration (PGA), spectral acceleration, and peak ground velocity). Many details are associated with the specific application of PSHA at a site, such as the types of earthquake sources, particulars of the recurrence behavior, effect of distance on ground motions, and the ground conditions. However, the same logical method (Figure 13.1) is the foundation of all PSHAs.

PSHA has for a long time served to provide input ground motions for the design of engineered structures, which requires estimates of hazard that are relevant to the lifespan of the structure. For example, building design typically considers return periods of 500 or 2,500 years, which are, respectively, equivalent to a 10 percent and 2 percent probability of exceedance in 50 years. In contrast, nuclear facilities and major hydrodam developments typically use hazard estimates with 10,000-year return periods or longer. Hazard estimates for these three return periods typically show large quantifiable differences across regions such as the western USA, Europe, Japan, and New Zealand, reflecting the long-term, tectonically driven differences in the expected future activity of earthquake sources across the regions. The differences are intuitively obvious, given that one would expect sites close to major plate boundary faults to experience more earthquakes in the long-term than sites further away. This is well illustrated by national PSH maps (e.g., New Zealand and USA; Figure 13.2(a) and (b)), and indeed, global maps (e.g., Global Seismic-Hazard Analysis Program (GSHAP), the predecessor of the Global Earthquake Model (GEM; globalquakemodel.org; Figure 13.2(c))), which show high hazard along the main plate boundary areas (areas of highest concentration of active faults and seismicity), and lower hazard away from the plate boundaries. This is useful information for engineering in particular, being the basis for development of design standards such as the New Zealand Loadings Standard NZS1170.5 (Standards New Zealand, 2004).

Response spectra (e.g., Figure 13.3) can be rapidly developed from a PSH model to provide seismic design loadings for a range of return periods and spectral periods. Spectral shapes differ according to the different mixes of earthquake magnitudes and distances surrounding the site of interest, which provides meaningful input to design loadings, including the selection of design earthquake scenarios and associated time histories. Spectra for given return periods (referred to as a uniform hazard spectra) can be disaggregated to identify the most likely (or most unlikely) earthquake scenarios for the site or region in question (Figure 13.4), and these scenarios are often used to select realistic time histories for input to seismic loading analysis, and for territorial authorities and others to plan for future earthquake hazards.

Many modern PSH models incorporate comprehensive epistemic (model or knowledge) uncertainties into every component of the model to account for all possible surprise events. The most recent version (version 3) of the Unified California Earthquake Rupture Forecast (UCERF) models, for instance (scec.usc.edu/scecpedia/UCERF3.0), allows virtually every possible

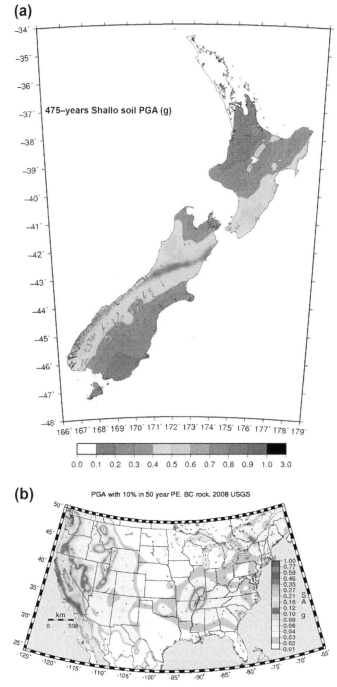

FIGURE 13.2   PSH maps produced from the (a) New Zealand national seismic hazard model; (b) US national seismic hazard model; and (c) Global Seismic Hazard Analysis Program (GSHAP) model. Each map shows the peak ground accelerations (PGAs) expected for a 500-year return period on soft rock sites. *Map sources are Stirling et al. (2012), Petersen et al. (2008), and Giardini et al. (1999), respectively.*

(c)

Global seismic hazard map

FIGURE 13.2   cont'd

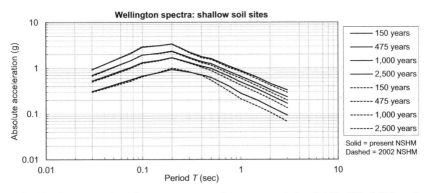

Wellington spectra: shallow soil sites

FIGURE 13.3   Examples of response spectra for return periods of 150, 475, 1,000, and 2,500 years, for shallow soil site conditions. The spectra are plotted as accelerations for a given spectral period, in which the PGA is plotted at 0.03 s due to the inability to plot 0.0 s in the log scale. These spectra are for Wellington city, New Zealand, and are derived from the 2002 and 2010 versions of the national seismic hazard model for New Zealand (Stirling et al., 2002, 2012).

combination of rupture geometry on the California fault sources, and uses seismological and geodetic data to define a range of distributed (background) seismicity rates.

Prior to PSHA, seismic-hazard assessment was based on deterministic methods, in which the location and predicted ground motions of the most important earthquake sources were the basis for hazard estimation (i.e., no recurrence information considered). Although conceptually simple and useful, deterministic methods can also lead to the overestimation of the hazard in the case of a source of extremely long recurrence interval being used to define the

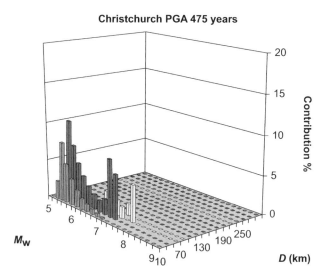

FIGURE 13.4 Example of a disaggregation for the city of Christchurch derived from the New Zealand national seismic-hazard model (Stirling et al., 2012). Despite the model being produced prior to the 2010−2012 Christchurch earthquake sequence, the disaggregation plot identifies two relevant classes of earthquakes that dominate the hazard of the city: earthquakes of M_w 5−6.0 at distances of <10 km to the city and M_w 6.0−7.5 at distances of 10−50 km. These classes of earthquakes encompass all the major earthquakes of the Canterbury earthquake sequence. PGA, peak ground acceleration.

hazard at the site, or underestimation of hazard by ignoring local sources with relatively short recurrence intervals for moderate earthquakes that may still produce damaging ground motions at a site. By disaggregating the hazard curve, a realistic deterministic scenario can be selected for use in scenario-based studies and analysis.

13.3 NATURE OF RECENT CRITICISMS OF PSHA

Many of the recent criticisms of PSHA imply that specific PSH models failed to anticipate the level of acceleration in recent, devastating earthquakes such as the Tohoku and Christchurch earthquakes (e.g., Stein et al., 2011). Although it is indeed the case that the PSH models did not provide any indication that the events were going to occur in the year 2011, criticisms of this nature have been made without full appreciation of the construction and intended use of PSH models. In short, people always want to see high levels of hazard on maps where the subsequent major earthquakes and associated strong ground motions occur, and when they do not they conclude that the models must be wrong. The reason for these apparent discrepancies is that the recurrence intervals or rates of occurrence of the causative earthquakes are fundamentally important drivers of the hazard estimates.

As an example, the Christchurch earthquake produced rock PGAs of about $1g$ in an area of formerly low seismicity where the New Zealand PSH model (Stirling et al., 2012) showed 500-year PGA motions of about $0.3g$. As I have mentioned, the 500-year return period is one of the most frequently used portrayals of PSH, but in low seismicity areas, the 500-year motions will almost invariably underestimate the motions produced by rare, damaging earthquakes if they occur there.

The ways of extracting strong motion information from a PSH model to adequately represent such rare events are threefold: First, the hazard can be calculated for return periods that are appropriately long (e.g., $\geq 2,500$ years) so as to produce higher levels of hazard. Important facilities such as hospitals usually consider a 2,500-year, return period hazard; Second, scenario (deterministic) hazard estimates can be provided by considering the motions that would be produced by specific local sources near a site, irrespective of the recurrence interval of the sources; Third, time-dependent or "time-varying" hazard models can be developed to produce high PSH if adequate knowledge of enhanced earthquake probabilities exists. In the absence of such time-varying information (the typical case), the PSH model will typically be based on a Poissonian probability model (time-independent earthquake rate or probability). The obvious limitations in the above are the choice of return period (people will often consider long return periods too conservative), choice of scenario earthquakes (people will be dubious of hazard estimates that are based on poorly defined sources of long/unknown recurrence interval), and data quantity/quality/uncertainty issues in time-varying hazard modeling. These limitations necessitate very careful recommendations being made in the PSHA process, and full documentation of the associated model caveats.

If a PSH model can be augmented with time-varying (forecasting) components, then the model can provide hazard estimates relevant to time periods of days to decades (e.g., Gerstenberger et al., in press). These models require two types of data: (1) a detailed knowledge of the earthquake history and prehistory of well-studied faults, so the earthquake recurrence interval and elapsed time since the last earthquake faults can be determined and (2) high-quality earthquake catalogs that allow temporal and spatial characteristics of earthquake occurrence to be deciphered and modeled with time-varying rate or probability models (e.g., Rhoades et al., 2010; Kossobokov, 2014; Schorlemmer and Gerstenberger, 2014; Wu, 2014; Zechar et al., 2014). In all efforts to establish such time-varying models, the resulting models are generally associated with large epistemic uncertainty and aleatory (random) variability. Furthermore, no one model is presently capable of providing a prospective short-term forecast of a large earthquake sequence that suddenly occurs (as was the case for the Canterbury earthquake sequence). Not surprisingly, the ability to provide actual earthquake forecasts for end-user application and policy is in the early stages of development. To attain reliable earthquake forecasts, there needs to be some significant advances in relevant scientific research and monitoring/ detection.

Legitimate criticisms can be leveled at elements of the PSHA inputs for regions relevant to the recent major earthquakes mentioned above. In the case of the Tohoku earthquake, the magnitude of the earthquake was considerably greater than the maximum magnitude of the Japanese PSH model, which was retrospectively seen as a clear deficiency in the model. However, in the Christchurch, New Zealand earthquake, the causative fault was unknown prior to the earthquake, due to the long recurrence interval and resulting lack of topographic expression and seismicity along the fault. Without a prior knowledge of a fault source, it was impossible to estimate where the strongest near-field ground motions would occur. PGAs $>2g$ were produced at some strong motion stations at soft soil sites around Christchurch during the earthquake, due to the city being close to the fault source, and a likely combination of near-fault effects such as high stress drop and hanging wall effects.

In the national seismic-hazard model for New Zealand (Stirling et al., 2012), the Christchurch earthquake was to an extent accounted for by the distributed seismicity model. Distributed seismicity models are typically made up of a set of sources at grid points that have activity rates defined on the basis of the spatial distribution of seismicity. They are designed to model the seismicity expected to occur away from the known fault sources in the area, and if correctly parameterized, will account for earthquakes on unknown sources. In the case of the New Zealand seismic-hazard model, the earthquake magnitudes, recurrence intervals, and ground motions were to an extent accounted for in the national seismic-hazard model in three ways.

- In the Christchurch area, the maximum magnitude of M_w 7.2 was defined in the distributed seismicity model several years prior to the occurrence of the M_w 7.1, September 4, 2010, Darfield earthquake, the main shock of the Canterbury sequence.
- The recurrence interval for Darfield-sized earthquakes estimated from this distributed seismicity model is of the order of 11,000 years (Stirling et al., 2012), which is of an order similar to the emerging recurrence estimates for the Greendale Fault derived from paleoseismology (uncertain recurrence estimate of 16,000 years; Quigley et al., 2010; Figure 13.5).
- The ground motions were to an extent accounted for in prior national seismic-hazard models, as illustrated by the response spectra in Figure 13.6. The New Zealand Loadings Standard (NZS1170) spectrum for the 2,500-year return period is not dissimilar to (i.e., depending on spectral period) the recorded motions for the Christchurch earthquake from strong motion stations around Christchurch (Figure 13.6). While the NZS1170

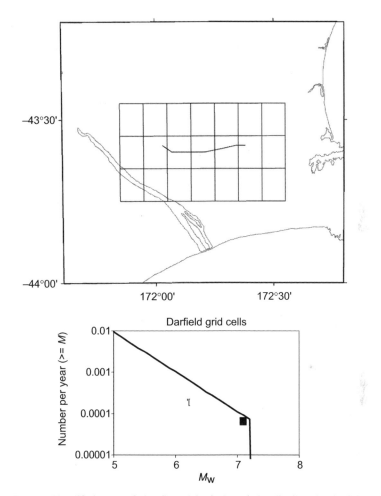

FIGURE 13.5 Simplified trace of the Greendale fault and the distributed seismicity grid cells of the New Zealand national seismic-hazard model in the immediate vicinity. The accompanying magnitude—frequency distribution shows the combined earthquake rates for these cells (cumulative number of events $\geq M$; magnitude range in the figure is limited to the magnitudes normally associated with damaging ground motions), and the 1/16,000-year rate estimate for the Greendale fault from paleoseismic data. *Figure reproduced from Stirling et al. (2012).*

spectra are based on a previous version of the national seismic-hazard model (Stirling et al., 2002), the estimated hazard for Christchurch did not change greatly in the later model (Stirling et al., 2012). Comparison of national seismic-hazard models at 2,500-year (and longer) return periods to the Christchurch earthquake is reasonable, given the likely long recurrence interval of the causative fault source, which had no evidence of prior ruptures in the epicentral area.

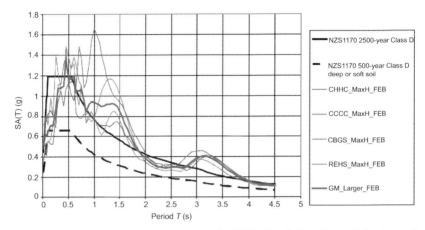

FIGURE 13.6   Examples of response spectra for Christchurch New Zealand, for deep soil site conditions. The thin solid lines are spectra recorded at selected strong motion stations in the city during the M_w 6.2 February 22, 2011, Christchurch earthquake, and the thicker dashed and solid lines are NZS1170 response spectra for 500- and 2,500-year return periods (Standards New Zealand, 2004). *Figure reproduced from McVerry et al. (2012).*

13.4 ADVANCES IN PSH INPUTS

The PSH models are only as good as the input data and methods of source parameterization, ground-motion estimation, and probability estimation. However, as long as the associated uncertainties and limitations are fully expressed in the model commentary, and appropriate advice provided, the models can still be useful. The use of deterministic models and appropriate parameterization of distributed seismicity models are two examples of solutions used to compensate for known or suspected deficiencies in PSH models.

Improvements to the performance of PSH models require improvements to PSHA inputs and components. It is therefore worthwhile to review some of the major advances in PSHA inputs and components being made at present, and identify issues and priorities for future research focus.

13.4.1 Ground-Motion Prediction

The aleatory variability in ground-motion prediction remains a large issue, despite major efforts to develop new GMPEs. The next generation attenuation (NGA) project (peer.berkeley.edu/nga) has involved some of the world's key GMPE developers producing a suite of GMPEs from the same quality-assured, strong motion data set. The models have incorporated more input parameters in an effort to improve ground-motion prediction, particularly with respect to source geometry. However, the aleatory variability, generally measured as the standard deviation of a log-normal distribution of ground motions, does not appear to have been reduced from those of earlier models (e.g., Watson-Lamprey, 2013). This may mean that the new parameters are not important as predictor variables, or they are important, but their impact on the GMPEs is being counteracted by the capturing of a more complete range of aleatory variability in the recently acquired strong motion data.

13.4.2 Monitoring

Efforts to improve the recording of input data by seismic and GPS networks are of fundamental importance to PSHA. Seismic networks (e.g., Geonet.org.nz) are making large improvements to the detection threshold of earthquakes (the minimum magnitude for a complete record of earthquakes), and the ability to observe temporal and spatial changes in seismicity. GPS is being increasingly used to provide input to source models (e.g., distributed seismicity models and subduction interface models). The generally short temporal coverage of GPS data is compensated for by a large spatial coverage, and as such, it can be a complement to other source models. This is illustrated in Figure 13.7 where I use GPS-based strain rates to develop a magnitude−frequency distribution for the plains surrounding Christchurch New Zealand (the Canterbury Plains). GPS results show that the order of 2-millimeters per year deformation rate is unaccounted for by the known active faults across the plains, and is presumably accommodated by earthquakes on unknown fault sources beneath the plains. The use of the GPS-derived 2-millimeters per year deformation rate to develop distributed seismicity rates for these sources produce seismicity rates that are a factor of two higher than the rates estimated from the observed seismicity prior to the Canterbury earthquake sequence.

Lastly, remote sensing techniques such as synthetic aperture interferometry are seeing improvements in applicability and resolution over time, and these will allow greater ability to detect the coseismic deformation field from earthquakes (e.g., Taubenböeck et al., 2014).

13.4.3 Active Faults: Detection and Characterization

Active fault data are the only PSHA input data set that is able to extend the earthquake record back in time to prehistory (Meghraoui and Atakan, 2014). Great improvements in the ability to detect and characterize active faults for input to PSHA have been seen in the last 10 years. Fault mapping has improved significantly through accumulated experience and the availability of new tools (e.g., LIDAR). Greater ability to map the surface geometry of faults and distributions of displacement has led to the improved characterization of fault sources in PSH models. The use of different disciplines and data sets together for fault characterization, particularly with respect to mapping fault ruptures in three dimensions, has yielded a great deal of understanding of rupture complexity and detail. Furthermore, increased age constraints on paleoearthquakes have made it

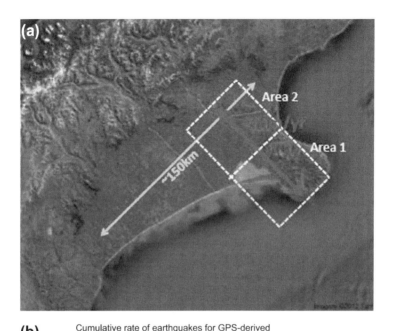

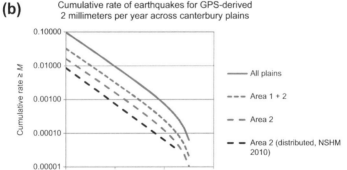

FIGURE 13.7 Magnitude–frequency distributions developed from GPS-measured 2 millimeters per year across the Canterbury Plains, New Zealand, for the entire plains (extent of yellow arrow); the 50-km-wide corridor across the full width of plains (Area 1 + Area 2), and for the 50 × 50-km Area 2. Also shown is the equivalent distribution derived for Area 2 from the distributed seismicity model of the New Zealand national seismic hazard model (Stirling et al., 2012). The city of Christchurch lies at the northeastern end of the boundary between Area 1 and Area 2.

possible to establish conditional probabilities and associated uncertainties for future earthquakes on specific faults.

13.4.4 Supercomputing to Consider all Possibilities

PSH models are increasingly drawing on diverse data sets and methods, and using high-end computing resources. The Californian UCERF3 model incorporates hundreds of thousands of logic tree branches in its comprehensive source model, and the use of supercomputers allows the model to be run through the four steps of PSHA (Figure 13.1). Furthermore, physics-based, seismic-hazard modeling efforts such as CyberShake (SCEC.org), use super-computers to run multiple realizations of earthquake scenarios from multiple sources, with shaking at the site computed directly from source, path, and site effects for each earthquake. The millions of calculations required would not be possible without access to major computing resources, and this was not possible as recently as a decade ago. Today, plausible scenarios such as the linking of fault sources to produce extended ruptures, and the range of un-certainties in magnitude—frequency statistics for the myriad of sources are able to be considered without the limitation of CPU demand. Already, exciting scientific results have emerged from the UCERF3 modeling efforts, such as the finding that the seismicity on faults cannot be modeled by the Gutenberg—Richter relationship, as this produces a poor fit to the paleoseismic data in California (Edward Field personal communication, 2013).

13.4.5 Forums for Scientific Debate

Some recent efforts have focused on providing scientific forums to openly debate some of the criticisms leveled at PSHA and the associated input. The American Geophysical Union and Seismological Society of America have held "Earthquake Debates" sessions on several occasions over the last 5 years. Furthermore, the Powell Center for Analysis and Synthesis (www.powell.usgs. gov) has recently supported a series of workshops that have brought together PSHA experts and critics from around the world to address issues associated with maximum magnitude estimation, testability of PSHA, and development of global, seismic-source models face to face (nexus.globalquakemodel.org/ Powell). These meetings have been very positive, as people have been working together on common ground rather than talking past each other in the literature.

13.5 FUTURE NEEDS

13.5.1 Correct Use of PSH Models

I have indicated that many of the criticisms of PSHA in the recent literature have been without full appreciation of the drivers of PSH models, and how the models should best be used. Comparing PSH estimates at short return periods to the motions produced by rare, damaging earthquakes (e.g., our earlier Christchurch earthquake example) is inappropriate in the context of PSH model evaluation. Appreciation of the recurrence behavior of the relevant earthquake sources in a region is an essential part of the PSH process, resulting in the selection of appropriate return periods and/or scenario earthquakes to quantify the hazard.

13.5.2 Earthquake Forecasting during Seismic Quiescence

The frequently expressed desire to enhance PSH models to enable them to be used for short-term forecasting presents a major scientific challenge. As is, PSHA is inappropriate for short-term forecasting for the simple fact that it uses the past as a proxy for the future. In other words, spatial differences in the level of seismic hazard are a direct consequence of the past distribution of historical seismicity, GPS strain, and active faults. Although this reflects the long-term distribution of hazard, it does not identify where and when major earthquakes will next occur, especially in low seismicity areas. What is missing is the ability to achieve forecasting without the forecasts being entirely based on the presence of existing earthquake sequences, and the spatial density and longer term activity of sources. For instance, none of the various forecasting models (e.g., Short Term Earthquake Probability (STEP), Epidemic Type Aftershock Sequence (ETAS) and Every Earthquake a Precursor According to Scale (EEPAS)) could have provided advanced warning of the Canterbury earthquake sequence of 2010–2012, as the area had essentially no seismicity in the preceding decades. In order to achieve short-term forecasting, future research efforts need to be focused on improving the ability to monitor and detect microseismicity and crustal deformation, and identification of reliable earthquake precursors not currently known or observed.

In the New Zealand context, efforts are now in place to advance the national seismic-hazard model through the incorporation of short-term forecasting models. The Canterbury earthquake sequence initiated this work through an immediate need to provide hazard estimates for rebuilding Christchurch. The resulting hazard estimates are an ensemble of time-varying and time-independent earthquake probability models (e.g., Gerstenberger et al., in press). The ensemble PSH model shows that the resulting hazard will be well above that calculated prior to the Canterbury earthquake sequence (Stirling et al., 2012) for some decades to come.

13.5.3 Reduction in Aleatory Uncertainty in Ground-Motion Prediction

The aleatory uncertainty in ground-motion estimation is very large (the standard deviations in GMPEs are typically about 0.5 in natural log units of ground motion), and does not seem to have been reduced in the complex GMPEs available today. In other words, a very large range in the potential ground motions still exists that could be produced at a single site due to earthquakes of the same magnitude and distance. In contrast, the differences between GMPEs (epistemic uncertainties) do appear to have been reduced in recent years, at least within the NGA project.

13.5.4 Testability of PSHA

Finally, efforts need to be supported in the objective testing of PSH models, as to date PSH models have largely been developed in the absence of any form of verification (e.g., Panza et al., 2014). The Collaboratory for the Study of Earthquake Predictability (http://www.cseptesting.org/ and Schorlemmer and Gerstenberger, 2014) has been developing testing strategies and methods for a wide variety of applications, and collaborative work has also been focused on developing ground motion-based tests of the New Zealand and US national seismic-hazard models.

The Global Earthquake Model (GEM), a worldwide seismic hazard and loss modeling initiative, is including testing and evaluation as an integral part of the overall model development. This work is mainly focused on the evaluation of components of the PSH model, such as GMPEs. The Yucca Mountain seismic-hazard modeling project developed innovative approaches to consider all viable constraints on ground motions for long return periods for nuclear waste repository storage, prior to the cancellation of the project in 2008 (Hanks et al., 2013). The need to verify the hazard estimates for return periods of 10^4-10^6 years advanced the use of geomorphic criteria to test the hazard estimates. The rationale is that "fragile geologic features" (Figure 13.8) provide evidence for nonexceedance of ground motions for long return periods, and prior to the cancellation of the Yucca Mountain project appeared to be showing new constraints on PSH models at long

FIGURE 13.8 Example of an ancient fragile geologic feature at Yucca Mountain, Nevada, that has been studied to obtain constraints on past ground motions for return periods equivalent to the age of the feature (10^4-10^5 years). The cowboy hat at the base of the feature gives an indication of scale.

return periods (e.g., Anderson et al., 2011). Specifically, ancient fragile geologic features observed at Yucca Mountain were inconsistent with extremely strong ground-motion estimates produced for the very long return periods considered for repository design. The Yucca Mountain project therefore revealed the extent to which PSH estimates become driven by the "tails" of statistical distributions, namely, the log-normal distribution of predicted ground motions from GMPEs. Had the Yucca Mountain project continued, fragile geologic features would have played a major part in reducing the design ground-motion estimates for repository design. Similar observations have been made elsewhere in the USA and in New Zealand (e.g., Stirling and Anooshehpoor, 2006; Anderson et al., 2011), and fragile geologic features are now recognized as the only viable criteria for constraining ground-motion estimates for long return periods (Anderson and Biasi, 2014; Stirling et al., 2014).

13.5.5 Complex PSHA on Normal Computers

If the future of PSHA is in the development of complex PSH models such as UCERF3, and in the development of physics-based PSH models, the reliance of these models on supercomputer resources will be a significant barrier to their widespread utility. Significant efforts will therefore need to be focused on making these models usable on standard computers, or uptake will be extremely limited in everyday PSHA for end-user applications.

13.6 CONCLUSIONS

PSHA is a simple, logical method that bases future earthquake hazard on the location, rate, and predicted shaking intensities of past earthquakes. The PSHA method provides useful solutions for end users, mainly as input to engineering design. As such it is here to stay until another alternative method is available that is shown to do better than PSHA. Work focused on the pursuit of such models should be supported by the funding institutions, whether they be governmental or private industry. Furthermore, input to PSHA must continue to be improved, which requires that all the current developments described above continue to be supported. Finally, recent experience has taught me that the critics of PSHA can make considerable contributions to doing things better when they are brought together with the PSHA scientists to participate in forums such as the Powell Center for Analysis and Synthesis. In recent forums, we found that scientists who had been very critical of PSHA in the literature were able to be understood and appreciated by the PSHA scientists in the face-to-face Powell setting. In turn, the responsibilities and practical needs of PSH scientists to provide timely solutions to end users were better appreciated by the critics. Forums like these must continue to be supported.

ACKNOWLEDGMENTS

I wish to thank Danielle Hutchings Mieler for her peer review of this chapter, and James Dolan for relevant discussions. Funding from the New Zealand National Hazards Research Platform is gratefully acknowledged. Lastly, I attribute many of the opinions expressed in this chapter to insights gained during the four workshops run by Ross Stein and myself at the Powell Center for Analysis and Synthesis over the period August 2012—September 2013. I therefore thank the Powell Center (especially Jill Baron) for supporting our workshop proposal, and the scientists who participated in the workshops: Ross Stein, Seth Stein, John Adams, Ned Field, Mark Petersen, Morgan Page, Nicola Litchfield, Oliver Boyd, Danijel Schorlemmer, Dave Jackson, Peter Bird, Yan Kagan, Mark Leonard, Oona Scotti, Laura Peruzza, Alex Allmann, Martin Kaesar, Harold Magistrale, Yufang Rong, Marco Pagani, Graeme Weatherill, Damiano Monelli, Anke Friedrich, Gavin Hayes and Margaret Boettcher.

REFERENCES

Anderson, J.G., Biasi, G.P., 2014. Precarious rocks and constraints on seismic hazard. In: Shroder, J., Wyss, M. (Eds.), Earthquake Hazard, Risk, and Disasters. Elsevier, London, pp. 377—403.

Anderson, J.G., Brune, J., Biasi, G., Anooshehpoor, A., Purvance, M., 2011. Workshop report: applications of precarious rocks and related fragile geological features to U.S. national hazard maps. Seismol. Res. Lett. 82 (3), 431—441.

Castanos, H., Lomnitz, C., 2002. PSHA: is it science? Eng. Geol. 66, 315—317.

Cornell, C.A., 1968. Engineering seismic risk analysis. Bull. Seismol. Soc. Am. 58, 1583—1606.

Gerstenberger, M., McVerry, G., Rhoades, D., Stirling, M. Seismic hazard modeling for the recovery of Christchurch, New Zealand. Earthquake Spectra, in press.

Giardini, D., Grunthal, G., Shedlock, K., Zeng, P., 1999. The GSHAP global earthquake hazard map. Ann. Geofis. 42, 1225—1230.

Hanks, T.C., Abrahamson, N.A., Baker, J.W., Boore, D.M., Board, M., Brune, J.N., Cornell, C.A., Whitney, J.W., 2013. Extreme Ground Motions and Yucca Mountain. U.S. Geological Survey. Open-File Report 2013—1245, 105 pp. http://dx.doi.org/10.3133/ofr20131245.

Kossobokov, V.G., Nekrasova, A., 2012. Global seismic hazard assessment program maps are erroneous. Seism. Instrum. 48 (2), 162—170.

Kossobokov, V.G., 2014. Times of increased probabilities for the occurrence of catastrophic earthquakes: 25 years of the hypothesis testing in real time. In: Shroder, J., Wyss, M. (Eds.), Earthquake Hazard, Risk, and Disasters. Elsevier, London, pp. 477—504.

McVerry, G.H., Gerstenberger, M., Rhoades, D., Stirling, M., 2012. Spectra and Pgas for the assessment and reconstruction of Christchurch. In: Proceedings of the 2012 Annual Conference of the New Zealand Society of Earthquake Engineering.

Meghraoui, M., Atakan, K., 2014. The contribution of paleoseismology to earthquake hazard evaluations. In: Shroder, J., Wyss, M. (Eds.), Earthquake Hazard, Risk, and Disasters. Elsevier, London, pp. 237—271.

Panza, G., Kossobokov, V., Peresan, A., Nekrasova, A., 2014. Why are the standard probabilistic methods of estimating seismic hazard and risks too often wrong? In: Shroder, J., Wyss, M. (Eds.), Earthquake Hazard, Risk, and Disasters. Elsevier, London, pp. 359—376.

Petersen, M., Frankel, A., Harmsen, S., Mueller, C., Haller, K., Wheeler, R., Wesson, R., Zeng, Y., Boyd, O., Perkins, D., Luco, N., Field, E., Wills, C., Rukstales, K., 2008. Documentation for the 2008 Update of the United States National Seismic Hazard Maps. U.S. Geol. Surv. Open-File Rept. 2008-1128, 61 pp.

Quigley, M., Van Dissen, R., Villamor, P., Litchfield, N., Barrell, D., Furlong, K., Stahl, T., Duffy, B., Bilderback, E., Noble, D., Townsend, D., Begg, J., Jongens, R., Ries, W., Claridge, J., Klahn, A., Mackenzie, H., Smith, A., Hornblow, S., Nicol, R., Cox, S., Langridge, R., Pedley, K., 2010. Surface rupture of the Greendale fault during the M_w 7 1 Darfield (Canterbury) earthquake, New Zealand: initial findings. Bull. N. Z. Natl. Soc. Earthquake Eng. 43 (4), 236−242.

Rhoades, D.A., Van Dissen, R.J., Langridge, R.M., Little, T.A., Ninis, D., Smith, E.G.C., Robinson, R., 2010. Re-evaluation of the conditional probability of rupture of the Wellington−Hutt Valley segment of the Wellington fault. Bull. N. Z. Natl. Soc. Earthquake Eng. 44, 77−86.

Schorlemmer, D., Gerstenberger, M., 2014. Quantifying improvements in earthquake rupture forecasts through testable models. In: Shroder, J., Wyss, M. (Eds.), Earthquake Hazard, Risk, and Disasters. Elsevier, London, pp. 405−429.

Standards New Zealand, 2004. Structural Design Actions—Part 5: Earthquake Actions—New Zealand. New Zealand Standard NZS 1170.5. Dept. Building and Housing, Wellington, New Zealand.

Stein, S., Geller, R., Liu, M., 2011. Bad assumptions or bad luck: why earthquake hazard maps need objective testing. Seismol. Res. Lett. 82 (5), 623−626.

Stirling, M.W., 2012. Earthquake hazard maps and objective testing: The hazard mapper's point of view. Opinion. Seismological Research Letters 83, 231−232.

Stirling, M.W., McVerry, G.H., Berryman, K.R., 2002. A new seismic hazard model for New Zealand. Bull. Seismol. Soc. Am. 92, 1878−1903.

Stirling, M.W., Anooshehpoor, R., 2006. Constraints on probabilistic seismic hazard models from unstable landform features in New Zealand. Bull. Seismol. Soc. Am. 96, 404−414.

Stirling, M.W., McVerry, G.H., Gerstenberger, M., Litchfield, N.J., Van Dissen, R., Berryman, K.R., Langridge, R.M., Nicol, A., Smith, W.D., Villamor, P., Wallace, L., Clark, K., Reyners, M., Barnes, P., Lamarche, G., Nodder, S., Pettinga, J., Bradley, B., Rhoades, D., Jacobs, K., 2012. National seismic hazard model for New Zealand: 2010 update. Bull. Seismol. Soc. Am. 102 (4), 1514−1542.

Stirling, M.W., Rood, D., Barrell, D., 2014. Using fragile geologic features to place constraints on long-term seismic hazard. Seismol. Res. Lett. http://www2.seismosoc.org/FMPro?.

Taubenböck, H., et al., 2014. The capabilities of earth observation to contribute along the risk cycle. In: Shroder, J., Wyss, M. (Eds.), Earthquake Hazard, Risk, and Disasters. Elsevier, London, pp. 25−53.

Watson-Lamprey, J., 2013. Incorporating the effect of directivity in the intra-event standard deviation of the NGA West 2 ground motion prediction equations. Abstracts for Annual Meeting of the Seismological Society of America. Seismol. Res. Lett. http://www2.seismosoc.org/FMPro.

Wu, Z., 2014. Duties of earthquake forecast. In: Shroder, J., Wyss, M. (Eds.), Earthquake Hazard, Risk, and Disasters. Elsevier, London, pp. 431−448.

Wyss, M., et al., 2012. Errors in expected human losses due to incorrect seismic hazards estimates. Nat. Hazards 62 (3), 927−935. http://dx.doi.org/10.1007/s11069-012-0125-5.

Zechar, D., Herrmann, M., Van Stiphout, T., Wiemer, S., 2014. Forecasting seismic risk as an earthquake sequence happens. In: Shroder, J., Wyss, M. (Eds.), Earthquake Hazard, Risk, and Disasters. Elsevier, London, pp. 167−182.

Chapter 14

Precarious Rocks: Providing Upper Limits on Past Ground Shaking from Earthquakes

John G. Anderson [1], Glenn P. Biasi [1] and James N. Brune [2]

[1] *Nevada Seismological Laboratory & Department of Geological Sciences and Engineering, University of Nevada, NV, USA*, [2] *Seismological Laboratory, University of Nevada Reno, Reno, NV, USA*

ABSTRACT

Precariously balanced rocks (PBRs) are naturally occurring geological features that could be easily toppled by strong earthquake shaking. They bring two forms of information valuable for understanding seismic hazard. In the specific regions where PBRs occur, they provide direct information about strong ground motions not exceeded during their lifetimes. Their more general application is to provide useful limits on ground motion variability. This article explains how precarious rocks may be interpreted, and why they are important for understanding strong ground motions in engineering application.

14.1 INTRODUCTION

Precariously balanced rocks (PBRs) address a need in the engineering design of critical facilities where the design must consider extreme events and ground motions with long return times, to include thousands or tens of thousands of years. Engineers find hazard curves to be a useful way to understand this information. Hazard curves express the rate at which ground motions are equaled or exceeded as a function of the amplitude of the motion. In principle, they could be measured by running an instrument at the site for a sufficiently long time, and counting how often the ground motion of interest is exceeded. Obviously, project time frames are too short for such an empirical approach. The instrumental earthquake catalog can support inference of hazard curves to return times of only years to perhaps a few decades, which is still not even close to being sufficient. As a result, estimates of hazard curves for ground motion must be developed from other information including geological appraisals of earthquake magnitudes and rates of nearby faults and models of the consequent ground motion. PBRs are important because they allow

Earthquake Hazard, Risk, and Disasters. http://dx.doi.org/10.1016/B978-0-12-394848-9.00014-6

a partial test of estimates of their hazard curves in the range of large amplitudes and very long return times (Section 14.2.1). If the estimated hazard curve implies that shaking strong enough to topple the PBR occurs frequently over a time interval equal to the age of the rock, the model used to estimate the hazard curve needs to be improved, and the hazard curve updated.

In some cases, contradictions can be resolved by a more detailed examination of the geological appraisal of nearby faults (Section 14.4.2). In these cases, the evidence from the PBRs results in a modified hazard estimate for any nearby location that is strongly affected by the adjustment of the fault model.

PBRs have also supported a better understanding of uncertainty in ground motion estimates (Section 14.4.3). At long return times, the common assumption of lognormal uncertainties in ground motion estimates leads to rare large ground motions that can erroneously imply that PBRs should have been toppled. These same ground motion overestimates can sometimes control the design and make earthquake engineering difficult and/or expensive. If the uncertainty is lower than estimates using traditional approaches, the ground motions are closer to the median estimate, and this effect is less severe. The universal implication is that traditional approaches to estimating ground motions need to be modified to treat the uncertainty properly.

As a prelude to discussing these critical applications, Section 14.2.1 presents the theoretical basis for the subsequent discussion, Section 14.2.2 describes data that need to be collected to utilize PBRs for these purposes, and Section 14.3 briefly reviews the geographical distribution of most of the PBRs that the authors have studied.

14.2 PBRS ARE PHYSICAL OBJECTS THAT PUT LONG-TERM BOUNDS ON PAST GROUND MOTIONS

Since as early as the late 1800s, seismologists have used the effects of ground motions on precarious features as an aid in understanding the causes and effects of earthquakes. Several early studies are summarized by Purvance et al. (2008a). Oldham (1899) made effective use of free-standing tombstones and monuments because their geometry makes them naturally sensitive to strong ground motions. Tombstones that have been undisturbed by earthquakes can be used to put an upper bound on the strength of the ground motions in the time interval since the tombstone was erected.

PBRs are naturally occurring objects with ground motion response properties similar to those of tombstones (Figure 14.1). Their advantage for hazard estimation is their longevity. In southern California, PBRs have been shown to have been fragile for thousands of years or longer. This makes PBRs a unique resource for limiting estimates of ground motion at long return times. The long ages of precarious rocks mean in particular that they can test the output of a probabilistic seismic hazard analysis (PSHA) for ground motions with very long return times.

FIGURE 14.1   **Examples of precariously balanced rocks.** These rocks are all present in seismically active regions of southern California. (a) Lovejoy Buttes. Colored tape has been applied to assist in modeling the three-dimensional shapes of the rocks. The pedestal height of approximately 1.5 m suggests that these rocks have been vulnerable for $\geq$10,000 years (Brune et al., 2014). The San Andreas fault is 14 km away at the base of the mountains at the skyline. (b) "Benton" rock, 2.5 m tall, between the San Jacinto and Elsinore faults. (c) "Tooth", also between the San Jacinto and Elsinore faults. (d) Generalized two-dimensional geometry of a precarious rock, from Shi et al. (1996). The properties of these rocks are given in Table 14.1.

14.2.1 Essential Theory

This section reviews three concepts that are essential for the application of PBRs to seismic hazard. First, we consider the physical relationships governing the toppling of rigid bodies. The response of PBRs can be understood in terms of the balance between rotational accelerations imparted by ground motion and the period-dependent net rotation of the rock. We then consider the properties of strong ground motions tested by PBRs, including their

TABLE 14.1 Properties of Precarious Rocks in Figure 14.1

Photographs	Rock Name	Lat Long	Height	α_1 (deg rad)	R_1	α_2 (deg rad)	R_2	Age	V_{S30} m/s
Figure 14.1(a)	Lovejoy Buttes (R02564)	34.59– 117.85	1 m	15 (0.26)	0.44 m	20 (0.35)	0.44 m	>10.5 K	>700
Figure 14.1(b)	Benton	33.59– 116.92	2.5 m	11 (0.19)	1.3 m	19 (0.33)	1.3 m	>10 K	>700
Figure 14.1(c)	Tooth	33.22– 116.47	1.5 m	10.3 (0.18)	0.82 m	10.9 (0.19)	0.82 m	~10 K	>700

Parameters are explained in Section 14.2.1.1.

amplitude and frequency dependence. Finally, we consider the fundamental output of a PSHA, the hazard curve, and how PBRs contribute to their evaluation.

14.2.1.1 Physics of Toppling Rocks

Conceptually, the toppling of a precarious rock by ground shaking seems straightforward. In fact, however, the PBR response to ground motion is nonlinear, and depends on both the frequency and amplitude of the inputs. Early literature on toppling criteria has been reviewed by Purvance et al. (2008a). To illuminate rock response to ground motions, we look briefly at their mathematical relationships, and then at some practical conclusions.

From the geometry (Figure 14.1(d)), a static horizontal force applied through the center of mass will just begin to lift the rock from one rocking point if it equals $mg \tan \alpha$, where g is the acceleration of gravity, m is the mass of the rock, and α is the angle from the vertical to the rocking point O in the direction of the force. If the force F is written in terms of mass times acceleration, $ma = mg \tan(\alpha)$, the static overturning acceleration, a, equals $g \tan(\alpha)$. Note that this result does not depend on the rock size or its rocking arm lengths R_1 and R_2. We will find later however that the rock size does affect the dynamic response of the rock.

The equations for rock response to dynamic ground motion are more complicated. The restoring force $F = mg \tan(\alpha - \theta)$ decreases as the rock rotates through an angle θ, and goes through zero when θ equals α, whereupon the rock topples unless subsequent ground motions rotate the rock back. This decrease in the restoring force as the rock rotates causes an increase in the period (i.e., the time it takes to settle back to its rocking points) as the rocking amplitude increases. The equation of motion of the rocking about rocking point O_i can be written in terms of the torque $I_i\ddot{\theta}$ just balancing static and dynamic restoring forces (Shi et al., 1996) as

$$I_i\ddot{\theta} + mgR_i \sin(\alpha_i - \theta) = mR_iA_y(t) \sin(\alpha_i - \theta) - mR_iA_x(t) \cos(\alpha_i - \theta),$$
(14.1)

where $\theta = \theta(t)$ is the rotation angle of the rock relative to the static position. I_i is the moment of inertia of the rock around rocking point O_i. Relative to the static case, Eqn (14.1) decomposes ground accelerations into $A_x(t)$, the horizontal acceleration of the ground, and $A_y(t)$, the vertical acceleration of the ground. DeJong (2012) developed a formulation equivalent to Eqn (14.1) using conservation of energy. For realistic ground motions, the PBR rocking problem can only be solved numerically. Numerical solutions typically conserve angular momentum during the transition from one rocking point to the other, so the differences in the elevation of the rocking points (represented by the angle β in Figure 14.1(d)) are handled naturally. Equation (14.1) neglects energy losses associated with physical damage during rocking. Approximate and exact solutions to Eqn (14.1) for simple input motions are available (e.g., Housner, 1963; Shi et al., 1996; Purvance et al., 2008a; DeJong, 2012). We will only show a few characteristics of the solutions.

Rectangular two-dimensional (2D) shapes provide a simple device to understand the frequency dependence of rock response to ground motion (Figure 14.2). Base width W is the same for each, and as height H increases from 1 to 10, α decreases from 1.0 to 0.1. By symmetry, the moment of inertia is the same for both rocking points: $I = \frac{4}{3}mR^2$. Figure 14.3 shows the aspects of the response of a block with $\alpha = 0.25$ when the input is a half cycle of a sinusoidal wave, with period T_P and peak acceleration A_{max}. Figure 14.3 shows the combinations of T_P and A_{max}, which topple a rock with $H = 150$ cm. For a general symmetrical shape, the moment of inertia of inertia is $I = c_I mR^2$, where c_I depends on the geometry. For a circular cylinder, $c_I = 1.356$, so the results in Figure 14.3 are approximately correct for this case as well. For a rock with a narrow base, as in Figure 14.1(b) and (c), c_I is smaller, in the range of approximately $1.05-1.2$. In the limit of all the mass concentrated in one point, $c_I = 1.0$. For m, R, and α held constant, a rock with a smaller value of c_I is slightly more easily toppled.

The results in Figure 14.3 help to illustrate important features of the physics of toppling PBRs. As the pulse period increases, the peak acceleration necessary to cause the rock to topple decreases asymptotically toward the static overturning acceleration $g \tan \alpha = 0.25g$. Smaller peak accelerations are unable to initiate rocking. As the pulse period decreases, the peak acceleration needed to cause toppling increases dramatically. An analog for the system modeled in Figure 14.3 is pulling out a rug from beneath a rigid object, with the intent of causing it to topple. No matter how sharply the rug is jerked (accelerated) sideways, if the net movement is small compared to the horizontal distance

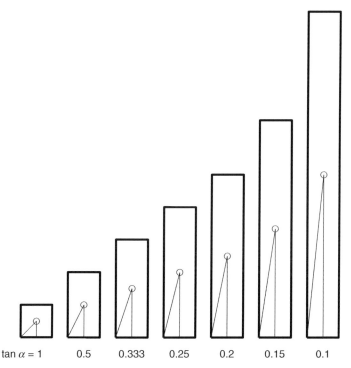

$\tan \alpha = 1$ 0.5   0.333   0.25   0.2   0.15   0.1

FIGURE 14.2   Rectangular blocks with a range of fragilities. For each block, the open circle is drawn at the center of gravity. The angle α is the angle between a vertical line through the center of gravity and a line from the center of gravity to a rocking point. A static horizontal acceleration equal to $g \tan \alpha$ is the minimum requirement to topple the block.

between the rocking point and the center of gravity, it may seem intuitive that the input would not suffice to cause the object to fall over. The dashed green line in Figure 14.3 tests this intuitive notion. The dashed green line shows the combination of T_P and A_{max} for which the final horizontal motion is $W/2$, the distance between the center of mass and the rocking point. The actual boundary differs from this intuitive criterion systematically, in an informative way. For long period pulses, gravitational forces tending to restore the rock to the equilibrium are important for the PBRs but not included in the rug analogy, so the actual toppling occurs when the ground displacement is greater than $W/2$. At short periods, both the solution and the intuitive model increase steeply as period decreases, but the lines are offset. The offset shows that a brief impulse with a small displacement is able to transfer enough angular momentum to cause the rock to topple even when the net displacement is smaller than $W/2$.

Since the static force to topple a rock depends only on the rock shape, one might expect that the size of the rock does not matter. However, the ground motions in earthquakes are transient and oscillatory, and not static. For that reason, the size of the rock is important. The boundary of the stable zone in

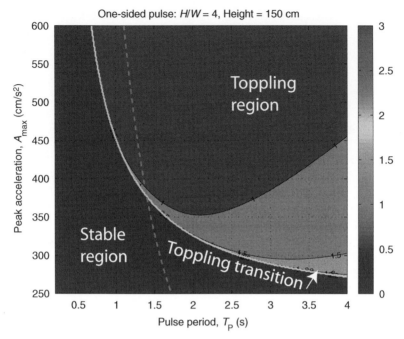

FIGURE 14.3   Contour plot showing elements of the response to a half cycle of a sinusoidal pulse of a rectangular block with height $H = 150$ cm and $W = 37.5$ cm, so $\tan \alpha = 0.25$. The horizontal acceleration has period T_P and peak acceleration A_{max}. The input function is thus $A_x(t) = A_{max} \sin\left(\frac{2\pi t}{T_P}\right)$ for $0 \leq t \leq T_P/2$. The vertical acceleration, $A_y = 0$. The dark blue region below the yellow line shows the range of A_{max} and T_P within which this single-sided pulse does not cause the rock to be toppled. Above the yellow line, colors indicate the time, after the start of the pulse, when the block rotation exceeds α. After this time, the gravitational force dominates to complete the fall of the block onto its side. When the pulse ends, at time $t = T_P/2$, the ground under the precarious rock has moved a horizontal distance $D_x = \frac{A_{max} T_P^2}{4\pi^2}$. The dashed green line shows $A_{max} = 2\pi^2 W/T_P^2$, the combination of T_P and A_{max} for which the final ground displacement is exactly equal to $W/2$, the horizontal separation of the center of gravity and the rocking point. Contours are obtained by the numerical integration of Eqn (14.1).

Figure 14.3 has been transferred to Figure 14.4 for comparison with the equivalent boundary for rocks of other heights. As the height increases, with α held constant, the boundary of the stable zone is shifted to longer periods (e.g., Housner, 1963; Purvance et al., 2008b). A decrease in c_I, for α held constant, has the same effect in Figure 14.4 as decreasing the height, but the effect of c_I is small compared to the effect of the range of heights in the figure. Anooshehpoor et al. (2004) suggested a rule of thumb that peak acceleration in a typical ground motion needs to exceed $1.3g \tan \alpha$ for rocks with heights in the range of about $1-2$ m. In Figure 14.4, this approximate threshold is about 320 cm/s^2. So long as the pulse has a period of over about $1.5-2.5$ s, this approximate threshold will be satisfied. Pulses meeting this criterion are more common in large-magnitude earthquakes (e.g., $M > 6$),

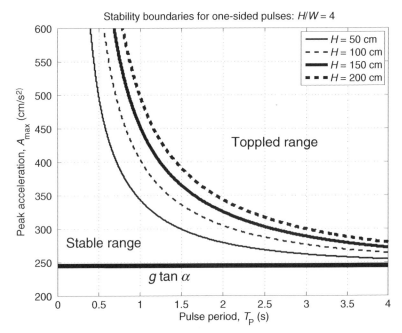

FIGURE 14.4   **Effect of height on the lowest boundary of the zone of stability.** The solid heavy line for $H = 150$ cm is the same as in Figure 14.3.

and are not expected for earthquakes with $M < 5$. Thus, small-magnitude earthquakes are not expected to topple large rocks, even though they sometimes have very high values of peak acceleration (e.g., Anderson, 2010). The implicit relation between ground motion pulse length and toppling figured into early preliminary definitions by Brune (1996) of precarious and semiprecarious rocks. By his definition, a rock is precarious if it can be toppled by a horizontal earthquake acceleration $<0.3g$, and a rock is semiprecarious if it can be toppled by a horizontal earthquake acceleration between $0.3g$ and $0.5g$. However, these definitions were developed based on rocks that were generally $1-2$ m high so the terms could be used to relate rock fragility to earthquakes common in the southwest United States. Also, the definition was originally based on ground motions similar to that of the 1940 El Centro earthquake.

14.2.1.2 Strong Ground Motions

Ground motions from earthquakes (e.g., Figures 14.5 and 14.6) are oscillatory time series. These motions have been modeled by band-limited white noise (e.g., Boore, 1983), but real earthquake records are normally more complex. Actual records have arrivals of P-waves, S-waves, and surface waves that originate at different parts of the fault.

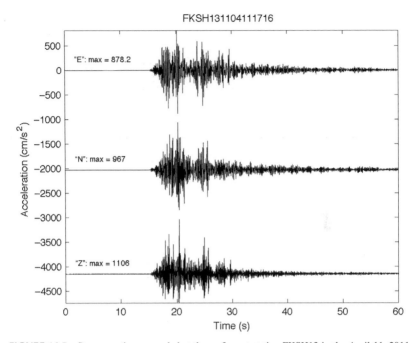

FIGURE 14.5   Strong motions recorded at the surface at station FKSH13 in the April 11, 2011, Fukushima Hamadori earthquake (M_w 6.7). Peak component amplitudes are given in units of centimeters per square second. These ground motions were observed by KiK-net, operated by the National Research Institute for Earth Science and Disaster Prevention (NIED).

The response of precarious rocks, such as those in Figures 14.1 and 14.2, to complex ground motions such as those in Figures 14.5 and 14.6 can be anticipated in probabilistic terms. Figure 14.7, based on Purvance et al. (2008a) using numerous shake table experiments and numerical simulations, shows an estimate of the probability of a rock being toppled as a function of PGA, the peak acceleration and SA(1 Hz), the peak acceleration response of a single-degree-of-freedom oscillator with undamped natural frequency of 1 Hz and 5% damping to the ground motions (e.g., Kramer, 1996). The vertical asymptote is controlled by the angle α. The horizontal asymptote is controlled by the size of the rock and the character of the ground motions. For instance, the record in Figure 14.5 has a very high peak acceleration, whereas the peak velocity in Figure 14.6 is not very high (unfavorable for toppling), but the period of the velocity pulse is long (favorable for toppling). The statistical model developed by Purvance et al. (2008a) considers records with these and other characteristics, since the character of the motion at any particular site is not easily predictable. For an accelerogram with peak acceleration and a response spectral amplitude plotting in the deep blue region, the rock will not topple, whereas if the records are in the deep red region, the rock is certain to topple. If the parameters from a randomly chosen accelerogram plot in the

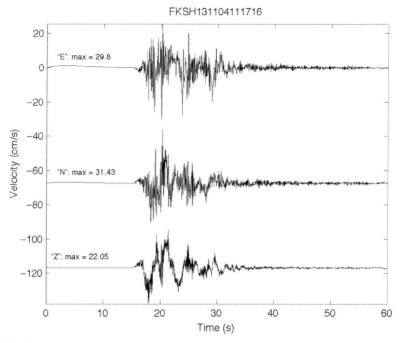

FIGURE 14.6 Velocity corresponding to the strong motions in Figure 14.5 recorded at the surface at station FKSH13 in the April 11, 2011, Fukushima Hamadori earthquake (M_w 6.7). Peak component amplitudes are given in units of centimeters per second.

rainbow of colors between these extremes, the rock has a chance of toppling or surviving, depending on the details of the ground motions. The uncertainty can only be reduced by calculating the response of a model of the rock to the actual accelerogram. The calculation is best performed based on a three-dimensional (3D) version of Eqn (14.1), as done by Hanks et al. (2013) in a study of seismic hazard at Yucca Mountain, Nevada.

14.2.1.3 Hazard Curves

The primary output of a PSHA is an estimated hazard curve, a key engineering tool for the quantitative understanding of seismic hazard and subsequently reducing risk (Tolis, 2014). The abscissa of an estimated hazard curve is an amplitude of ground motion, and the ordinate is the exceedance rate of the ground motion. There are of course many measures of the amplitude of ground motions, for instance, peak acceleration, peak velocity, and response spectral amplitudes (e.g., Kramer, 1996). Figure 14.8 shows an example of two estimated hazard curves for the Lovejoy Buttes rock shown in Figure 14.1(a). The two parameters considered in Figure 14.8 are peak acceleration and the pseudoacceleration response spectral amplitude of a 5 percent damped single-degree-of-freedom oscillator with an undamped

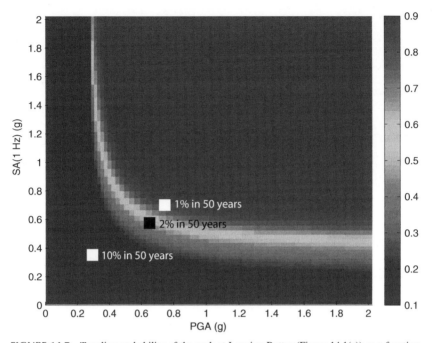

FIGURE 14.7   Toppling probability of the rock at Lovejoy Buttes (Figure 14.1(a)) as a function of peak acceleration and peak response of a 1 Hz oscillator to that acceleration. The probabilities, illustrated by different colors according to the scale on the right, are based on a statistical study of observed strong motion records, leading to a model that can be applied depending on key parameters of any rock (Purvance et al., 2008a,b). Square points, based on Figure 14.8, show the 2008 ground motions from the United States Geological Survey National Seismic Hazard Maps (Petersen et al., 2008) at the selected probabilities. These points are obtained from hazard curves, which are discussed in Section 14.2.1.3.

natural period of 1.0 s (SA(1s)). The procedure for estimating a seismic hazard curve has been summarized conceptually by Anderson et al. (2011) and Stirling (2014) and is described in more detail by, for example, McGuire (2004).

Essential components of the input used to estimate hazard curves are a model of the seismicity and a ground motion prediction equation (GMPE). The seismicity model would ideally include a description of locations, magnitudes, and occurrence rates of all the possible earthquakes that might affect the site. For the site of interest, the GMPE estimates the ground motion for each earthquake in the hazard model.

Estimated hazard curves such as those in Figure 14.8 are the result of an effort to predict the outcome of a hypothetical physical experiment that could measure the true hazard curve that exists for that location. The hypothetical experiment is to operate a strong motion instrument at the site of the precarious rock for, say, 10^4 years or more, with the duration controlled by the

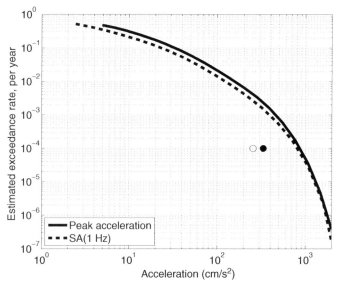

FIGURE 14.8 Seismic hazard curves for peak acceleration and for the acceleration spectral response at 1 Hz, SA(1 Hz) for Lovejoy Buttes, California (34.591 N, 117.849 W), corresponding to the 2008 release of the US Geological Survey's National Seismic Hazard Map (Petersen et al., 2008). The points in Figure 14.7 representing the Lovejoy Buttes rock are determined from these curves by finding the exceedance rate that corresponds to the three chosen probabilities, and finding the corresponding accelerations. The circles in this figure also represent the rock in Figure 14.1(a), with an assumed age of 10,000 years and for which $g \tan \alpha = 0.26g$. The open circle is shown at the quasistatic toppling acceleration of $0.26g$, and the solid point is at an acceleration a factor of 1.3 greater, as discussed in the text. The solid point should be compared only with the hazard curve for peak acceleration. Although the location of this point below the hazard curve is indicative of an inconsistency, the probabilities contoured in Figure 14.7 are needed for a more reliable analysis. This case is discussed in Section 14.4.3.

smallest rate of interest. With that long of a time sample of records, assuming the overall physics of the situation does not change, the number of exceedances of the selected ground motion parameters could simply be counted and then divided by the duration of the experiment to obtain the measured hazard curve. Time intervals with durations of $10^4 - 10^5$ years are still short compared to the rate of change of tectonic configurations of a region, so the thought experiment is meaningful to that level.

Within this perspective, it is essential to clearly recognize the difference between the hazard curve (or "true hazard curve") for the site and an estimate of that hazard curve.

14.2.2 Data Needs

Precarious rocks can be used to test models of the seismic hazard. Essential data needs for the tests are rock geometry, age of the rock combined with some understanding of how it was formed, and a knowledge of the site amplification effects of shallow geology at the site. This section elaborates on those data requirements.

The first type of essential information is the geometry of the rock, including all the parameters used in Eqn (14.1). Several methods are available to obtain this information about the rock geometry.

The angles α_1 and α_2 are determined by the center of mass and the rocking points. For some rocks with relatively simple shapes, the rocking points are obvious from field inspection, and the center of mass can be determined fairly accurately from photographs. If the rock is tabular, a simple analytic 2D approximation can be used. For more complex shapes, several computer codes are available to reconstruct 3D solid objects from 2D orthographic views. We use a computer program called PhotoModeler (Eos Systems Inc., Vancouver, British Columbia, Canada), which requires about 40 photographs, with a large number of identifiable targets on the rock (Figure 14.1(a)), as well as carefully placed targets on the base. A few hours of work on the computer yields a quite precise shape of complexly shaped rocks, and this in turn allows an accurate estimate of the center of mass and rocking points.

An independent estimate of parameters is provided by the direct measurement of force versus tilt angle, or, in simple cases, by measuring force through the center of mass along with the measurement of density. Simple tests with laboratory scale models have verified the procedures. Taking direct measurements as true values, Brune et al. (2011) estimated that the uncertainty of estimating α is about 0.05 radians using PhotoModeler, and about 0.06 radians using "expert opinion" based on photographs from several directions. Estimates using an analytic 2D approximation based on photographs are on average biased, on average about 0.05−0.1 radian low, depending on how close the rock is to being ideally tabular, and because the 2D analyses depend on the photograph, which may not be at the best viewing angle to determine α. These accuracies are sufficient for approximate tests of various current seismic hazard estimates and for illustrating the geographical distribution of PBRs. An approximate verification of the procedures is provided by the behavior of rocks and solids of various shapes on simple shake tables in the laboratory and on the UNR engineering department shake table. The details of the procedures are described in various earlier publications cited in Anooshehpoor et al. (2004), Purvance et al. (2008a), and Brune et al. (2011).

The second data need is for the age of the rock and geomorphology of the site. Many of the rocks in the desert Southwest, for example, Mojave Desert, are covered with desert "rock varnish", which, depending on how thick or dark it is, can be thousands of years old, indicating no surface weathering for that period (Brune and Whitney, 1992). Other geomorphic evidence in the desert Southwest indicates a very reduced rock erosion rate since the end of the last ice age, near or earlier than the beginning of the Holocene. Based on this, Bell et al. (1998) used a simple one-parameter model commonly used in dating glacial moraines and alluvial fans, in this case assuming that the PBRs were exhumed relatively rapidly prior to the Holocene, and since that time have been exposed to cosmic rays that generate cosmogenic isotopes in the outer layer of the rocks and in the rock pedestals. Knowing the concentration of the isotopes gives the time since exhumation. Bell et al. (1998) found ages of thousands to tens of thousands of years, in all cases greater than or equal to the age constraint given by rock varnish (usually about 10,000 years). Balco et al. (2011) used a more complex inversion model with three erosion rate parameters. Perhaps surprisingly, they also found very rapid exhumation rates at some time prior to the Holocene (5 centimeters per year, effectively instantaneous). Thus, not so surprising, they found age dates similar to that of Bell et al. (1998), that is, thousands to tens of thousands of years. These times are much longer than typically used for PSHA analysis (e.g., 2 percent in 50 years or recurrence times of 2475 years), and thus, the age dating provides assurance that the PBRs can be used to test seismic hazard estimates. Recent work has shown that cosmogenic PBR ages correlate well in many cases with pedestal height (Brune et al., 2014). This new development has expanded the number of rocks about which some estimate of age can be offered.

Finally, in order to use PBRs to develop upper limits on estimated past ground motions, an estimate is required of the shallow geologic properties of the site. Because PBRs develop in granitic terrains, this is proving to be perhaps the best constrained element of PBR science. Recent geophysical measurements in granitic terrains of southern California (Yong et al., 2013) indicate typical V_{S30} values of ≥ 650 m/s. As a result, V_{S30} values at PBR sites are similar or higher than values associated with rock site conditions.

14.3 DISTRIBUTION OF PBRS

PBRs are relatively rare geological features. Most of the PBRs in southern California occur in granitic areas with many remnant core stones. This section describes some of the geological considerations for their existence, and looks at their distribution in southern California.

14.3.1 Circumstances for Creation and Preservation

PBRs are not found everywhere. The most basic observation is that they occur in landscapes where erosion has exposed the bedrock geological formations. Jointed bedrock is favorable. Weathering in this case occurs primarily along the joints leaving harder core stones in between the joints. In granitic terrain, as described by Bell et al. (1998) and Brune et al. (2007b), the soft material along joints can be eroded leaving core stones behind, some of which are precarious.

Arid regions, such as those of the southwestern United States, are the most favorable location. Rapid erosion and weathering rates are not consistent with the existence of old precarious rocks, since it causes a rapid destruction of precarious rocks. Similarly, in some regions with heavy rainfall, deep soil forms on eroding topography, and precarious rocks do not outcrop.

In the southwestern United States, at the end of the last glacial maximum, the climate became more arid, and erosion rates dropped sharply. Many of the precarious rocks in this region are coated with desert varnish, a mineral deposit that thickens very slowly on the surface of rocks exposed to the air. A thick coating of desert varnish is evidence that the rocks have been in basically their present position since their formation.

Even in the deserts of the southwestern United States, old precarious rocks are generally not found in the high mountains, where winter snow may gradually creep downslope or form avalanches. Precarious features in these locations may be informative for ground motions in a recent earthquake, but do not provide constraints over long time intervals.

14.3.2 PBRs Show That Hazard Even in Active Areas is Not Homogeneous

Figure 14.9 shows a map of PBRs that have been found in southern California for which $\alpha \leq 0.3$ based on preliminary estimates of α using the 2D approximation. The distribution is not at all homogeneous. Several factors influence the distribution:

- Access, as rocks are usually discovered when hiking in granitic terrain starting on developed roads. Rocks far from sufficiently developed roads are much less likely to be found.
- Appropriate geology, as discussed above.
- Earthquake hazard. Most of the PBRs shown in Figure 14.9 occur at a distance of at least 10−15 km from the nearest active fault. Exceptions are rocks near trans-tensional stepovers, where observations and numerical modeling indicate that ground motions are greatly reduced (Brune, 2003; Lozos et al., 2011, 2013).

As Figure 14.9 shows, most PBR locations include multiple similarly vulnerable features. Anderson et al. (2011) illustrate this point by showing the distribution of α at two of the better fields of precarious rocks. Thus, we can say that precarious rocks are not just freak anomalies that have been very lucky to survive multiple earthquakes. It brings up the issue of the increase in the confidence one gains in setting constraints on past ground motion by the presence of multiple rocks in a small area. Consider the situation where there are n rocks in a field of PBRs. We can hypothesize a trial ground motion vector $\vec{\mathbf{m}}$ with two components: the axes of Figure 14.7. For the trial ground motion, let p_i be the probability that the ith rock in a field would be toppled by an

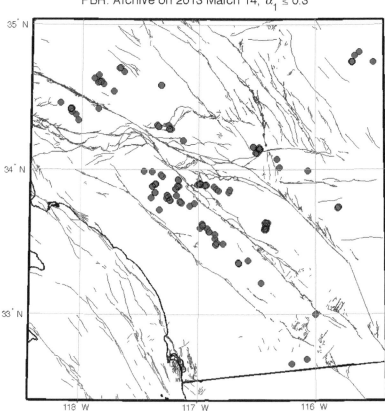

FIGURE 14.9 Map of precarious rocks in southern California, compared to locations of active faults. All rocks on this map have an estimated $\alpha < 0.3$. Many estimates of α are based on two-dimensional profiles, and are thus subject to the uncertainties discussed in Section 14.2.2. The points very close to active faults, such as San Andreas, San Jacinto, Cleghorn, and Garlock, represent rocks with either very young ages, proximity to fault stepovers, trans-tensional regimes, the potential that the rock is unstable only in an azimuthal direction in which polarized shaking is weak from the earthquakes that control the hazard, or uncertain activity rates.

earthquake, based on results comparable to those in Figure 14.7. On the assumption that these rocks are spatially separated by enough distance to be regarded as independent trials, the probability of all these rocks surviving is given by

$$p_A = 1 - \prod_{i=1}^{n} (1 - p_i) \tag{14.2}$$

The chance of all the rocks surviving, p_A, is necessarily smaller than the chance of any one of the rocks surviving. In this way, the confidence in the limit of ground motion given by $\vec{\mathbf{m}}$ is increased. Purvance et al. (2008b) showed that for an area with several PBRs of similar fragilities that individually had a high probability of being toppled, the odds against all of them accidentally surviving are very high.

14.4 APPLICATIONS

14.4.1 General: Test Hazard Models

The goal of a PSHA is to estimate the hazard curve, and the most important application of precarious rocks is to test those estimates. Where contradictions occur, the subsequent efforts can study the models used to estimate the hazard curve, and explore how the input models can be improved.

There are two main ways that precarious rocks can be used to test hazard models. The simpler approach, starting with a description of historical earthquakes and the assumption that the rock survived these historical events, tests models for the ground motion. If one can add information about the age of the rock, and if the rock is very old, then one can also make meaningful tests of estimated hazard curves such as those in Figure 14.8.

Figure 14.8 shows what Hanks et al. (2013) refer to as a "point in hazard space" beside the estimated hazard curve. PGA points on the abscissa are static and dynamic toppling accelerations for the Lovejoy Buttes rock. The Lovejoy Buttes rock is estimated to be >10,000 years old. The ordinate for this point is plotted at rate = (1/age). The point falls below the estimated hazard curves for these accelerations. This suggests that there may be an inconsistency between the Lovejoy Buttes rock and the United States Geological Survey (USGS) hazard curve for that location. Using this line of reasoning, Anderson and Brune (1999b) suggested that "the probability of exceeding a ground motion capable of toppling a precarious rock during a time period equal to the age of the rock is equal to the confidence level at which the inputs to the PSHA can be rejected."

Purvance et al. (2008a,b) described a more sophisticated and appropriate approach for this problem, based on vector-valued PSHA (Bazzurro and Cornell, 2002). The vector-valued hazard for PGA and SA(1 s) is a generalization of the scalar hazard represented in Figure 14.8. The derivative of the exceedance rate curve in Figure 14.8 would estimate the occurrence rates of PGA and of SA(1 s). The vector-valued probability estimates the occurrence rates of a vector of ground motions, where in this case the vector has the two components PGA and SA(1 s).

Figures 14.3, 14.4, and 14.7 all indicate that the toppling of a PBR depends on at least two properties of the ground motion. Thus, Purvance et al. (2008a) estimate the joint probability of rock toppling on two distinct parameters. The first is PGA. For the second, they found results for three alternative parameters, all of which depend on lower frequency components of the strong motion: peak ground velocity (PGV), and response spectra for oscillator periods of 1 and 2 s (SA(1 s), SA(2 s)). Then, Purvance et al. (2008b) multiplied the rock fragility (Figure 14.7) with the estimated vector-valued probability for the same pair of parameters (PGA and SA(1 s) in their case) to obtain overturning rates for the vector pair of ground motion parameters. An integral over all combinations yields an overall estimate of the total overturning rate.

To carry this out correctly, one needs to know how PGA and SA(1 s) are correlated, since obviously both need to be sufficiently high during the same earthquake, whereas the earthquakes dominating the hazard curve for PGA at a given exceedance rate are not necessarily the same as the earthquakes that dominate the SA(1 s) hazard curve at the same exceedance rate. However, in the case of the Lovejoy Buttes rock, the deaggregations of the estimated hazard indicate that the PGA and SA(1 s) maxima are likely to be from the same large earthquake on the San Andreas fault. This allows us to conclude that this rock is almost certainly inconsistent with the combination of peak acceleration and SA(1 s) that is predicted by this map to occur with a probability of 1 percent in 50 years. The rock is marginally consistent at 2 percent in 50 years.

14.4.2 Test Fault Activity Rates

There are several instances where precarious rocks have led to the reevaluation of the activity rates of faults that were input into a seismic hazard analysis. Brune et al. (2007a) reported a number of PBRs around Silverwood Lake, about 7 km northwest of the San Andreas fault in Cajon Pass.

Anooshehpoor et al., (SCEC, 2008) estimated the fragilities of several of these rocks at Silverwood Lake and also several in the nearby Grass Valley area ($\sim$11 km from the San Andreas fault). The Silverwood Lake rocks were inconsistent with the corresponding estimated hazard curves. Brune et al. (2007a) suggested several explanations, for example, local low stress drops on the San Andreas Fault, and the possibility that the intersection of the San Andreas and San Jacinto fault may be a releasing stepover with relatively low normal stresses, and consequent low ground motions. Schlom et al. (SCEC, 2008) suggested that the rocks at Grass Valley indicate that the Cleghorn fault is not as active as assumed in the 2008 hazard maps. Southeast of this area, Anooshehpoor et al. (SCEC, 2009) tested six PBRs less than about 1 km from the Pinto Mountain fault. The fragilities were inconsistent with the 2008 seismic hazard maps, in part because of being so close to the fault. This might indicate that the Pinto Mountain fault is not as active as assumed for the hazard maps.

14.4.3 Test Ground Motion Estimates

A contradiction between a precarious rock and an estimated hazard curve can also be caused by an inappropriate GMPE. The field of precarious rocks at Lovejoy Buttes, at distances of 14−16 km from the San Andreas fault, is an example of this possibility. The estimated ages of the rocks at Lovejoy Buttes are >10,500 years. The rock in Figure 14.1(a), which is a representative of this field, is located 14.4 km from the San Andreas fault. On this part of the San Andreas fault, the most recent earthquake occurred in 1857, with M_w 7.9 (e.g., Sieh, 1978; Zielke et al., 2010). Paleoseismic estimates indicate that similarly sized ground rupturing earthquakes have a recurrence interval of about 135 years (Scharer et al., 2011). Thus, a rock that is >10,000 years old would have survived at least approximately 65−75 major earthquakes. As noted in Section 14.4.1, one estimate of the hazard curves for Lovejoy Buttes (Figure 14.8) is inconsistent with the PBR. Since the activity rate on the San Andreas is well established, and the San Andreas fault is the only major fault that affects Lovejoy Buttes, the contradiction would seem to be caused by an inappropriate treatment of the highest ground motions.

The input of the GMPE to develop the estimated hazard curve comes in two parts. The first is the model of the median ground motion at the site, which in this case we designate as μ_{SAF}. The second is the variability of the ground motion relative to the median value, which for the moment we represent with the generic symbol σ.

There are no strong motion data from $M_W \sim 8$ strike-slip earthquakes at the distance of Lovejoy Buttes from the San Andreas fault, so estimates of μ_{SAF} given by the GMPEs are extrapolated from smaller magnitudes, mostly recorded at larger distances. The extrapolations are, of course, guided by physics-based models, but there are uncertainties in the physics needed for the extrapolation. The median values of PGA estimated from the GMPEs used to develop the hazard curve in Figure 14.8 are $\mu_{SAF} = 0.271g$ (Boore and Atkinson, 2008), $\mu_{SAF} = 0.223g$ (Campbell and Bozorgnia, 2008), and $\mu_{SAF} = 0.294g$ (Chiou and Youngs, 2008). Thus, the median estimates could be consistent with the survival of the PBRs at Lovejoy Buttes.

On the other hand, the value used for σ could be responsible for the inconsistency. Figure 14.10 illustrates how a change in σ can affect the outcome of a PSHA. The probability density function shows that as σ is reduced from the original value given by the model to 70 percent and 40 percent of the original value, the distribution of expected values of PGA gets narrower. Taking $\alpha = 0.26$ for Lovejoy Buttes, and using the 30 percent rule as the threshold of toppling, the threshold of toppling is about $0.34g$. For the full value of $\sigma = 0.521$ given in the selected GMPE (Campbell and Bozorgnia, 2008), over 10,000 years, one expects $0.34g$ to be exceeded about 15 times. As noted in Section 14.2.1.2, the toppling of a PBR is not a one-dimensional problem, but given typical correlations of peak acceleration and other measures of strong ground motion, the chance of the rock's survival would seem small. For $\sigma = 0.365$, Figure 14.10 predicts about eight exceedances, and for $\sigma = 0.15$, the chance of one exceedance drops below 20 percent. Thus, the estimated chance of exceeding the toppling acceleration is very sensitive to σ, and a small enough value effectively could eliminate the current contradiction to PBRs at Lovejoy Buttes. Note that in this example,

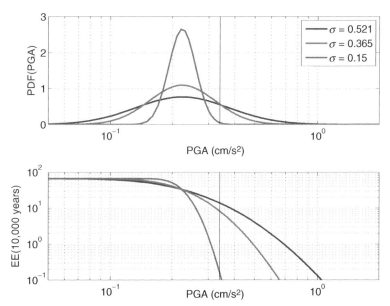

FIGURE 14.10 The upper frame gives the probability density function for peak acceleration based on a median value of $0.223g$ and three values of σ. The vertical black line at $0.34g$ is the estimated PGA needed to topple the rock in Figure 14.1(a), approximately $1.3g \tan \alpha$. From these distributions, the role of sigma can be seen. For $\sigma = 0.52$, it would not be unusual to draw at random a ground motion to be larger than necessary to topple the rock. For $\sigma = 0.15$, it should occur only very rarely. Drawing at random from these distributions corresponds to individual earthquakes. If σ is indeed large, one of the many large earthquakes on the San Andreas fault near Lovejoy Buttes would have toppled the rock. The lower frame quantifies this expectation. The lower frame gives the expected number of exceedances (EE) of PGA in 67 occurrences of the event (based on an average 150 year recurrence interval and rock age of 10,000 years). The median value of PGA and maximum considered value of σ are based on the motion expected from an M_w 8 earthquake on the San Andreas fault at Lovejoy Buttes, based on the ground motion prediction equation of Campbell and Bozorgnia (2008). The smaller values of σ are reduced to 0.7 and 0.4 of the original value.

as in a PSHA, σ accounts for the variability of the ground motion in time, from multiple realizations of earthquakes on the same fault and recorded at the same station.

An understanding of sigma suggests reasons why GMPEs might overestimate the size of this parameter. Let Y_{es} represent the amplitude of a ground motion parameter of interest, such as peak acceleration, from earthquake e at station s. The mean prediction of the GMPE can be represented as $\mu = \widehat{Y}_{es}(Z_{es})$ where Z_{es} is a vector of predictor variables (magnitude, distance, etc.). The actual datum can then be written, in the notation of Anderson and Uchiyama (2011) as

$$\log Y_{es} = \widehat{Y}_{es}(Z_{es}) + \delta_{es}^{I} = \widehat{Y}_{es}(Z_{es}) + E_e + S_s + \delta_{es}^{III} \qquad (14.3)$$

where E_e represents a log average event term (since some events of the same magnitude may radiate more energy than others) and S_s represents a log average station term due to the differences in geological structure. Superscripts in residuals δ_{es}^I and δ_{es}^{III} designate the level of residual processing. Variance values are found as the mean of the squares of the various residual terms: $(\sigma^I)^2$ from δ_{es}^I, $(\sigma^{III})^2$ from δ_{es}^{III}, $\sigma_{stations}^2$ from E_e, and σ_{events}^2 from S_s. Note that the total variance is $(\sigma^I)^2 = \sigma_{events}^2 + \sigma_{stations}^2 + (\sigma^{III})^2$. Standard deviations are the square roots of the variances. (Another common notation for σ^I is σ_T, for σ_{events} is τ, and for $\sigma_{stations}$ is ϕ_{SS}, as, e.g., in Abrahamson and Youngs, 1992; Al Atik et al., 2010; Lin et al., 2011.)

Handling the variability associated with site effects is of particular concern. We illustrate their role in the estimates of sigma using a study by Anderson et al. (2013). Figure 14.11(a) and (c) shows raw peak accelerations and peak velocities as observed at 82 stations that recorded the Fukushima Hamadori, Japan, earthquake (M_w 6.7) at distances of <100 km. The mean bias due to the event term (E_e) has been removed. The solid lines show the prediction of one GMPE (Boore and Atkinson, 2008). Observed values of PGA are scattered about equally above and below the prediction line. Since there is only one event, the standard deviation based on these data alone is $\sigma^I = \sqrt{\sigma_{stations}^2 + (\sigma^{III})^2}$. Numerical values of σ^I are given in the figure legends. Figure 14.11(b) and (d) show the same data after applying station-specific site corrections. The corrections had previously been determined by Kawase and Matsuo (2004). The standard deviation is σ^{III} in these cases. The data are clustered noticeably more tightly around the predictions after the site correction is applied. The standard deviations drop by 28 percent for PGA and 36 percent for PGV. The reduced values are comparable to, or smaller than, estimates of "single-station sigma" (e.g., Atkinson, 2006) obtained in other studies (e.g., Rodriguez-Marek et al., 2012). This example illustrates a general point that site variability is a major contributor to the total variance in GMPEs.

Anderson and Brune (1999a) pointed out that PSHA for a single station can inaccurately estimate the hazard if it uses a value of σ that includes the variability introduced by the site variability between stations. The site conditions do not change materially from one earthquake to another. However, most GMPEs do not estimate S_S; the data needed to determine S_S are often not available. At the sites of PBRs, where bedrock is outcropping at the surface, the site amplification is likely to be small, and any uncertainty model that incorporates the potential of high site amplifications can lead to an over-estimate of the hazard, consistent with results such as in Figure 14.8.

The "ergodic assumption", as defined by Anderson and Brune (1999a), is made by including the variability of the site conditions in the value of σ used with GMPEs in the seismic hazard calculation at a single site, that is, treating the spatially variable uncertainty due to station and path as if it is a function of

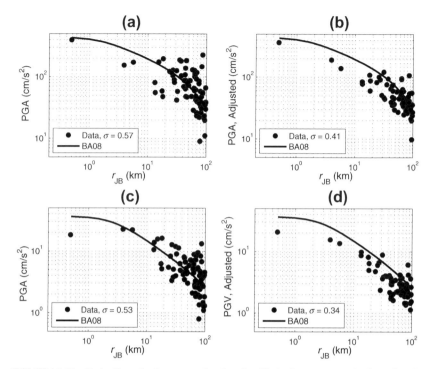

FIGURE 14.11   Reduction of σ by accounting for site effects. Data are mean horizontal peak acceleration and velocity at 82 stations that recorded the April 11, 2011, Fukushima Hamadori earthquake (M_w 6.7) at distances of <100 km, as reported by Anderson et al. (2013). For peak acceleration, the value of σ is reduced from approximately 0.57 in Frame (a) to approximately 0.42 in Frame (b). For peak velocity, it is reduced from 0.53 in Frame (c) to 0.34 in Frame (d). The reference curves in all frames are the ground motion prediction equation of Boore and Atkinson (2008). Frames (a) and (c) use $V_{S30} = 400$ m/s, representing a network average. Frames (b) and (d) use $V_{S30} = 760$ m/s, which is the reference value to which the data have also been adjusted using the approach described by Kawase and Matsuo (2004a,b). Data have also been adjusted for the event terms found by Anderson et al. (2013). *Modified from Anderson et al. (2013).*

time at the PBR site. Unfortunately, this is often done in PSHA, including the estimated hazard curves for Lovejoy Buttes (Figure 14.8) and corresponding toppling probabilities in Figure 14.7.

Uncertainty in the station response cannot be ignored, but site response can be measured. If it is not measured, but rather included in the estimate of σ, the PSHA calculations effectively make the assumption that the spatial variability of the site response represents a random variation in time for repeated earthquakes at the same distance and source for the same magnitude. Of course, it does not make sense if a large part of σ is representing variable site effects and path effects, because these are fixed and should not contribute to a long tail on the probability distribution. If this contribution to the hazard has not been measured, or if it is intended to be included in the hazard curve, the uncertainty due to the site and path effects should be represented by branches of a logic tree, rather than as an aleatory uncertainty. The precarious rocks at Lovejoy Buttes (Brune, 1999) were particularly important in the recognition of the ergodic assumption used in many probabilistic seismic hazard analyses (Anderson and Brune, 1999a).

Merely recognizing the existence of the ergodic assumption falls short of demonstrating that this is the cause of the discrepancy at Lovejoy Buttes and, by analogy, at some of the other PBR sites. Estimates of the value of sigma for ground motions from repeat ruptures recorded on a single station are hard to obtain, but some preliminary results by Yagoda-Biran et al. (2014) suggest that the value of 0.20 is consistent with very limited data from California. We thus speculate, as did Anderson and Brune (1999a), that the decrease in σ that can be achieved by eliminating the ergodic assumption would eliminate the inconsistency at Lovejoy Buttes.

The universal implication is that a site-specific site response study may afford a PSHA pathway to reduce uncertainty in ground motion, leading to better estimates of hazard and potentially lower design ground motions.

14.4.4 Interpreting Inconsistent Rocks

Some estimated hazard curves may, inherently, be inconsistent with PBRs. An example is the estimated hazard curves in Figure 14.8. The explanation of this condition raises a significant issue.

The 2008 USGS map estimates ground motion for a site condition characterized by $V_{S30} = 760$ m/s. However, this site condition, although being a constraint, does not uniquely characterize the site. For a given input motion, if one considers the set of all possible site profiles that have $V_{S30} = 760$ m/s, there will be a range of ground motions at the surface. For instance, one possibility is that the site has relatively sharp boundaries causing resonances, although a different possibility is that the upper 30 m is a uniform layer of hard rock. By design, the USGS National Seismic Hazard Map includes those uncertainties in the hazard calculation since the users of the map will rarely have a budget to support a site-specific study. Thus, the cost of not knowing more about the site effects is that a bias toward a higher hazard estimate is introduced. A geotechnical firm that estimates the seismic hazard for a specific site may measure the site properties, and the results of those measurements may decrease σ, with the consequence that the estimate of the hazard curve could decrease at small annual exceedance rates (e.g., $<10^{-3}$). The 2008 USGS methodology includes those uncertainties in the estimate of σ_T, rather than as in a logic tree as described in the previous section. A precarious rock at that site may then be inconsistent with the estimated hazard curve in the USGS maps, but a site-specific study might resolve the inconsistency.

14.5 SUMMARY

Precarious and semiprecarious rocks are valuable observations for seismic hazard analysis. In the limited areas where they are available, they can be used to place approximate limits on the amplitude of ground motion that have occurred since the rock formed. In desert regions, those times may be large compared to the recurrence interval of earthquakes, in which case, the rocks provide a direct test of PSHA inputs. Some of the lessons learned from reconciling contradictions can be applicable globally.

ACKNOWLEDGMENTS

Many colleagues have contributed greatly to the data, experimental methods, and analytical understanding of PBRs. We recognize especially Rasool Anoosheh-poor, Matt Purvance, John Whitney, and Tom Hanks. We also recognize Rich and Sue Brune for their assistance in discovery and documentation. The ground motions in Figures 14.5 and 14.6 were provided by the National Research Institute for Earth Science and Disaster Prevention (NIED) of Japan. This research was supported in part by the US Geological Survey Awards G12AP20086 and G13AP00065. This research was also supported by the Southern California Earthquake Center. SCEC is funded by NSF Cooperative Agreement EAR-1033462 and USGS Cooperative Agreement G12AC 20038. The SCEC contribution number for this paper is 1910.

REFERENCES

Abrahamson, N.A., Youngs, R.R., 1992. A stable algorithm for regression analyses using the random effects model. Bull. Seismol. Soc. Am. 82, 505–510.

Al Atik, L., Abrahamson, N., Bommer, J.J., Scherbaum, F., Cotton, F., Kuehn, N., 2010. The variability of ground motion prediction models and its components. Seismol. Res. Lett. 81, 794–801.

Anderson, J.G., 2010. Source and site characteristics of earthquakes that have caused exceptional ground accelerations and velocities. Bull. Seismol. Soc. Am. 100, 1–36.

Anderson, J.G., Brune, J.N., 1999a. Probabilistic seismic hazard analysis without the ergodic assumption. Seismol. Res. Lett. 70, 19–28.

Anderson, J.G., Brune, J.N., 1999b. Methodology for using precarious rocks in Nevada to test seismic hazard models. Bull. Seismol. Soc. Am. 89, 456–467.

Anderson, J.G., Uchiyama, Y., 2011. A methodology to improve ground-motion prediction equations by including path corrections. Bull. Seism. Soc. Am. 101, 1822–1846.

Anderson, J.G., Brune, J.N., Biasi, G., Anooshehpoor, A., Purvance, M., 2011. Workshop report: applications of precarious rocks and related fragile geological features to U.S. National Hazard Maps. Seismol. Res. Lett. 82, 431–441.

Anderson, J.G., Kawase, H., Biasi, G.P., Brune, J.N., 2013. Ground motions in the Fukushima Hamadori, Japan, Normal faulting earthquake. Bull. Seismol. Soc. Am. 103, 1935–1951.

Anooshehpoor, A., Brune, J.N., Zeng, Y., 2004. Methodology for obtaining constraints on ground motion from precariously balanced rocks. Bull. Seismol. Soc. Am. 94, 285–303. http://dx.doi.org/10.1785/0120020242.

Anooshehpoor, R., Purvance, M.D., Brune, J.N., 2008. Field-testing precariously balanced rocks in the vicinity of San Bernardino, California: seismic hazard ramifications. In: 2008 Southern California Earthquake Center Annual Meeting Poster 1-059.

Anooshehpoor, A., Brune, J.N., Purvance, M.D., 2009. Field-test of precariously balanced rocks near Yucca Valley, California: seismic hazard ramifications. In: 2009 Southern California Earthquake Center Annual Meeting Poster 1-038.

Atkinson, G.M., 2006. Single-station sigma. Bull. Seismol. Soc. Am. 96, 446–455.

Balco, G., Purvance, M.D., Rood, D., 2011. Exposure dating of precariously balanced rocks. Quat. Geochronol. 6, 295–303.

Bazzurro, P., Cornell, C.A., July 21–25, 2002. Vector-valued probabilistic seismic hazard analysis (VPSHA). In: Proc. 7th U.S. National Conf. on Earthquake Engineering, Boston, Massachusetts. Paper No. 61.

Bell, J.W., Brune, J.N., Liu, T., Zreda, M.G., Yount, J.C., 1998. Dating precariously balanced rocks in seismically active parts of California and Nevada. Geology 26, 495–498.

Boore, D.M., Atkinson, G.M., 2008. Ground-motion prediction equations for the average horizontal component of PGA, PGV, and 5%-damped PSA at spectral periods between 0.01 s and 10 s. Earthquake Spectra 24, 99–138.

Boore, D.M., 1983. Stochastic simulation of high-frequency ground motions based on seismological models of the radiated spectra. Bull. Seismol. Soc. Am. 73, 1865–1894.

Brune, J.N., 1996. Precariously balanced rocks and ground-motion maps for southern California. Bull. Seismol. Soc. Am. 86, 43–54.

Brune, J.N., 1999. Precarious rocks along the Mojave section of the San Andreas fault, California: constraints on ground motion from great earthquakes. Seismol. Res. Lett. 70, 29–33.

Brune, J.N., 2003. Precarious rock evidence for low near-source accelerations for transtensional strike-slip earthquakes. Phys. Earth Planet. Inter. 137, 229–239.

Brune, J.N., Biasi, G., Brune, R.J., 2007a. "Precariously balanced rocks at Silverwood Lake, seven kilometers from the San Andreas fault in Cajon Pass: What's going on?", abstract. In: 2007 Southern California Earthquake Center Annual Meeting. www.scec.org.

Brune, J.N., Purvance, M.D., Anooshehpoor, A., 2007b. Gauging earthquake hazards with precariously balanced rocks. Am. Sci. 95, 36–43.

Brune, J.N., Whitney, J., January–March 1992. Precariously balanced rocks with rock varnish: paleoindicators of maximum ground acceleration? SSA 1992 meeting, [abs.]. Seismol. Res. Lett. 63, 351.

Brune, J.N., Brune, R., Biasi, G.P., Anooshehpoor, R., Purvance, M.D., 2011. Accuracy of nondestructive testing of PBRs to estimate fragilities, abstract. In: 2011 Southern California Earthquake Center Annual Meeting. www.scec.org.

Brune, R.J., Brune, J.N., Grant-Ludwig, L., 2014. Preliminary precariously balanced rock (PBR) age dates based on various models of erosion. SSA Abstr., 14–198.

Campbell, K.W., Bozorgnia, Y., 2008. NGA ground motion model for the geometric mean horizontal component of PGA, PGV, PGD and 5% damped linear elastic response spectra for periods ranging from 0.01 to 10 s. Earthquake Spectra 24, 137–171.

Chiou, B.S.-J., Youngs, R.R., 2008. An NGA Model for the Average Horizontal Component of Peak Ground Motion and Response Spectra. Earthquake Spectra 24, 173–215.

DeJong, M.J., 2012. Amplification of rocking due to horizontal ground motion. Earthquake Spectra 28, 1405–1421.

Hanks, T.C., Abrahamson, N.A., Baker, J.W., Boore, D.M., Board, M., Brune, J.N., Allin Cornell, C., Whitney, J.W., 2013. Extreme Ground Motions and Yucca Mountain. Open File Report 2013–1245, U.S. Department of the Interior, U. S. Geological Survey, Reston, Virginia, 105 pp.

Housner, G.W., 1963. The behavior of inverted pendulum structures during earthquakes. Bull. Seismol. Soc. Am. 53, 403–417.

Kramer, S.L., 1996. Geotechnical Earthquake Engineering. Prentice Hall.

Kawase, H., Matsuo, H., 2004a. Amplification Characteristics of K-NET, KiK-NET, and JMA Shindokei Network Sites based on the Spectral Inversion Technique. 13th World Conference on Earthquake Engineering, Vancouver, Canada. Paper No.454, 2004.8.

Kawase, H., Matsuo, H., 2004b. Relationship of S-wave Velocity Structures and Site Effects Separated from the Observed Strong Motion Data of K-NET, KiK-net, and JMA Network. Japan Association of Earthquake Engineering, 126−145 (in Japanese).

Lin, P.-S., Chiou, B., Abrahamson, N., Waling, M., Lee, C.-T., Cheng, C.-T., 2011. Repeatable source, site, and path effects on the standard deviation for empirical ground-motion prediction models. Bull. Seismol. Soc. Am. 101, 2281−2295.

Lozos, J.C., Oglesby, D.D., Brune, J.N., Olsen, K.B., 2011. "The effects of intermediate fault segments on rupture behavior and ground motion in strike-slip stepovers", abstract. In: 2011 Southern California Earthquake Center Annual Meeting. www.scec.org.

Lozos, J.C., Olsen, K.B., Oglesby, D.D., Brune, J.N., 2013. Rupture and ground motion models on the Claremont-Casa Loma stepover of the San Jacinto fault, incorporating complex fault geometry, stresses, and velocity structure, abstract. In: 2013 Southern California Earthquake Center Annual Meeting. www.scec.org.

McGuire, R.K., 2004. Seismic Hazard and Risk Analysis. Earthquake Engineering Research Institute, Oakland, CA.

Oldham, R.D., 1899. Report on the great earthquake of 12th June 1897. Mem. Geol. Survey India 29, 1−397.

Petersen, M.D., Frankel, A.D., Harmsen, S.C., Mueller, C.S., Haller, K.M., Wheeler, R.L., Wesson, R.L., Zeng, Y., Boyd, O.S., Perkins, D.M., Luco, N., Field, E.H., Wills, C.J., Rukstales, K.S., 2008. Documentation for the 2008 Update of the United States National Seismic Hazard Maps: U.S. Geological Survey Open-File Report 2008-1128, 61 pp.

Purvance, M.D., Anooshehpoor, R., Brune, J.N., 2008a. Freestanding Block Overturning Fragilities: Numerical Simulation and Experimental Validation. In: Earthquake Engineering and Structural Dynamics, vol. 37. John Wiley & Sons, Ltd, pp. 791−808.

Purvance, M.D., Brune, J.N., Abrahamson, N.A., Anderson, J.G., 2008b. Consistency of precariously balanced rocks with probabilistic seismic hazard estimates in southern California. Bull. Seismol. Soc. Am. 98, 2629−2640.

Rodriguez-Marek, A., Montalva, G.A., Cotton, F., Bonilla, F., 2012. Analysis of single-station standard deviation using the KiK-net data. Bull. Seismol. Soc. Am. 101, 1242−1258. http://dx.doi.org/10.1785/0120100252.

Scharer, K.M., Biasi, G.P., Weldon II, R.J., 2011. A reevaluation of the Pallett Creek earthquake chronology based on new AMS radiocarbon dates, San Andreas fault, California. J. Geophys. Res. 116, B12111. http://dx.doi.org/10.1029/2010JB008099.

Schlom, T.M., Ludwig, L.G., Kendrick, K.J., Brune, J.N., Purvance, M.D., Rood, D.H., Anooshehpoor, R., 2008. An initial study of precariously-balanced rocks at the Grass Valley site, and their relevance to quaternary faults in and near the San Bernardino Mountains, CA. In: 2008 Southern California Earthquake Center Annual Meeting Poster 1-063. www.scec.org.

Shi, B., Anooshehpoor, A., Zeng, Y., Brune, J.N., 1996. Rocking and overturning of precariously balanced rocks by earthquakes. Bull. Seismol. Soc. Am. 86, 1364−1371.

Sieh, K., 1978. Slip along the San Andreas fault associated with the great 1857 earthquake. Bull. Seismol. Soc. Am. 68, 1421−1448.

Stirling, M.W., 2014. The Continued Utility of Probabilistic Seismic Hazard Assessment. In: Shroder, J., Wyss, M. (Eds.), Earthquake Hazard, Risk, and Disasters. Elsevier, London, pp. 359−376.

Tolis, S.V., 2014. To What Extent Can Engineering Reduce Seismic Risk. In: Shroder, J., Wyss, M. (Eds.), Earthquake Hazard, Risk, and Disasters. Elsevier, London, pp. 531—541.

Yagoda-Brian, G., Anderson, J.G., Miyake, H., Koketsu, K., 2014. Between-event variance for large "repeating earthquakes", submitted to Seismological Research Letters.

Yong, A., Martin, A., Stokoe, K., Diehl, J., 2013. ARRA-funded VS30 measurements using multi-technique approach at strong-motion stations in California and central-eastern United States. U.S. Geological Survey Open-File Report, 2013—1102, 60 p. and data files. http://pubs.usgs.gov/of/2013/1102/.

Zielke, O., Arrowsmith, J.R., Ludwig, L.G., Akciz, S.O., 2010. Slip in the 1857 and earlier large earthquakes along the Carrizo Plain, San Andreas fault. Science 327, 1119—1122. http://dx.doi.org/10.1126/science.1182781.

Chapter 15

Quantifying Improvements in Earthquake-Rupture Forecasts through Testable Models

Danijel Schorlemmer [1,2] and Matthew C. Gerstenberger [3]

[1] *GFZ German Research Centre for Geosciences, Potsdam, Germany,* [2] *University of Southern California, Los Angeles, CA, USA,* [3] *GNS Science, Lower Hutt, New Zealand*

ABSTRACT

The philosophy of Karl Popper suggests that scientific hypotheses should be evaluated in experiments and the results should be used to improve the underlying models. With these improved models, new hypotheses should be formulated and, again, put under test in experiments. Because earthquake-rupture forecast models are societally relevant products of seismological research and they influence public policy making, rigorous testing and evaluation of these models are a must. Recently, interest has been reinvigorated in developing tests for earthquake forecast models in the seismological community with the application of many forecast experiments and long-term testing underway. In this chapter, we describe some philosophies behind testing, the types of earthquake forecasts that are currently in development or under testing, and introduce test metrics and procedures. The difficulties of forecast and experiment development are laid out. In particular, we highlight the importance of testing centers for unbiased and fully prospective experiments, that is, testing earthquake forecasts against future observations. We show how the seismological community is embracing this philosophy and how it is applied to earthquake-rupture models and also to other aspects of seismic hazard assessment.

15.1 INTRODUCTION

George Box, a statistician and important contributor to the field of time series analysis once stated "Remember that all models are wrong; the practical question is how wrong do they have to be to not be useful" (Box and Draper, 1987). The challenge comes in understanding, or, more importantly, in quantifying how useful a model may be. Models are approximations of real systems, and how good they are depends on how the system is described in the

Earthquake Hazard, Risk, and Disasters. http://dx.doi.org/10.1016/B978-0-12-394848-9.00015-8

model and also on the amount and quality of data. The usefulness of a model is an important concept in the field of earthquake forecasting, including seismic hazard assessment. The only true way to understand the amount of useful information in a forecast is by comparing the forecast to observed earthquakes; although seemingly easy in concept, this can become a difficult challenge in practice, mostly due to the relatively slow rate of occurrence of large earthquakes. As with much of the earth sciences, and, in fact, all of the sciences, earthquake forecasting work has at its core the concept of inductive reasoning. As discussed by Karl Popper (2004), this is the process of using controlled experiments to draw inductive conclusions from what is observed in the experiment. Examples of this are the development of an earthquake-recurrence model for a particular fault based on paleoseismic data, or the use of the past rate of magnitude 5.0 events in the catalog to forecast the number of magnitude 5.0 earthquakes that are expected in the next year. To tie this back to George Box, it is well understood that models such as these are not completely correct; however, what is not well understood is just how much useful information these models contain.

The key step in understanding the amount of useful information a forecast may contain is to develop the forecast as a testable hypothesis and scrutinize it via testing. Here, it is important to distinguish between a testable hypothesis and a falsifiable hypothesis. A clearly formulated hypothesis that lacks ambiguity can be rejected by a test using even a single observation that violates the hypothesis. This is a falsifiable hypothesis. In the early days of earthquake forecasting research that aimed at predicting the occurrence of the next big event, the focus was on developing falsifiable hypotheses. Unfortunately, this was largely a failure without a successful prediction. Research has since moved toward probabilistic forecasting. If well specified, a probabilistic forecast can be rigorously tested; however, it cannot be falsified with a single observation. In testing of probabilistic forecast models, the result is unlikely to be clear because there is always a chance of incorrectly rejecting a model; however, by testing over many observations, a researcher can begin to develop an understanding of just how useful a model is. Without such testing, it becomes easy for a belief to become accepted as the truth. Back to Karl Popper, he has stated that if a hypothesis is not falsifiable, it is based on belief and is not scientific; here, we are relaxing his definition to require a more loosely defined testable hypothesis that allows for a quantitative understanding of the performance of the model. Therefore, every theory needs to be tested or subjected to attempts at falsification and scrutinized based on the results. As a consequence, the theory (i.e., model) will be improved or replaced, and the cycle starts again. Of course, in practice, such things are never so simple, and later, we will discuss ideas of what may constitute a useful test.

Although quantitative statistical testing of earthquake forecast models has been happening for decades (e.g., Evison and Rhoades, 1993, 1997;

Kossobokov et al., 1997; Kagan and Jackson, 1994, 2000; Vere-Jones, 1994, 1999), the Regional Earthquake Likelihood Model (RELM) experiment (Field, 2007) was the first attempt at a comprehensive and comparative test that included multiple models and had a goal of providing results that would be directly applicable to seismic hazard modeling. Other early forecasting research was often aimed at "earthquake prediction". As we discuss in the next section, several well-known—but unsuccessful—predictions of large earthquakes have been made in the past. In the absence of successful and useful predictions, work in this field, including seismic hazard, has largely shifted to the development of probabilistic earthquake forecast models. These models specify the probability for a range of earthquakes to occur, in a region and in some period; importantly, the probability of these earthquakes is almost always much less than 100 percent, and is most often less than 1 percent.

In this chapter, we will briefly discuss the history of earthquake forecasting and prediction research, how this has led to current work on earthquake testing centers, and how it relates to the ideas of George Box and Karl Popper. We will then describe the basics of how one can set up a forecast test, including the critical step of defining a testable earthquake forecast. Finally, we will discuss some of the statistical tests that are currently in use and provide some examples of their application.

15.2 SOME REMARKS ABOUT EARTHQUAKE PREDICTION

Earthquake prediction research went through an enthusiastic period in the 1960s and 1970s. This period was characterized by an intense search for earthquake precursors in the hope of identifying signals or anomalies that predict an impeding large earthquake and allow for societally relevant countermeasures. The enthusiasm to solve the earthquake prediction problem, amplified by the successful evacuation in Haicheng 1975 (Hammond, 1976), began to fade starting in 1976 when an M7.8 earthquake destroyed the city of Tangshan in China (Lomnitz and Lomnitz, 1978). No scientist issued a warning, and the earthquake occurred in a region previously not identified as active. With time, reports about the ambiguity of the Haicheng prediction success were published (Wang et al., 2006). Following the failure of the 1984 United States Geological Survey prediction for a moderate sized event to occur at Parkfield before 1993 (Bakun and Lindh, 1985), pessimism spread in the seismological community. Voices were raised that earthquakes are inherently unpredictable (Geller et al., 1997a) and triggered a widely perceived debate (e.g., Aceves and Park, 1997; Geller et al., 1997b; Wyss, 1997). A list of proposed earthquake precursory signals was evaluated by the International Association for Seismology and Physics of the Earth's Interior, but no signal showed sufficient evidence to be used as a precursor (Wyss and Booth, 1997). As a consequence, earthquake prediction work was increasingly replaced by seismic hazard assessments. However, earthquake prediction research did not come to a complete halt, and significant programs are still in place in Russia (Sobolev and Chebrov, 2014; Kossobokov, 2014) and China (Wu, 2014). One aspect of prediction research has not changed over time: unscientific or exaggerated claims of precursor observations or of successful predictions remain common. The promise made in the 1970s that researchers will be able to predict earthquakes within a decade (Press, 1975) still remains unfulfilled.

Given the societal need for earthquake-hazard information, the scientific community mostly moved from the quest to predict earthquakes to calculating statistical forecasts of distributions of seismicity. Predictions refer to a single earthquake given as location, time, and magnitude in such narrow bounds that the occurrence or nonoccurrence of this earthquake renders the prediction right or wrong; forecasts, on the contrary, provide statistical descriptions of future distributions of seismicity such that a larger set of earthquakes is necessary to evaluate their forecasting power (Rundle et al., 2013). Traditionally, the earthquake forecasting community has been split between those forecasting seismic hazard for long periods of time (e.g., 50 years) and those forecasting earthquake occurrence on the short to medium term (e.g., days to decades).

15.3 FORECAST MODELS

A variety of earthquake forecast models has been developed that cover many different types of input data and attempts to capture different hypotheses of geological processes or to provide statistical descriptions of earthquake occurrences. Most models use past seismicity as their primary or sole input, while others take into account geologic data such as fault information, fault slip rates, strain rates, or recurrence estimates based on paleoseismic data. Some others also include geodetic measurements, and finally, there are models that use electromagnetic signals, water level measurements, radon emissions, and atmospheric observations.

Another important characteristic that distinguishes among models is the period covered by the forecast. Although a useful prediction model will aim to predict the next significant earthquake within a short period, such as on the order of days, forecast models span a large range of time horizons with single models often capable of producing forecasts for any number of periods. Seismic hazard models typically cover many years to decades and are often considered stationary, expected to reflect long-term seismicity. Other models attempt to forecast seismicity for the next year and may be based on aftershocks and other longer-term clustering (Rhoades and Evison, 2005). Short-term models try to capture seismicity changes that occur within the time frame of months or even, as in the case of forecasting aftershocks, hours to days (e.g., Gerstenberger et al., 2005).

15.4 GRIDDED RATE-BASED FORECASTS

Also important is how the forecast is specified. A forecast will typically cover a region (e.g., a country or smaller tectonic region). A common method for specifying the spatial component of a forecast is by using a grid that covers the region. Because the rates of earthquakes strongly depend on magnitudes, the cells within the grid are subdivided into a range of magnitude bins, [M4.95, M5.05), [M5.05, M5.15), and so on, up to some larger bin of a large magnitude, for example, $M \geq 7.95$. Such a forecast will be for a single period, and all bins within the grid are assigned an expected number of earthquakes. We call this a gridded rate-based forecast, and it is the type used in the RELM experiment (Field, 2007). RELM has developed and tested a suite of different forecast models, each covering the area of California in $0.1° \times 0.1°$ cells of which each was separated into 41 magnitude bins covering the magnitude range from 4 to 9 in 0.1 magnitude unit steps (Schorlemmer and Gerstenberger, 2007), totaling 314,962 bins (see Figure 15.1 for examples).

15.5 ALARM-BASED AND REGIONAL FORECASTS

Another type of forecast is produced by alarm-based models that predict the occurrence of one or more earthquakes for a specific space-time volume. Such models are generally based on some type of precursory signal. Alarm-based models generally use a threshold value of occurrence probability above which an alarm is issued and below which no target earthquake is expected. One prominent example of an alarm-based forecasting method that has been developed several decades ago is the M8 model (Kossobokov, 2014). Choosing a seismicity rate threshold, any aforementioned gridded rate-based forecast can be turned into an alarm-based forecast. Varying the threshold creates the alarm function of a particular rate forecast.

Alternative modeling approaches may provide an expected number of events for a larger region or polygons of differing sizes. One particular example of this type of model is a fault-based forecast that provides some expected number of events for a single fault, or a collection of faults.

Some more sophisticated models are spatially continuous and may provide a forecast for any exact location that is required; however, in practice, these models are generally specified on a grid for application and for testing.

15.6 PROBABILISTIC SEISMIC HAZARD ANALYSIS AND HYBRID MODELS

Hybrid models are a type of model that combines multiple models into a single forecast. The submodels may be subsets of each other representing similar processes, time frames, and magnitudes, or they may be completely independent of each other. The submodels are combined using some sort of weighting scheme that may be quantitative or subjective.

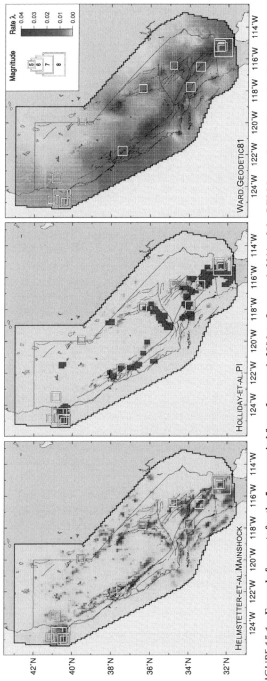

FIGURE 15.1 **Example forecasts for the 5-year period from January 1, 2006, to January 1, 2011, of the RELM experiment.** For each cell, the forecasts are summed up over all magnitude bins ($M \geq 4.95$), and shown in color. White frames indicate target earthquakes that occurred during the testing period, and their size is scaled to magnitude (left). The smoothed seismicity model with adaptive kernel by Helmstetter et al. (2007) is concentrated along the active faults and in areas of high seismic activity in the past (center). The forecast model based on the Pattern Informatics method by Holliday et al. (2007) focuses on the most probable areas for future earthquakes, defined as areas with strongly fluctuating seismicity (right). The geodetic forecast model by Ward (2007) is based on fault slip rates mapped into smoothed geologic moment rates.

Probabilistic seismic hazard analysis (PSHA) models provide a particularly interesting example of a heavily used hybrid forecast model that combines multiple hypotheses into one algorithm. Typically, a PSHA source model consists of two submodels: a fault-based model and a smoothed seismicity model based on an earthquake catalog (e.g., Stirling et al., 2012). The fault-based model is primarily derived from paleoseismic data, using varying models of earthquake occurrence, and is intended to represent longer-term processes and the occurrence of larger earthquakes; the smoothed seismicity model uses seismicity rates derived from an earthquake catalog of hypocenters, magnitudes, and time, and spatially smoothed locations. Because the number of mapped faults does not contain all potential faults for any region, the smoothed seismicity model is used to model the occurrence of smaller events below some magnitude where it is assumed that all faults are known. The sources that dominate the forecast depend on the period for which the forecast is generated, with faults becoming more dominant for longer periods. Because there is no strict model that defines how the fault model and smoothed seismicity model should be combined, the combination is typically optimized by using global positioning system (GPS)-derived moment rates or slip rates.

Other models that combine submodels consistently across all magnitudes and space have been developed more recently. Examples of this are the recently developed seismic hazard model for Canterbury, New Zealand (Gerstenberger et al., 2011), which combines models representing clustering at three different time scales (Figure 15.2), and recent work by Rhoades (2013) and Marzocchi et al. (2012).

15.7 SOME COMMON ASSUMPTIONS AND QUESTIONS POSED BY EARTHQUAKE FORECAST MODELS

As described above, most earthquake forecast models are derived from past observations of earthquake occurrence or other measures such as GPS-calculated strain rates. For earthquake hazard, most often, the occurrence of large earthquakes is estimated from information about active-fault size and recurrence interval. The occurrence of smaller earthquakes (e.g., $M \lesssim 6.5$) is typically calculated from the most recent decades of recorded seismicity. In both cases, the assumption is made that the future will behave like the past. As with the development and application of any forecast model, many questions are raised: How should we interpret the observations? Have we already observed all or the most dominant features of seismicity? Is the spatial distribution of seismicity stationary? Or to what extent is it stationary? Does a model describing past seismicity forecast future seismicity well? These questions can in part be answered by rigorous testing of earthquake forecast models and their forecasts against future observations; only prospective testing against earthquakes that occurred after the forecast was issued can prevent any bias, including the potential for unconscious bias in the model and forecast creation.

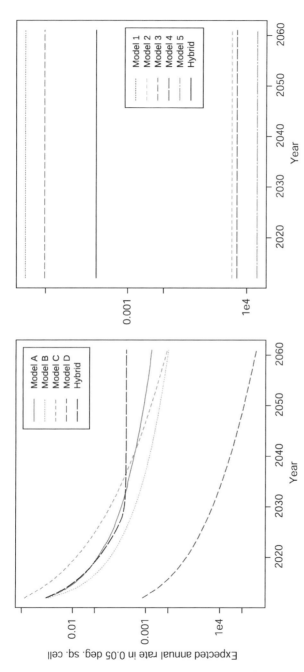

FIGURE 15.2 Multiple models representing time-dependent seismicity and long-term stationary seismicity were combined using expert elicitation into a hybrid model for Canterbury, New Zealand (Gerstenberger et al., 2011). The variability among the different models for 50 one-year forecasts for a central Christchurch location for the 50-year period following the Canterbury earthquake sequence (2011) is shown (left). Time-dependent models: four component models (Model A—Model D) representing different scales of earthquake clustering. The hybrid model combines both time-dependent and stationary contributions (right). Stationary models: five component models (Model 1—Model 5) were used. Here, the hybrid model only contains the stationary contributions.

15.8 THE ROLE OF TESTING EARTHQUAKE OCCURRENCE MODELS

As with any scientific hypothesis, the most specific measure of the usefulness of a model for describing a system, or of its predictive power, comes through testing of the model against future observations. A model that matches past data perfectly may fail in predicting the future; one example when this may occur is when a model has been "overfit" to the past data, resulting in optimization of parameters that may be unimportant or may poorly fit future data. Only prospective tests can identify such an overfit. This highlights the thoughts of George Box: It is important to assess the usefulness of a model. Therefore, in an ideal world, every model should be tested rigorously and its performance carefully assessed, before the model is used in a societally relevant manner; however, when dealing with the realities of available earthquake data and the relatively low rate of occurrence of large earthquakes, it is rarely possible to test a model in an ideal manner.

To create a testable forecast, there must be a clear statement of precisely: (1) the magnitude range for the expected rates or probabilities of event(s); (2) the enclosed region in which the event(s) will occur; (3) the period during which the event(s) will occur. Careful attention must be paid to the definitions to avoid ambiguity in the testing. For example, this has been seen in practice; the magnitude type (e.g., M_w or M_L) should be specified with the forecast. Without such a specification, there is the potential for an inconsistency due to one magnitude estimate occurring outside the forecast range, while another is within. The forecast should typically express the probability for one or more events to occur, and the probability will almost always be less than 1. Defining the forecast in such a way removes ambiguity about the result following the occurrence of one or more events. One objection modelers often have is that if an event occurs only slightly beyond any of the specified boundaries, the model will be penalized for being incorrect when the feeling is the model is "mostly" correct. While this may be the case, such a definition allows the modelers to understand how the model is performing poorly and to adjust and improve their hypothesis to potentially match future observations more precisely.

An instructional example of a model without a precise definition is that of the characteristic earthquake model. Schwartz and Coppersmith (1984) suggested that characteristic earthquakes are the maximum-size earthquakes of a fault segment as defined by displacement and recurrence time data; Wesnousky (1994) considered them as maximum-size earthquakes of fault zones or segments with generally quiescent periods in between. A precise and agreed-upon definition of this has been lacking within the earthquake-hazard community. This has resulted in multiple interpretations of which parameters describe a characteristic earthquake and how similar their observations need to be to constitute characteristic earthquakes. This has hindered progress on the understanding of the validity of this model.

Ultimately, there are two main aims when testing a model: (1) to understand if there is useful predictive information in forecasts calculated by a model; and (2) what parts of the model forecast earthquakes well, what parts of the model forecast earthquakes poorly, and how we can improve the performance of the model. Most often, when we wish to understand the information contained in a model, we compare it to a null hypothesis. The null hypothesis will typically be an existing and applied model, such as a model that forecasts earthquakes to be a simple Poissonian distribution in time but at locations of past earthquakes. By testing against other models, we can see which models have improved predictive power and what the details are that contribute to this.

Tests of predictive power can only be conducted in a prospective manner, meaning that a forecast should exclusively be tested against earthquakes in the future to exclude any biases of the modeler, even if they are unconscious. Testing against past data is helpful for model calibration or as sanity checks of a model but cannot replace a truly prospective test. Models successfully tested in a retrospective sense may be highly optimized to match past data, but their predictive power may be low. However, truly prospective tests pose many problems, as will be discussed in the Problems and Solution section. Having said this, retrospective testing does have a place in model development. It is easier to do than prospective testing, and although it is likely to produce a biased result, if a model cannot pass a retrospective test, it is unlikely to be worth the effort to test it prospectively.

15.9 DEVELOPING TESTS OF EARTHQUAKE FORECAST MODELS

Testing earthquake forecasts or predictions can be a trivial task or a massive undertaking. This mainly depends on the design and purpose of the experiment, and is also influenced by the degree of reproducibility of the experiment. On the one hand, if one wants to test a single prediction, waiting to see if the predicted earthquake occurs or not is sufficient as long as the length of the forecast period is reasonable. The prediction will be rejected if the predicted earthquake did not occur; using the results, the modeler can then aim to understand the reasons for the failure and seek improvements in the model. On the other hand, if the aim is to have a more complete understanding of the model behavior over multiple earthquakes, to test a large suite of models comparatively, or to test models that generate multiple forecasts (e.g., at regular intervals), a dedicated infrastructure can become necessary to operate such an experiment. Furthermore, in this case, reproducibility becomes even more important for maintaining transparency, recovering from operational problems, etc. It is generally not possible for an outsider to reproduce such an operation without a massive effort. Therefore, the credibility of the experiment and the confidence people have in the results will strongly depend on the level of reproducibility of the results and the transparency of the methods and procedures involved.

A number of decisions must be made before conducting a test of an earthquake forecast. These decisions will guide how the test is done and should ensure that useful and informative results are produced through the testing. The decisions are discussed in detail in the following sections and are outlined here:

- What is the goal of the test and what does one hope to learn? Is it to understand the overall predictive power of a model? Is it a relative comparison of models? Is it to understand the strengths and weaknesses of models?
- Are appropriate observations available to perform the tests and what is their source? Are the uncertainties quantified?
- Can the observations be used without any additional interpretation?
- Are the data sufficiently complete for testing (i.e., are necessary observations missing from the data)?
- How many earthquakes are required to provide a meaningful result, and how long will this take to achieve?
- What is the testing region?
- What are the minimum and maximum magnitudes and how will magnitude be binned?
- What is the time length of the forecast?
- How many time lengths will the test cover (e.g., if the forecast is for 1 day, how many 1-day periods will the test cover)?
- Will the test be retrospective, prospective, or both?
- What testing method(s) will be used?
- Is a reference model required, and if so, what is it?
- What would a meaningful result be?
- What are the appropriate statistical tools to allow one to achieve a meaningful result?
- How will this result be communicated?

For example, a test can be designed to select the better performing model between two alternate hypotheses, or it may be designed to understand the level of absolute information that a single model can provide. A researcher may be interested in how a model will perform in an applied situation (e.g., using unrevised and potentially erroneous data), or they may be interested in whether or not a particular scientific hypothesis is supported by observations. In many cases, testing of an earthquake forecast will be reliant on using observed earthquake catalog data, such as time, location, and magnitude; however, other forecasts may be more easily tested using secondary data such as observed intensity (e.g., recorded ground-motion amplitude data or Modified Mercalli Intensity (MMI)). If one wishes to test against hypocentral locations, the location uncertainties are typically small enough to allow for a meaningful test result; however, if one wishes to test a fault-based forecast, there is no consistent observation of fault rupture that allows for robust testing without the observation itself being a very uncertain model.

In all cases, it is important to understand how complete the observed data sets are and how long it will take to record enough observations to have a statistically powerful and also a meaningful and useful result. For example, a typical PSHA model, which can provide a forecast for a 10 percent probability of ground-motion exceedance in 50 years (i.e., 475-year return period), is likely not directly testable in the lifetime of any one researcher, or even several generations of researchers. Therefore, a meaningful result might be obtained by breaking the model down into specific assumptions, such as the characteristic earthquake model, and testing them on a global scale; the result of this test can then be used to inform the development of the seismic hazard model. This topic is discussed in more detail in the Problems and Solution section.

Several specific statistical tests are described below, but the particular test, or tests, used should be chosen based on the goals of the experiment including whether or not multiple models are being compared, whether one is testing an alarm-based prediction or a gridded forecast, or one wishes to know the absolute level of information in a forecast.

15.10 TESTING METHODS

The choice of testing procedures and metrics is dependent on the type of models to be tested and should take into account any model peculiarities. Furthermore, the suite of applied tests should attempt to cover all the aspects of the model forecast. However, not only does the occurrence of forecast earthquakes need to be considered but the nonoccurrence of forecast earthquakes is also important information. Besides the spatial component, models should also forecast the number and the magnitude distribution of events as correctly as possible. Only if the applied suite of tests covers most aspects of forecasts will it become difficult for a modeler to unfairly optimize their model for the tests.

For the purpose of this chapter we use the following categorization: gridded rate-based tests and alarm-based tests are applied to forecast models that forecast earthquakes as point sources (hypocenter location) and per space-magnitude bin. For testing, the observed earthquake hypocenters are associated with the respective bin in the forecast testing region. Thus, these tests are relatively simply to implement. Another type of proposed testing is one based on fault ruptures. Such forecasts would include the location and length of the fault that ruptured and require the equivalent observation. These tests are difficult to implement and are discussed in the subsection on fault-based testing.

15.11 GRIDDED RATE-BASED TESTING

RELM started testing rate-based forecasts with a small set of tests, the N-, L-, and R-tests as described by Schorlemmer et al. (2007). Whereas the N- and L-tests check for consistency of the forecast with the observation in the number domain and likelihood domain (i.e., the probability for the occurrence of the observation in the model considering magnitude, location, and time), respectively, the R-test performs a pairwise comparative test to identify the model with a higher forecasting power. This set of tests was later complemented by the S- and M-tests (Zechar et al., 2010a). These tests can be used to test individual components of the L-test, and they consider the consistency of model forecasts with the observation in the space and magnitude domain; these tests allow for an easier understanding of the results by making it clear which factors influence the results. In addition, the CL-test measures the consistency in the likelihood conditional on the number of observed events (Werner et al., 2011). The T- and W-tests were added for model comparisons and to overcome the ambiguity inherent in the R-test result (Rhoades et al., 2011), eventually replacing the R-test completely. Finally, residual-based tests, the deviance and Pearson residuals, were introduced by Clements et al. (2011). The Pearson residuals allow for the identification of spatially localized overprediction or underprediction of models, whereas the deviance residuals spatially measure the performance of one model relative to another.

Common to all these tests is the fact that the forecast rates are given per spatio-magnitude bin and for a fixed period. The size of the bins and the length of the testing period do not change the application of the test, and both parameters are driven by the purpose of the models (and tests). It is assumed that all bins are independent and that earthquakes are Poissonian distributed; these are incorrect assumptions, and their effect is the subject of ongoing work (Schorlemmer et al., 2010b). In general, bins are defined on a regular spatial grid (e.g., a $0.1° \times 0.1°$ grid) with a regular subdivision of magnitudes (e.g., 0.1 magnitude units). Although a regular spatial grid is generally appropriate for models, sometimes, models forecast rates for unevenly shaped areas. Such areas can mostly be described with circles or polygons.

Even though it is simple to assign earthquake hypocenters to bins, the uncertainties of location and magnitude need to be taken into account. Earthquakes that are located close to the border of a cell or even outside the testing region, may, given the uncertainties, have truly occurred in a neighboring cell or within the testing region, respectively. The procedure to take these uncertainties into account has been introduced by Schorlemmer et al. (2007).

For stationary models, the fact that forecasts have to be made for a fixed period is not an inconsistency, but it fully complies with the model. For time-varying models that attempt to capture changes in earthquake activity, fixed-length forecast periods can be a strong limitation. In particular, during aftershock sequences following larger events, the estimated rate changes during, for example, a 1-day testing period are not negligible. Also, the test results of such a 1-day period in which the large events occur are meaningless because the models have no chance to recalculate their forecast to account for the expected aftershocks. Making the testing periods shorter to account for short-term changes in earthquake activity (e.g., 30 min) is not a good solution because the computational load is increased dramatically while most periods during an experiment contain no earthquakes at all. A possible solution is to adjust the length of the testing periods to the seismic activity, in the extreme case defining the testing periods as the times between two events.

15.12 ALARM-BASED TESTING

The simplest test metrics for alarm-based models are based on the contingency table in which the number of hits (earthquakes in the alarm zone), correct negatives (no earthquake in the zone without any alarm), misses (earthquakes in the zone without any alarm), and false alarms (no earthquake in the alarm zone) are collected. From this table, various simple metrics (e.g., hit rate, miss rate) can be derived (Mason, 2003). One is the receiver operating characteristics (ROC), based on the hit rate and the false-alarm rate. However, the ROC is considered a weak test as it uses a spatially uniform model as the reference. The Molchan diagram (Molchan, 1991; Molchan and Kagan, 1992; Harte and Vere-Jones, 2005) is an extension of the ROC such that different reference models can be selected. An extension of the Molchan diagram is the area skill score by Zechar and Jordan (2008) that allows for testing the entire alarm function.

15.13 FAULT-BASED TESTING

Some models, in particular the ones used in PSHA, not only forecast the hypocenter of earthquakes or earthquake rates on a grid but the exact parts or segments of one or multiple faults that will rupture. Testing such models poses some difficulties: All faults and their respective segments need to be clearly defined. The more difficult task is to assign an observed event to one or more of the predefined segments. Not only should the location match but also the rupture planes should match with the segments. Unfortunately, such information is not calculated on a regular basis and could thus be taken from an independent data provider. Even worse, any solution is nonunique, and much interpretation of available data is necessary to derive the extent, location, and orientation of a rupture; a process that cannot be automated and will as such always create dissent among the participants of an experiment. One possible solution requires modelers to change their model to forecast only observables, namely, earthquake hypocenters and focal mechanisms or moment tensors. To be able to use these observables, the models need to smooth the rates of all faults and their segments to a grid (also including location uncertainties that will make earthquakes appear to occur at off-fault locations). As such, fault-based forecasts could be treated in the same way as gridded rate-based forecasts and the testing methods could be applied. The focal mechanisms could additionally be employed to ensure that the earthquake-rupture orientation matches with the orientation of the fault or fault segments. On the one hand, such an experiment requires some significant modifications of fault-based models; on the other hand, fault-based models are otherwise untestable, which is scientifically speaking not satisfactory.

15.14 STRUCTURED TESTING

An example of an experiment that attempted to test a large suite of models using a structured and transparent framework is the RELM experiment (Field, 2007). The plan of testing RELM models included testing a suite of models that had each generated a single 5-year forecast, and a suite of 1-day models over many periods. This experiment required: (1) retrieving input data on a daily basis (i.e., earthquake catalogs); (2) generating forecasts from the models; (3) testing these forecasts against observations; (4) processing the results; and (5) doing all of this in a transparent and reproducible manner. It became quickly clear that an automated testing facility was needed. One can hardly imagine that every modeler would process, on a daily basis, the input data, generate a new forecast and submit this forecast in time to an independent group of testers. Such an experiment may succeed for a week or so, but will undoubtedly fail over a period of 5 years or longer. Any mistake, or any submission of forecasts after the daily deadline, could render the experiment a failure. But, even if this setup would work for 5 years or longer, there may be scientific challenges: a modeler may change his or her model after analyzing its performance but continue to submit the forecasts. In such a case, the model would become a moving target and the results could not be attributed to a particular model implementation, again rendering the experiment a failure. As a consequence, each model should ideally run independently from the daily scientific working of the modeler. This ensures that the model is not changed during the duration of the experiment. To accomplish this, RELM established the concept of a testing center, which is an independent and automated facility for processing and testing earthquake forecasts (Schorlemmer and Gerstenberger, 2007). The testing center releases the burden of real-time processing; if an independent modeler wants to create a forecast for day d, the forecast has to be created before day d starts and also submitted to the testers before day d. This is a challenge for even the most organized scientist. In our domain of earthquakes, earthquake catalogs are not produced in real time at a sufficient quality level. It can take hours to days and sometimes months until automatically detected earthquakes will be reviewed by a seismologist and potentially relocated. Therefore, in a practical sense, 1-day forecasts for day d cannot be created on the day before d, but once all events in the period before day d have been reviewed. Therefore, a model needs to be installed in a testing center where the center can ensure that the model cannot be changed and then the testing center can invoke the forecast generation for day d weeks after that day but providing the model only input data of the period before day d (Schorlemmer and Gerstenberger, 2007). Thus, truly prospective tests can be performed without having to organize them in a real-time fashion with the aforementioned drawbacks and problems.

In addition to these advantages, a testing center offers many more improvements over simple testing without the dedicated infrastructure. The input data for one experiment can be downloaded centrally and the very same data set can be provided to all participating models. This way, the testing center ensures that all models use the same input data and that no additional information is used in generating forecasts. The testing center stores each downloaded data set, thus allowing the replay of a test of a particular day. All models can be run automatically, and the entire testing process can be implemented to run without human interaction. By storing all input data sets and all results of each testing run, the testing center provides all means to fully reproduce any test result. This feature in particular is important for a truly scientific experiment. In general, earthquake catalogs are provided as is, and they are subject to change without notice. Networks recompute magnitudes or relocate events long after a reviewed location and magnitude were estimated. Therefore, it is in general impossible to retrieve the very same catalog that was provided some time ago by the network. The testing center however stores each input data set (here a catalog) for each test run, and the very same catalog can be used for a replay of the test run or any other external analysis. The testing center also requires that each model that depends on random numbers retrieves them through an interface from the testing center, to ensure that the same random numbers will be used when replaying a test run. If not, the test run will not be fully reproducible. The design of the testing center has strictly followed the guidelines of *transparency, controlled environment, comparability,* and *reproducibility* as provided by Schorlemmer and Gerstenberger (2007).

15.15 TESTING CENTERS: COLLABORATORY FOR THE STUDY OF EARTHQUAKE PREDICTABILITY

During the RELM experiments, it was decided to implement a dedicated testing center to make sure that the aforementioned rules and guidelines could be followed. All the modelers submitted their forecasts to the testing center before January 1, 2007, and the forecasts covered the 5-year period starting January 1, 2007. Eleven forecasts were submitted that forecast mainshocks only, and six forecasts were submitted that forecast both mainshocks and aftershocks. In late 2006, a group at the Southern California Earthquake Center (SCEC) inherited the RELM project (Jordan, 2006) and implemented the first testing center (Schorlemmer and Gerstenberger, 2007; Zechar et al., 2010b) as part of a new project called Collaboratory for the Study of Earthquake Predictability (CSEP). CSEP developed the infrastructure for testing the RELM models and worked with three other international testing centers so that the CSEP infrastructure became an international standard. Starting September 1, 2007, the first version of the testing center became operational, and continuous testing of the RELM models started.

Simultaneously, the CSEP testing center also started testing two 1-day RELM models, the Short-Term Earthquake Probabilities model (Gerstenberger et al., 2005) and the Epidemic-Type Aftershock Sequence (ETAS) model (Zhuang et al., 2008). These two models were the first models to be installed as fully functional software code that operates without human interaction. Both models generate daily forecasts for the next 1-day period in California and are tested every day. To ensure seamless operation, the testing was designed as a four-tier system as described by Zechar et al. (2010b). The first tier is the development system on which the development of testing center codes takes place and which is also used to integrate models into the work flow of the testing center. Every 24 h the second tier, the certification system, runs self-tests of the software codes to make sure that any changes in the codes did not break the system. Weeks before a new version of the testing center software is released, the software codes are not changed anymore and are tested for functionality on the certification system. With each new release, the codes are moved to the operational system, the third tier, where they run without any human interaction. All the results computed at this system are transferred directly to the fourth tier, the web server for publishing.

CSEP expanded over time into different areas and included many more different models. The first off-spin in 2007 was the testing center at GNS Science in New Zealand (Gerstenberger and Rhoades, 2010), covering experiments for the New Zealand area. In 2008, another testing center at the Earthquake Research Institute of the University of Tokyo, Japan, was started (Tsuruoka et al., 2012). At this testing center, several experiments were set up covering three different testing areas: Kanto area, main Japanese Islands, and Japan including offshore area. Finally, in 2009, a testing center at ETH Zurich was implemented covering experiments for the area of Italy (Schorlemmer et al., 2010a). Over time, the testing center at SCEC (located at the University of Southern California) started experiments in the Western Pacific region (Eberhard et al., 2012), for oceanic transform faults, and a global experiment. Overall, time horizons of implemented models encompass 1 day, 3 months, 1 year, and 5 years.

15.16 PROBLEMS AND SOLUTIONS

A number of challenges must be faced when testing a model. For the purposes of this discussion, we break those challenges into four categories: (1)

developing a testable model; (2) data issues; (3) hypothesis versus models; and (4) test application.

15.16.1 Developing a Testable Model

The need for a testable model has been outlined in the section on The Role of Testing Earthquake Occurrence Models. However, the challenge involved in doing so should not be underestimated. To go from an idea or hypothesis to a testable model forces researchers to pin down the details of their idea and to examine its implications in ways they may not have considered before. Often, a useful way to develop and fully understand one's assumptions is to try and formulate one's hypothesis into a testable model that rigorously specifies the exact boundaries and conditions of the forecast. For other modelers, the challenge in making their model testable may come in that data are not available to test their particular model. Sometimes, the only available observables are other models (e.g., fault models as a substitute for the limited knowledge about faults). Particular problems are tests that require time frames that are too long to be reasonable for any useful scientific endeavor. All these aspects should be considered when developing a model.

15.16.2 Data Issues

As discussed in the section, Testing Methods, when developing an experiment to test a forecast or a model, one must decide what data are appropriate to use as the observations for the test. In most cases, the observations will be an earthquake catalog. As with any data, an earthquake catalog is only an interpretation of what has occurred and will contain uncertainties, and the data quality will need to be assessed. To ensure that a test is as unbiased as possible, the minimum magnitude of complete recording will need to be at least as low as the minimum magnitude being tested against. This may require an extensive study as performed by Schorlemmer et al. (2010a). If the magnitude range being tested includes magnitudes that may not be completely recorded, it is impossible to know if the test result is truly representative of the performance of the model. Additionally, any data reported in an earthquake catalog will have some associated uncertainty, such as in the magnitude and location. Although in many cases these uncertainties may be small relative to the resolution of the test, they can cause edge effects that may significantly affect the test results (Werner and Sornette, 2008). However, handling these uncertainties in an efficient manner and in one that does not overly complicate the results (Schorlemmer et al., 2007) is not an easy task and is one that has not yet been solved. In other cases, a catalog may contain more egregious errors, such as in depth, which a researcher must be aware of and adapt the test conditions appropriately.

In general, as long as the researchers are aware of any data quality issues, these can be handled and a test can be conducted. However, the more difficult challenge comes in the amount of available earthquake data. Two issues must be considered related to the number of observed earthquakes that a test will require: (1) statistical power and (2) how representative the observed earthquakes are of a sample from a longer period (in particular when testing stationary models). Statistical power is the probability that the test will correctly reject the model and is largely controlled by the number of observed earthquakes the model is tested against. For example, when comparing two models, the power can be thought of as the ability of a test to distinguish between the models, and the power is therefore dependent on the forecast of each of the models. By creating simulated catalogs from each model's forecast, and examining how often the other forecast fails the test, the power can be estimated (Zechar et al., 2010a). The higher the power, the more information is contained in the test. Understanding how representative a test result from any earthquake sample is of a test result from a longer period is a function of the spatial stationarity of seismicity and the spatial variability of a given model. Just how long of a testing period will be required is model dependent and is not yet well understood, but is an area of active research.

A particularly interesting case is for PSHA models that are capable of producing forecasts over any number of time frames, and are typically used to generate a forecast for relatively long lengths such as 475 or 2500 years. For such a model, it is clear that it will never be tested over the periods for which it is creating forecasts. A difficult challenge for PSHA modelers is to develop tests of their models that can provide informative and unbiased results, and can indicate how well a model performs and when improvements are made. Typically, this problem is approached by investigating some of the basic assumptions in the model (e.g., magnitude distributions).

Furthermore, one of the most difficult challenges in testing earthquake models comes from the length of time required to test a model, and is due to the nature of scientists. As scientists and modelers, we are constantly changing, adapting, and hopefully improving our ideas. This means that a model we were working on 1 or 2 years ago may now be of lesser interest to us, or we may have altered the model in a subtle but important way. However, to fully test the ideas, it may take many years; if each time we develop or change our ideas, we forget about our past models, we may never be fully able to quantify if our changes are truly creating improved earthquake forecasts. However, having each version of such a model implemented in a testing center will provide a detailed track record over time and may help better understand the relative improvements or declines in the performance between the different versions.

In the context of how long it takes to test a model, it is important to discuss the role of testing large earthquakes versus testing of small earthquakes. Because of the power-law nature of the frequency-magnitude distribution of earthquakes (Gutenberg and Richter, 1944), many more small earthquakes occur than large earthquakes. This can easily be exploited to shorten the time required to test earthquake forecast models; however, for several reasons, this is typically not done. Mainly this is avoided because large magnitudes are of most interest to researchers and end users, and how well the occurrence of small earthquakes predicts the occurrence of large earthquakes is not yet well understood. Second, models such as PSHA have independent models for forecasting small and large earthquakes, and by testing on smaller magnitude earthquakes, over a short period, little is learned about the ability to forecast large earthquakes.

A slightly different data consideration is illustrated in the case of testing forecasts based on Coulomb stress change modeling (e.g., Cocco et al., 2010; Woessner et al., 2011; Toda and Enescu, 2011). When a forecast is made using Coulomb modeling, one of the inputs required to create the forecast is a slip model that indicates the extent of the fault that ruptured and the distribution of slip on the fault. At present, the science behind the slip models is rapidly evolving and a potential for multiple slip models exists for any earthquake. Additionally, the optimal slip models will commonly use aftershock locations to help constrain the fault location. This presents two challenges: (1) the best slip models are not available until some time after the mainshock, this can be days or weeks; and (2) if testing retrospectively, the use of these slip models will introduce a potentially significant bias into the testing by using a model that was developed using data that the forecast is being tested against. A seemingly simple solution to this is to use automated slip models that are generated coseismically, or immediately after the event. However, these models are likely to be only first-order results and will not allow for a full understanding of the Coulomb forecast model's potential. The same is true for any forecast model that requires input data during the experiment that is based on interpretation or careful selection specific to the forecast model.

15.16.3 Hypothesis versus Model

Finally, we must remember that when models become complicated requiring complex calculations and lengthy computer programs, what we are actually testing is that particular model implementation of the hypothesis (or hypotheses). Each computer code may be different in subtle, or not-so-subtle, ways, and as any scientific programmer knows, it will also contain errors. Such differences may become important in testing, or may even be highlighted and better understood via testing. One such example of the variability that comes with implementation is the ETAS model (Ogata and Zhuang, 2006), which has been developed as a model by many different researchers, each adding their own influence to the model. Although a particular test may be described as testing the ETAS model, it will be testing a slightly, but often significantly, different hypothesis than when an ETAS test is done by another modeler.

15.16.4 Test Application

Throughout the forecasting community, a consensus has not yet evolved on the best and most informative suite of tests. The community is still working on understanding exactly what each of the tests is indicating and how to draw conclusions from the test result. A particular challenge is to understand how the results can be taken up into model improvements. Additionally, it is still possible to obtain conflicting results when applying a suite of tests. As the understanding of the tests improves, these conflicts will hopefully be better understood, including their implications for the models tested. Until these conflicts are better understood, they can present a challenge in communicating the results. Even when the results are clearer, consideration is required on how the results are presented to nonspecialists. This is a difficulty the entire community is currently working to overcome.

15.17 OUTLOOK AND CONCLUSIONS

In the last 10 years, emphasis has been renewed on earthquake forecasting research, and a particularly vigorous effort has been made to develop and apply statistical tests that can measure the amount of information contained in any forecast and compare that to other forecasts. It is through this type of statistical testing that we can best improve our ability to forecast earthquakes and work toward improvements in estimates of seismic hazard. Many challenges remain. Due to the relatively low occurrence rate of earthquakes, new methods are required to strengthen our understanding of the ability of forecast and hazard models to provide useful information.

A first important step in improving testing was made by the RELM initiative that collectively tested a suite of models over an initial 5-year period and will continue testing for an indefinite period (Schorlemmer et al., 2010b; Zechar et al., 2013). Coming out of RELM were two important impacts: First, a shift has occurred in the thinking of the earthquake forecast community, and also in the greater seismology community, with a stronger focus on testing of models and an expectation that it be done. Second, some useful testing results and learning about the test themselves have been achieved. Several classes of models were tested in the experiment, including smoothed seismicity, geodetic, geological, and earthquake simulator. In this particular 5-year test, the Helmstetter et al. (2007) model of a smoothed seismicity adaptive kernel was the clear winner in almost any of the tests applied (Zechar et al., 2013). Also, it was learned that a purely geodetic-based model performed well in the tests and had significant information. Interestingly, Rhoades (2013) demonstrated that when creating optimized hybrid models based on this 5-year testing period, models that performed relatively poorly in the RELM experiment could provide a useful contribution when included in a hybrid model. The performance of the Helmstetter et al. (2007) model has been considered for various seismic hazard models, and it has been included in the forthcoming UCERF3 model for California (Field et al., 2013).

The final result of a PSHA model is a forecast of the expected ground motions. Because of the long time frames of PHSA models, their testing is most often split into testing of the earthquake-rupture forecast components and the ground-motion components separately to derive the most information possible about the model's performance. However, ultimately what matters in application is how the combined forecast performs. To this end, recent works (e.g., Stirling and Petersen, 2006; Stirling and Gerstenberger, 2010; Beauval et al., 2008) have developed tests of the final hazard output using either historical intensity data, such as MMI, or more recently recorded strong ground-motion data. These tests have proved to be insightful; however, they are limited to small data sets that are lacking larger ground motions that are of more interest to the PSHA community. An alternative to this approach is directly testing the models that predict ground motion or intensity for a given earthquake magnitude and location, and are not concerned with the probability of earthquake rupture. Traditionally, this work has been retrospective, but recent work within the framework of the Global Earthquake Model (www.globalquakemodel.org) is designing a CSEP-style framework for prospectively testing ground-motion prediction equations and intensity prediction equations.

REFERENCES

Aceves, R.L., Park, S.K., 1997. Cannot earthquakes be predicted? Science 278, 488.

Bakun, W.H., Lindh, A.G., 1985. The Parkfield, California, earthquake prediction experiment. Science 229, 619–624.

Beauval, C., Bard, P.Y., Hainzl, S., Gueguen, P., 2008. Can strong-motion observations be used to constrain probabilistic seismic-hazard estimates? Bull. Seismol. Soc. Am. 98 (2), 509–520. http://dx.doi.org/10.1785/0120070006.

Box, G.E.P., Draper, N.R., 1987. Empirical Model Building and Response Surfaces. John Wiley & Sons, New York, NY.

Clements, R.A., Schoenberg, F.P., Schorlemmer, D., 2011. Residual analysis methods for space-time point processes with applications to earthquake forecast models in California. Ann. Appl. Stat. 5 (4), 2549–2571. http://dx.doi.org/10.1214/11-AOAS487.

Cocco, M., Hainzl, S., Catalli, F., Enescu, B., Lombardi, A.M., Woessner, J., 2010. Sensitivity study of forecasted aftershock seismicity based on Coulomb stress calculation and rate- and state-dependent frictional response. J. Geophys. Res. 115, B05307. http://dx.doi.org/10.1029/2009JB006838.

Eberhard, D.A.J., Zechar, J.D., Warmer, S., 2012. A prospective earthquake forecast experiment in the western Pacific. Geophys. J. Int. 190 (3), 1579–1592. http://dx.doi.org/10.1111/j.1365-246X.2012.05548.x.

Evison, F.F., Rhoades, D.A., 1993. The precursory earthquake swarm in New Zealand — hypothesis tests. N. Z. J. Geol. Geophys. 36 (1), 51–60.

Evison, F.F., Rhoades, D.A., 1997. The precursory earthquake swarm in New Zealand: hypothesis tests II. N. Z. J. Geol. Geophys. 40 (4), 537–547.

Field, E.H., 2007. Overview of the working group for the development of regional earthquake likelihood models (RELM). Seismol. Res. Lett. 78 (1), 7−16.

Field, E.H., et al., 2013. Uniform California earthquake rupture forecast, version 3 (UCERF3): the time-independent model. USGS Open-File Report. 1165.

Geller, R.J., Jackson, D.D., Kagan, Y.Y., Mulargia, F., 1997a. Earthquakes cannot be predicted. Science 275, 1616−1617.

Geller, R.J., Jackson, D.D., Kagan, Y.Y., Mulargia, F., 1997b. Cannot earthquakes be predicted? Responses. Science 278, 488−490.

Gerstenberger, M.C., Wiemer, S., Jones, L.M., Reasenberg, P.A., 2005. Real-time forecasts of tomorrow's earthquakes in California. Nature 435, 328−331. http://dx.doi.org/10.1038/03622.

Gerstenberger, M.C., Rhoades, D.A., 2010. New Zealand earthquake forecast testing centre. Pure Appl. Geophys. 167 (8−9), 877−892. http://dx.doi.org/10.1007/s00024-010-0082-4.

Gerstenberger, M.C., Cubrinovski, M., McVerry, G.H., Stirling, M.W., Rhoades, D.A., Bradley, B., Langridge, R.M., Webb, T.H., Peng, B., Pettinga, J., Berryman, K.R., Brackley, H.J., 2011. Probabilistic assessment of liquefaction potential for Christchurch in the next 50 years. GNS Sci. Report, 15, 25 pp.

Gutenberg, B., Richter, C.F., 1944. Frequency of earthquakes in California. Bull. Seismol. Soc. Am. 34, 185−188.

Hammond, A.H., 1976. Earthquakes: an evacuation in China, a warning in California. Science 192, 538−539.

Harte, D., Vere-Jones, D., 2005. The entropy score and its uses in earthquake forecasting. Pure Appl. Geophys. 162 (6−7), 1229−1253. http://dx.doi.org/10.1007/s00024-004-2667-2.

Helmstetter, A., Kagan, Y.Y., Jackson, D.D., 2007. High-resolution time-independent grid-based forecast for $M \geq 5$ earthquakes in California. Seismol. Res. Lett. 78 (1), 78−86. http://dx.doi.org/10.1785/gssrl.78.1.78.

Holliday, J.R., Chen, C., Tiampo, K.F., Rundle, J.B., Turcotte, D.L., Donnellan, A., 2007. A RELM earthquake forecast based on pattern informatics. Seismol. Res. Lett. 78 (1), 87−93. http://dx.doi.org/10.1785/gssrl.78.1.87.

Jordan, T.H., 2006. Earthquake predictability, brick by brick. Seismol. Res. Lett. 77 (1), 3−6.

Kagan, Y.Y., Jackson, D.D., 1994. Long-term probabilistic forecasting of earthquakes. J. Geophys. Res. 99 (B7), 13685−13700. http://dx.doi.org/10.1029/94JB00500.

Kagan, Y.Y., Jackson, D.D., 2000. Probabilistic forecasting of earthquakes. Geophys. J. Int. 143 (2), 438−453. http://dx.doi.org/10.1046/j.1365-246X.2000.01267.x.

Kossobokov, V.G., Healy, J.H., Dewey, J.W., 1997. Testing an earthquake prediction algorithm. Pure Appl. Geophys. 149 (1), 219−232. http://dx.doi.org/10.1007/BF00945168.

Kossobokov, V.G., 2014. Times of increased probabilities for the occurrence of catastrophic earthquakes: 25 years of the hypothesis testing in real time. In: Shroder, J., Wyss, M. (Eds.), Earthquake Hazard, Risk, and Disasters. Elsevier, London, pp. 477−504.

Lomnitz, C., Lomnitz, L., 1978. Tangshan 1976: a case history in earthquake prediction. Nature 271, 109−111. http://dx.doi.org/10.1038/271109a0.

Marzocchi, W., Zechar, J.D., Jordan, T.H., 2012. Bayesian forecast evaluation and ensemble earthquake forecasting. Bull. Seismol. Soc. Am. 102 (6), 2574−2584. http://dx.doi.org/10.1785/0120110327.

Mason, I.B., 2003. Binary events. In: Jolliffe, I.T., Stephenson, D.B. (Eds.), Forecast Verification. Wiley, Hoboken, pp. 37−76.

Molchan, G.M., 1991. Structure of optimal strategies in earthquake prediction. Tectonophysics 193, 267−276.

Molchan, G.M., Kagan, Y.Y., 1992. Earthquake prediction and its optimization. J. Geophys. Res. 97, 4823–4838.

Ogata, Y., Zhuang, J., 2006. Space-time ETAS models and an improved extension. Tectonophysics 413, 13–23.

Popper, K., 2004. Conjectures and Refutations: The Growth of Scientific Knowledge (Reprinted Ed.). Routledge, London, ISBN 0-415-28594-1.

Press, F., 1975. Earthquake prediction. Sci. Am. 232 (5), 14–23. http://dx.doi.org/10.1038/scientificamerican0575-14.

Rhoades, D.A., Evison, F.F., 2005. Test of the EEPAS forecasting model on the Japan earthquake catalogue. Pure Appl. Geophys. 161, 1271–1290.

Rhoades, D.A., Schorlemmer, D., Gerstenberger, M.C., Christophersen, A., Zechar, J.D., Imoto, M., 2011. Efficient testing of earthquake forecasting models. Acta Geophys. 59 (4), 728–747. http://dx.doi.org/10.2478/s11600-011-0013-5.

Rhoades, D.A., 2013. Mixture models for improved earthquake forecasting with short-to-medium time horizons. Bull. Seismol. Soc. Am. 103 (4), 2203–2215. http://dx.doi.org/10.1785/0120120233.

Rundle, J.B., Holliday, J.R., Graves, W.R., Donnellan, A., Turcotte, D.L., 2013. Challenges in web-based real time earthquake forecasting (RTEF): localizing forecast probabilities in space and time (abstract). Seismol. Res. Lett. 84 (2), 324.

Schorlemmer, D., Gerstenberger, M., 2007. RELM testing center. Seismol. Res. Lett. 78 (1), 30–36.

Schorlemmer, D., Gerstenberger, M., Wiemer, S., Jackson, D.D., Rhoades, D.A., 2007. Earthquake likelihood model testing. Seismol. Res. Lett. 78 (1), 17–29.

Schorlemmer, D., Christophersen, A., Rovida, A., Mele, F., Stucchi, M., . Marzocchi, W., 2010a. Setting up an earthquake forecast experiment in Italy. Ann. Geophys. 53 (3). http://dx.doi.org/10.4401/ag-4844.

Schorlemmer, D., Zechar, J.D., Werner, M.J., Field, E.H., Jackson, D.D., Jordan, T.H., The RELM Working Group, 2010b. First results of the regional earthquake likelihood models experiment. Pure Appl. Geophys 167 (8–9), 859–876. http://dx.doi.org/10.1007/s00024-010-0081-5.

Schwartz, D.P., Coppersmith, K.J., 1984. Fault behaviour and characteristic earthquakes: examples from Wasatch and San Andreas fault zones. J. Geophys. Res. 89, 5681–5698.

Sobolev, G., Chebrov, V., 2014. The experience of real time earthquake predictions in Kamchatka. In: Shroder, J., Wyss, M. (Eds.), Earthquake Hazard, Risk, and Disasters. Elsevier, London, pp. 449–475.

Stirling, M., Gerstenberger, M., 2010. Ground motion-based testing of seismic hazard models in New Zealand. Bull. Seismol. Soc. Am. 100 (4), 1407–1414. http://dx.doi.org/10.1785/0120090336.

Stirling, M., Petersen, M., 2006. Comparison of the historical record of earthquake hazard with seismic-hazard models for New Zealand and the continental United States. Bull. Seismol. Soc. Am. 96 (6), 1978–1994. http://dx.doi.org/10.1785/0120050176.

Stirling, M., McVerry, G., Gerstenberger, M., Litchfield, N., Van Dissen, R., Berryman, K., Barnes, P., Wallace, L., Villamor, P., Langridge, R., Lamarche, G., Nodder, S., Reyners, M., Bradley, B., Rhoades, D., Smith, W., Nicol, A., Pettinga, J., Clark, K., Jacobs, K., 2012. National seismic hazard model for New Zealand: 2010 update. Bull. Seismol. Soc. Am. 102 (4), 1514–1542. http://dx.doi.org/10.1785/0120110170.

Toda, S., Enescu, B., 2011. Rate/state Coulomb stress transfer model for the CSEP Japan seismicity forecast. Earth Planets Space 63 (3), 171–185. http://dx.doi.org/10.5047/eps.2011.01.004.

Tsuruoka, H., Hirata, N., Schorlemmer, D., Euchner, F., Nanjo, K.Z., Jordan, T.H., 2012. CSEP testing center and the first results of the earthquake forecast testing experiment in Japan. Earth Planets Space 64, 661–671. http://dx.doi.org/10.5047/eps.2012.06.007.

Vere-Jones, D., 1994. Statistical models for earthquake occurrence: clusters, cycles and characteristic earthquakes. In: Bozdogan, H. (Ed.), Proc. 1st US/Japan Conf. on Frontiers of Statistical Modeling: An Informational Approach. Kluwer, pp. 105–136.

Vere-Jones, D., 1999. Probabilities and information gain for earthquake forecasting. Comput. Seismol. 30, 248–263.

Wang, K., Chen, Q.-F., Sun, S., Wang, A., 2006. Predicting the 1975 Haicheng earthquake. Bull. Seismol. Soc. Am. 96 (3), 757–795. http://dx.doi.org/10.1785/0120050191.

Ward, S.N., 2007. Methods for evaluating earthquake potential and likelihood in and around California. Seismol. Res. Lett. 78 (1), 121–133. http://dx.doi.org/10.1785/gssrl.78.1.121.

Werner, M.J., Sornette, D., 2008. Magnitude uncertainties impact seismic rate estimates, forecasts and predictability experiments. J. Geophys. Res. 113 (B8), B08302. http://dx.doi.org/10.1029/2007JB005427.

Werner, M.J., Helmstetter, A., Jackson, D.D., Kagan, Y.Y., 2011. High-resolution long-term and short-term earthquake forecasts for California. Bull. Seismol. Soc. Am. 101 (4), 1630–1648. http://dx.doi.org/10.1785/0120090340.

Wesnousky, S.G., 1994. The Gutenberg-Richter or characteristic earthquake distribution, which is it? Bull. Seismol. Soc. Am. 84, 1940–1959.

Woessner, J., Hainzl, S., Marzocchi, W., Werner, M.J., Lombardi, A.M., Catalli, F., Enescu, B., Cocco, M., Gerstenberger, M.C., Wiemer, S., 2011. A retrospective comparative forecast test on the 1992 Landers sequence. J. Geophys. Res. 116, B05305. http://dx.doi.org/10.1029/2010JB007846.

Wu, Z., 2014. Duties of earthquake forecast: Cases and lessons in China. In: Shroder, J., Wyss, M. (Eds.), Earthquake Hazard, Risk, and Disasters. Elsevier, London, pp. 431–448.

Wyss, M., 1997. Cannot earthquakes be predicted? Science 278, 487–488.

Wyss, M., Booth, D.C., 1997. The IASPEI procedure for the evaluation of earthquake precursors. Geophys. J. Int. 131, 423–424.

Zechar, J.D., Jordan, T.H., 2008. Testing alarm-based earthquake predictions. Geophys. J. Int. 172 (2), 715–724. http://dx.doi.org/10.1111/j.1365-246X.2007.03676.x.

Zechar, J.D., Gerstenberger, M.C., Rhoades, D.A., 2010a. Likelihood-based tests for evaluating space-rate-magnitude earthquake forecasts. Bull. Seismol. Soc. Am. 100 (3), 1184–1195. http://dx.doi.org/10.1785/0120090192.

Zechar, J.D., Schorlemmer, D., Liukis, M., Yu, J., Euchner, F., Maechling, P.J., Jordan, T.H., 2010b. The collaboratory for the study of earthquake predictability perspective on computational earthquake science. Concurr. Comput. 22 (12), 1836–1847. http://dx.doi.org/10.1002/cpe.1519.

Zechar, J.D., Schorlemmer, D., Werner, M.J., Gerstenberger, M.C., Rhoades, D.A., Jordan, T.H., 2013. Regional earthquake likelihood models I: first-order results. Bull. Seismol. Soc. Am. 103 (2A), 787–798. http://dx.doi.org/10.1785/0120120186.

Zhuang, J., Christophersen, A., Savage, M., Vere-Jones, D., Ogata, Y., Jackson, D.D., 2008. Differences between spontaneous and triggered earthquakes: their influences on foreshock probabilities. J. Geophys. Res. 113, B11302. http://dx.doi.org/10.1029/2008JB005579.

Chapter 16

Duties of Earthquake Forecast: Cases and Lessons in China

Zhongliang Wu

Institute of Geophysics, China Earthquake Administration, Beijing, People's Republic of China

ABSTRACT

The duties of earthquake forecasting are accomplished by a system with the output of the earthquake forecast serving the disaster risk management of the society. The system, with different degrees of complexity in different places, consists of monitoring, analysis, predictive modeling, and forecast-oriented, decision-making procedures. The performance of the system is determined by the scientific and technical capability and the public understanding of these capabilities. This chapter analyzes the duties of earthquake forecasting using China as a case example, which provides a "field map" of the Chinese system, highlighting the scientific products and their performance and the underlying ideas for the production, improvement, and use of these scientific products.

16.1 INTRODUCTION: (MIS)UNDERSTANDING EARTHQUAKE FORECAST/PREDICTION IN CHINA

Earthquake forecast/prediction is both a scientific issue and a social issue. The entanglement of these two issues makes earthquake forecast/prediction one of the more complicated topics, in both natural science and social science. For over half a century, the term "seismosociology" has been a session topic of international symposia or a topic of academic publications. Rapid developments in both science and society have been providing this field with rich material and new questions. On the other hand, however, the systematic study in this field in a comprehensive perspective is, to a large extent, just at its beginning.

In this chapter, we discuss the duties of earthquake forecast/prediction, a special topic in "seismosociology", taking the Chinese approach as an example. This approach has been investigated, introduced, and commented upon by several works, either focusing on a specific aspect (Editorial Committee of China Today, 1993; Chen and Wang, 2010) or in the international context

Earthquake Hazard, Risk, and Disasters. http://dx.doi.org/10.1016/B978-0-12-394848-9.00016-X
431

(Bormann, 2011; Jordan et al., 2011). The Chinese approach is characterized by its long history, large-scale practice, active and complicated interaction with a changing society, and many aspects of crosscultural misunderstanding. To some extent, if the Chinese approach were understood, then other approaches would be easily understood.

Earthquake forecast/prediction has been one of the key issues in the earthquake science in China since the 1950s. In 1956, *Study of Seismicity in China and Preparedness Against Seismic Disasters* was enrolled into the *National Twelve-Year Long-term Project of the Development of Science and Technology (1956—1967)*. In this document, four topics related to earthquakes have been proposed: (1) Development of seismograph networks and seismological instrumentation; (2) Seismic intensity zonation and study of the regional features of seismicity; (3) Study of the effect of seismic strong ground motion on buildings and development of earthquake engineering; and (4) Study of earthquake forecast/prediction. A large-scale study of earthquake forecast/prediction was started after the 1966 Xingtai earthquake that was the start of the seismically active decade from 1966 to 1976 on the Chinese mainland, during which nine major earthquakes occurred (Ma et al., 1982). Many of the outside reports concentrated on the short- to intermediate-term earthquake prediction (e.g., Hammond, 1976). However, the Chinese wording "earthquake forecast/prediction (study)" has actually a much broader sense as compared to the "earthquake forecast/prediction (study)" in English, which is a cultural difference similar to the difference between the Chinese "Lung" and the western "dragon".[1] Table 16.1 shows the comparison of the related concepts. In this chapter, we will not discuss seismic-hazard assessment, which is related to seismic intensity zonation, although in China it is also understood as a kind of long-term earthquake forecast/prediction. Remarkably, however, it is worth pointing out that Chinese have noticed for a long time that the "long-term forecast/prediction" in the sense of seismology (i.e., for the time range of 100 years starting from the present) and the "long-term forecast/prediction" in the sense of physics (for the time range centered at 100 years after ±10 years) are different issues, and the self-organized criticality of seismicity (even if being seismologically correct) does not exclude the long-term predictability of earthquakes.

---

1. Note: Probably due to this "broadband" understanding of the word "earthquake forecast/prediction", excepting for a few cases in fundamental studies, in the Chinese literature generally the words "forecast" and "prediction" are not differentiated sharply and are used in a "hybrid" manner, although in some cases "forecast" means the scientific product and "prediction" means the message sent to the public, and in some other cases "forecast" indicates probabilistic assessment of hazard, whereas "prediction" indicates alarms based on an apparent probability increase. In the whole text of this chapter, therefore, we use "forecast/prediction" to reflect this situation.

TABLE 16.1 Chinese Wording (Direct Translation) versus Related Research Field

Chinese Wording (Direct Translation)	Research Field Related
Long-term earthquake forecast/prediction	Seismic-hazard assessment
Intermediate- and long-term earthquake forecast/prediction	Time-dependent seismic-hazard assessment
Intermediate-term earthquake forecast/prediction	Intermediate-term medium-range earthquake forecast
Forecast of large-scale seismic tendency (within 3 years)	Intermediate-term medium-range earthquake forecast
Annual earthquake forecast/prediction	Intermediate-term medium-range earthquake forecast
Short- and imminent-term earthquake forecast/prediction	Earthquake prediction or operational earthquake forecast
Aftershock forecast/prediction	Earthquake interaction

The Law of the People's Republic of China on Protecting Against and Mitigating Earthquake Disasters was adopted on December 29, 1997, and came into force as of March 1, 1998. On December 27, 2008, this Law, amended and adopted at the Sixth Meeting of the Standing Committee of the Eleventh National People's Congress, was promulgated, which went into effect on May 1, 2009. During the discussion on the revision of the Law, a debate occurred on whether earthquake forecast/prediction should be kept as a governmental duty, given the present scientific capacity, with the result that earthquake forecast/prediction was still kept as a governmental duty. Evidently the "broadband" understanding of earthquake forecast/prediction in Chinese (from long-term to short- and imminent-term further to aftershocks) plays an important role in reaching this consensus. According to the legislative regulations, the China Earthquake Administration (CEA) provides forecast-based scientific information to the government, and the government (with necessary evaluation and approval procedures) is responsible for releasing the forecast/prediction information to the public.

Related to the cultural differences, another barrier of communication occurs with scientific publication in Chinese. The first professional journal in geophysics in China was *Acta Geophysica Sinica* starting in 1948. The publication had a gap from 1967 to 1973 during the Cultural Revolution when scientific research was disrupted. In 1979, the first volume of *Acta Seismologica Sinica* was published. Most of the Chinese journals in earthquake and engineering seismology started their publication in the 1980s. In recent years, more and more Chinese seismologists have been publishing their papers in international journals using English as the working language. But many more papers are still published in Chinese with English abstracts. Institutions of the CEA and local earthquake administrations publish their own journals (Chen et al., 2003). There has been a debate on whether it is a good thing to have so many journals and whether to form a few "united"/optimized ones (such as *Earth*, *Planets*, and *Space* in Japan) is worth trying. Besides academic journals, the Seismological Press, founded in 1976, publishes books on earthquake science, and edits the *China Earthquake Yearbook* annually, which includes the records of academic events and earthquakes (Editorial Committee of China Earthquake Yearbook, 1990). Since 1988 (Zhang, 1988), *Earthquake Cases* (in Chinese with English abstracts) have been summarized and published regularly by the Seismological Press. Overcoming the language barrier, in the 1970s, some American Chinese seismologists translated selected papers published in *Acta Geophysica Sinica* into English, to form a few volumes of *Chinese Geophysics* published by the American Geophysical Union. The Chinese National Committee for IASPEI (1987, 1991, 1995, 1999),[2] regularly summarizes the publications and introduces them to international geophysical communities. Despite these efforts, there are still numerous works in China, although playing an active role contributing to the study of earthquake forecast/prediction, to which not enough attention is paid by international colleagues.

16.2 EARTHQUAKE FORECAST/PREDICTION FOR DIFFERENT TIME SCALES: EXAMPLES OF SCIENTIFIC PRODUCTS AND THE MECHANISM OF THEIR GENERATION AND QUALITY CONTROL

16.2.1 The Key Regions Subject to Enhanced Monitoring and Preparedness

Since the 1990s, a national research project has been organized by the CEA, aiming at the identification of "Key Regions Subject to Enhanced Monitoring and Preparedness" (Research Group of "Researches on Earthquake Risk Regions and Losses Prediction of China Continent During from 2006 to 2020", 2007). This work used geological, seismological, and geophysical methods, aided by numerical geodynamic modeling, with no significant difference from what is being used by the international seismological communities. But the application is somehow Chinese characterized, targeting at the identification of the regions

---

2. Liu, J., Zhang, X.D., Jiang, Z.S., Jiang, H.K., Zhang, Y.X., Zhou, L.Q., Li, M.X., Xue, Y., Wang, Y.L., 2007. Earthquake prediction: studies and experiments (2003−2006). In: Chinese National Committee for the International Union of Geodesy and Geophysics (Eds.), The 2003−2006 China National Report on Seismology and Physics of the Earth's Interior for the General Assembly of the IUGG, Perugia, Italy, July 2−13, 2007.

that are probable for earthquakes over M_S7 in the (seismically active) western China and earthquakes over M_S6 in the (populated and economically developed) eastern China, for the period of 1.5 decades. The target periods for such identifications are three five-year plans for economy and social development, or three successive terms of the government. Among the 17 Key Regions identified in western China for the period 2006—2020, the south-to-mid Longmenshan fault zone was identified as a probable location for earthquakes over M_S7. This expectation was first fulfilled by the May 12, 2008, Wenchuan $M_S8.0$ earthquake (with the epicenter or nucleation point located in the mid-Longmenshan fault zone, and the rupture spanning the mid-to-north Longmenshan fault zone). In addition, the April 20, 2013, Lushan $M_S7.0$ earthquake (rupturing the south Longmenshan fault zone) further fulfilled the forecast. In contrast, almost all other approaches, for example, the Annual Consultation that will be discussed in the later sections, and the consultation on the short-term earthquake hazard based on the monitoring of precursory anomalies, failed to identify the Longmenshan fault zone as seismically hazardous at different time scales. Continuing this approach, after the 2008 Wenchuan earthquake, the CEA organized a working group on the long- to intermediate-term forecast/prediction of major to great earthquakes (Working Group of M7, 2012).

16.2.2 The Annual Forecast

Started in 1972 and officially formalized in 1975, the Annual Consultation Meeting on the Likelihood of Earthquakes in the Next Year has been one of the regular activities organized by the CEA. At the turn of the year, seismologists use their methods of precursor/anomaly identification and earthquake forecast/prediction to identify the regions with increased probability of earthquakes in the coming year. A group of experts conduct the evaluation of the proposed precursors/anomalies and the forecast/prediction. By combining tectonic, seismic, and other geophysical information, this group of experts draws conclusions about the seismic tendency in the next year and identifies the regions with a higher seismic risk. This working group writes a report to the CEA, and the CEA in turn reports these results to the central government. Wu (1997) and Zhao et al. (2010, the appendix) introduced the history and flow chart of the Annual Consultation Meeting.

During the Consultation Meeting, three problems are addressed and proposed to be discussed with higher priorities: (1) Is the present a seismically active or inactive period? (2) What will be the overall level of seismic activity in the next year? Will there be any major to great earthquakes with a magnitude >7.0 occurring in the whole Chinese mainland the next year, and will there be any strong earthquakes with a magnitude >6.0 occurring in densely populated eastern China the next year? and (3) Which regions will have a higher probability of earthquakes larger than a magnitude of 6.0 (in the western part) and/or a magnitude of 5.0 (in the eastern part) in the coming year?

Evaluation of the performance of the Annual Consultation (Shi et al., 2001; Zhuang and Jiang, 2012) shows that on average the Annual Consultation, with both successful and unsuccessful (including failures to forecast and false alarms) estimates, significantly outperforms random forecasts, while this apparent success is, on the other hand, dependent on the probabilistic estimation of background seismicity. A unique scientific merit of the Annual Consultation Meeting lies in that it is a real forward forecast/prediction practice that has been persistently conducted for four decades. The method or philosophy for the assimilation of different kinds of data for the assessment of seismic hazard and the role of the empirical judgment of the expert panel in reaching such an assessment are subject to further investigation (Wu et al., 2007).

16.2.3 The Evaluation of Earthquake Forecast/Prediction

During the decade from 1966 to 1976, a seismically active period on the Chinese mainland, Chinese seismological agencies carried out extensive studies and experiments on earthquake forecast/prediction (Ma et al., 1982; Mei et al., 1993; Chen and Wang, 2010), leaving the record of the successful but controversial prediction of the February 4, 1975, Haicheng, Liaoning, $M_S7.3$ earthquake. The foreshock swarm used to predict the Haicheng earthquake was selected by the Subcommittee of Earthquake Prediction of the International Association of Seismology and Physics of the Earth's Interior (IASPEI) as one of the three significant precursors (Wyss, 1991; Wyss and Dmowska, 1997). But other precursors as well as the forecast-based decision-making process for evacuation were questioned (Jackson, 2004; Wang et al., 2006). The precursor-hunting experiment has lasted up to the present time, with both successes and failures. Retrospectively investigated, there were several precursor-like anomalies before the 2008 Wenchuan earthquake (for a review, see Ma and Wu, 2012), but none of these anomalies was so conclusive as to lead to the intermediate-term or short-term alarm of the earthquake.

The Department for Seismological Monitoring and Earthquake Forecast of the CEA has established years ago a mechanism for evaluating forward forecast/predictions. Those who report/claim an alarm for a future earthquake are required to fill out a simple form (called "forecast/prediction card") stating the time range, spatial range, and magnitude range of the predicted earthquake, and the abstracted scientific basis for such a predictive argument. These "forecast/prediction cards" are used for both the test of the forecast/predictions and the decision making to reach the final conclusion for assessing time-dependent seismic hazard by assimilating different sources of information.

For evaluating the would-be earthquake precursors, Chinese seismologists have used for a long time the "R value" proposed by Xu (1989), which considers both the hit rate and the false alarm rate, being similar to the presently well-known receiver operating characteristic test (the ROC diagram).

In 1999, the National Committee for the Evaluation of Earthquake Prediction was founded. In 2008, the CEA invited several experts, who were not only Chinese but also those from abroad, to evaluate the forecast made by a Russian group[3] indicating that there would be a strong earthquake near Beijing around the time of the Olympic Games.

16.2.4 The Monitoring of Potential Precursory Anomalies

Monitoring of potential earthquake precursors in the CEA had been through four disciplines, namely, seismological, geodetic, geoelectric and geomagnetic, and ground fluid or seismohydrology/hydroseismology. Quality control is conducted by the Coordination Groups for Observation and Interpretation. Except for the ground fluid observation system, the related observation networks all have dual missions: (1) monitoring earthquakes and geophysical fields for earthquake emergency and seismological research, as well as data service to Earth science studies; and (2) monitoring of potential anomalies in seismicity, seismic waves, and geophysical fields for earthquake forecast/prediction.

Since the turn of the century, infrastructures for seismological, geodetic, and other geophysical observation and monitoring have been developed and modernized considerably.[4] Infrastructure for global positioning system monitoring was constructed jointly by the CEA and other geodetic and/or geoscience agencies through the Crustal Movement Observation Network of China. This project has caused much attention in the international scientific communities (e.g., Wang et al., 2001). Almost in parallel, the "China Digital Seismological Observation System" was completed and began operation at the end of 2000, which includes the National Digital Seismograph Network, 20 regional digital seismograph networks, and a mobile digital seismograph network consisting of 100 portable digital seismographs. Between 1999 and 2001, the Beijing Capital-Circle Digital Seismograph Network (covering Beijing Municipality, Tianjin Municipality, and Hebei Province, with real-time data transmission) was established, consisting of 107 seismic stations. The "China Digital Earthquake Observation Networks Project", in which two

---

3. The paper, "Fundamental Peculiarities of the Entropy Model of Energy Processes in Seismic Areas and Earthquake Prediction", by Sibgatulin, V.G., Peretokin, S.A., Khlebopros, R.G., published in 2007 by *Earth Science Frontiers*, indicated that there would be an earthquake with a magnitude 6–7 near Beijing around August 2008. The forecast-related paragraph was eliminated by the editor in the proof-reading stage, considering the sensitivity of the place and time (in the sense of the 2008 Beijing Olympic Games). This paper caused attention within the CEA. Six experts abroad were invited to review the paper, mainly the predictive conclusion. The evaluation turned down the forecast, and it was shown that the evaluation was correct.
4. Chen, J.M., Earthquake disaster reduction in developing China, Keynote Presentation, the 14th World Conference on Earthquake Engineering (WCEE), October 12–17, 2008, Beijing, China, electronic version available at: http://www.iitk.ac.in/nicee/wcee/article/14_K002.pdf.

major components are a geophysical-/geochemical-anomaly monitoring network and a digital seismograph network, was launched in June 2004, constructed from 2004 to 2008, and passed the acceptance inspection in April 2008 (just one month before the Wenchuan earthquake). The project has significantly enhanced the density of seismic stations and geophysical/ geochemical monitoring stations. After the project, the monitoring capability can be characterized by four geophysical/geochemical-anomaly monitoring stations and more than eight seismic stations for every 0.1 million km^2, and 95 percent of the monitoring equipment being digital. The completeness of earthquake catalog was improved from $M_L 4.5$ to $M_L 2.5$ nationwide, and to $M_L 1.5$ in the regions near megacities. Considering the huge territory of China, however, this deployment is still neither region-balanced (e.g., most areas of the Tibetan plateau still have a very low monitoring capability) nor sufficient (e.g., even if in the seismically active central-China north-south seismic belt, to capture the high-resolution picture of the spatiotemporal variation of some geophysical fields such as the geomagnetic field or the gravity field, is still difficult), and accordingly construction has been going on since the 2008 Wenchuan earthquake.

16.3 WITHIN THE LIMIT OF THE CAPABILITY OF EARTHQUAKE FORECAST/PREDICTION: ROLES OF TIME-DEPENDENT SEISMIC-HAZARD ASSESSMENT IN SEISMIC RISK MANAGEMENT

In 2010, the CEA published the Open File *Guidelines for Strengthening Seismological Monitoring and Earthquake Forecast/Prediction*,[5] highlighting the "five-combinations policy", namely, (1) the combination of long-term, intermediate-term, and short- and imminent-term forecast/prediction, and sequence type and aftershock forecast/prediction; (2) the combination of seismological monitoring, earthquake forecast/prediction, and earthquake early warning; (3) the combination of monitoring, research, forecast/prediction test, and forecast/prediction application; (4) the combination of CEA's work and the works of the other institutions/agencies as well as the public; and (5) the combination of earthquake forecast/prediction with disaster reduction countermeasures. This reflects the "holistic view" of the Chinese approach, that on the one hand, "earthquake forecast/prediction" is understood in a "broadband" perspective; on the other hand, even if with the limits of the present capacity of earthquake forecast/prediction, further promoting the research and development works and the application for disaster risk reduction is not only necessary but also feasible.

5. http://www.cea.gov.cn/manage/html/8a8587881632fa5c0116674a018300cf/_content/11_03/28/ 1301300870907.html.

16.3.1 From Long-term Preparedness to Annual Enhanced Preparation

The 10- to 15-year estimation of the "Key Regions Subject to Enhanced Monitoring and Preparedness" has shown its relatively sound scientific basis and potential for application for social sustainability. At the end of the application, however, a comprehensive investigation is still needed of how well this scientific product has contributed to the regional capacity building for earthquake disaster risk reduction. Since 2011, the CEA has been organizing a major project supported by the National Social Science Foundation of China for the sociological investigation and evaluation of the earthquake disaster risk reduction in the Key Regions. The result of this investigation will have significant impacts on policy making.

Between the 1.5 decade seismic-hazard assessment and the annual seismic-hazard forecast, there is also an activity in China, aiming at a three-year estimation of seismic tendency for the whole of the Chinese mainland. This estimation has less contact with society, and mainly serves the Annual Consultation by providing a measure of the "background" degree of earthquake activity. The performance of this practice has not yet been evaluated systematically.

The Annual Consultation, with a hit rate of about 20—30 percent, and statistically outperforming random forecast, helps to a large extent the local preparedness. In the regions that are expected to have a higher probability of earthquakes, seismological monitoring is strengthened, and some engineering/social countermeasures are taken accordingly. At present, however, no standard procedure exists, either for the short-term, prediction-oriented enhanced monitoring or for the targeted preparedness. Countermeasures vary from place to place and from time to time, to some extent reasonably due to the regionalized characteristics of economy and seismicity. Remarkably, due to the size of the Chinese territory, a major earthquake is never a rare event for the whole country. Therefore, for those regions that are not listed in the annual seismically-risky regions, the nationwide Annual Consultation plays the role of reminding the local governments and the society of the "visible" seismic hazard.

In principle, earthquake forecast/prediction can be decomposed into three components: scientific, technique, and engineering. The scientific component deals with the physical predictability of earthquakes and the methods for feasible earthquake forecast/prediction. The technique component, in turn, is to make full use of the results of the scientific component to implement the forecast/prediction scheme within the limits of the scientific capacity, that is, to provide the forecast/prediction with different spatiotemporal resolutions and with different uncertainties. The engineering component is to consider both the technical issues and the social issues to pursue the optimization of the social benefit. The philosophy of the identification of the Key Regions, the 3-year estimation of seismic tendency, and the Annual Consultation is to seek an optimization in solving the problem of earthquake preparedness by balancing the scientific, the technique, and the engineering components, so that earthquake science may be able to make the maximum contribution to the society within the limit of its capacity.

16.3.2 Enhanced Monitoring Campaign for Special Places and Times

In China, the forecast/prediction-oriented monitoring has a special activity. Evaluation of such an activity is still to be conducted. For some special times and places, such as the time and place being the focus of the society, or the time and place where there are some anomalies observed, or there are some concerns from the society about the short-term earthquake hazard, the CEA will organize some enhanced monitoring activities. Falsification and verification of the observed anomalies is one of the missions of this activity. Capturing some more anomalies based on denser deployment of observational sensor networks is another mission, which is not completely independent of the former. Since the identification of earthquake anomalies needs a long time for the baseline of comparison, it is unclear as to how effectively this kind of enhanced monitoring contributes to the assessment of time-dependent seismic hazard. On the other hand, the case-specific special-time, space-oriented monitoring activity at least enhances the alert level of the existing monitoring systems and avoids unnecessary social disorder.

One dramatic example was that, based on the analysis of seismicity and geophysical anomalies, the seismological agency announced to the organization committee of the 1990 Beijing Asian Games that there would probably be small earthquake/s occurring during the Asian Games but that these would be unlikely destructive. Just before the opening ceremony there occurred a felt earthquake. No panic resulted due to the information provided beforehand by seismological agencies (although hard to be considered as a forecast/prediction in the standard sense).

Denser networks of enhanced monitoring accumulate scientific data, no matter whether or not they do have the "information gain" for the assessment of short-term seismic hazard. During some earthquakes in Yunnan Province, southwestern China, even if the enhanced monitoring system could not further constrain the time window of the "target" earthquake, the near-earthquake-source recordings (such as seismic, strong motion, and geodetic recordings) provided earthquake studies with interesting and, in some cases, unique data.[6]

---

6. For example, the April 12, 2001, Shidian, Yunnan, $M_S5.9$ earthquake, with an epicenter distance 1.1 km recording horizontal PGA 500gal; and the October 27, 2001, Yongsheng, Yunnan, $M_S6.0$ earthquake, with an epicenter distance 10 km, recording horizontal PGA 167.7gal. Courtesy of Dr Qiao Sen, Earthquake Administration of Yunnan Province.

16.3.3 The Haicheng Debate: A Historical Perspective

Before the February 4, 1975, Haicheng $M_S7.3$ earthquake, mainly based on foreshocks and other geophysical precursors as well as on the macroanomalies reported to seismological agencies from the public, the local government evacuated the residents. In spite of the countermeasures taken, still 1328 fatalities resulted (Zhang, 1988), which invoked doubt about the reality of the prediction (e.g., Geller, 1997). However, this earthquake occurred in a densely populated area with almost completely unreinforced buildings, and it was in the wintertime of northeastern China with temperatures typically below $-10\,°C$. A retrospective scenario estimate indicates that the evacuation could have saved from 8,000 to 24,000 lives (Wyss and Wu, 2014).[7]

Historically, this unique event has both the chance component and the deterministic component. The performance of foreshocks ranks the best among the proposed precursors, but only a minor ratio of earthquakes have the foreshocks that could be used in practice for the alarm. The origin time of the earthquake was about 7:30 pm local time, which made people still have patience to "wait for" the quake in the cold windy outdoor places. People believing reports of anomalous animal behavior were easily persuaded to evacuate, and what was dramatic was that there were many reports of anomalous animal behavior, another unique feature of the Haicheng earthquake. In this sense, the success was to a large extent due to luck, or a lucky interaction among the seismological agencies, the local government, the society, and most importantly, the earthquake as well as its preparation process. On the other hand, however, at that time, only the Chinese believed that earthquake forecast/prediction can be accomplished and applied "in that way", which turns out that only the Chinese were able to capture such a success, even if it was by chance.

The single success in forecasting/predicting an earthquake does not mean that the entire earthquake forecast/prediction effort is a success. In 1976, based on almost exactly the same method (and philosophy), Chinese seismologists missed the tragic Tangshan $M_S7.8$ earthquake. Until now, Haicheng is still a record that has not been broken worldwide. In a longer historical perspective, both the successes and the failures of earthquake forecast/prediction are treasures of human civilization, reflecting the permanent dream and persistent pursuit of science to challenge the seemingly impossible.

16.3.4 Aftershock Probability for Rescue and Reconstruction

The assessment of the type of earthquake sequence and the likelihood of strong aftershocks has a relatively sound scientific basis and has played a positive role in assisting the rescue and relief as well as the reconstruction.

---

7. Wyss, M., Wu, Z. L., 2014. How many lives were saved by the evacuation before the *M7.3* Haicheng earthquake of 1975? *Seismol. Res. Lett.*, 85, 126–129.

After the May 12, 2008, Wenchuan $M_S8.0$ earthquake, estimates of the duration, maximum magnitude, rate, and most-likely locations of the strong aftershocks were made, which was shown to be consistent with the observed aftershock activity. The estimates were based on the presently available knowledge of seismicity, such as the Omori's law, the Gutenberg–Richter's law, and the change of the Coulomb failure stress. Scientists of the Commission on Geophysical Risk and Sustainability of the International Union of Geodesy and Geophysics (IUGG) and the US Geological Survey were invited to join in the predictive estimate of the aftershock tendency. These estimates helped the rescue and reconstruction activities, which was especially noticeable considering the landslides and landslide lakes threatening the relief and reconstruction. However, on May 19, a false alarm of a "magnitude 7 aftershock" broadcast on television caused widespread social disorder in Chengdu, Xi'an, and other cities near the Wenchuan earthquake. The process of this unbelievable social panic has been puzzling till the present, and might be a puzzle forever.

16.4 DECISION-MAKING ISSUES OF EARTHQUAKE FORECAST/PREDICTION

In discussing the duties of earthquake forecast/prediction, the decision-making issues of earthquake forecast/prediction have to be taken into account. Such decision-making issues include two levels of problems: (1) how to analyze comprehensively the information from different disciplines—all have some clues leading to the assessment of time-dependent seismic hazard, but all have large uncertainties—to get the predictive conclusions so as to communicate with the government; and (2) how to suggest to the government what decisions to take responding to the assessment conclusions for the reduction of earthquake disasters. Both levels of problems are complicated and risky. It seems that the spirit of forensic seismology (Bowers and Selby, 2009), the discipline that was mainly related to the monitoring of a comprehensive nuclear test ban treaty, may also apply to the field of earthquake forecast/prediction.

16.4.1 Decision Making for Predictive Conclusions

The estimate of the tendency of seismic activity, at different spatiotemporal scales, is basically organized in a matrix. Academically, the studies on earthquake forecast/prediction are divided into $4 + 1$ subjects: seismology; geomagnetism and geoelectricity; ground deformation and gravity; and underground fluid and geochemistry, plus comprehensive analysis. Geographically, the whole of the China mainland is divided into four regions for study: the north- and northeast region; the northwest region; the southwest region; and the east- and south region. The candidate precursory anomalies under consideration are almost the same as those studied in other countries/regions, such as seismicity patterns, geomagnetic and geoelectric field anomalies, ground deformation and gravity variations, and changes of the level and chemical contents of underground water, among others. Case studies of earthquakes played an important role in the accumulation of experience (e.g., Zhang, 1988).

Such a matrix structure reflects one of the basic ideas in the study of earthquake forecast/prediction in China. One of the fundamental questions related to earthquake forecast/prediction is "In the case that earthquake forecast/prediction is still an unsolved scientific problem, is it possible to obtain *some* useful information about earthquake hazard based on the present (incomplete) knowledge and (incomplete) data?" The Chinese answer is "yes".

Other questions are "Having different kinds of observations, is it possible to draw *some* useful conclusions by a comprehensive consideration based on experience and qualitative knowledge?" Or "Will the comprehensive analysis enhance the information gain of these observations, even further leads to the emergence of an increased forecast/prediction capability?" The Chinese answer is "yes". The answer to the second question reflects the long tradition of Chinese thinking consistent with the ideas of nonlinear dynamics, while the answer to the first question reflects the Chinese idea about the nature of decision making.

In the comprehensive analysis that combines the information from different disciplines, Chinese seismologists prefer the combination of quantitative calculation and qualitative experience. Some of the calculation tools, such as fuzzy logic, pattern recognition, artificial neural networks, and expert systems, are used for learning and forecasting/predicting. On the other hand, experienced (senior) experts play an important role in the forming of the final conclusions, although the underlying physics of some of the empirical "field knowledge" of earthquake tendency (i.e., the relation between the observation of experts and the predictive conclusions they draw, and the relation between such logic and the process of earthquake preparation) are still not fully understood, even by the experts themselves.

16.4.2 Monitoring and Modeling for Prediction: Lessons from the Wenchuan Earthquake

In the comprehensive analysis of potential anomalies to reach the predictive conclusions, one of the long-time used straightforward methods is to investigate at how many observing sites anomalies appear, and those places having a sufficient number of observing sites (or more often, a sufficient portion of observing sites) exhibiting anomalies are identified as the places for an enhanced probability of future earthquakes. The shortcoming of this seemingly straightforward idea becomes clear when considering the probable cause of the anomalies in the perspective of geodynamics that connects the anomalies to the preshock variation of the tectonic stress field. Due to the tensor characteristic of tectonic stress, the distribution of anomalies (if they do exist and are associated with, for example, the increase of the stress level) will not have spatial homogeneity—rather they are controlled by the seismotectonic stress field.

The May 12, 2008, Wenchuan $M_S8.0$ earthquake in southwestern China (Chen and Booth, 2011), occurring along the mid-to-north Longmenshan fault zone and rupturing two parallel faults with an about 240- and 72-km length, respectively, provided earthquake forecast/prediction study with a vivid lesson to understand the above mentioned importance of geodynamics in controlling the potential anomalies. According to field investigations and geological studies, the tectonic context that the western Sichuan plateau is being thrust over the Chengdu basin at the Longmenshan fault zone determined the behavior of observed precursory anomalies. Earthquake-forecast-oriented Chinese literature (e.g., Department for Earthquake Monitoring and Prediction of the China Earthquake Administration, 2009) described the preparation process of the Wenchuan earthquake by the "three units model": the deformation unit (the western Sichuan plateau), the locking unit (the Longmenshan fault zone), and the resisting unit (the Chengdu basin). Anomalies could be observed in the "deformation unit". Counting the observing sites at which anomalies appeared, however, there was no conclusion reached that along the Longmenshan fault a major-to-great earthquake was approaching.

"Monitoring and modeling for prediction" is by no means a new concept in seismology: The IASPEI has even a commission named "Commission on Seismic Source—Monitoring and Modeling for Prediction". However, at the practice level, such a concept has not become an operational tool in the decision-making to estimate the increase of the probabilities of earthquakes. As a result, there were several cases of earthquakes that occurred near the margin of the regions delineated by the expert panel as the regions with increased seismic hazard.

16.4.3 Social Pressures in the Predictive Decision Making

In the decision-making process, whether an earthquake forecast/prediction conclusion is to be reported to the government, or more often, whether the forecast/prediction is to be released to the public by the government, social pressures have occurred that depended on the economic and social tolerance of false alarms or failures to predict. When dealing with the short- to intermediate-term earthquake forecast/prediction, which has very large uncertainties and is often connected to emergency management countermeasures such as evacuation actions, this social pressure becomes even greater. Chen Zhangli, former director general of the China Seismological Bureau (1996–2001) the predecessor of the CEA, recorded and discussed some interesting (and to a great extent unique) case examples for the leaders of the national and local seismological agencies to make hard decisions based on the short-term forecast/prediction with significant uncertainties (Chen, 2007).

Releasing, or not releasing, that is a question dealing with an earthquake forecast/prediction. In recent years, more and more Chinese seismologists have started to think about this question in an alternative perspective (e.g., Wu et al., 2013). The most important issue is the coupling of earthquake forecast/prediction with disaster risk reduction countermeasures. With the present limit of the capacity of forecast/prediction, if the forecast/prediction can be used for the "if-then" prediction of strong ground motion and economic impact serving the risk management and capacity building, then the forecast/prediction, even if with significant uncertainties, will be very useful. Otherwise, if over-simply, the government and the public are exposed directly to a forecast/prediction, with large uncertainties, of a "probable explosion of a bomb" (note that news media prefer to talk about the trinitro toluene-yield equivalence of an earthquake, and even if a moderate earthquake can be related to an A-bomb), then the decision making for risk management will automatically change to the decision making for emergency. In this case, a *natural disaster* (earthquake) may be converted to a *technological accident* (false alarms or failures to predict), just due to the incorrect understanding of the role of the forecast/prediction for the reduction of earthquake disaster risk.

16.5 CONCLUDING REMARKS AND DISCUSSION: EARTHQUAKE FORECAST/PREDICTION AS A BRANCH OF "MODERN" SCIENCE AND TECHNOLOGY

In this chapter, we take the Chinese approach as an example for discussing the duties of earthquake forecast/prediction. The duties include the duty of the government, duty of the seismological agency (in China, the CEA), and duty of scientific community. The duties are accomplished by a series of scientific products serving decision making. Quality control of the scientific products is achieved through an organized and evolving system. Although so many distinctive differences are associated with the Chinese approach, some of the conclusions may make a more general sense for understanding the duties of earthquake forecast/prediction.

In the perspective of the development and application of earthquake science, two relatively clear clues of thinking can be figured out, although the cases and lessons are very complicated. The first clue of thinking is the *transition* of a series of *scientific capabilities*, that is, (1) the capability of understanding (the mechanism of earthquakes); (2) the capability of observing and modeling (the potential indications of future earthquakes); (3) the capability of monitoring (the potential indications of future earthquakes); and (4) the capability of forecast/prediction (translating the monitoring results to the forecast/prediction of future earthquakes).

The second clue of thinking is the *transition* of a series of *application capabilities*, that is, (1) the capability of predicting strong ground motion and destruction in the case of a scenario earthquake; (2) the capability of predicting the seismic risk of the society; (3) the capability of decision making for risk management; and (4) the capability of enhancing the capability for the reduction of earthquake disasters through a series of countermeasures of risk management.

Earthquake forecast/prediction studies as a branch of "modern" science needs the concept of the transition from basic research, to development research, to applied research, and further to application, enhancing the technical readiness level (TRL) of earthquake forecast/prediction. Earthquake forecast/prediction application as a component of "modern" technology needs the concept of systems engineering, which facilitates the transition from earthquake forecast/prediction to the reduction of earthquake disasters. The present shortcoming of earthquake forecast/prediction studies lies in the fact that from basic research to application, there is still no clear concept of the TRL, and the "short circuit" from basic research to application has caused many misunderstandings, both in the public and in scientific communities. The present shortcoming of earthquake forecast/prediction application lies in the fact that the role of forecast/prediction-based evacuation is to a large extent overemphasized. Before these two shortcomings are overcome, the duties of earthquake forecast/prediction cannot be correctly understood or properly accomplished.

REFERENCES

Bormann, P., 2011. From earthquake prediction research to time-variable seismic hazard assessment applications. Pure Appl. Geophys. 168, 329–366.

Bowers, D., Selby, N.D., 2009. Forensic seismology and the comprehensive Nuclear-Test-Ban treaty. Annu. Rev. Earth Planet Sci. 37, 209–236.

Chen, Z.L., 2007. Earthquake Prediction: Practice and Reflection. Seismological Press, Beijing in Chinese.

Chen, Y., Booth, D.C., 2011. The Wenchuan Earthquake of 2008: Anatomy of a Disaster. Science Press in Cooperation with Springer, Beijing.

Chen, Q.F., Wang, K.L., 2010. The 2008 Wenchuan earthquake and earthquake prediction in China. Bull. Seismol. Soc. Am. 100, 2840–2857.

Chen, Y.T., Wu, Z.L., Xie, L.L., 2003. Centennial national and institutional reports: seismology and physics of the Earth's interior – China (Beijing). In: Lee, W.H.K., Kanamori, H., Jennings, P.C., Kisslinger, C. (Eds.), International Handbook of Earthquake and Engineering Seismology, Part B. Academic Press, Amsterdam, pp. 1317–1321 (Chapter 79.14).

Chinese National Committee for IASPEI (Ed.), 1987. National Report on Seismology and Physics of the Earth's Interior (1983–1986). Seismological Press, Beijing in Chinese.

Chinese National Committee for IASPEI (Ed.), 1991. National Report on Seismology and Physics of the Earth's Interior (1987–1990). Seismological Press, Beijing in Chinese.

Chinese National Committee for IASPEI (Ed.), 1995. National Report on Seismology and Physics of the Earth's Interior (1991–1994). Meteorological Press, Beijing.

Chinese National Committee for IASPEI (Ed.), 1999. National Report on Seismology and Physics of the Earth's Interior (1995–1998). Meteorological Press, Beijing.

Department for Earthquake Monitoring and Prediction of the China Earthquake Administration (Ed.), 2009. Science Report on the Wenchuan Magnitude 8.0 Earthquake. Seismological Press, Beijing in Chinese.

Editorial Committee of China Earthquake Yearbook (Ed.), 1990. China Earthquake Yearbook (1949–1981). Seismological Press, Beijing in Chinese (*China Earthquake Yearbook* has been published annually since 1982).

Editorial Committee of China Today (Ed.), 1993. China Today: Earthquake Hazard Reduction Undertaking. Publishing House of China Today in Chinese.

Geller, R.J., 1997. Earthquake prediction: a critical review. Geophys. J. Int. 131, 425–450.

Hammond, A.L., 1976. Earthquakes: an evacuation in China, a warning in California. Science 192, 538–539.

Jackson, D.D., 2004. Earthquake prediction and forecasting. In: IUGG (Ed.), State of the Planet: Frontiers and Challenges. AGU, Washington, DC, pp. 225–348.

Jordan, T.H., Chen, Y.T., Gasparini, P., Madariaga, R., Main, I., Marzocchi, W., Papadopoulos, G., Sobolev, G., Yamaoka, K., Zschau, J., 2011. Operational earthquake forecasting: state of knowledge and guidelines for utilization. Ann. Geophys. 54 (4). http://dx.doi.org/10.4401/ag-5350.

Ma, T.F., Wu, Z.L., 2012. Precursor-like anomalies prior to the 2008 Wenchuan earthquake: a critical-but-constructive review. Int. J. Geophys. http://dx.doi.org/10.1155/2012/583097, 583097.

Ma, Z.J., Fu, Z.X., Zhang, Y.Z., Wang, C.M., Zhang, G.M., Liu, D.F., 1982. Earthquake Prediction: Nine Major Earthquakes in China (1966–1976). Seismological Press, Beijing in Chinese; English version: 1990. Springer-Verlag, Berlin.

Mei, S.R., Feng, D.Y., Zhang, G.M., Zhu, Y.Q., Gao, X., Zhang, Z.C., 1993. Introduction to Earthquake Prediction in China. Seismological Press, Beijing in Chinese.

Research Group of "Researches on Earthquake Risk Regions and Losses Prediction of China Continent During from 2006 to 2020" (Ed.), 2007. Researches on Earthquake Risk Regions and Losses Prediction of China Continent During from 2006 to 2020. Seismological Press, Beijing in Chinese.

Shi, Y., Liu, J., Zhang, G., 2001. An evaluation of Chinese annual earthquake predictions, 1990–1998. J. Appl. Probab. 38A, 222–231.

Wang, Q., Zhang, P.-Z., Freymueller, J.T., Bilham, R., Larson, K.M., Lai, X., You, X., Niu, Z., Wu, J., Li, Y., Liu, J., Yang, Z., Chen, Q., 2001. Present-day crustal deformation in China constrained by global positioning system measurements. Science 294, 574–577. http://dx.doi.org/10.1126/science.1063647.

Wang, K., Chen, Q.-F., Sun, S., Wang, A., 2006. Predicting the 1975 Haicheng earthquake. Bull. Seismol. Soc. Am. 96, 757–795. http://dx.doi.org/10.1785/0120050191.

Working Group of M7, 2012. Study on the Mid- to Long-term Potential of Large Earthquakes on the Chinese Continent. Seismological Press, Beijing in Chinese.

Wu, F.T., 1997. The annual earthquake prediction conference in China (national consultative meeting on seismic tendency). Pure Appl. Geophys. 149, 249–264.

Wu, Z.L., Liu, J., Zhu, C.Z., Jiang, C.S., Huang, F.Q., 2007. Annual consultation on the likelihood of earthquakes in continental China: its scientific and practical merits. Earthquake Res. China 21, 365–371.

Wu, Z.L., Ma, T.F., Jiang, H., Jiang, C.S., 2013. Multi-scale seismic hazard and risk in the China mainland with implication for the preparedness, mitigation, and management of earthquake disasters: an overview. Int. J. Disaster Risk Reduction 4, 21–33.

Wyss, M. (Ed.), 1991. Evaluation of Proposed Earthquake Precursors. AGU, Washington, DC.

Wyss, M., Dmowska, R. (Eds.), 1997. Earthquake prediction, state-of-the-art. Pure Appl. Geophys. 149 (special issue), 1–264.

Wyss, M., Wu, Z.L., 2014. How many lives were saved by the evacuation before the *M7.3* Haicheng earthquake of 1975? Seismol. Res. Lett. 85, 126–129.

Xu, S.X., 1989. The evaluation of earthquake prediction ability. In: Department of Science, Technology and Monitoring, State Seismological Bureau (Ed.), The Practical Research Papers on Earthquake Prediction Methods (Seismicity Section). Seismological Press, Beijing, pp. 586−589 (Chinese).

Zhang, Z.C. (Ed.), 1988. Earthquake Cases in China (1966−1975). Seismological Press, Beijing in Chinese (Note: *Earthquake Cases* series has been published regularly by Seismological Press since 1988. For some important earthquakes, there are also special books systematically collecting and summarizing the research results. Note that the contents of *Earthquake Cases* not only include precursory anomalies and retrospective summary of the forecast/prediction but also background information of the earthquake such as intensity distribution, focal mechanism, and seismotectonics.).

Zhao, Y.Z., Wu, Z.L., Jiang, C.S., Zhu, C.Z., 2010. Reverse tracing of precursors applied to the annual earthquake forecast: retrospective test of the annual consultation in the Sichuan-Yunnan region of southwest China. Pure Appl. Geophys. 167, 783−800. http://dx.doi.org/10.1007/s00024-010-0077-1.

Zhuang, J., Jiang, C., 2012. Scoring annual earthquake predictions in China. Tectonophysics 524/525, 155−164.

Chapter 17

The Experience of Real-Time Earthquake Predictions on Kamchatka

Gennady Sobolev[1] and Victor Chebrov[2]

[1] *Institute of Physics of the Earth, RAS, Moscow,* [2] *Kamchatka Branch of Geophysical Survey, RAS, Petropavlovsk-Kamchatsky, Russia*

ABSTRACT

A Kamchatka Expert Council was established as a branch of the national Russian Expert Council (KB REC) in 1998. The authors of various methods of earthquake prediction submit their forecasts to KB REC. All predictive messages are documentarily recorded. KB REC's Messages about the seismic situation in Kamchatka are delivered to the Administration and EMERCOM of Kamchatka. Twenty-one earthquakes with $M_w \geq 6.0$ occurred in KB REC's area of responsibility in 1998–2012. A seismic situation was assessed by nine different methods. Most of the applied prediction methods give earthquake waiting periods of about 1 month. Eight predictions came true with regard to three parameters of the expected earthquake: area, time, and magnitude; seven predictions were erroneous with regard to one or two parameters; three of six missed events had depths >550 km. The experience of the documentarily recorded medium-term earthquake predictions made within 15 years using a suite of methods is unique. It allows an assessment of the level of this domain of science.

17.1 INTRODUCTION

After the Kamchatka earthquake of December 5, 1997, which had a magnitude $M_w = 7.8$ and was the strongest regional seismic event over the last 50 years, the Russian Geophysical Survey, with the support from the Administration of Kamchatka, made a decision to start an experiment on studying the prospects and possibilities for real-time forecasting of the earthquakes. The authors of various methods of earthquake prediction expressed their wish to participate in this project and submit their forecasts to the Kamchatka Branch of the Russian

Earthquake Hazard, Risk, and Disasters. http://dx.doi.org/10.1016/B978-0-12-394848-9.00017-1
449

Expert Council on Earthquake Prediction (KB REC), which incorporated 17 specialists in the different fields of geophysics, including some designers of the prediction methods.

The authors of the predictions were not required to formalize the procedure of calculations and the prognostic estimates. The following philosophy was assumed. The scarce experience in the real-time prediction of the earthquakes, which has been gained so far, does not provide the grounds for drawing negative conclusions concerning any of the suggested methods. The size of the spatial domain and time interval of the prediction can (and should) vary depending on the magnitude, location, and source depth of the seismic event to be forecasted. The authors were free to update their methods according to the new experience gained. Thus, it was not the purpose of the experiment to statistically estimate the issued forecasts. The only task was to (purely empirically) reveal the most promising methods and outline the ways to their refining. The authors were not required to explain the physics of the considered anomaly. Only the single criterion of truth was assumed: whether the forecast has proven to be correct in terms of the place, time, and magnitude.

At the same time, prognostic estimates were important for the Administration and, EMERCOM of Kamchatka, since these estimates provide the basis for taking preventive measures to mitigate the adverse consequences of the expected earthquake that could induce ground shaking with the intensity of >7. The KB REC thoroughly discussed all the obtained forecasts; analyzed the significance of anomaly relative to background noise, graphs, and maps presented by the author; took into account the experience of retrospective predictions; compared the given forecast with the estimate of seismic situation that was provided by the members of the KB REC and external experts. If the forecast merited attention, it was used as the basis (with the allowance for the other data) for issuing the reports on the current seismic situation in Kamchatka for the Administration and EMERCOM.

The procedure we report on has similarities with the prediction efforts in China (Wu, 2014), but is different from the method used by Kossobokov (2014) to predict great earthquakes worldwide, and very different from the approach to forecast future seismicity in California described by Schorlemmer and Gerstenberger (2014).

We underline that personal opinions of the authors of this chapter as to the methods for assessing the validity of the precursors, prognostic anomalies, probability, and significance of a successful forecast have nothing in common with the content of the chapter. It is inexpedient to discuss these opinions here. In this chapter, we present the results of the unique experiment, which was carried out by the other researchers and evaluated by the independent experts of the KB REC.

17.2 SEISMICITY AND SYSTEM OF OBSERVATIONS

Kamchatka and the Commander Islands are classed as regions where contemporary geodynamic processes achieve maximum intensity on the planet. According to data on general seismic zoning in Russia, earthquakes may occur in the subduction zone in Kamchatka, on average, once every 100–110 years with $M_w \geq 9.0$, once every 26 years with $M_w \geq 8.0$, and once every 3–4 years with $M_w \geq 7.0$. A seismic catalog with a minimum representative magnitude $M_l \geq 3.5$ exists since 1962 and includes >50,000 events. Figure 17.1 shows epicenters of the earthquakes with $M_w \geq 6.0$ that occurred from January 1, 1962, to December 31, 2012 (before 1976, M_w values have been obtained from M_s).

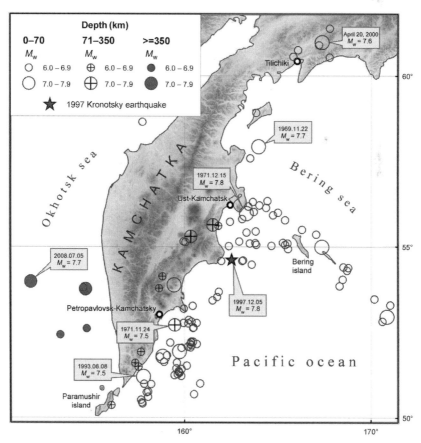

FIGURE 17.1 **Kamchatka earthquake epicenters map ($M_w \geq 6$) for 1962–2012.** Asterisk—epicenter of the Kronotsky earthquake on December 5, 1997, $M_w = 7.8$.

The system for the integrated monitoring of geodynamic processes in the zone of Kurile−Kamchatka and Aleutian island arc junction includes networks of seismic, geophysical, geodetic, and hydrogeochemical observations, as well as technical and software tools for data acquisition, processing, and storage.

Seismological observations. As of December 31, 2012, the seismic station network in Kamchatka consists of 70 seismic signal-recording points (Chebrov et al., 2013). Data from all seismic stations are available in digital format on a real-time basis.

In addition to seismic stations, the system for integrated monitoring of changes in geological environment, which is owned by the Kamchatka Branch of the Geophysical Survey (KB of GS) of the RAS (Russian Academy of Sciences), includes 62 points for other types of observations (Chebrov et al., 2013). One part of this system is shown in Figure 17.2.

The hydrogeochemical observation network (HGGH) consists of four hydrogeological stations including 12 water points for direct measurements at the point and collection of samples with subsequent analysis in laboratory conditions. Measurement and sample collection frequency is once in 3 days. The parameters measured in the field are water flow rate, water and air temperature, and atmospheric pressure. The following is determined in water samples under laboratory conditions: pH, concentrations of chlorine, hydrocarbonate, sulfate, sodium, potassium, calcium, magnesium, boric acid, and silicic acid. The following is determined in gas samples under laboratory conditions: methane, nitrogen, oxygen, carbon dioxide, argon, helium, hydrogen, and hydrocarbon gases. These observations have been made since 1977 (Khatkevich and Ryabinin, 2006).

GPS observations. There are 20 permanent points with data transmission in 1- and 30-s modes. High-precision measurements are made using double-frequency receivers. The observations have been made since 1996 (Levin, 2009).

Tiltmeter observations have been made since 2010 at eight permanent points based on APPLIED GEOMECHANICS 701-2A inertia-free tiltmeters. Sampling frequency is 100 Hz (Levin et al., 2006).

The gas-geodynamic observation network consists of nine points (pits with a depth down to 4 m, 60-m dry well, flowing wells with a depth between 60 and 1757 m), where the following is measured: temperature, pressure, and humidity at 10-min intervals; concentration of hydrogen (H_2), radon (Rn), thoron (Tn), and carbon dioxide (CO_2) at 10 (30)-min intervals. The observations have been made since 1997 (Firstov and Shirokov, 2005).

Water level observations (WLO) include two 665- and 800-m-deep piezometric wells where the water level and atmospheric pressure are measured every 5 min. The observations have been made since 1987 (Kopylova, 2001).

Electrotelluric observations are made at four points where the difference in potentials between electrodes is measured in mutually perpendicular directions

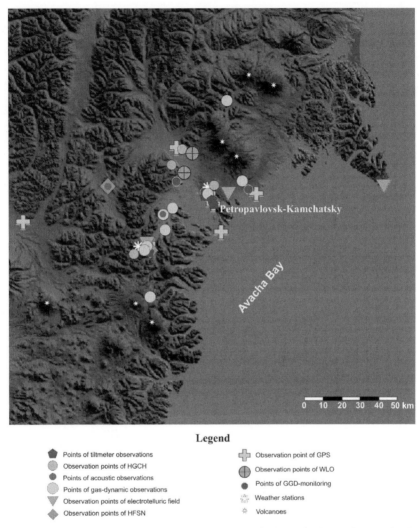

FIGURE 17.2   **Integrated monitoring points (without seismic stations) near Petropavlovsk-Kamchatsky.**

at 1-min intervals (one point—1 s). The observations have been made since 1990 (Moroz et al., 2004).

Observations of high-frequency seismic noises (HFSN) are made at two points. The measured parameters are seismic noise in a narrow frequency band with the center frequency of about 30 Hz. Sampling frequency is 100 Hz. Earth tide wave O_1 is derived in a signal envelope. The observations have been made since 1992 (Saltykov et al., 2008).

The listed types of observations are conducted by operatives from the Kamchatka Branch of the Geophysical Survey of RAS. A number of surveys are carried out by other organizations.

OAO Kamchatgeologiya. *Hydrogeodeformation monitoring* (GGD) is according to results of measurements of underground water level, temperature, electrical conductivity, and atmospheric pressure at seven points with wells (50−800 m in depth). Sampling interval is 10 min. The observations have been made since 2001.

Institute of Volcanology and Seismology of Far Eastern Branch of RAS. *Monitoring of geoacoustic emission (GAE) level and electric component of the Earth's electromagnetic field.* This has four observation points with deep wells. Measurements in the wells are made in a number of frequency ranges at depths of 1012 and 270 (Well 1); 730 (2); 200 (3); 600 (4) m. Sampling frequency is 32 Hz. The observations have been made since 2000 (Gavrilov et al., 2008).

Institute of Cosmophysical Research and Radio Wave Propagation of Far Eastern Branch of RAS. *Vertical ionospheric sounding* at 15-min intervals since 1967, *observations of 1- to 10-kHz natural Vertical low frequency ionospheric sounding (VLF) electromagnetic waves* since 1996, *geomagnetic field variations* with a 1-Hz frequency since 1967, *atmospheric electric conductivity* with a 1-Hz frequency since 1997, *vertical component of atmospheric electric field* with a 1-Hz frequency since 1989.

The created system for integrated monitoring of geodynamic processes has helped to accumulate experience in the detection of precursors. This chapter describes only the results of studies of medium-term precursors, with an emphasis on precursor detection in real time.

17.3 REAL-TIME PREDICTIONS FOR 1998−2012

The experience of studying the Kronotsky earthquake of December 5, 1997, $M_w = 7.8$ (see the section "Precursors and Prediction of the 1997 Kronotsky earthquake") has demonstrated that prediction of seismic events in Kamchatka is not a hopeless task. In this regard, to provide on-line seismic hazard assessment and prediction of earthquakes and volcanic explosions, a Kamchatka Expert Council was established as a branch of the national Russian Expert Council (KB REC) in 1998 (Chebrov et al., 2011). The KB REC consists of 17 specialists in different fields of geophysics. In a normal mode, meetings of the KB REC are held every week. The normal mode means the absence of strong earthquakes with $M_w \geq 6.0$ or large volcanic explosions. In other instances, meetings are held more frequently, taking into account the receipt of new on-line information about seismic and volcanic process development. At the meetings of the KB REC, medium-term expert judgments for the next 6 months are updated every 3 months. *All predictive messages are documentarily recorded.* Following the results of the meetings, KB REC's Messages about seismic activity in Kamchatka are prepared; these messages

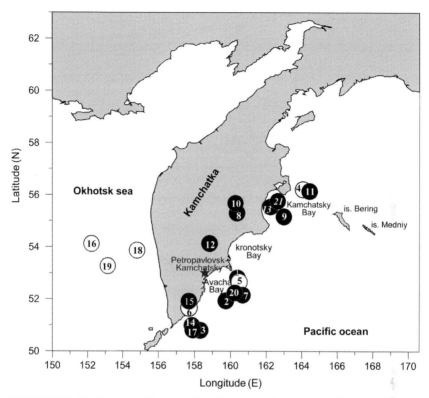

FIGURE 17.3   Earthquakes with $M_w \geq 6.0$ in Kamchatka from May 1, 1998–December 31, 2012. Nos of earthquakes corresponds to the numbers in the table. Black circles—predicted events, open circles—missed events. The depths of three from six missed events (Nos 16, 18, and 19) were >550 km (see the table).

are then delivered to the Russian Expert Council, the Administration, EMERCOM of Kamchatka, and IPE (Institute of Physics of the Earth) of RAS, Moscow. If the rise in seismic hazard was forecasted by several methods, the administrations of the Ministry for Civil Defense and Emergencies initiate certain activities such as training of the personnel and examination of medicine and food stocks and fire-extinguishing equipment.

From May 1, 1998, to December 31, 2012, 21 earthquakes with $M_w \geq 6.0$ occurred in KB REC's area of responsibility. They are shown in Figure17.3 and itemized under Nos 1–21 in Table 17.1.

Column 3 indicates the methods used for the assessment of a seismic situation for this period. The main reason that not all of the nine medium-term prediction methods were applied prior to all earthquakes was the absence of operatives in Kamchatka because of business trips and vacations. The Monitoring of seismicity kinetic parameters (DT) method has only been applied since 2006. Column 4 lists the predictions that came true with regard to three

TABLE 17.1 Real-Time Predictions for 1998–2012

No	Date M_W H (km)	Coordinates φ (°N)	λ (°E)	Methods Used for Assessment of Seismic Situation	Successful Predictions	Partially Successful Predictions
1	June 01, 1998 $M_W = 6.4$ $H = 31$	52.81	160.37	M6, CODA, HFSN, WLO, HGCH	HGCH (May 21, 1998)	–
2	March 08, 1999 $M_W = 6.9$ $H = 7$	51.93	159.72	M6, CODA, Z-test, HFSN, HGCH	–	Z-test (June 01, 1998)
3	September 18, 1999 $M_W = 6.0$ $H = 40$	50.99	157.84	M6, CODA, Z-test, HFSN, HGCH	HFSN (September 16, 1999)	HGCH (August 13, 1999)
4	August 02, 2001 $M_W = 6.3$ $H = 25$	56.21	164.05	CODA, RTL, Z-test	–	–
5	October 08, 2001 $M_W = 6.5$ $H = 31$	52.62	160.46	CODA, RTL, Z-test, HFSN, HGCH	–	–
6	October 16, 2002 $M_W = 6.2$ $H = 108$	51.66	157.68	M6, CODA, RTL, Z-test, HFSN, HGCH	–	–
7	March 15, 2003 $M_W = 6.1$ $H = 4$	52.15	160.66	M6, CODA, RTL, Z-test, HFSN, WLO, HGCH	–	HFSN (January 10, 2003), (February 05, 2003)

8	June 16, 2003 $M_w = 6.9$ $H = 190$	55.30	160.34	M6, CODA, HFSN	—	HFSN (May 30, 2003) Z-test (April 30, 2003)
9	April 14, 2004 $M_w = 6.2$ $H = 48$	55.16	162.97	M6, CODA, RTL, Z-test, HFSN	—	RTL (October 24, 2003) Z-test (March 05, 2004)
10	June 10, 2004 $M_w = 6.9$ $H = 208$	55.68	160.25	M6, CODA, HFSN, WLO	—	HFSN (May 13, 2004) WLO (May 21, 2004) M6 (November 06, 2003) Z-test (May 28, 2004)
11	April 12, 2006 $M_w = 6.0$ $H = 1$	56.14	164.42	M6, CODA, RTL, Z-test, DT, GAE	GAE (March 23, 2006), (March 30, 2006), (April 07, 2006) DT (February 12, 2006)	—
12	May 22, 2006 $M_w = 6.2$ $H = 213$	54.13	158.81	M6, CODA, DT, HFSN, WLO, HGCH, GAE	WLO (April 20, 2006)	GAE (April 21, 2006), (April 28, 2006) Z-test (March 24, 2006)
13	August 17, 2006 $M_w = 5.7$ $H = 82$	55.59	162.13	M6, CODA, RTL, Z-test, DT, GAE	DT (May 29, 2006)	RTL (March 24, 2006)
14	August 24, 2006 $M_w = 6.5$ $H = 38$	50.75	157.97	M6, CODA, RTL, Z-test, DT, HFSN, HGCH, GAE	HFSN (July 04, 2006), (August 04, 2006), (August 18, 2006)	RTL (March 24, 2006) Z-test (August 11, 2006)
15	June 30, 2007 $M_w = 6.4$ $H = 129$	51.92	157.67	M6, CODA, RTL, Z-test, DT, HFSN, WLO, HGCH	—	RTL (January 19, 2007) Z-test (November 19, 2007)

Continued

TABLE 17.1 Real-Time Predictions for 1998–2012—cont'd

No	Date M_W H (km)	Coordinates φ (°N)	λ (°E)	Methods Used for Assessment of Seismic Situation	Successful Predictions	Partially Successful Predictions
16	July 05, 2008 $M_W = 7.7$ $H = 665$	53.98	151.74	M6, CODA, DT, HFSN	—	—
17	July 24, 2008 $M_W = 6.2$ $H = 40$	50.61	158.04	M6, CODA, RTL, Z-test, DT, HFSN, WLO, HGCH, GAE	—	Z-test (January 25, 2008), (May 29, 2008) RTL (May 30, 2008)
18	November 11, 2008 $M_W = 7.3$ $H = 564$	53.77	154.69	M6, CODA, DT, HFSN	—	—
19	December 10, 2009 $M_W = 6.3$ $H = 597$	53.27	153.13	M6, CODA, DT, HFSN	—	—
20	July 30, 2010 $M_W = 6.3$ $H = 38$	52.22	160.46	M6, CODA, RTL, Z-test, DT, HFSN, WLO, HGCH, GAE	Z-test (June 28, 2010)	HFSN (June 09, 2010) DT (March 12, 2010) RTL (March 20, 2009)
21	February 20, 2011 $M_W = 6.1$ $H = 48$	55.73	162.48	M6, CODA, RTL, Z-test, DT, GAE	Z-test (December 03, 2010)	RTL (November 06, 2009)

RTL - Region, time, length method.

parameters of the expected earthquake: area, time, and magnitude. Column 5 of the table lists the predictions that are erroneous with regard to one or two parameters. Given below is a brief description of the essence of the given predictions; the names of authors are given in brackets. The relevant predictive statements were made at the regularly held sessions of the KB REC. They applied to the stated months following the session date given in the table. Sessions at which no predictive statements were made were held, but they are not listed in the following.

In the following, the predictive statements made at sessions of the KB REC are ordered by the method on which the statement was based.

Hydrogeochemical monitoring, HGCH (Khatkevich, Ryabinin). According to retrospective data since 1977, anomalous variations of given hydro-geochemical parameters were observed before earthquakes with $M \geq 6$. The most frequent of them were variations of chloride ion (Cl^-) and sodium ion (Na^+) concentrations. The distances from wells to epicenters were up to 350 km, and the lead time varied from 1 week to 5–6 months.

In the case of the prediction as of May 21, 1998, the most expressed anomalies related to a reduction of chloride ion (Cl^-), Ca ion, SO_4 ion, and sodium ion (Na^+) concentrations were recorded in well GK-15 at the Paratunka station (52.83°N–158.13°E). The authors predicted that the intensity of shaking in Petropavlovsk-Kamchatsky would be 5–6 points on the 12-point scale with a waiting time up to 6 months. Such an intensity in the city may be caused by earthquakes with $M = 6.0$ in the case of epicentral distance up to 70 km and $M = 7.0$ in the case of epicentral distance up to 150 km (Gusev and Shumilina, 2000). Earthquake # 1 on June 1, 1998, with $M_w = 6.4$ occurred at the epicentral distance of 120 km from the city and caused shaking with an intensity of 4–5 points. Taking into account the fact that the author was not a seismologist, this prediction was estimated by the KB REC as true in spite of error in intensity. *In the case of the prediction as of August 13, 1999*, the anomalies related to the reduction of (Cl^-), (HCO_3^-), and (Na^+) concentrations were recorded in well GK-1 at the Pinachevo station (52.83°N–158.13°E). The authors predicted that the intensity of shaking in Petropavlovsk-Kamchatsky would be 5 points and above with the waiting time of 1 month. Earthquake # 3 on September 18, 1999, with $M_w = 6.0$ occurred at the epicentral distance of 230 km from the city and caused shaking with an intensity of 3–4 points. An error was made in the estimation of the expected shock intensity and, therefore, magnitude and/or area.

HFSN (Saltykov) (see the section "Precursors and Prediction of the 1997 Kronotsky earthquake"). According to retrospective data for 1992–2006, when persistent stabilization of the phase shift between the HFSN tidal component and earth tide wave O_1 is detected, within at least three weeks, the probability of an earthquake with $M \geq 4.0 + 0.008 \times \Delta$ increases. Monitoring covers seismically active areas with a maximum distance from observation points $\Delta \leq 400$ km. The area is set by a geographical name or coordinates of its vertices and selected depending on the value of the phase shift. The earthquake occurrence waiting time was up to 1 month.

The prediction as of September 16, 1999, is based on an anomaly at the Nachiki station (53.1°N–157.8°E). An earthquake with $M \geq 6.0$ was predicted at a distance up to 250 km from this station with the waiting time of 1 month. Earthquake # 3 occurred on September 18, 1999, with $M_w = 6.0$ at the epicentral distance of 237 km from the Nachiki station.

In the case of the prediction as of January 10, 2003, based on observations at the Karymshina station (52.8°N–158.15°E), an earthquake with $M \geq 6.0$ at a distance up to 250 km from the station was expected within 1 month. In accordance with the *repeated prediction as of February 5, 2003*, based on data from the Nachiki station, an earthquake with $M \geq 6.0$ at a distance up to 250 km from the station was expected within 1 month as well. Earthquake # 7 with $M_w = 6.1$ occurred on March 15, 2003, at a distance of 223 km from the Nachiki station and 187 km from the Karymshina station. An error was made in waiting time estimation.

In accordance with the prediction as of May 30, 2003, based on observations at the Nachiki station, an earthquake with $M \geq 6.0$ at a distance up to 250 km from this station was expected within 1 month. Earthquake # 8 with $M_w = 6.9$ occurred on July 16, 2003, at a distance of 294 km from the Nachiki station. An error was made in distance estimation and in the time window.

In accordance with the prediction as of May 13, 2004, based on observations at the Nachiki station, an earthquake with $M \geq 6.0$ at a distance of up to 250 km from the station was expected until the end of May. Earthquake # 10 with $M_w = 6.9$ occurred on June 10, 2004, at a distance of 327 km from the Nachiki station. An error was made in the estimation of the prediction validity period and area of the expected earthquake.

The prediction as of July 4, 2006, was extended on August 4, 2006, and August 18, 2006, based on observations at the Nachiki and Karymshina stations. According to the wording as of August 18, 2006, an earthquake with $M > 5.0$ at $\Delta < 120$ km or $M > 4.0 + 0.0083\ \Delta$ at $120 < \Delta < 300$ km was expected within a week. Earthquake # 14 with $M_w = 6.5$ occurred on August 24, 2006, at a distance of 263 km from the Nachiki station and 231 km from the Karymshina station.

According to the prediction as of June 9, 2010, based on observations at the Nachiki and Karymshina stations, an earthquake with $M > 5.0$ at $\Delta < 120$ km or $M > 4.0 + 0.0083\ \Delta$ at $120 < \Delta < 300$ km was expected within a month. Earthquake # 20 with $M_w = 6.3$ occurred on July 30, 2010, at a distance of 207 km from the Nachiki station and 171 km from the Karymshina station. An error was made in the estimation of the magnitude and prediction validity period.

17.3.1 Water Level Observations WLO (Kopylova)

According to observational data since 1987, water level monitoring in well E1 (53.26°N−158.48°E, depth 665 m) revealed variations (from weeks to months) preceding earthquakes with magnitudes of at least 6 with the epicenters located at distances up to 370 km. Predominant water level lowering occurring at an increased rate was observed before earthquakes (Kopylova, 2001).

According to the prediction as of May 21, 2004, an earthquake with $M \geq 5.6$ was expected within the span from the first few weeks to 3.5 months in the Avacha Bay and Kronotsky Bay (southern part of Kamchatka) areas at a distance within 200−250 km from well E1. Earthquake # 10 with $M_w = 6.9$ occurred on June 10, 2004, at a distance of 359 km from well E1. An error was made in the estimation of the earthquake occurrence area.

According to the prediction as of April 20, 2006, an earthquake with $M \geq 5.5 \pm 0.5$ was expected in Kamchatka within the first months. It occurred on May 22, 2006 (# 12), at a distance of 235 km from well E1 and had $M_w = 6.2$.

M6 algorithm (Shirokov) (see the section "Precursors and Prediction of the 1997 Kronotsky earthquake").

According to the prediction as of November 6, 2003, an earthquake with $M = 6.8−7.3$ was expected in the Kamchatka sector of the Pacific island arc within 8 months. Earthquake # 10 with $M_w = 6.9$ occurred on June 10, 2004, in central Kamchatka. An error was made in the determination of the location.

Well geoacoustic observations. GAE (Gavrilov). GAE responses to an external electromagnetic field are studied based on data of acoustic measurements at the depths of 1012 and 270 m in well G-1 (53.05°N, 158.66°E) at the Khlebozavod point. According to the results of combined acoustic and electromagnetic measurements since 2000, it has been established that the structure of GAE responses to the daily variations of an external electromagnetic field in the area of the well changes before earthquakes with $M \geq 5.5$ at a distance of up to 550 km from the well. The lead time varies from several days to 1 month (Gavrilov et al., 2008). The physical reasons for the different reactions of GAE to the external electromagnetic field are associated with changes in fluid saturation (and electric conductivity) of the rocks surrounding the well at final stages of earthquake nucleation.

The prediction as of March 23, 2006, was extended on March 30, 2006, and April 7, 2006. An earthquake with $M = 5.0−6.0$ was expected in the area with coordinates 48.0°−57.0°N, 156.0°−165.0°E by April 21, 2006. Earthquake # 11 with $M_w = 6.0$ occurred on April 12, 2006.

According to the prediction as of April 21, 2006, and its extension as of April 28, 2006, an earthquake with $M = 5.0−6.0$ was expected in the area with coordinates 48.0°−57.0°N, 156.0°−165.0°E by May 4, 2006. Earthquake # 12 with $M_w = 6.2$ occurred on May 22, 2006. An error was made in the estimation of the magnitude and prediction validity period.

17.3.2 Monitoring of Seismicity Kinetic Parameters, DT (Tomilin)

Based on laboratory studies of acoustic emission and retrospective field observations, time intervals Δt between successive earthquakes and interval variation coefficients $V_{\Delta t}$ are used as predictive signs. Regularities in the breach of Poisson's law are sought. The macrofailure process (earthquake) occurs at different scale levels. It has been established by experiment that correlation relationships exist between seismic source development time, earthquake nucleation area size, and the value of released energy (Tomilin et al., 2005).

The statistic criteria of failure-source formation that have been proposed as part of the hierarchical model allow localization of a spatial area of focus nucleation, determine the point of time when this area goes into an unstable state, and estimate potential energy release in it. Thus, a principal possibility of predicting the location, time, and energy of an impending seismic event emerges, irrespective of the scale of this event. The DT method is used for predicting the area and increase in the probability of an earthquake with $M \geq 6.0$; the waiting time is not established yet.

According to the prediction as of February 12, 2006, an earthquake with $M = 5.5−6.2$ was expected in the area with coordinates 54.5°−56.5°N, 163.5°−167°E. Earthquake # 11 with $M_w = 6.0$ occurred on April 2, 2006 (54.58°N, 163.62°E).

According to the prediction as of May 29, 2006, an earthquake with $M = 5.5−6.3$ was expected in the area with coordinates 51.0°−57.0°N, 161.0°−163.8°E. Earthquake # 13 with $M_w = 5.7$ occurred on August 17, 2006 (55.58°N, 162.12°E).

Z-test (the method was proposed in Wyss (1986) and Wyss and Habermann (1988); the authors of the predictions in Kamchatka: *Saltykov, Kravchenko*).

A Kamchatka Region catalog with energy classes $K \geq 8.5$ has been used to reveal seismic regime anomalies. The catalog is cleared from aftershocks. The area under study is limited by latitudes $\varphi = 50.5°N$ and 56.5°N, longitudes $\lambda = 156.0°E$ and 167.0°E, and depth from 0 to 100 km.

In order to distinguish space−time blocks with a significant change of seismic-event flow intensity in a seismically active zone, the area under study was scanned by variable radius cells (cylinder) with a depth from 0 to 70 km. The radius varied from 30 to 55 km at 1-km intervals, which was determined by the density of epicenters (at least 300 earthquakes shall be accounted for one design cylinder).

Seismic flow rate in an "elementary" time span was determined for each cylinder. Then, calculations were performed using the following equation:

$$Z = R_1 - R_2 / \left(\sigma_1^2 / n_1 + \sigma_2^2 / n_2 \right)^{1/2},$$

where R_1 is the average rate in the first time interval (from T_0 to t); R_2 is the average rate in the second time interval (from t to T_e); t is the current time ($T_0 < t < T_e$); T_0 is the catalog beginning time; T_e is the calculation end time; σ_1 and σ_2 are the standard deviations in the specified intervals; n_1 and n_2 are the counts. The greater Z is, the more significant the difference in the seismic flow at intervals t_1 and t_2.

However, a greater Z value indicates only the reliability of quiescence identification, but not its value. In addition to the method (Wyss and Habermann, 1988), for anomalous intervals with $Z > 3$, a value of reduction of seismic flow rate SRD(t) was calculated (Saltykov and Kugaenko, 2000):

$$\mathrm{SRD}(t) = 1 - R_2 / R_1$$

and the maximum value of SRD = max(SRD(t)) and corresponding time t were determined.

Scanning results in a map of maximum values of statistically significant ($Z > 3$) seismic flow rate reduction SRD within the time interval from 1 to 5 years counted back from the estimated date. SRD = 1 corresponds to absolute seismic quiescence, SRD = 0.875—seismicity level reduction by 8 times, SRD = 0.75—by 4 times. The calculations are performed for successive dates at 30-day intervals, starting from the beginning of the catalog (January 1, 1962). Then, a map is constructed; it indicates the grid nodes corresponding to seismic flow reduction by 2 times (SRD = 0.5), by 4 times (SRD = 0.75), by 8 times (SRD = 0.875), and absolute quiescence (SRD = 1). The distinguished volumes where seismicity reduction exceeds a specified level are combined together and further considered as a single area of seismic quiescence. Based on the retrospective use of the method in a seismically active area in Kamchatka since 1970, an area of the expected earthquake with $M \geq 6.0$ is indicated within the interval up to 2.5 years.

According to the prediction as of June 1, 1998, an earthquake with $M > 7$ was expected in the area within 51°–52°N, 157°–159°E. Earthquake # 2 occurred on August 3, 1999 (51.93°N, 159.72°E), and had a magnitude $M_w = 6.9$. An error was made in the estimation of the earthquake occurrence area and magnitude.

According to the prediction as of April 30, 2003, an event with $M > 6.0$ was expected to the south of Avacha Bay or in Kronotsky Bay by October 2003. This event (# 8) occurred on June 16, 2003, in central Kamchatka and had a magnitude $M_w = 6.9$. An error was made in area estimation.

According to the prediction as of May 03, 2003, an event with $M > 6.0$ was expected in the region 51°–54°N during 2004. Event # 9 occurred on April 14,

2004 (55.16°N, 162.97°E), and had a magnitude $M_w = 6.2$. An error was made in area estimation.

According to the prediction as of May 28, 2004, an event with $M > 6.0$ was expected in the region 51°−54°N during 2004−2005. Event # 10 occurred on June 10, 2006 (55.68°N, 160.25°E), and had a magnitude $M_w = 6.9$. An error was made in area estimation.

According to the prediction as of March 24, 2006, an event with $M > 6.0$ was expected in the region 51°−54°N during 2006. Event # 12 occurred on May 25, 2006 (54.13°N, 158.81°E), and had a magnitude $M_w = 6.2$. An error was made in area estimation.

According to the prediction as of August 11, 2006, an event with $M > 6.0$ was expected in the region 51°−54°N during half of the year. Event # 14 occurred on August 24, 2006 (50.75°N, 157.97°E), and had a magnitude $M_w = 6.5$. An error was made in area estimation.

According to the prediction as of January 19, 2007, an event with $M > 6.0$ was expected in the region 51°−54°N during 1 year. Event # 15 occurred on May 30, 2007 (51.92°N, 157.67°E), and had a magnitude $M_w = 6.4$.

According to the prediction as of January 25, 2008, an event with $M > 6.0$ was expected in the region 51°−54°N, 158.5°−164.5°E during 2008. Event # 17 occurred on July 24, 2008 (50.61°N, 158.04°E), and had a magnitude $M_w = 6.1$. An error was made in area estimation.

According to the prediction as of June 28, 2010, an event with $M > 6.0$ was expected in the region 51°−55°N, 158.5°−164.5°E during 1 year. Event # 20 occurred on July 30, 2010 (52.22°N, 160.46°E), and had a magnitude $M_w = 6.3$.

According to the prediction as of December 03, 2010, an event with $M > 6.0$ was expected in the region 51°−56°N, 158.5°−164.5°E during 1 year. Event # 21 occurred on February 20, 2011 (55.73°N, 162.48°E) and had a magnitude $M_w = 6.1$.

RTL method (authors of the method are Sobolev and Tyupkin (1997) and Sobolev (2001) (see the section "Precursors and Prediction of the 1997 Kronotsky earthquake"), authors of the predictions are (*Saltykov, Kravchenko*).

A Kamchatka Region catalog with energy classes $K \geq 8.5$ is used to reveal seismic regime anomalies. The catalog is cleared from aftershocks. The area under study is limited by latitudes $\varphi = 50.5°N$ and $56.5°N$, longitudes $\lambda = 156.0°E$ and $167.0°E$, and a depth from 0 to 100 km. In order to distinguish a seismic quiescence area, the studied region is covered with a grid with a step of 0.0625° by latitude and 0.125° by longitude.

The RTL-analyzer software product used for work in Kamchatka (Ivanov and Saltykov) helps to obtain a visual image of RTL parameter values at a given point in time at all grid nodes. Moving through time at certain intervals (15 days), one can observe a dynamic pattern of the seismic quiescence development and extinction process and therefore determine a space−time position of the quiescence. Each node of the grid within the area of the visually distinguished quiescence is attributed to a minimum RTL parameter value for the duration of the existence of the anomaly. The grid nodes with RTL values < -3 are combined together based on an adjacency principle and determine the seismic quiescence area. The distinguished anomaly is characterized by a minimum RTL parameter value for the entire time of its existence and the duration of the quiescence corresponding to the time interval during which RTL parameter values were < -3.

This method was retrospectively verified in Kamchatka only for earthquakes with $M \geq 7$ (Sobolev, 2001), including the Kronotsky earthquake in 1997. On this basis, the authors of RTL predictions for 1998–2012 (Saltykov, Kravchenko) indicated the magnitude of the expected earthquakes as $M > 7.0$ in all cases. The magnitudes of the occurred earthquakes were < 7.0. For this reason, all predictions in the table are characterized as partially true.

The following areas were indicated in the *prediction as of October 24, 2003*: (1) South of Kamchatka, (2) Gulf of Kamchatka. Earthquake # 9 occurred on April 14, 2004, in the Gulf of Kamchatka and had a magnitude $M_w = 6.2$.

The following areas were indicated in the *prediction as of March 24, 2006*: (1) South of Kamchatka, (2) Gulf of Kamchatka. Earthquake # 13 occurred on August 17, 2006, in the Gulf of Kamchatka and had a magnitude $M_w = 5.7$. Another earthquake # 14 occurred on August 24, 2006, in the south of Kamchatka and had a magnitude $M_w = 6.5$.

The following areas were indicated in the *prediction as of January 1, 2007*: (1) South of Kamchatka, (2), Gulf of Kamchatka. Earthquake # 15 with $M_w = 6.4$ occurred on May 30, 2007, in the south of Kamchatka.

The following areas were indicated in the *prediction as of May 30, 2008*: (1) South of Kamchatka, (2) Areas to the east of the Kronotsky Peninsula and southwest of the Bering Island. Earthquake # 17 with $M_w = 6.2$ occurred on July 24, 2008, in the south of Kamchatka.

The following areas were indicated in the *prediction as of March 20, 2009*: (1) South of Kamchatka, including the south of Avacha Bay. Earthquake # 20 with $M_w = 6.3$ occurred on July 30, 2010, in Avacha Bay.

The following areas were indicated in the *prediction as of November 06, 2009*: (1) South of Kamchatka, including the Avacha Bay, (2) Kamchatka Bay. Earthquake # 21 with $M_w = 6.1$ occurred on February 20, 2011, in Kamchatka Bay.

17.4 DISCUSSION

When assessing the experience of the real-time predictions made in 1998–2012 in Kamchatka, we should note the following circumstances: The real seismic hazard is posed by earthquakes in the Pacific focal zone with $M \geq 7.5$. Such earthquakes may cause oscillations with an intensity of $J \geq 7$ on the 12-point scale and Peak gravity acceleration (PGA) $\geq 100 \ cm^2/s$ on the east coast of Kamchatka.

KB REC's messages sent to the Russian Expert Council, Administration, EMERCOM of Kamchatka and IPE of RAS never contained a prediction of the earthquake with $M \geq 7.5$ or with an intensity of $J \geq 6$ points. Special attention was paid to the estimation of J in Petropavlovsk-Kamchatsky with a population of about 180,000 people and a well-developed infrastructure. *All predictive messages contained the conclusions of the absence of hazardous earthquakes. Fortunately, this proved to be true.* No earthquakes posed a hazard for the population and infrastructure between 1998 and 2012. The only earthquake with $M \geq 7.5$ (No. 16 in the table) had a depth of 665 km and intensity of $J \approx 4-5$ points. It was not predicted by any of the methods. Thus, the predictions listed in the table referred to moderate earthquakes with $6 \leq M \leq 7$. It is characteristic that no prediction was given with regard to a seismic-situation assessment by the CODA method (refer to the table). This is connected with the fact that this method, in the opinion of the authors, facilitates the prediction of earthquakes with $M \geq 7$ only.

Taking into account the relatively low reliability and efficiency of predicting all three parameters of the expected earthquake (area, time, and magnitude), a seismological-situation assessment was given in KB REC's messages sent to the Russian Expert Council, Administration, EMERCOM of Kamchatka and IPE of RAS, instead of these parameters. *As an example, we give an extract from KB REC's message as of February 10, 2012. It was the only one where a relatively high J value (up to 6 points) was expected.*

The assessment of a seismic hazard for the next month (until March 10, 2012)

At the present time, anomalies in the observed geophysical fields are noted (HFSN, DT, Z-test, HGCH, WLO, GAE), which allows us to draw a conclusion about the increased probability of earthquakes with $M = 6-7$ by $5-10$ times in the area of the southern Kamchatka, including Avacha Bay. The expected intensity of shocks in Petropavlovsk-Kamchatsky does not exceed $5-6$ points on the 12-point MSK-64 scale.

All data (except for seismological data) have been obtained near Petropavlovsk-Kamchatsky, so the prediction specifically refers to the southern part of Kamchatka (including, in particular, Avacha Bay), which is indicated in prediction conclusions for certain methods. The maximum of the magnitude range is limited because no geodetic anomalies and precursor variations have been revealed using the CODA method.

The average frequency of earthquakes with magnitude $M \geq 6.0$ for the south of Kamchatka ($50.5°-53.5°N$, $155°-162°E$) is 0.7 per year or 0.06 per month (based on catalog data for $1962-2012$). In the case of the described prediction (alarm time $-$ 1 month), the probability of the expected earthquake increases to $0.3-0.6$.

In reality, an earthquake with $M = 6.9$ in the south of Kamchatka ($50.93°N-157.34$ E) occurred on February 28, 2013. This case, along with the table data, indicates that most of the applied prediction methods, except for the Z-test and RTL methods, *give unjustifiably short earthquake waiting periods of about 1 month.* We point to two possible reasons. The retrospective assessments of method capabilities were too optimistic. Moreover, the dynamics of medium properties were not taken into account. An example is given below of chloride ion concentration variations in a well over 28 years.

It follows from Figure 17.4 that, within the period until 1998, variations were smoother, and anomalous concentration reduction was observed at longer time intervals as compared to the situation after 1998. As described in the section "Precursors and Prediction of the 1997 Kronotsky earthquake", this event with $M_w = 7.8$ occurred on December 5, 1997. It obviously changed the stress condition in the seismically active area of Kamchatka (Sobolev, 2010). The seismic activity has reduced. The annual number of earthquakes with $K \geq 8.5$ (without aftershocks) within 1963–1997 (before the Kronotsky earthquake) was equal to N/year = 518.37 and, after that, within 1997–2012, it was equal to N/year = 469.38, that is, activity reduction was 1 percent. Such changes as in Figure 17.4 might also be in the structure of other predictive signs. However, they were not taken into account in the real-time predictions (table), the authors of which used retrospective experience before 1998 as the basis.

It is still too early to assess the reliability and efficiency of the predictions presented in the table, in terms of both individual methods and as a whole. It was suggested by Molchan (1991) to use the equation

$$\varepsilon = n + \tau,$$

where n is the relative number of failures to predict and τ is the relative alert time. The lower the coefficient ε, the more efficient the precursor; the n term can be considered as a measure of reliability.

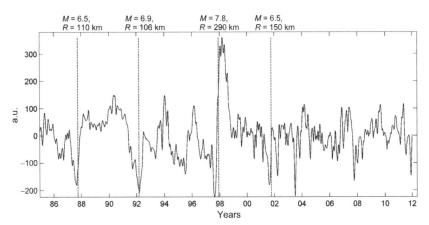

FIGURE 17.4   Variations of chloride ion concentrations in water from well GK-1 (52.83°N–158.13°E) (depth-1261 m). Vertical dashed lines indicate the time of the earthquakes, before which the clear bay-like reduction of chloride ion concentration was observed. The upper abscissa scale indicates earthquake magnitudes and distances to the well.

The authors of the predictions in Kamchatka were not set a task of accumulating uniform statistics. It follows from the data presented in the table that a seismic situation was assessed by not all methods before 21 earthquakes. For example, in the WLO and GAE methods, miserly statistics constituted only 7 cases. Criteria for issuing predictions changed several times in the course of the accumulation of experience. In some cases, the focal area of the predicted earthquake was not determined. A limit of waiting time was not even established in the DT method. It can be said on a qualitative level that the reliability of the methods used leaves much to be desired. In addition to the data presented in the table, it should be noted that, following a number of predictions, the expected earthquakes did not occur, and the number of such "false alarms" was twice that of predictions that came true. As far as we know, the experience of the documentarily recorded medium-term earthquake predictions made within 15 years using a suite of methods is unique. It allows an assessment of the level of this domain of science. In general, the prediction is still a long way off efficient practical use. However, it should be noted that this conclusion relates to the prediction of earthquakes with a moderate magnitude $M < 7$, which does not pose a real hazard to the population and infrastructure. It is not improbable that several methods will give clear predictive anomalies in the case of nucleation of earthquakes with $M > 7.5$ in Kamchatka. To some extent, this was before the Kronotsky earthquake of 1997 with $M = 7.8$.

17.5 PRECURSORS AND PREDICTION OF THE 1997 KRONOTSKY EARTHQUAKE

On December 5, 1997, the Kronotsky earthquake with $M_w = 7.8$ occurred in Kamchatka (54.64°N−162.55 E). Before this event, precursors had been detected using five methods, and predictions had been made in real time. It was the most powerful earthquake within the Kamchatka seismic area during the period represented in the instrumental catalog since January 1, 1962. These predictions were made by different researchers independent of each other. Below, we summarize the observations by each method and the statements resulting concerning future earthquakes in the region. These predictions were erroneous with regard to one of expected parameters: time, place, or magnitude. In the terms of the present chapter, they can be considered as partly successful predictions.

17.5.1 The RTL Method

The analysis using the RTL method (Sobolev and Tyupkin, 1997; Sobolev, 2001) made it possible to identify an area of seismic quiescence prior to the Kronotskoe earthquake. The RTL method uses three functions to measure the state of seismicity at a given location as a function of time $R(x, y, z, t)$ assigns a decreasing weight to each earthquake in the catalog as a function of the epicentral distance from the point of interest, $T(x, y, z, t)$ decreases the weight of each event as a function of the difference from the time of interest, and $L(x, y, z, t)$ weighs the contribution to the algorithm by the rupture length of each event.

These functions are defined as

$$R(x, y, z, t) = \left[\sum \exp(-r_i/r_0) \right] - R_{ltr}$$

$$T(x, y, z, t) = \left[\sum \exp(t - t_i) \Big/ t_0 \right] - T_{ltr}$$

$$L(x, y, z, t) = \left[\sum (l_i/r_i)^p \right] - L_{ltr}$$

In these formulas, x, y, z, and t are the coordinates, the depth, and time, respectively. r_i is the epicentral distance of current events from the location selected for analyses, t_i is the occurrence time of the past seismic events, and l_i is the length of rupture. The R_{ltr}, T_{ltr}, L_{ltr} are the long-term averages of these functions. By subtracting them, they eliminate the linear trends of the corresponding functions. r_0 is a coefficient that characterizes the diminishing influence of more distant seismic events; t_0 is the coefficient characterizing the rate at which the preceding seismic events are "forgotten" as the time of analysis moves on; and p is the coefficient that characterizes the contribution of the size of each preceding event. With $p = 1$, 2, or 3, this quantity is proportional to rupture length, square of rupture, or the energy, respectively. R, T, and L are dimensionless functions. They are further normalized by their standard deviations, σ_R, σ_T, and σ_L, respectively. The product of the above three functions is calculated as the RTL parameter, which describes the deviation from the background level of seismicity and is in units of the standard deviation, $\sigma = \sigma_R \sigma_T \sigma_L$.

Figure 17.5 shows how quiescence was developing in 1994−1997 prior to the Kronotskoe earthquake when the RTL curve was below the background level. The thin arrow in Figure 17.5 indicates the time (August 27, 1996) when

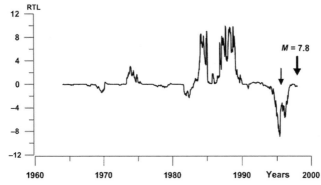

FIGURE 17.5   **RTL graph prior to the Kronotsky earthquake with** $M_w = 7.8$**.** The thin arrow indicates the time of prediction.

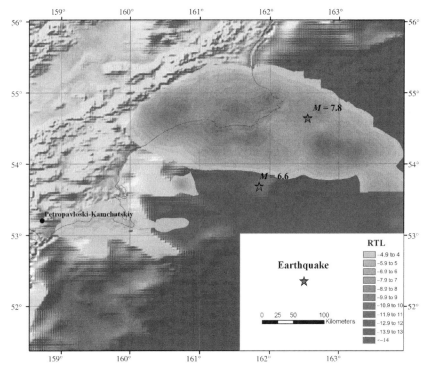

FIGURE 17.6   **The map of seismic quiescence 1.7 years prior to the Kronotsky earthquake with $M_w = 7.8$ and main aftershock with $M_w = 6.6$.**

a report was sent to the National Earthquake Prediction Council of the Russian Ministry Emergency Situations. It was suggested, based on the chart in Figure 17.5 and the map of seismic quiescence in Figure 17.6, that an earthquake with $M \sim 7$ was expected in Northern Kamchatka inside of the quiescence area (Figure 17.6). The alarm interval was defined as 2 years. The expected earthquake did happen 1.7 years after the prediction issue, and the epicenter was in the zone $\sigma < -9$.

Seismic quiescence phenomena were used early to predict the May 7, 1986, Andreanof Island earthquake with magnitude $M_s = 7.7$ (Kisslinger, 1988). The author applied the difference from the RTL technique. The magnitude estimate (M_s 7–7.5) was based on the size of the quiescent zone and the seismic history of the Adak region. The magnitude was underestimated as in our case.

17.5.2 The HFSN Method

The prediction was also made based on HFSN observations (Saltykov et al., 1998). This original method provides for the use of earth tides as a reference signal for the purpose of studying seismic emission regularities in the

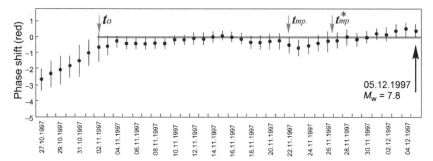

FIGURE 17.7   The phase variations of the HFSN prior to the Kronotsky earthquake with $M_w = 7.8$.

frequency band of about 30 Hz. HFSN synchronization with external tidal excitation occurring before large earthquakes is observed as a stabilized phase shift between tidal wave O_1 and the periodical component of the HFSN envelope with the same period—$T = 25.8$ h. The HFSN monitoring for detecting precursors of large regional earthquakes with $M \geq 6.0$ yielded the following stable positive results: of the nine earthquakes that have occurred within 400 km of the observing stations during the period 1992—1997, it was only in a single case that no precursor was identified.

Information parameter: phase shift between an earth tide wave and the corresponding component of the HFSN envelope. Precursor pertains to the phase shift stabilization at a certain level within three weeks. The warning was made 9 days before the Kronotsky earthquake on December 5, 1997. Figure 17.7 (correct) shows the phase variations of the HFSN component associated with the action of tide wave O_1 before the Kronotsky earthquake, according to data from the Nachiki station (53.1°N—157.8°E). The horizontal line shows the stabilization of this phase. Vertical lines correspond to a confidence 1σ interval relative to background noise. The arrows indicate the following: t_0—start of precursor manifestation (time of synchronization with a tide wave), t_{mp}—alarm time determined by the method, t^*_{mp}—alarm time in accordance with a prediction conclusion.

Before the Kronotsky event, precursors had been detected and predictions had been made in real time using three more methods.

17.5.3 The CODA Method

Parameter α was introduced in order to describe the anomalous properties of an individual earthquake coda (Gusev and Lemzikov, 1985). The α value is an anomaly (deviation from a regional average value) for a derivative of the logarithm of a coda-record envelope of current amplitude with respect to time.

$$\alpha = \frac{d}{dt}[\lg A(t) - \lg a(t)],$$

where $A(t)$ is the current amplitude of a coda envelope and $a(t)$ is the support function describing an average regional shape of a coda-record envelope. Positive α values indicate a relatively smoother coda-envelope amplitude decay in comparison with a support curve, and negative ones are indicative of a relatively steeper decay. By the retrospective experience gained by 1994, based on the CODA method, a precursor of a strong earthquake may be considered as expressed bay-like synchronous anomalies of α at a pair of adjacent stations with a lead time to the expected main shock of 0.8–1.6 years (Gusev, 1997).

In April 1996, based on these data and retrospective experience, a medium-term prediction was given to indicate that an event with $M > 7.5$ would occur in the area of the Gulf of Kamchatka in the period before December 1996 (Abubakirov et al., 1998). The alarm was canceled in the beginning of 1997.

17.5.4 The M6 Method

The methodological approach of the M6 algorithm for a prediction problem solution is based on identifying the signs space–time regularity in a seismic process, which are most noticeable at a final stage of strong earthquake nucleation (Shirokov, 2001).

The following procedures are implemented on the basis of seismological data analysis.

Identification of seismic quiescence zones in a wide range of energy classes and seismic focus depths. Identification of anomalies in the distribution of earthquakes by depth and energy classes. Identification of the anomalies of various duration, which are related to diurnal distribution of earthquakes in the seismically active volume under study. Identification of anomalies in seasonal distribution of earthquakes. Analysis of earthquake swarm sequences and identification of anomalies in space–time distribution of swarm events. Identification of periodic, cyclic components in the seismic process with periods of up to tens of years, including study of the influence of monthly (29.53 days) and long-term (with the periods of 8.85 and 18.613 years) lunar components on seismicity. Prediction parameters are magnitude $M \geq 6.0$, seismic focal zone in the Kamchatka area, prediction time—about 1 month. Based on the M6 algorithm, in October 1997, an increased probability of an earthquake higher than latitude 52.0°N with a magnitude of about 6.0 was predicted to occur before the end of 1997 (Shirokov, 2001). The prediction came true after 34 days with an error in magnitude.

17.5.5 Geodetic Method

On December 4, 1997, a medium-term prediction was given based on the results of geodetic measurements. At the end of November, repeated linear measurements were made at the Ust–Kamchatsk testing site (Fedotov et al.,

1999). The comparison of the measurement results with the data obtained in 1996 has shown that line reduction took place throughout the testing site. Maximum compression was observed for the line perpendicular to the seismic focal zone. The compression in this line was close to 4×10^{-6}. The obtained data indicated potential nucleation of a strong earthquake with $M > 7.0$ in the area of the Gulf of Kamchatka.

17.6 CONCLUSION

The experience of studying the Kronotsky earthquake of December 5, 1997, $M_w = 7.8$ has demonstrated that prediction of seismic events in Kamchatka is not a hopeless task. In this regard, to provide on-line seismic hazard assessment and prediction of earthquakes and volcanic explosions, a Kamchatka Expert Council was established as a branch of the national Russian Expert Council (KB REC) in 1998. The authors of various methods of earthquake prediction submit their forecasts to KB REC. *All predictive messages are documentarily recorded.* Following the results of the meetings, KB REC's Messages about seismic activity in Kamchatka are delivered to the Russian Expert Council, the Administration, EMERCOM of Kamchatka and IPE of RAS, Moscow.

From May 1, 1998, to December 31, 2012, 21 earthquakes with $M_w \geq 6.0$ occurred in KB REC's area of responsibility. A seismic situation was assessed by nine different methods. Most of the applied prediction methods give earthquake waiting periods of about 1 month. Eight predictions came true with regard to three parameters of the expected earthquake: area, time, and magnitude; seven predictions were erroneous with regard to one or two parameters; three of six missed events had depths of >550 km. The number of false alarms was twice that of the predictions that came true.

The experience of the documentarily recorded medium-term earthquake predictions made within 15 years using a suite of methods is unique. It allows an assessment of the level of this domain of science. A number of predictive methods in Kamchatka has significantly increased over the years that have passed since the partially successful prediction of the 1997 Kronotsky earthquake, $M_w = 7.8$. The reasonable assessment of prediction reliability and efficiency will enable preventive measures to reduce potential loss, which corresponds to the hazard degree of the future seismic event. In general, the prediction is still a long way off efficient practical use.

REFERENCES

Abubakirov, I.R., Gusev, A.A., Guseva, E.M., 1998. Manifestation of preparation process of the 1997 Kronotskoye earthquake in temporal variations of Coda decay rate of small earthquakes. In: Kronotskoye Earthquake of December 5, 1997 on Kamchatka: Precursors, Properties, Effects, Petropavlovsk-Kamchatskii: KGARF, pp. 112–120 (in Russian).

Chebrov, V.N., Saltykov, V.A., Serafimova, Yu.K., 2011. Earthquake prediction on Kamchatka. In: The Results of the Kamchatka Branch of Russian Expert Council for Seismic Hazards and Risk Estimation in 1998−2009. M.: Svetoch Plus, p. 304 (in Russian).

Chebrov, V.N., Droznin, D.V., Kugaenko, Yu. A., et al., 2013. The system of detailed seismological observations in Kamchatka in 2011. Volcanol. Seismol. 7 (1), 16−36.

Fedotov, S.A., Chernyshev, S.D., Matvienko, Yu.D., Zharinov, N.A., 1999. Prediction of Kronotsky earthquake December 5, 1197, $M = 7.8−7.9$, Kamchatka, and of strong aftershocks $M > 6$. Volcanol. Seismol. 20 (6), 597−613.

Firstov, P.P., Shirokov, V.A., 2005. Dynamics of molecular hydrogen and its relation to deformational processes at the Petropavlovsk−Kamchatskii. Geodynamic test site: evidence from observations in 1999−2003. Geochem. Int. 43 (11), 1056−1064.

Gavrilov, V., Bogomolov, L., Morozova, Yu., Storcheus, A., October/December 2008. Variations in geoacoustic emissions in a deep borehole and its correlation with seismicity. Ann. Geophys. 51 (5/6), 737−753.

Gusev, A., Lemzikov, V., 1985. Properties of scattered elastic waves in the lithosphere of Kamchatka: parameters and temporal variations. Tectonophysics 112, 137−153.

Gusev, A.A., Shumilina, L.S., 2000. Modeling the intensity−magnitude−distance relation based on the concept of an incoherent extended earthquake source. Volcanol. Seismol. 21, 443−463.

Gusev, A., 1997. Temporal variations of the Coda decay rate on Kamchatka: are they real and precursory? J. Geophys. Res. 102 (B4), 8381−8396.

Khatkevich, Yu.M., Ryabinin, G.V., 2006. Geochemical and ground-water studies in Kamchatka in the search for earthquake precursors. Volcanol. Seismol. 4, 34−42.

Kisslinger, C., 1988. An experiment in earthquake prediction and the 7 May 1986 Andreanof Islands earthquake. Bull. Seismol. Soc. Am. 78, 218−229.

Kopylova, G.N., 2001. Variations of water level Elizovskaya 1 well, Kamchatka due to large earthquakes: 1987−1998 observations. Volcanol. Seismol. 2, 39−52.

Kossobokov, V.G., 2014. Times of increased probabilities for the occurrence of catastrophic earthquakes: 25 years of the hypothesis testing in real time. In: Shroder, J., Wyss, M. (Eds.), Earthquake Hazard, Risk, and Disasters. Elsevier, London, pp. 477−504.

Levin, V.E., 2009. GPS monitoring of recent crustal movements in Kamchatka and the commander islands during 1997−2007. Volcanol. Seismol. 3 (3), 60−70.

Levin, V.E., Magus'kin, M.A., Bakhtiarov, V.F., et al., 2006. The multisystems geodetic monitoring of recent crustal movements in Kamchatka and the commander is. Volcanol. Seismol. 3, 54−67.

Molchan, M., 1991. Structure of optimal strategies in earthquake prediction. Tectonophysics 193, 267−276.

Moroz, Yu.F., Moroz, T.A., Nazarets, V.P., et al., 2004. Electromagnetic field in studies of geodynamic processes, in complex seismological and geophysical researches of Kamchatka. In: The 25th Anniversary of Kamchatkan Experimental & Methodical seismological Department, Petropavlovsk-Kamchatskii: Kamchatskii Pechatnyi Dvor, pp. 152−170 (in Russian).

Saltykov, V.A., Kugaenko, Yu. A., 2000. Seismic quiescences before two large Kamchatka earthquakes of 1996. Volcanol. Seismol. 22 (1), 87−98.

Saltykov, V.A., Kugaenko, Yu.A., Sinitsyn, V.I., Chebrov, V.N., 2008. Precursors of large Kamchatka earthquakes based on monitoring of seismic noise. Volcanol. Seismol. 2 (2), 94−107.

Saltykov, V.A., Sinitsyn, V.I., Chebrov, V.N., 1998. The use of the high-frequency seismic noise for intermediate-term prediction of strong earthquakes. In: Kronotskoye Earthquake of December 5, 1997 on Kamchatka: Precursors, Properties, Effects, Petropavlovsk−Kamchatskii: KGARF, pp. 99−105 (in Russian).

Schorlemmer, D., Gerstenberger, M.C., 2014. Quantifying improvements in earthquake rupture forecasts through testable models. In: Shroder, J., Wyss, M. (Eds.), Earthquake Hazard, Risk, and Disasters. Elsevier, London, pp. 405–429.

Shirokov, V.A., 2001. Short_Term forecasting of the times, locations, and sizes of the Kamchatka m 6–7.8 earthquakes from combined seismological data. In: Geodinamika i vulkanizm Kurilo–Kamchatskoi ostrovoduzhnoi sistemy (The geodynamics and Volcanism of the Kuril–Kamchatka island arc system), Petropavlovsk-Kamchatskii: IVGiG DVO RAN, pp. 95–116 (in Russian).

Sobolev, G., 2001. The examples of earthquake preparation in Kamchatka and Japan. Tectonophysics 338, 269–279.

Sobolev, G.A., 2010. The present-day seismicity variations in the Kuril–Kamchatka seismic zone. Volcanol. Seismol. 4 (6), 367–377.

Sobolev, G.A., Tyupkin, Yu.S., 1997. Low-magnitude seismicity precursors of large earthquakes in Kamchatka. Volcanol. Seismol. 18 (4), 433–446.

Tomilin, N.G., Damaskinskaya, E.E., Pavlov, P.I., 2005. Statistical kinetics of fracture and the prediction of seismic phenomena. Fiz. Tverd. Tela 47 (5), 955–959 (in Russian).

Wu, Z.L., 2014. Duties of earthquake forecast: cases and lessons in China. In: Shroder, J., Wyss, M. (Eds.), Earthquake Hazard, Risk, and Disasters. Elsevier, London, pp. 431–448.

Wyss, M., 1986. Seismic quiescence precursor to the 1983 Kaoiki ($M_s = 6.6$) Hawaii earthquake. Bull. Seismol. Soc. Am. 76, 785–800.

Wyss, M., Habermann, R.E., 1988. Precursory quiescence. Pageoph 126, 319–332.

Chapter 18

Times of Increased Probabilities for Occurrence of Catastrophic Earthquakes: 25 Years of Hypothesis Testing in Real Time

Vladimir G. Kossobokov

The Abdus Salam International Centre for Theoretical Physics e SAND group, Trieste, Italy;
Institute of Earthquake Prediction Theory and Mathematical Geophysics, Russian Academy
of Sciences, Moscow, Russian Federation; Institut de Physique du Globe de Paris, France;
International Seismic Safety Organization, ISSO

ABSTRACT

Earthquake prediction is an uncertain profession. Many methods for earthquake forecast/
prediction have been proposed and some of these methods may be reliable. Some of those
might be even useful in mitigating seismic risks and reducing losses due to catastrophic
earthquakes and associated phenomena. Regretfully, most of currently known earthquake
forecast/prediction methods cannot be adequately tested and evaluated just because of
lack of a precise definition and/or shortage of data for a reliable verification. A rare
exception is the pattern recognition algorithm M8, which was designed in 1984 for
prediction of great, magnitude 8, earthquakes, hence its name. This computer-coded
algorithm was originally conceived for application targeting other magnitude ranges,
so that by 1986 it was already tested in retrospective applications aimed at earthquakes,
down to magnitude 5. Since then the M8 algorithm has been used for systematic
monitoring of seismic activity in a number of seismic regions worldwide. After
successful early forecasts of the 1988 Spitak (Armenia) and the 1989 Loma Prieta
(California) earthquakes, a rigid test to evaluate the efficiency of the reproducible
intermediate-term middle-range earthquake prediction technique has been designed.
Since 1991 every 6 months the algorithm M8, along and in combination with its
refinement MSc, has been applied in a real-time prediction mode to seismicity of the
entire Earth to outline the areas where magnitude 8.0+ and 7.5+ earthquakes are most
likely to occur before the next update. Each of the four statistics achieved to date in the
Global Test proves with confidence above 99 percent rather high efficiency of the M8 and
M8-MSc predictions limited to intermediate-term middle- and narrow-range accuracy.

Earthquake Hazard, Risk, and Disasters. http://dx.doi.org/10.1016/B978-0-12-394848-9.00018-3

The null hypothesis of random recurrence in earthquake-prone areas has been rejected, at least for magnitude 8.0+ and 7.5+ earthquakes. The results of this global experimental testing are indirect confirmations of: (1) earthquake predictability; (2) the existence of dynamic features that are common in different tectonic environments; and (3) diverse behavior in the course of durable phase transitions in complex hierarchical, nonlinear system of blocks and faults of the naturally fractal lithosphere of our planet Earth.

18.1 INTRODUCTION

The catastrophic nature of earthquakes has been known for centuries due to the resulting devastation by many of them. The evident abruptness along with irregularity and infrequency of seismic extreme occurrences facilitated formation of a common perception that earthquakes are *random unpredictable phenomena*.

It is common knowledge that usually forecast/prediction of extreme events is not an easy task: By definition, an extreme event is a rare one in a series of kindred phenomena. Generally speaking, it implies investigating a small sample of case histories with a help of delicate statistical methods applied to data of different quality collected in various conditions. Many extreme events are correlated and/or clustered being apparently far from independent and follow some "strange" distribution like a mono- or multi-fractal that is hardly a uniform one. Evidently, such an "unusual" situation complicates search and definition of precursory behaviors to be used for forecast/prediction purposes.

Moreover, making forecast/prediction claims quantitatively probabilistic in the frames of the most popular objectivists' viewpoint on probability requires a long series of "yes/no" forecast/prediction outcomes, which cannot be obtained without an extended rigorous test of the candidate method. The set of errors of types I and II, in particular, "success/failure" scores and space-time measure of alarms, and other additional information obtained in such a test supply researchers with data necessary to judge the candidate's potential as a forecast/prediction tool and, eventually, to find its improvements. Obviously, this is to be done first in comparison against random guessing that results do permit evaluation of confidence measured in terms of statistical significance.

Note that an application of the forecast/prediction tools could be very different in a variety of different costs and benefits, and, therefore, requires determination of optimal strategies. Specific costs and benefits may suggest modifying the forecast/prediction tools for more adequate "optimal" applications.

Evidently, all these general considerations apply to prediction of seismic extremes, i.e., strong catastrophic earthquakes. The perception of earthquakes' occurrence as a model random process was formed naturally at the times when comprehensive catalogs of earthquakes were absent or too short for any scientifically reliable conclusion (Kanamori, 1981). Nowadays the situation has changed. The data on earthquake occurrences and sizes accumulated over the period of instrumental record of about a century provide the necessary base for tackling the earthquake prediction problem from the prospective of a knowledgeable statement in advance of seismic events.

Almost 50 years ago Charles Richter has written a one-third page note (Richter, 1964) commenting on an early, and possibly the first quantitatively described observation of a general increase in seismic activity prior to large earthquakes, pattern Σ (Keilis-Borok and Malinovskaya, 1964). He noted "a creditable effort to convert this rather indefinite and elusive phenomenon into a precisely definable one", marked as important a confirmation of "the necessity of considering a very extensive region including the center of the approaching event", and outlined "difficulty and some arbitrariness, as the authors duly point out, in selecting the area which is to be included in each individual study". However at that time, as mentioned above, the information database for earthquake prediction research remained sparse and fragmentary and, therefore, did not allow meaningful testing of hypotheses in any systematic way about phenomena that were claimed to be precursory to large earthquakes.

18.2 DEFINITION AND CLASSIFICATION OF EARTHQUAKE PREDICTIONS

The United States National Research Council, Panel on Earthquake Prediction of the Committee on Seismology suggested the following definition (Allen et al., 1976, p. 7):

An earthquake prediction must specify the expected magnitude range, the geographical area within which it will occur, and the time interval within which it will happen with sufficient precision so that the ultimate success or failure of the prediction can readily be judged. Only by careful recording and analysis of failures as well as successes can the eventual success of the total effort be evaluated and future directions charted. Moreover, scientists should also assign a confidence level to each prediction.

We follow this definition as a consensus necessary precondition of any scientific prediction study; although we recognize that a confidence level could be assigned rather to a method than to a single specific prediction. We also recognize that predictions could be different in their accuracy related to magnitude range, geographical area, and time interval.

Recently, the International Commission on Earthquake Forecasting (Jordan et al., 2011) has suggested the following distinction of prediction and forecast:

A prediction is defined as a deterministic statement that a future earthquake will or will not occur in a particular geographic region, time window, and magnitude range, whereas a forecast gives a probability (greater than zero but less than one) that such an event will occur.

Jordan et al., 2011, p. 319

However, when coming to practical application, the same authors advise as follows:

Recommendation G2: Quantitative and transparent protocols should be established for decision-making that include mitigation actions with different impacts that would be implemented if certain thresholds in earthquake probability are exceeded.

Jordan et al., 2011, p. 363

Apparently, according to their definition and recommendation Jordan et al. (2011) guide us to make a conclusion that forecasting may become useful when and only when formulated as prediction. Note that any prediction (including earthquake predictions discussed in this chapter) can get an attributed number "greater than zero but less than one" that may or may not characterize the uncertainty of its deterministic statement. In the absence of a solid theory of earthquake probabilities there is no scientific merit to distinguish forecasts and predictions. Therefore we consider probabilistic forecasting as a fuzzy, uncertain prediction. On the other hand, earthquake statistics and their scaling are rather instructive for classification of predictions.

First of all, the famous Gutenberg—Richter relationship suggests limiting magnitude range of prediction to about one unit. Otherwise, the statistics of outcomes would be essentially related to dominating smallest earthquakes and may be misleading when attributed to the largest earthquakes of the targeted magnitude range. This is usually ignored. Moreover, the intrinsic uncertainty of earthquake sizing (see e.g., Bormann, 2012) allows self-deceptive picking additional cases of justification "just from below" the magnitude range of target earthquakes. Of course, these might be important encouraging evidence but, by no means, can be a "helpful" additive to the statistics of a rigid testing.

Usually, earthquake prediction is classified in respect to duration of expectation time while overlooking term-less identification of earthquake-prone areas. This basic zero step of hierarchical classification is of particular importance for strategic hazard assessment as well as for testing temporal predictability of catastrophic earthquakes. An outstanding example of term-less prediction by pattern recognition is given in Figure 18.1. Gelfand et al. (1976) defined 73 D-intersections of morphostructural lineaments in California and Nevada as earthquake-prone for magnitude 6.5+ events. Since the date of publication their prediction has been confirmed by all the 15, magnitude 6.5+ earthquakes including the recent most April 4, 2010, M7.2 Baja California earthquake. Each of these earthquakes occurred in a narrow vicinity of the D-intersections (union of yellow circles in Figure 18.1). Moreover, Kossobokov (2012) has reminded that the Puente Hills thrust fault was "predicted" (coincides exactly with the lineament drawn in 1976) decades in advance it was "rediscovered" by the 1995 Northridge earthquake (Shaw and Shearer, 1999).

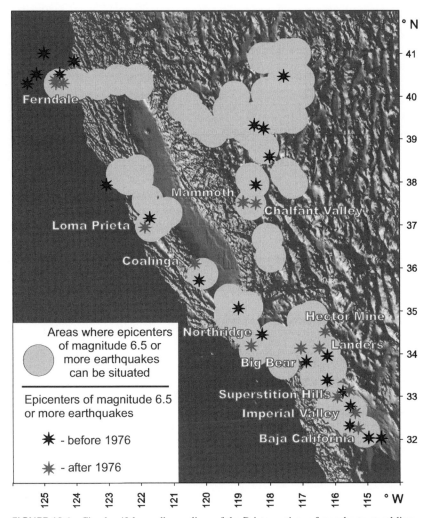

FIGURE 18.1   Circular 40-km radius outlines of the D-intersections of morphostructural lineaments in California and Nevada and epicenters of magnitude 6.5+ earthquakes before (black stars)
and after (red stars with names) publication of (Gelfand et al., 1976).

Contrary to the pattern recognition approach to term-less earthquake prediction the probabilistic ones fail in assessing earthquake hazard. After
the disastrous January 12, 2010, M7.3 Port-au-Prince, Haiti earthquake,
(Kossobokov and Nekrasova, 2010, 2012) have shown that Global Seismic
Hazard Assessment Program (GSHAP) maps are misleading and, in fact, do
not predict location of strong earthquakes. Instead of a promised "10 percent
of exceedance in 50 years" (Giardini et al., 1999) the maps have not anticipated the observed ground shaking at epicenters of about 50 percent of 1,181

strong, magnitude 6.0 or larger earthquakes in 2000–2009. The percentage of the GSHAP errors grows with magnitude, so that shaking at each epicenter of magnitude 7.5 or larger earthquakes (same as for the top dozen deadliest earthquakes) in 2000–2009 was violating GSHAP prediction with the average underestimation of about 2 units of macroseismic intensity. Emphasizing the need for objective testing Stein et al. (2011) have demonstrated erroneous character of the regional, presumably, improved version of probabilistic earthquake hazard assessment map for Japan.

Regretfully, the spatial accuracy of an earthquake prediction method is usually not taken into consideration (Jordan et al., 2011). The forecasts are often made for a "cell" (Schorlemmer et al., 2010; Lee et al., 2011) or "seismic region" (McCann et al., 1979; Kagan and Jackson, 1991, 1995) whose area is not linked to the size of the target earthquake. This might be another source for making a wrong choice in parameterization of a forecast/prediction method and, eventually, for unsatisfactory performance in real-time applications.

Summing up, prediction of time and location of an earthquake of a certain magnitude range can be classified into the categories listed in Table 18.1 according to its temporal and spatial accuracy. Note that a wide variety of possible combinations exist that is much larger than the usually considered "short-term exact" one. In principle, such an accurate statement about anticipated seismic extreme might be futile due to the complexities of the Earth's lithosphere, its blocks-and-faults structure, and evidently nonlinear dynamics of the seismic process. The observed scaling of source size and preparation zone with earthquake magnitude (Dobrovolsky et al., 1979) implies exponential scales for territorial accuracy of predictions similar to the temporal ones. Naturally, the spatial accuracy of prediction is linked to the source zone linear dimension, l. It varies from exact pinpointing the source to long-range uncertainty of about a few tens of l.

From the viewpoint of such a classification, the earthquake prediction problem might be approached by a hierarchical, step-by-step refinement technique, which accounts for a multiscale dynamic escalation of seismic

TABLE 18.1 Classification of Earthquake Prediction Accuracy

Temporal, in Years		Spatial, in Source Zone Size l	
Long-term	10	Long-range	Up to 100
Intermediate-term	1	Middle-range	5–10
Short-term	0.01–0.1	Narrow	2–3
Immediate	0.001	Exact	1

activity to the main rupture (Keilis-Borok, 1990; Kossobokov et al., 1990, 1999b). Such a technique (Kossobokov and Shebalin, 2003) starts with recognition of earthquake-prone zones for earthquakes of a number of magnitude ranges, then follows with a dynamic determination of long- and intermediate-term areas and times of increased probability (TIPs), and, finally, may come out with an exact short-term or even immediate alert.

The earthquake prediction algorithms described below are in complete agreement with the consensus definition (Allen et al., 1976) and essentially provide predictions of at least intermediate-term, middle-range accuracy. In contrast, probability mappings like those originally accepted for testing at Collaboratory for the Study of Earthquake Predictability (http://www. cseptesting.org/; Jordan, 2006) are not earthquake predictions in this sense. As pointed out by Jordan et al. (2011) an earthquake probability model may become useful for practical applications, when someone specifies exactly the probability cutoff and the expected magnitude range for a given mapping so as to produce a rigid prediction statement.

18.3 EARTHQUAKE PREDICTION ALGORITHMS M8 AND MSC

Scholz (1997) describes predicting earthquakes as an easy task of: (1) deployment of precursor detection instruments at the site of expected earthquake; (2) recognition of the precursors; and (3) publicly predicting the earthquake through approved channels. We do it systematically since 1984, although in a clear understanding that: (1) some "precursor detection instruments" are already deployed worldwide and their record is available for general use; (2) some "precursors" are already recognized and can be detected; and, finally, (3) over the last 25 years many strong earthquakes have been "publicly predicted". The most difficult of completing "an easy" task remains convincing "all your colleagues to agree" on predictability of earthquakes (Ismail-Zadeh and Kossobokov, 2011).

In our practice, a routine nowadays, we make use of the USGS/NEIC Global Hypocenters' Data Base System, which Preliminary Determinations of Epicenters (PDE) and Quick Earthquake Determinations (QED) provide an up-to-date global catalog of seismic events, and adjust our precursor detection algorithms to the different level of completeness in PDE and QED.

Our earthquake prediction algorithms are based on a simple general scheme illustrated in Figure 18.2: a seismically active territory is covered with a sample of areas, usually, circles of investigation (CIs) (a), each area has its own "history" of seismic events of different magnitude (b), each seismic "history" is described in terms of specified precisely definable moving counts (c), which combination is subject to pattern recognition of "precursor" signifying whether or not the incoming period is a TIP, for the occurrence of anticipated target earthquake (d).

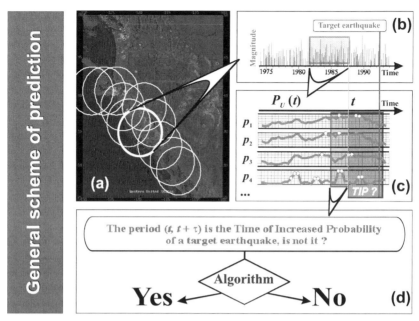

FIGURE 18.2 General scheme of an earthquake prediction tool. See text for description of sub-parts a–d.

For more than 25 years we use the M8 algorithm (now available from IASPEI Software Library—Kossobokov, 1997) as an intermediate-term, middle-range approximation. This earthquake prediction method was designed by retroactive analysis of dynamics of seismic activity preceding the great, magnitude 8.0 or more, earthquakes worldwide, hence its name. Its prototype (Keilis-Borok and Kosobokov, 1984) and the original version (Keilis-Borok and Kossobokov, 1987) were tested first retroactively, then prospectively. The original version of M8, explicitly defined in (Healy et al., 1992; Kossobokov, 1997), is subject to the ongoing real-time experimental testing started officially in 1992.

M8 algorithm. Prediction is aimed at earthquakes from magnitude range $MM_0+ = (M_0, M_0 + \Delta m)$, where $\Delta m < 1$. The earthquake magnitude scale we use reflects the size of earthquake sources. Overlapping CIs of the fixed diameter $D(M_0)$ scan seismic locus in the region under study. The sequence of earthquakes with aftershocks removed is considered within each CI. Sequences in different CIs are normalized to about the same prefixed average annual number of earthquakes $\overline{N}$ by selecting the lower magnitude cutoff $\underline{M} = M_{min}(\overline{N})$.

For a given sequence several functions are computed in the trailing time window $(t - s, t)$ and magnitude range $(\underline{M} \leq \text{magnitude} < M_0)$. These functions include: (1) the number of earthquakes $N(t)$ of magnitude $\underline{M}$ or greater in time window $(t - s, t)$; (2) the deviation of $N(t)$ from longer-term trend, $L(t)$; (3) linear concentration $Z(t)$ estimated as the ratio of the average source diameter to the average distance between sources; and (4) the maximum number of aftershocks $B(t)$. Each of the functions N, L, and Z is calculated twice with $\underline{M} = M_{\min}(\overline{N})$ for $\overline{N} = 20$ and $\overline{N} = 10$. As a result, the earthquake sequence is given a robust description by seven functions $N1$, $N2$, $L1$, $L2$, $Z1$, $Z2$, and B. "Anomalously large" values are identified for each function using the condition that they are higher than $Q\%$ of the encountered values. An alarm or a TIP is diagnosed for τ years from the moment of time t when at least six out of seven functions, including B, show up "anomalously large" values within a narrow time window $(t - u, t)$. To make prediction more stable this condition is required for two consecutive moments, $t - 0.5$ and t years. In course of a real-time monitoring, the alarm may extend beyond or be terminated before τ years in case the updating causes changes in determination of the magnitude cutoffs and/or the percentiles of the encountered functions.

The following standard values of parameters indicated above are prefixed in the algorithm M8: $D(M_0) = \{\exp(M_0 - 5.6) + 1\}^{\circ}$ in degrees of meridian (this is 384, 560, 854, and 1333 km for $M_0 = 6.5$, 7.0, 7.5, and 8, respectively, about 5–10 times the length of the target earthquake source), $s = 1$ year for B and 6 years for the other six functions, $u = 3$ years, $Q = 75$ percent for B and 90 percent for the other six functions, and $\tau = 5$ years. Usually, the average diameter of the source, l, is estimated by $\frac{1}{N} \sum_{\{i\}} 10^{\beta(M_i - \alpha)}$, where N is the number of main shocks in $\{i\}$, $\beta = 0.46$ to meet the condition of proportionality to the linear dimension of source, and $\alpha = 0$ (which does not restrict generality), while the average distance, r, between them is set proportional to $\sqrt[3]{\frac{1}{N}}$. The usage of more accurate estimate of the linear concentration of main shocks may improve the performance of the algorithm. The ultimate unambiguous description of the M8 algorithm with all the prefixed parameters and rules of data processing is published as a computer code in the IASPEI Software Library (Kossobokov, 1997).

From a general viewpoint (Figure 18.3), the algorithm M8 uses traditional description of a dynamical system adding to a common phase space of rate (N)

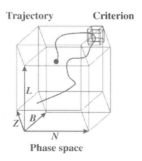

Trajectory   Criterion

Phase space

FIGURE 18.3 The M8 algorithm criterion in the seismic expanded normalized phase space.

and rate differential (*L*) dimensionless concentration (*Z*), and a characteristic measure of clustering (*B*). The algorithm recognizes criterion, defined by extreme values of the phase space coordinates, as a vicinity of the seismic dynamical system singularity. When the trajectory of an area of investigation enters the criterion, the probability of an extreme event increases to the level sufficient for its effective provision. By analogy with a delay time from maximum entropy to catastrophe in dynamical systems, the probability of an extreme event may remain that high for some time after the trajectory departs from the criterion.

The middle-range accuracy of the M8 algorithm is far from ideal. Therefore, the Mendocino Scenario algorithm for reducing the area of alarm was designed (Kossobokov et al., 1990) by retroactive analysis of the detailed regional seismic catalog prior to the Eureka earthquake (1980, *M* = 7.2) near Cape Mendocino in California, hence its name abbreviated to MSc.

MSc algorithm. Given a TIP diagnosed for a certain territory **U** at the moment **T**, the algorithm is designed to find within **U** a *smaller* area **V**, where the predicted earthquake can be expected. An application of the algorithm requires a reasonably complete catalog of earthquakes with magnitudes $M \geq (M_0 - 4)$ (where M_0 is the minimum magnitude of the target event), which is lower than the minimal threshold usually used by M8. In case this condition is not fulfilled, we assume that the dynamics of earthquakes available in the database inherit the behavior from the lower levels of seismic hierarchy. The detection of the MSc criteria in such a case is more difficult, if possible, and might result in additional failures to predict.

The essence of MSc can be summarized as follows. Territory **U** is coarse-grained into small squares of $s \times s$ size. Let (i,j) be the coordinates of the centers of the squares. Within each square (i,j) the number of earthquakes $n_{ij}(k)$, aftershocks included, is calculated for consecutive, short time windows, *u* months long, starting from the time $t_0 = (\mathbf{T} - 6$ years) onward, to allow for the earthquakes that contributed to the TIP's diagnosis; *k* is the sequence number of a time window. In this way the timespace considered is divided into small boxes (i,j,k) of the size $(s \times s \times u)$. *"Quiet" boxes* are singled out for each small square (i,j); they are defined by the condition that $n_{ij}(k)$ is below the *Q* percentile of n_{ij}. The clusters of *q* or more quiet boxes connected in space or in time are identified. Area **V** is the areal projection of these clusters.

The standard values of parameters adjusted for the case of the 1980 Eureka earthquake are as follows: *u* = 2 months, *Q* = 10 percent, *q* = 4, and *s* = 3*D*/16, *D* being the diameter of the circle used in algorithm M8.

Qualitatively, the MSc algorithm outlines an area of the territory of alarm, where the activity, from the beginning of seismic inverse cascade recognized by the first approximation prediction algorithm (e.g., by the M8 algorithm), is continuously high and infrequently drops for a short time, i.e., characteristic of *intermittency*. Such an alternation of activity must have a sufficient temporal and/or spatial span. The phenomenon, which is used in the MSc algorithm, might reflect the second, possibly, shorter-term and, definitely, narrow-range stage of the premonitory rise of seismic activity near the incipient source of the main shock. By this analogy, the M8 algorithm diagnoses loss of stability at an early stage, when the behavior of the equilibrium state is replaced by oscillations and period doubling, and the MSc algorithm determines the stage when "a cycle loses its skin" forming a strange attractor.

Thus, in our practice, the M8 algorithm provides intermediate-term prediction in the first, middle-range, approximation, and the MSc algorithm, if the data permit, narrows down the area covered by the alarm to narrow or even define the exact location of the incipient target earthquake. Naturally, both apply to the null approximation delivered by identifying earthquake-prone zones, e.g., in terms of "seismic regions", "active fault zones", "D-intersections or knots".

18.4 REAL-TIME PREDICTIONS BY THE M8-MSC ALGORITHMS

The 1988 Spitak (Armenia), M6.8 earthquake was the first tragic confirmation of the potential efficiency of the M8-MSc monitoring achieved in the real-time prediction mode (Kossobokov, 1986; Keilis-Borok and Kossobokov, 1988). The CI centered at coordinates 42° N and 45° E was the only one out of ten CI's covering the entire Caucasus, where an M8 TIP targeting magnitude range of M6.5+ has been diagnosed in July 1988. The MSc algorithm has identified in the second narrow-range approximation the 96 by 120 km rectangle. The epicenters of the December 7 main shock and its first aftershocks filled about a quarter of the M8-MSc prediction area.

The results of the experimental monitoring of the former Soviet Union seismic regions (1986—1990) were encouraging: six out of seven target large earthquakes were predicted at the M8 approximation. Regretfully, the collapse of the Soviet Union has discontinued most of this series of real-time experiments.

The prediction of the 1989 Loma Prieta (California), M7.1 earthquake (Kossobokov, 1986; Keilis-Borok et al., 1990) did stimulate a design of the global and regional test experiments in earthquake-prediction. The M8 algorithm TIP targeting M7.0+ in California has emerged in 1985 with a single alerted CI out of the eight considered. The area of alarm expanded to the south in 1988 when another TIP was diagnosed. At the beginning of 1989 the data permitted the MSc algorithm to outline the reduced area of alert to two rectangular shapes, one of which pointing on Mammoth Lakes was disregarded later when QED was substituted with PDE data. The epicenter of October 18 (17), 1989, Loma Prieta earthquake did confirm the M8-MSc prediction centered at a segment of San Andreas fault behind Monterey Bay. It should be noted that this alarm was among other predictions covering the magnitude range from 6.4 to 8 presented by Prof. Vladimir I. Keilis-Borok to the National Earthquake Prediction Evaluation Council (NEPEC), in May 1989. The predictions were subject of discussions at a specially dedicated meeting of NEPEC (Updike, 1989) and resulted in a recommendation to the United States Geological Survey to perform a systematic testing of the Soviet earthquake prediction methods, starting with the apparently simplest and then most documented algorithm M8.

18.5 GLOBAL TEST OF THE M8-MSC PREDICTIONS

Following the recommendation of NEPEC John H. Healy (USGS, Menlo Park), James W. Dewey (USGS/NEIC, Golden), and Vladimir G. Kossobokov (RAS, Moscow) collaborated in setting up a decisive test in a real-time application of the M8 algorithm (Healy et al., 1992). By 1992 all the components necessary for such a reproducible real-time prediction experiment, i.e., an unambiguous definition of the algorithms and the database, were specified in publications. These components are:

- Algorithm M8 (Keilis-Borok and Kossobokov, 1984, 1987, 1990);
- Algorithm MSc (Kossobokov et al., 1990); and
- The National Earthquake Information Center *Global Hypocenters Data Base* (*1989*) sufficiently complete since 1963.

This allowed a systematic application of the M8 and MSc algorithms since 1985, when requiring 12 years of data for a stable evaluation of the seven functions contained in the algorithm and additional 10 years to form a reasonable evaluation of their percentiles for determination of "anomalously large" values. For the catalog preprocessing, which includes elimination of possible duplicates and identification of aftershocks, we used the algorithm designed by P. Shebalin (Shebalin, 1992). The results of retrospective simulation of prediction in advance by the M8 algorithm targeting M7.5+ earthquakes in the Circum Pacific area, 1985−1990 were encouraging. Moreover, in the course of preparation of the US Open-File Report (Healy et al., 1992) the real-time update on January 1991 issued the M8 prediction for M7.5+ for the territory of Costa Rica and Panama. The April 22, 1991 Limon (Costa Rica), M7.6 earthquake confirmed this prediction.

Since 1992 the input catalog and M8-MSc predictions were updated every 6 months with up to 1-month delay, and the target earthquake occurrences were testing the hypothesis of random guessing proportionate to the empirical seismic rates in different regions worldwide, i.e., *Seismic Roulette* null hypothesis.

Seismic Roulette. When using the literal measure of territory in km^2 one may overestimate statistical significance of the obtained results by equalizing the areas of high and low seismic activity, at the extreme, areas where earthquake happens and do not happen. The actual, empirical distribution of earthquake locations is the best present-day knowledge estimate of where earthquakes may occur. Our recipe of using earthquake-oriented measure μ, and counting probability p of random guessing is the following: Choose a sample catalog representative of seismic locus. Count how many events from the catalog are inside the territory considered; this will be your denominator. At a given time, count how many events from the catalog are inside the area of alarm; this will be your numerator. Integration of the average ratio over the time of prediction experiment provides the estimate of an a priori probability p. The significance level of the prediction results can be estimated as $B(n, N, p)$, where B is the cumulative binomial distribution function that provides the probability of observing n or fewer successes in N trials, with the probability of success on a single trial denoted by p. The confidence level of the prediction results is dual to the significance level and equals $1 - B(n - 1, N, p)$. The higher is the confidence level achieved in a test, the lower is the probability of getting by pure chance an outcome as good as or better than in the test.

This simple recipe has a nice analogy that justifies using statistical tools available since Blaise Pascal (1623−1662).

Seismic Roulette: Consider a roulette wheel with as many sectors as the number of target events in your sample earthquake catalog, a sector for each event. Make your bet according to prediction: determine which events are inside area of alarm, and put one chip in each of the corresponding sectors. Nature turns the wheel.

If seismic roulette is not perfect, one can win systematically. This may require a switch from the original algorithm that loses systematically to its "antipodal" version (Molchan, 2003).

As a result of the Global Test of the M8-MSc predictions, the worldwide performance of earthquake prediction algorithms M8 and M8-MSc can be characterized by the following Table 18.2.

For both magnitude ranges, the significance level estimates are far below 1 percent and, therefore, suggest rejecting the hypothesis of random guessing with confidence higher than 99 percent.

Thus, having in mind the intrinsic seismic uncertainties we may conclude with certainty that *Seismic Roulette* is not perfect. It should be stressed that: (1) we use the most conservative measure of the alarm volume accounting for empirical distribution of epicenters per se (without any hypothetical expansion to a subjectively chosen *parent universe*); and (2) to drive any of the estimates of confidence below 95 percent, the Global Test should encounter at least 10 failures to predict without a single successful confirmation in any of the magnitude ranges M8.0+ or M7.5+.

Error diagram. Besides statistical significance, the important characteristic of a prediction method is its effectiveness measured by costs and benefits. Molchan (1997, 2003) has introduced a theory of prediction strategies based on statistical analysis of the prediction errors of the two types. In a nutshell, the lower envelope, Γ, of the error points set $\{n^o, \tau^o\}$, where n^o is the ratio of

TABLE 18.2 Worldwide Performance of Earthquake Prediction Algorithms M8 and M8-MSc

| | | | | Large Earthquakes | | | | |
| | | | | Measure of Alarms, % | | Confidence Level, % | |
Test Period	Total	M8	M8-MSc	M8	M8-MSc	M8	M8-MSc
Magnitude 8.0+							
1985–present	21	16	10	32.84	16.62	99.99	99.90
1992–present	19	14	8	29.80	14.78	99.99	99.63
Magnitude 7.5+							
1985–present	68	40	16	28.73	9.32	99.99	99.96
1992–present	56	30	10	23.14	8.31	99.99	98.36

Note: Confidence level tells how sure one can be that the achieved performance is not arisen by chance.

failures to predict to the total number of targets and τ^o is the ratio of alarm time to the total period considered, characterizes the effectiveness of the predictions. Denoted as *error diagram*, Γ demonstrates how far from a random guessing are the predictions resulting from the algorithm. A trade-off between n^o and τ^o depends on a choice of adjustable parameters of the prediction algorithm and a loss function $\gamma = \gamma(n^o, \tau^o)$, which balance costs and benefits. It may be very different when different preparedness problems and corresponding measures are considered in response to prediction. The point where γ and Γ touch each other determines both the minimal achievable loss and the optimal set of adjustable parameters of the prediction method.

The error diagram Γ is a desirable signature of the prediction method. Even a single error point, A, attributed to some prediction method implies Γ to be the broken line connecting the three points O (0, 1), A (n^o, τ^o), and P (1, 0) that can be used to define the optimal strategy. On the other hand, constructing the lower envelope Γ assumes extended variation of the adjustable parameters and therefore delivers a demonstration of the level of stability of the algorithm in retrospective analysis. It should be noted also that in practical applications error diagrams must account for a finite number of targets available (Kossobokov, 2006) as well as for an appropriate measure of space-time (Molchan, 2010).

Molchan and Romashkova (2010) have performed an exhaustive independent analysis of the M8 algorithm predictions targeting M8.0+ in 1985−2009. Their estimates of significance were obtained after a substantial variation of the magnitude scales, spatial measure and regionalization are slightly different than that presented here due to a huge margin of reserved uncertainty and an apparent overestimate of the number of degrees of freedom when splitting space into a set of nonintersecting areas. Nevertheless, Molchan and Romashkova (2010) conclude that their "results argue in favor of nontriviality of the M8 prediction algorithm".

Figure 18.4 shows an example of semi-annual maps that summarize the M8-MSc predictions for a current half-year period. We take liberty of picking up the one from the password protected Web site (http://www.mitp. ru/en/restricted_global/2010a/2010am8.html) that displays the global situation as on January 1, 2010 in advance the February 27, 2010 offshore Maule, Chile M8.8 mega-thrust (http://www.mitp.ru/en/predictions/2010feb27.html). Although the M8-MSc predictions are intermediate-term, medium-range, and by no means imply "red alert", some colleagues have expressed a legitimate concern about maintaining necessary confidentiality. That is why the up-to-date predictions are posted on a Web site of restricted access provided to Test Observers (their number has grown from 100 members in 1999 to 200 in 2013). The format of this publication does not allow describing in detail the complete collection of all the M8-MSc prediction outcomes issued in the Global Test. Therefore, we discuss just a few typical case histories that

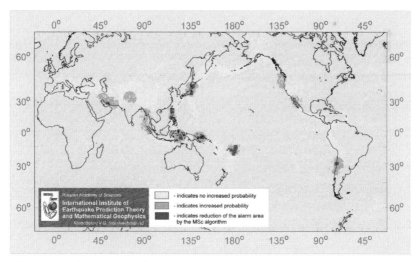

FIGURE 18.4   Global testing of the M8-MSc predictions: Regions of increased probability of magnitude 8.0+ as on January 1, 2010. Note: The diagnosis of TIP in circles of investigation is color-coded: yellow corresponds to absence of alarm; dark yellow, presence of alarm; red highlights most probable areas of occurrence according to the MSc algorithm. Predictions are shown through the mask of the global distribution of seismicity (and, of course, do not refer to the areas where earthquakes are not recorded in the period of instrumental observations) and limited to areas where the available information on earthquakes is enough to apply the standard version of the M8 algorithm. TIP, time of increased probability.

include confirmed and false diagnosis of TIPs as well as failures to predict target earthquakes:

An example of confirmed prediction—October 4, 1994 Shikotan, M8.3 earthquake. From the end of the 1970s to about 1990, the region of the southern Kuril Islands was in a steady state of seismic activity (Kossobokov et al., 1999a). In 1991, a swarm of earthquakes, seven of which were of magnitude 6 or more, lasted for 2 weeks and was followed by a magnitude 7.4 earthquake offshore Urup Island. This rise of seismic activity was first recognized by the M8 algorithm in July 1992 as a TIP for magnitude range M7.5+ in the region of the Eastern part of Hokkaido and the southern Kuril Islands. The first major earthquake confirming the prediction occurred on January 15, 1993 M7.6, in Kushiro-Oki (Hokkaido) at a depth of 102 km. Furthermore, in January 1994 the M8 algorithm, run to predict M8.0+ earthquakes, determined another TIP, now for a great earthquake in the region (Figure 18.5). The Shikotan Island M8.3 earthquake that followed on October 4, 1994 fitted exactly the temporal, territorial, and magnitude ranges of the predictions updated in July 1994 (using data through the first half of 1994). It is notable that algorithm MSc pinpointed the location of the October 4 event providing eventually the exact spatial accuracy of prediction: the refined

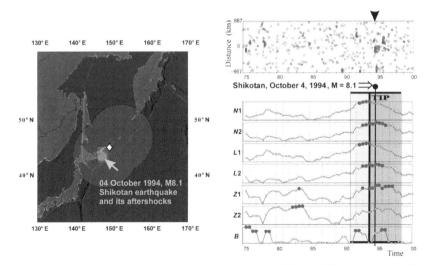

FIGURE 18.5   Global testing of the M8-MSc predictions, M8.0+: The October 4, 1994 Shikotan earthquake. Note: The highlighted circular areas of alarm in the first approximation determined by algorithm M8 and rectangular areas of alarm in the second approximation determined by algorithm MSc are shown on the right. A remarkable swarm of seven magnitude 6 earthquakes (diamond) occurred about a month in advance of the 1994 great shock in the southern Kuril Islands. On the right, space-time occurrences of earthquakes in the circle of investigation in projection on plate boundary are given on top of the M8 seven function's plots. "Anomalously large" values are marked with solid dots, TIP—with dark gray rectangle, and the origin time of the 1994 great earthquake—with the vertical line. TIP, time of increased probability.

210 × 160 km area of alarm coincides remarkably with the aftershock zone of the Shikotan earthquake. Moreover, in August 1994, about 5 weeks in advance the great shock a swarm of seven magnitude 6.0+ earthquakes (marked with diamond in Figure 18.4) occurred in about 200 km to the northeast of its future epicenter.

The same TIP was confirmed in progress when using data through the first half of 1995, whereas the MSc algorithm second approximation migrated to the northeast and expanded slightly to the size of 210 × 210 km, most of which hosted the epicenter and the first aftershocks of the December 3, 1995 M8.0 Iturup earthquake (Kossobokov et al., 1999a). Similar to the 1994 Shikotan earthquake, the 1995 Iturup, one has a remarkable model example of a clear foreshock sequence; it started on November 24, 1995 with a magnitude 6.6 and escalated until the very main shock on December 3. This swarmlike sequence consists of 64 events of magnitude 4 or above, of which 19 have $M \geq 5$ and four have $M \geq 6$.

Failures to predict. Although a comprehensive analysis of the M8-MSc predictions in the Global Test is not yet performed, it appears that failures may have systematic characters. In particular, we observe the emerging two types of failures to predict illustrated in Figure 18.6: So far, all the five M8.0+ earthquakes that were not predicted in the course of the Global Test are either in the area of scoring near but lower than critical (June 23, 2001; August 15, 2007) or in the chain of correlated earthquakes (Keilis-Borok et al., 2004) connected with the M8-MSc prediction (September 25, 2003; November 15, 2006; January 13, 2007).

Both kinds of failures to predict reflect the physically continuous character of seismic hazard distribution, which was intentionally violated in the design of

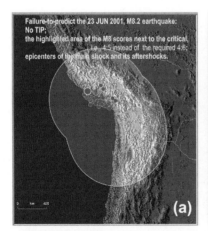
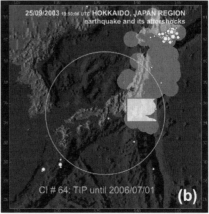

FIGURE 18.6   Global testing of the M8-MSc predictions, M8.0+: The examples of the two kinds of failures to predict—(a) scores lower than next to required for the diagnosis of TIP (left) and—(b) connected to the M8-MSc TIP location by a chain of correlated earthquakes (right). TIP, time of increased probability.

the algorithms subject to rigid unambiguous testing. Naturally, different ways exist for transforming the binary diagnosis of M8 and MSc into a continuous fuzzy measure of dynamically changing hazard, for example, such as was done by David Vere-Jones and coauthors (Harte et al., 2003). In addition, the extent of the preparation zone for a target earthquake estimated by Dobrovolsky et al. (1979) is about 3—4 times larger than the diameter $D(M_0)$ of a CI used in application of the M8 algorithm. Therefore, another possibility exists for introducing a gradual decay of alarm level to the temporal limit and outside its area, which could define a probabilistic forecasting. We do not do it for the reason of yet a small sample size that may help understanding an appropriate shape of seismic hazard distribution within the earthquake preparation area.

False diagnosis of TIPs. A similar character of physical continuity can explain that a number of TIPs expired without any target earthquakes, i.e., so-called "false alarms". For example, the July 9, 1994 Great Deep Bolivia M8.2 earthquake occurred more than 600 km away from the Pacific coast just outside the Global Test area, but next to a cluster of TIPs, which was in effect from the beginning of the Test in 1992 until the middle of 1995 when it was declared a "false alarm".

Another cluster of "false" TIPs in Japan lasted from the middle of 2001 through 2010 gradually migrating from southwestern to the northern regions. This could have been associated with the abovementioned failure to predict the September 25, 2003 Kushiro-Oki (Hokkaido, Japan) M8.3 earthquake and a series of earthquakes that started with the July 28, 2002 M7.3 deep event (depth 566 km) near Priamurye-Northeastern China border in the back of and outside the alerted section of subduction zone, followed by 12 shallow magnitude 7.0 or larger earthquakes in the area alerted in 2002—2010, and ended with the August 9, 2009 M7.1 deep earthquake (depth 292 km) beneath Izu Islands. Also we cannot exclude the possibility of a Tokai silent earthquake, initiated in 2001 that lasted for many years in the middle of this cluster of "false alarms", as a physically related phenomenon. (The area of this cluster of TIPs in the first half of 2010 can be found in Figure 18.4.)

The case history of this "false alarm" in Japan has continued with the March 11, 2011, M9.0 mega-thrust earthquake, which occurred off the Pacific coast of Tohoku region (Kossobokov, 2011). The Tohoku earthquake happened on the 70th day after the TIP was canceled. Its first aftershocks (white dots in Figure 18.7) filled one of the two M8-MSc prediction areas from its northeast to southwestern limit. The devastating mega-earthquake was preceded by a series of earthquakes starting with the March 09, 2011, Sankiru-Oki, M7.3 earthquake and signifying short-term forerunners some 51 h in advance of the main shock.

Premature removal of the TIP with tentative expiration date on July 01, 2011, as a result of regular updating of the Global Test predictions by algorithms M8 and MSc in January 2011, was caused by the fact that the threshold value of one of the seven functions used in declaring an alarm (specifically, Z1) had decreased slightly (by 1.5 percent of its value). Such a marginal change in

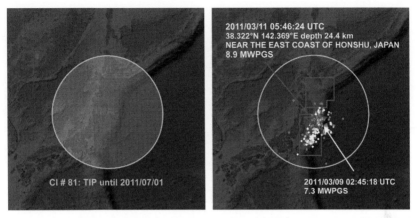

FIGURE 18.7 Global testing of the M8-MSc predictions, M8.0+: The March 11, 2011 off the Pacific coast of Tohoku region, Japan M_w 9.0 mega-earthquake. Note: On the left, the circle of investigation where on July 1, 2010 the M8 algorithm has diagnosed TIP until July 1, 2011 (yellow) and the areas determined by the MSc algorithm (red). This TIP was canceled prematurely in the update dated January 2011. On the right, the epicenters of the March 11, 2011 mega-thrust (big star), its first aftershocks (white dots), as well as of the March 9, 2011 earthquake (small blue star) and its aftershocks (blue dots) are plotted on the background of the July 2010 M8-MSc prediction outlines. TIP, time of increased probability.

one of the seven graphs involved in the diagnosis of TIP (Figure 18.8) could hardly be observed with a naked eye. However, the "black box" code of the M8 algorithm determined the change in scores of abnormally high values of functions from the required pair of six and six (including *B*) actual in July 2010 to six and five, and canceled the alarm. Note that, formally, the 2011 Tohoku mega-thrust is outside the M8.0+ range considered in the Global Test of M8-MSc predictions.

These case histories of "confirmed predictions", "failures to predict", and "false alarms" should not be considered as an attempt to exclude some errors from overall statistics of performance. To the contrary, we would like to demonstrate the imperfections that indicate the complexity inflicted by the fractal distribution of faults and the dynamics of the earthquake process. The richness of the observed patterns in seismic activity suggests finding further improvements in-line with joint approaches based on hierarchical, step-by-step techniques that account for multiscale escalation of seismic activity to the main rupture (Keilis-Borok, 1990; Kossobokov et al., 1990). On the other hand the improvements may benefit from using continuous fuzzy measures of time-dependent seismic hazard (Harte et al., 2003). Such a joint "all-shades-of-gray" dynamic hazard assessment may start with the recognition of earthquake-prone zones for earthquakes from a number of magnitude ranges, and then follow with determination of long- and intermediate-term areas and TIPs, and, finally, come out with an exact short-term or immediate alert. This may allow dynamic assessment and mitigation of seismic risks.

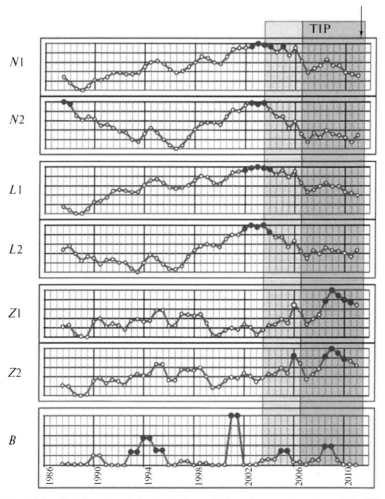

FIGURE 18.8   Global testing of the M8-MSc predictions, M8.0+: functions of the M8 algorithm determined in the circle of investigation. Note: Functions are normalized to arbitrary units ranging from minimal value of 0 to maximal value of 1. The anomalous values are marked with black dots, while the value of function $Z1$, critical in cancellation the TIP in January 2011, with a white dot of enlarged size. Time interval of $u = 3$ years is given by the light gray background, while the prematurely canceled TIP is marked gray. The origin time of the March 11, 2011 mega-earthquake is marked with the vertical arrow on the top. TIP, time of increased probability.

18.6   OTHER M8 ALGORITHM APPLICATIONS

As already mentioned, the design of the Global Test had been stimulated by the encouraging statistics of retrospective applications of the M8 algorithm in a number of seismic regions worldwide (Kossobokov, 1986; Keilis-Borok and Kossobokov, 1990). In hindsight, the algorithm aimed at different magnitude ranges from the original M8.0+ down to M5.0+ did recognize TIPs in advance 39 out of 44 target earthquakes. Our achievements in setting up regular experimental monitoring and testing in real time are rather modest. At the moment besides the Global Test we run on a regular basis the M8 algorithm and its modified versions in Italy, California, Vrancea region (Romania), and Armenia (Kossobokov et al., 2002; Peresan et al., 2005; Keilis-Borok et al., 2000; Popa et al., 2007; Antonyan et al., 2007) targeting earthquakes in M7.0+, M6.5+, M6.0+, and M5.5+ magnitude ranges.

The December 26, 2004 Sumatra–Andaman, M_w 9.1–9.3 giant earthquake happened to be the first indication that the algorithm designed for prediction of M8.0+ earthquakes can be rescaled for prediction of mega-earthquakes. An application of the M8 algorithm to seismic activity record through the middle of 2004 in a circle of investigation of 6,000-km diameter centered at epicenter of the 2004 Sumatra–Andaman earthquake did recognize a TIP for the period 2001–2005. Since 2005 and apart from the Global Test of the M8-MSc predictions, we have expanded application of the M8 algorithm targeting mega-earthquakes worldwide. In agreement with a well-known switch of scaling of the earthquake source length at about magnitude M7.8–8.0 (Bormann, 2012), the diameter of CIs has been chosen to follow a different proportion than that of the original formula for $D(M_0)$ and was set to 2,000 km for M8.5+ and 6,000 km for M9.0+. The decision of using much larger CIs eventually happened to be appropriate when the M8 algorithm is aimed at giant earthquakes like the December 26, 2004 M9.2 Sumatra–Andaman, March 28, 2005 M8.7 Nias, February 27, 2010 M8.8 Chile, and March 11, 2011 M9.0 Tohoku earthquakes, which ruptured the 1,300-, 400-, over 700-, and 480-km segments of subducting margins of the Indian and Pacific Oceans.

As concerning refining localization by the MSc algorithm, we did not manage to find such a suitable adjustment of the algorithm parameters that would pinpoint the locus of the incipient mega-earthquakes. It is hard to claim that the reason for that is the exclusive size of these extremes, which preparation may extend to the entire lithosphere of the Earth (Romashkova, 2009).

According to retrospective analysis of the spatial and temporal distribution of alarms diagnosed by the M8 algorithm aimed at M9.0+ starting in 1985, one of the small clusters of alarms (a goup of eight circles) falls into 1984–1989 and the region of the Mediterranean. Another compact cluster combines in 1994–1999, the five overlapping circles in the region of Cascadia in western USA. No mega-earthquakes were registered in these cases. In addition to that, the alarm area of the M8 algorithm diagnosis for M9.0+ in 2004 has had the global extent to aggregation of 126 out of total 262 circles with diameter 6,000 km (Kossobokov, 2005a; Romashkova et al., 2005).

Since all four mega-earthquakes of the twentieth century (Kamchatka, November 4, 1952, $M_w = 9.0$; Aleutian Islands, March 9, 1957, $M_w = 9.1$; Chile, May 22, 1960, $M_w = 9.5$; Alaska, March 28, 1964, $M_w = 9.2$) occurred in a short period of time, this cluster is unlikely in the model of independent identically distributed events (probability less than 1 percent); therefore, on January 20, 2005 in Kobe at the plenary Special Session "The Indian Ocean Disaster: risk reduction for a safer future" of the UN World Conference on Disaster Reduction we have stated the likely occurrence of the next mega-earthquake in the next 5−10 years (Kossobokov, 2005b). From 2005 on, out of the frames of the Global Test of predictions by the M8 and MSc algorithms, we update systematically in real-time prediction mode, the worldwide map of alarms diagnosed for the magnitude ranges M8.5+ and M9.0+.

By the year 2008 (Kossobokov et al., 2008) the area of alarm for M9.0+ decreased from 113 in 2005 to 46 circles of 6,000 km diameter. The centers of these 46 overlapping circles were forming the four clusters, of which the largest (35 circles) is associated with Sumatra−Andaman earthquake stretching from its source to the north and west, to Assam and Kashmir, and to the east, to Timor Island. The March 28, 2005 Nias, M8.7 mega-earthquake and two pairs of coupled great earthquakes—September 12, 2007 Southern Sumatra, M8.5 and M8.1 and April 11, 2012 Northern Sumatra, M8.6 and M8.2—have already confirmed indirectly the diagnosis of increased hazard of mega-earthquakes in this particular cluster of alarms.

One of the other three clusters of TIP's located in South America diagnosed in a circle with its center at (31°S, 70°W) and remained in all semiannual updates of predictions up to January 2010. Unlike the 2004 Sumatra−Andaman earthquake, the M8.8 mega-thrust on February 27, 2010 offshore Maule, Chile happened within the alarms determined in the frames of the Global Test as well. The epicenters of the main shock and its aftershocks fill the 700-km portion of the South American subduction zone, which is about one-half of the major seismic belt segment highlighted by TIPs diagnosed in the regular experimental testing aimed at magnitude M8.0+ and M7.5+. The regular update on January 1, 2010 missed the earthquake epicenter in the second approximation diagnosed by algorithm MSc. This failure of MSc algorithm appears natural, taking into account the linear extent of the event, which is about half of the area alerted in the first approximation (Kossobokov et al., 2010).

Upon the update of the M8 algorithm TIPs for M9.0+ earthquakes in January 2011, the areas of alarm included 44 overlapping circles: compared to 2008, the largest Sumatra cluster of 36 CIs remained nearly the same, the one in Chile has spread to six CIs, while the one in New Guinea has shrunk to a single CI at (8.75° S, 158° E), the cluster of TIPs in Cascadia disappeared, and the new one in Northwestern Pacific centered at (45.5° N, 150.5° E) has been switched on in July 2009. The epicenter of the March 11, 2011, M9.0 mega-thrust earthquake off the Pacific coast of Tohoku region has occurred well inside this area of TIP in Northwestern Pacific about a thousand kilometers from its center.

The existing catalogs of earthquakes in the period before a group of mega-earthquakes of the twentieth century are not full enough for the retrospective validation of the standard version of the M8 algorithm in the example of these events. At the same time, the results of diagnosis by means of a modified version of the algorithm with less demanding level of the catalog completeness, suggest the presence of precursory signs of a complex activation in advance of at least three of the four mega-earthquakes (namely, for the events of 1952, 1957, and 1964). In addition, as in the case of the 2004 Sumatra—Andaman event, the sequences of smaller size earthquakes at the approach and after mega-earthquakes of the entire period of instrumental observations demonstrate both common features and variety of dynamics in the epicenter vicinity of different sizes (Romashkova et al., 2005).

Naturally, we cannot get in our lifetime enough statistics required for the ultimate justification of applicability and efficiency of predictions in magnitude ranges M8.5+ and M9.0+. However, in the same fashion as the unique case history of retrospective diagnosis of the M8-type TIP in advance of the largest "starquake" in a series of the 111 flashes of energy release radiated from the neutron star with celestial coordinates 1806-20 (Kossobokov et al., 2000), the application aimed at mega-earthquakes appears to us very important both from theoretical and practical viewpoints.

18.7 DISCUSSION AND CONCLUSIONS

Thus, our efforts over 25 years since the first confirmation of the M8-MSc predictions issued in real time result in rejecting with certainty the hypothesis of random earthquake occurrence. Statistical validity and reliability of the results achieved in rigid real-time testing experiments confirm the following:

18.7.1 The Four Paradigms

- Some earthquakes are predictable. Seismic premonitory patterns exist and some of them are reliable.
- Formation of precursors at a scale of years involves fault systems by far larger than the size of the impending earthquake source.
- The general activation of seismic activity in advance of the main shock with its diverse variety of associated phenomena is similar in a wide range of tectonic environments and...
- ...its direct analogies exist in some other complex nonlinear systems (e.g., starquakes, magnetic disturbances, outcomes of elections, starts and ends of economic recessions, episodes of a sharp increase in the unemployment rate, and surges of homicides in a mega-city (Kossobokov and Soloviev, 2008)).

18.7.2 Seismic Roulette is Not Perfect

- The M8 and M8-MSc predictions can be used in a knowledgeable way. Their accuracy, confidence level, and efficiency established in the course the Global Test are already enough: (1) for undertaking earthquake low-key preparedness measures, which would prevent a considerable part of damage and human loss, although far from the total (Davis et al., 2012); as well as (2) for setting up data acquisition systems aimed at observation of local precursory phenomena in advance forthcoming potentially disastrous earthquakes for the purposes of further refinement of prediction.

- Methodologies linking intermediate-term, middle-range predictions with disaster management strategies and dynamic risk assessment do exist (Molchan, 2003; Davis, 2012).

18.7.3 Implications for Physics

- The M8-MSc predictions provide reliable empirical constraints on predictability, universality, and diversity of seismic processes for realistic modeling earthquake sequences.
- The case histories of specific TIPsdiagnosed in the course of testing experiments indicate that distributed seismic activity is a problem of statistical physics and...
- ...favor the hypothesis that earthquakes follow a general hierarchical process that proceeds via a sequence of inverse cascades to produce self-similar scaling (i.e., indicative of *intermediate asymptotic*), which then truncates at the largest scales bursting into direct cascades of aftershocks (Gabrielov et al., 1999).

Evidently, the algorithms M8 and MSc are neither optimal nor unique. Their efficiency and their accuracy could be improved by systematic monitoring of the alarm areas and by designing new earthquake prediction techniques based on in-depth analysis of seismic activity and associated geophysical observables. This requires more data of a different nature being analyzed systematically to establish reliable correlations between the occurrence of extreme events and observable phenomena.

Losses from natural disasters continue to increase mainly due to the lack of knowledge and poor understanding by the majority of the scientific community, as well as by decision makers and the population, the three components of risk, i.e., hazard, exposure, and vulnerability. Contemporary science, geophysics and seismology, in particular, are responsible for not coping with challenging changes of exposures and their vulnerability inflicted by growing population, its concentration, which result in a steady increase of losses due to natural hazards. Scientists owe to society for lack of knowledge, education, and communication. Some cases of recent disastrous earthquakes are on the limit of unacceptable fault committed by technocrats and their advisers.

We hope that the ongoing testing of intermediate-term, middle-range earthquake prediction algorithms have demonstrated in practice already that contemporary science can do a better job in disclosing natural hazards, assessing dynamic risks, and delivering useful information in advance of catastrophic events. The enormous progress in real-time retrieval and monitoring of distributed multitude of geophysical data appeals to multidisciplinary research aimed at predictability of natural catastrophes. Geoscientists must initiate shifting the minds of community from pessimistic disbelieve to optimistic challenging issues of dynamic hazard predictability.

REFERENCES

Allen, CR (Chairman), Edwards, W., Hall, W.J., Knopoff, L., Raleigh, C.B., Savit, C.H., Toksoz, M.N., Turner, R.H., 1976. Predicting Earthquakes: A Scientific and Technical Evaluation − with Implications for Society. Panel on Earthquake Prediction of the Committee on Seismology, Assembly of Mathematical and Physical Sciences, National Research Council. U.S. National Academy of Sciences, Washington, D.C.

Antonyan, ASh, Manukyan, A.V., Romashkova, L.L., Kossobokov, V.G., 2007. Re-establishing seismic monitoring aimed at intermediate-term prediction of strong earthquakes in Armenia. Geophys. Res. Abstr. vol. 9. Abstracts of the Contributions of the EGU General Assembly 2007, Vienna, Austria, April 15−20, 2007 (CD-ROM), EGU2007-A-06626.

Bormann, P. (Ed.), 2012. New Manual of Seismological Observatory Practice (NMSOP-2). IAS-PEI, GFZ German Research Centre for Geosciences, Potsdam doi:10.2312/GFZ.NMSOP-2 urn:nbn:de:kobv:b103-NMSOP-2. http://nmsop.gfz-potsdam.de.

Davis, C., Keilis-Borok, V., Kossobokov, V., Soloviev, A., 2012. Advance prediction of the March 11, 2011 great east Japan earthquake: a missed opportunity for disaster preparedness. Int. J. Disaster Risk Reduct. 1, 17−32 http://dx.doi.org/10.1016/j.ijdrr.2012.03.001.

Davis, C.A., 2012. Loss functions for temporal and spatial optimizing of earthquake prediction and disaster preparedness. Pure Appl. Geophys. 169, 1989−2010 http://dx.doi.org/10.1007/s00024-012-0502-8.

Dobrovolsky, I.R., Zubkov, S.I., Myachkin, V.I., 1979. Estimation of the size of earthquake preparation zone. Pure Appl. Geophys. 117, 1025−1044.

Gabrielov, A., Newman, W.I., Turcotte, D.L., 1999. An exactly soluble hierarchical clustering model: inverse cascades, self-similarity, and scaling. Phys. Rev. E 60, 5293−5300.

Gelfand, I., Guberman, Sh, Keilis-Borok, V., Knopoff, L., Press, F., Ransman, E., Rotwain, I., Sadovsky, A., 1976. Pattern recognition applied to earthquakes epicenters in California. Phys. Earth Planet Inter. 11, 227−283.

Giardini, D., Grünthal, G., Shedlock, K.M., Zhang, P., 1999. The GSHAP global seismic hazard map. Ann. Geofis. 42 (6), 1225−1228.

Global Hypocenters Data Base CD-ROM NEIC/USGS, Denver, CO, 1989.

Harte, D., Li, D.-F., Vreede, M., Vere-Jones, D., 2003. Quantifying the M8 prediction algorithm: reduction to a single critical variable and stability results. New Zealand J. Geol. Geophys. 46, 141−152.

Healy, J.H., Kossobokov, V.G., Dewey, J.W., 1992. A Test to Evaluate the Earthquake Prediction Algorithm, M8. U.S. Geological Survey. Open-File Report 92-401, 23 pp. with 6 Appendices.

Ismail-Zadeh, A.T., Kossobokov, V.G., 2011. Earthquake forecast M8 algorithm. In: Gupta, H. (Ed.), Encyclopaedia of Solid Earth Geophysics. Springer, Heidelberg, pp. 178−182. http://dx.doi.org/10.1007/978-90-481-8702-7.

Jordan, T., Chen, Y., Gasparini, P., Madariaga, R., Main, I., Marzocchi, W., Papadopoulos, G., Sobolev, G., Yamaoka, K., Zschau, J., 2011. ICEF report. Operational earthquake forecasting: state of knowledge and guidelines for utilization. Ann. Geophys. 54 (4). http://dx.doi.org/10.4401/ag-5350.

Jordan, T.H., 2006. Earthquake predictability, brick by brick. Seismol. Res. Lett. 77, 3−7.

Kagan, Y.Y., Jackson, D.D., 1991. Seismic gap hypothesis: ten years after. J. Geophys. Res. 96, 21,419−21,431.

Kagan, Y.Y., Jackson, D.D., 1995. New seismic gap hypothesis: five years after. J. Geophys. Res. 100 (B3), 3943−3959.

Kanamori, H., 1981. The nature of seismicity patterns before large earthquakes. In: Ewing, M. (Ed.), Series 4: Earthquake Prediction − An International Review. AGU Geophysical Monographs, Washington, D.C., pp. 1−19.

Keilis-Borok, V., Shebalin, P., Gabrielov, A., Turcotte, D., 2004. Reverse tracing of short-term earthquake precursors. Phys. Earth Planet Inter. 145 (1−4), 75−85.

Keilis-Borok, V.I., Kosobokov, V.G., 1984. A complex of long-term precursors for the strongest earthquakes in the world. Earthquakes and prevention of natural disasters. In: Proc. 27-th International Geological Congress, 4 August 14, 1984, Moscow. Colloquium C6, vol. 61. Nauka, Moscow, pp. 56−66.

Keilis-Borok, V.I., Kossobokov, V.G., 1987. Periods of High Probability of Occurrence of the World's Strongest Earthquakes. Computational Seismology 19: 45−53. Allerton Press Inc.

Keilis-Borok, V.I., Malinovskaya, L.N., 1964. One regularity in the occurrence of strong earthquakes. J. Geophys. Res. 69, 3019−3024.

Keilis-Borok, V., Kossobokov, V., Healy, J., Turcotte, D., 2000. Reproducible earthquake prediction and earthquake preparedness. In: Geophysical Research Abstracts. European Geophysical Society, 25th General Assembly (CD-ROM).

Keilis-Borok, V.I., 1990. The lithosphere of the Earth as a nonlinear system with implications for earthquake prediction. Rev. Geophys. 28, 19−34.

Keilis-Borok, V.I., Kossobokov, V.G., 1988. Premonitory Activation of Seismic Flow: Algorithm M8. Lecture Notes of the Workshop on Global Geophysical Informatics with Applications to Research in Earthquake Prediction and Reduction of Seismic Risk (15 November−16 December, 1988). ICTP, Trieste, 17 pp.

Keilis-Borok, V.I., Kossobokov, V.G., 1990. Premonitory activation of seismic flow: algorithm M8. Phys. Earth Planet Inter. 61, 73−83.

Keilis-Borok, V.I., Knopoff, L., Kossobokov, V., Rotwain, I.M., 1990. Intermediate-term prediction in advance of the Loma Prieta earthquake. Geophys. Res. Lett. 17 (9), 1461−1464.

Kossobokov, V., 2005a. Can we predict mega-earthquakes. Geophys. Res. Abstr. 7, 05832. Abstracts of the Contributions of the EGU General Assembly 2005, Vienna, Austria, April 24−29, 2005 (CD-ROM).

Kossobokov, V., 2011. Are mega earthquakes predictable? Izv. Atmos. Oceanic Phys. 46 (8), 951−961. http://dx.doi.org/10.1134/S0001433811080032.

Kossobokov, V., Romashkova, L., Nekrasova, A., 2008. Targeting the next mega-earthquake. Geophys. Res. Abstr. 10. Abstracts of the Contributions of the EGU General Assembly 2008, Vienna, Austria, April 13−18, 2008 (CD-ROM), EGU2008-A-07303.

Kossobokov, V.G., 1986. The test of algorithm M8. In: Gabrielov, A., Dmitrieva, O.E., Keilis-Borok, V.I., Kossobokov, V.G., Kuznetsov, I.V., Levshina, T.A., Mirzoev, K.M., Molchan, G.M., Negmatullaev, S.Kh, Pisarenko, V.F., Prozoroff, A.G., Rinehart, W., Rotwain, I.M., Shebalin, P.N., Shnirman, M.G., Shreider, S.Yu (Eds.), Algorithms of Long-term Earthquakes' Prediction. Centro Regional de Sismologia para America del Sur (CERESIS), Lima, Peru, pp. 42−52.

Kossobokov, V.G., 1997. User manual for M8. In: Healy, J.H., Keilis-Borok, V.I., Lee, W.H.K. (Eds.), Algorithms for Earthquake Statistics and Prediction, vol. 6. IASPEI Software Library, El Cerrito, CA. Seismol. Soc. Am.

Kossobokov, V.G., 2005b. 26 December 2004 greatest Asian quake: when to expect the next one? Statement at special session on the Indian Ocean disaster: risk reduction for a safer future. In: UN World Conference on Disaster Reduction, January 18−22, 2005, Kobe, Hyogo, Japan.

Kossobokov, V.G., 2006. Testing earthquake prediction methods: "The West Pacific short-term forecast of earthquakes with magnitude MwHRV ≥ 5.8". Tectonophysics 413, 25−31.

Kossobokov, V.G., 2013. Earthquake prediction: 20 years of global experiment. Nat. Hazards 69, 1155−1177. http://dx.doi.org/10.1007/s11069-012-0198-1 (Published online: April 21, 2012).

Kossobokov, V.G., Keilis-Borok, V.I., Smith, S.W., 1990. Localization of intermediate term earthquake prediction. J. Geophys. Res. 95 (B12), 19763−19772.

Kossobokov, V.G., Shebalin, P.N., Healy, J.H., Dewey, J.W., Tikhonov, I.N., 1999a. A real-time intermediate-term prediction of the October 4, 1994, and December 3, 1995, southern Kuril Islands earthquakes. In: Chowdhury, D.K. (Ed.), Computational Seismology and Geodynamics. The Union, Washington, D.C., pp. 57−63. Am. Geophys. Un. 4.

Kossobokov, V., Shebalin, P., 2003. Chapter 4. Earthquake prediction. In: Keilis-Borok, V.I., Soloviev, A.A. (Eds.), Nonlinear Dynamics of the Lithosphere and Earthquake Prediction. Springer, Heidelberg, pp. 141−207.

Kossobokov, V.G., Nekrasova, A.K., 2010. Global seismic hazard assessment program maps are misleading. Eos Trans. AGU 91 (52). Fall Meet. Suppl., Abstract U13A-0020.

Kossobokov, V.G., Romashkova, L.L., Nekrasova, A.K., 2010. Targeting the next megaearthquake: the February 27, 2010 Chile case. Eos Trans. AGU 91 (26). Meet. Am. Suppl., Abstract U41A-06.

Kossobokov, V.G., Maeda, K., Uyeda, S., 1999b. Precursory activation of seismicity in advance of the Kobe, 1995 earthquake. Pure Appl. Geophys. 155, 409−423.

Kossobokov, V.G., Romashkova, L.L., Panza, G.F., Peresan, A., 2002. Stabilizing intermediate-term medium-range earthquake predictions. J. Seismol. Earthquake Eng. 4 (2−3), 11−19.

Kossobokov, V.G., Soloviev, A.A., 2008. Prediction of extreme events: fundamentals and prerequisites of verification. Russ. J. Earth Sci. 10, ES2005. http://dx.doi.org/10.2205/2007ES000251.

Kossobokov, V.G., Keilis-Borok, V.I., Cheng, B., 2000. Similarities of multiple fracturing on a neutron star and on the Earth. Phys. Rev. E 61, 3529−3533.

Kossobokov, V.G., Nekrasova, A.K., 2012. Global seismic hazard assessment program maps are erroneous. Seismic Instrum. 48 (2), 162−170. http://dx.doi.org/10.3103/S0747923912020065.

Lee, Y., Turcotte, D.L., Holliday, J.R., Sachs, M.K., Rundle, J.B., Chen, C., Tiampo, K.F., 2011. Results of the regional earthquake likelihood models (RELM) test of earthquake forecasts in California. PNAS 108 (40), 16533−16538. http://dx.doi.org/10.1073/pnas.1113481108.

McCann, W.R., Nishenko, S.P., Sykes, L.R., Krause, J., 1979. Seismic gaps and plate tectonics: seismic potential for major boundaries. Pure Appl. Geophys. 117, 1082−1147.

Molchan, G., 2010. Space-time earthquake prediction: the error diagrams. In: Seismogenesis and Earthquake Forecasting: The Frank Evison, vol. II, pp. 53−63 (Pageoph Topical Volumes).

Molchan, G., Romashkova, L., 2010. Earthquake prediction analysis based on empirical seismic rate: the M8 algorithm. Geophys. J. Int. 183 (3), 1525−1537.

Molchan, G.M., 1997. Earthquake prediction as a decision making problem. Pure Appl. Geophys. 149, 233−247.

Molchan, G.M., 2003. Chapter 5. Earthquake prediction strategies: a theoretical analysis. In: Keilis-Borok, V.I., Soloviev, A.A. (Eds.), Nonlinear Dynamics of the Lithosphere and Earthquake Prediction. Springer, Heidelberg, pp. 209−237.

Peresan, A., Kossobokov, V., Romashkova, L., Panza, G.F., 2005. Intermediate-term middle-range earthquake predictions in Italy: a review. Earth-Sci. Rev. 69 (1−2), 97−132.

Popa, M., Cadichian, N., Romashkova, L.L., Radulian, M., Stanica, D., Kossobokov, V.G., 2007. Seismic monitoring aimed at intermediate-term prediction of strong earthquakes in the Vrancea region. Geophys. Res. Abstr. vol. 9. Abstracts of the Contributions of the EGU General Assembly 2007, Vienna, Austria, April 15−20, 2007 (CD-ROM), EGU2007-A-06563.

Richter, C.F., 1964. Discussion of paper by V.I. Keylis-Borok and L.N. Malinovskaya, 'One regularity in the occurrence of strong earthquakes'. J. Geophys. Res. 69, 3025.

Romashkova, L.L., 2009. Global-scale analysis of seismic activity prior to 2004 Sumatra−Andaman mega earthquake. Tectonophysics 470, 329−344.

Romashkova, L., Nekrasova, A., Kossobokov, V., 2005. Seismic cascades in advance and after 26 December 2004 Sumatra-Andaman and other mega-earthquakes. Eos Trans. AGU 86 (52). Fall Meet. Suppl., Abstract U11B-0834.

Scholz, C.H., 1997. Whatever happened to earthquake prediction. Geotimes 42 (3), 16−19.

Schorlemmer, D., Zechar, J.D., Werner, M.J., Jackson, D.D., Field, E.H., Jordan, T.H., RELM Working Group, 2010. First results of the regional earthquake likelihood models experiment. Pure Appl. Geophys. 167 (8−9), 859−876.

Shaw, J.H., Shearer, P.M., 1999. An elusive blind-thrust fault beneath metropolitan Los Angeles. Science 238, 1516−1518.

Shebalin, P.N., 1992. Automatic Duplicate Identification in Set of Earthquake Catalogues Merged Together. U.S. Geological Survey. Open-File Report 92−401, Appendix II.

Stein, S., Geller, R., Liu, M., 2011. Bad assumptions or bad luck: why earthquake hazard maps need objective testing. Seismol. Res. Lett. 82, 623−626.

Updike, R.G. (Ed.), 1989. Proceedings of the National Earthquake Prediction Evaluation Council. U.S. Geological Survey. Open-File Report 89−114.

Chapter 19

Review of the Nationwide Earthquake Early Warning in Japan during Its First Five Years

Mitsuyuki Hoshiba

Meteorological Research Institute, The Japan Meteorological Agency, Tsukuba, Japan

ABSTRACT

In 2007, the Japan Meteorological Agency (JMA) started a nationwide Earthquake Early Warning (EEW) service for the general public aiming to mitigate earthquake disasters by giving people enough time to take appropriate safety measures in advance of strong ground shaking. The warning announcement is broadcast by various ways, such as TV, radio, and cellular phone broadcasting messages. This chapter reviews the performance of the JMA EEW system over its first 5 years, during which time the $M_w 9.0$ Tohoku Earthquake occurred. During the earthquake, the system functioned as designed for the Tohoku district. However, it underpredicted the seismic intensity of the mainshock for the Kanto district due to the large extent of the fault rupture, and it issued some false alarms due to multiple simultaneous aftershocks. A questionnaire survey indicates that most people consider the EEW useful despite the occurrence of false alarms.

19.1. INTRODUCTION

Seconds before the strong ground shaking of a major earthquake hits a city, a warning message is issued. On receiving the warning, the railway companies give an emergency notice to all train drivers to stop trains immediately, elevators stop at the nearest floor and open their doors automatically, surgeons stand back from the operating table, and pupils duck under desks in schools. The warning is called "Earthquake Early Warning (EEW)". EEW has been researched worldwide in recent decades by many researchers (Allen et al., 2009).

In Japan, automatic shutdown systems were required on all kerosene stoves in the early 1970s. These systems, which automatically turn off the flame when they detect strong shaking, are integrated into each kerosene fan heater or stove for home use. Gas companies introduced automatic shutoff systems in

Earthquake Hazard, Risk, and Disasters. http://dx.doi.org/10.1016/B978-0-12-394848-9.00019-5

the late 1980s (Tokyo Gas, 2013); these automatically cut off the supply of gas in the case of strong shaking. They are a lesson learned from the 1923 Kanto earthquake ($M7.9$), in which Tokyo was devastated by big fires and more than 105,000 people were killed mostly by the fire. More recently, several chemical plants have introduced more sophisticated automatic shutdown systems (Takamatsu, 2009) to prevent disaster due to strong ground shaking, in which when a P wave is detected, the following S wave is predicted. This is based on the information observed at a site, and a warning is issued or an automatic control measure is applied to the site or a nearby location. This is the so-called on-site method of the EEW technique.

In order to halt bullet trains (Shinkansen) safely during earthquakes, an EEW technique called the Urgent Earthquake Detection and Alarm System (UrEDAS) was introduced in 1992 (Nakamura and Saita, 2007). In this system, the earthquake magnitude was first estimated from the predominant frequency of the P wave during the initial 3 s, and then the epicentral distance was determined from the measured amplitude, in which the estimated magnitude is applied to the attenuation relation inversely to estimate the epicentral distance (attenuation relations are usually applied to determine amplitude from the distance and magnitude). The UrEDAS was replaced in 2004–2005 with a new system (Yamamoto et al., 2011; Noda et al., 2012), in which the epicentral distance is estimated from the rate of increase of the P wave envelope during the initial 3 s ($B-\Delta$ method; Odaka et al., 2003); a sudden increase indicates a short epicentral distance, and a gradual increase means a long epicentral distance. In the EEW technique for the new system, the epicentral distance is estimated first and then the magnitude is determined from the amplitude, which is different from the UrEDAS. The new systems are adopted to the Shinkansen railways at present. The railway EEW systems are based on analyses of the seismic signals from a network of seismometers installed along the railway and along coastal line.

The nationwide EEW service has started in Japan in 2007 for the general public (it had started in 2004 for specific users), by the Japan Meteorological Agency (JMA) (Hoshiba et al., 2008; Kamigaichi et al., 2009; Doi, 2011). This is definitely different from earthquake prediction. The seismic waves which are detected at stations relatively near the hypocenter are rapidly analyzed, and distribution of strong ground shaking is quickly predicted, and then the information is delivered immediately to people (EEW users) living relatively far from the hypocenter. Even after an earthquake occurs, it is possible to warn people being at a certain distance from the hypocenter before strong ground shaking reaches them. Even though the interval between the delivery of EEW and the time strong shaking reaches people is relatively short (counted in seconds), EEW is a useful and powerful tool for mitigation of earthquake disaster by giving people enough time to take appropriate safety measures (e.g., stop trains, duck under desks) in advance of strong shaking.

This chapter reviews the 5 year performance of the operational stage of the JMA EEW. During this period, the 2011 off the Pacific coast of Tohoku earthquake ($M_w9.0$, hereafter the Tohoku earthquake) occurred on March 11, 2011. The JMA EEW system functioned as designed in the Tohoku district (Hoshiba et al., 2011); the details are described in this chapter. JMA obtained feedback from the general public with a questionnaire survey after the Tohoku Earthquake. The results and analyses of the survey are also reported here.

19.2. OPERATION OF JMA EEW

JMA is the national authority responsible for announcing earthquake information to the general public. The information is provided via TVs, radios, and Internet within 2−3 min just after the occurrence of the earthquake. The announcement contains the epicenter location, focal depth, magnitude, and distribution of observed seismic intensity. When a tsunami is predicted from the epicenter location, focal depth, and magnitude, JMA issues tsunami warnings (performance of JMA's tsunami warning during the Tohoku earthquake and its improvement plan were reported in Hoshiba and Ozaki (2012, 2013), and Ozaki (2012)). JMA has a long history of seismic observation, its maintenances, and experiences of communicating earthquake information to the general public. Based on this background, JMA started the EEW service.

The JMA EEW system relies mainly on a network technique in which an event's hypocenter and magnitude, M, are quickly determined by an automated analysis. The hypocenter is determined by a combination of several techniques (Hoshiba et al., 2008), using approximately 1100 stations from the JMA network and the Hi-net network of the National Research Institute for Earth Science and Disaster Prevention (NIED). M is calculated mainly from maximum displacement amplitudes (Kamigaichi et al., 2009). Using the quickly estimated hypocenter and M, seismic intensities are then predicted from an attenuation relation of ground motion and site amplification factors, in which extent of fault rupture, L (km), is taken into account based on an empirical relation, $L = 10^{0.5M-1.85}$. Details are described in Hoshiba and Ozaki (2013).

The JMA EEW announcements are divided into two grades: "forecasts" and "warnings" (Hoshiba et al., 2008; Kamigaichi et al., 2009; Doi, 2011) depending mainly on the predicted seismic intensity on the JMA scale (Figure 19.1). Definition of JMA seismic intensity is explained in the Appendix. EEW forecasts are issued to advanced users (e.g., railway companies, elevator companies, and other contractors) through provider companies when the predicted seismic intensity is 3 or greater on the JMA scale, when events are estimated to be $M3.5$ or larger, or when the observed acceleration exceeds 100 cm/s^2. In forecasts, the regions are particularly specified where seismic intensity 4 or greater is predicted. When intensity is predicted to be 5 lower or greater at any observation station in the seismic intensity networks, an EEW warning is issued to the general public in regions

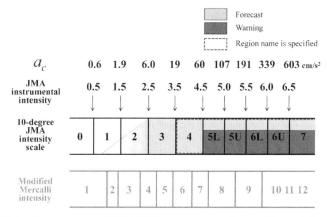

FIGURE 19.1   Two categories of JMA EEW message, warning, and forecast, in relation to JMA seismic intensity. The categories are determined mainly from the predicted intensity. When an intensity of 4 or larger is predicted, the region of strong shaking is especially specified in the EEW messages. The relation of a_c in (Eqn (19.A1) in appendix), JMA instrumental seismic intensity and the 10 degree JMA intensity scale is also indicated as well as the approximate relation of JMA and Modified Mercalli intensity scales. Letters "L" and "U" on 5 and 6 of the 10 degree JMA scale represents the delineations of "lower" and "upper", respectively.

where intensity 4 or greater is predicted. The warning announcement is broadcast by various ways, such as TV, radio, and cellular phone broadcasting messages. JMA EEWs are updated as the amount of available data increases with elapsed time; that is, M is revised as a function of time as the rupture grows. Accordingly, EEWs are issued repeatedly with improving accuracy. For example, a total of 15 EEW messages were issued for the Tohoku earthquake ($M_w9.0$). The time evolutional estimation of M is a unique point of JMA EEW, while many other researchers try to estimate eventual M deterministically from the first several seconds. The warning message is updated when the seismic intensity is predicted to be 5 lower or greater in regions where the intensity was previously estimated to be less than 4 (i.e. new regions which are subject to shaking above the threshold). In the updated warning, the newly added regions are described.

19.3. PERFORMANCE OF JMA EEW

In this section, the performance of the JMA EEW is reported during three periods of operation: from 2007 to February 2011 (before the Tohoku Earthquake), during the Tohoku Earthquake, and after the earthquake.

19.3.1. October 2007 to February 2011

The first 41 months correspond to the start of JMA EEW service to just before the $M_w9.0$ Tohoku Earthquake. During this period, the JMA EEW system issued warnings for 17 events and forecasts for 1,880 events. Table 19.1(a) summarizes the performance of the JMA EEW for events, for which an EEW warning was issued (that is, seismic intensity 5 lower or greater was predicted), and the hypocenter of the earthquakes are shown in Figure 19.2(a). Table 19.1(b) and Figure 19.2(b) indicate the events for which a warning was not issued but shaking of intensity 5 lower or greater was actually observed.

No large earthquakes occurred from October 2007 until the $M5.2$ event on April 28, 2008 near Okinawa islands. The predicted intensity was 5 lower, but the actual intensity was observed to be 4; that is, the shaking was slightly overpredicted. The next earthquake was an $M7.0$ event on May 8, 2008 off the coast of the Kanto district. For this event, a warning was issued much later than the first trigger (58.3 s) although the detection and forecast of the event took place much earlier (9.3 s). The late warning was issued much after people had already felt ground shaking, which caused widespread misunderstanding that more severe shaking was expected to start again. The delay of this warning was due to the growth of the estimated M caused by the fluctuation of gradual increase of displacement amplitude of later phases. After this experience, JMA introduced a criterion that a warning is updated only when the elapsed time is <60 s from the first trigger (This criterion caused a problem during the $M_w9.0$ Tohoku Earthquake as described later.).

The third event with a JMA EEW warning announcement was the $M7.2$ Iwate−Miyagi earthquake of June 14, 2008 (17 persons killed). Its epicenter was about 80 km north of Sendai, a major city of the Tohoku district with more than 1 million population, where intensities of 5 upper were observed at some sites. In Sendai, one person was killed, 26 wounded, and 10 houses were partially damaged (Fire Disaster Management Agency, Japan, report on June 18, 2010). The warning was issued 4.5 s after the first trigger, which gave approximately 15 s of lead time for Sendai. After this experience, people increasingly recognized EEW as a valuable measure for their safety.

During the first 41 months of EEW warnings, the predictions of seismic intensity were appropriate for most events within 1 degree of JMA intensity (Table 19.1(a)). For the $M4.1$ event (August 25, 2009; No. 11 in Table 19.1(a)), however, the intensity was seriously overpredicted: that is, a warning was issued but the observed intensity was 0 meaning that the shaking was too weak to be sensed by human. This false alarm was caused by a software error. One day before the event, engineers changing the data transmission software mistakenly set the amplitude parameter by three digits, which amplified observed shaking by 1,000 times. The error led to overestimation of M, and then overprediction of seismic intensity. After this false alarm, the director general of the seismological and volcanological department of JMA headquarters apologized to the general public at a press conference (JMA, 2009).

Table 19.1(b) lists the seven cases during the 41 months of "missed alarm", in which a warning was not issued although intensity 5 lower or greater was observed. For all seven cases, the predicted intensity was 4 and actual

TABLE 19.1(a) Summary of JMA EEW Warnings before the Tohoku Earthquake (October 2007 to February 2011) for which EEW Warnings were Issued (that is, Seismic Intensity 5 Lower or Greater was Predicted)

No.	Origin Time (JST)	Mag.	Seismic Intensity Predicted	Seismic Intensity Observed	Forecast[1] (s)	Warning[1] (s)	Remarks
1	2008/04/28 02:32	5.2	5L	4	4.6	10.6	
2	2008/05/08 01:45	7.0	5L	5L	9.3	58.3	
3	2008/06/14 08:43	7.2	6U	6U	3.5	4.5	Iwate-Miyagi earthquake
4	2008/06/14 09:20	5.7	5L	5L	3.6	8.4	Aftershock of no. 4
5	2008/06/14 12:27	5.2	5L	4	3.8	51.4	Aftershock of no. 4
6	2008/07/08 16:42	6.1	5L	5L	4.8	13.9	
7	2008/07/24 00:26	6.8	5L	6L	4.1	20.8	
8	2008/09/11 09:20	7.1	5U	5L	7.8	9.7	

No.	Origin time	Magnitude	Predicted seismic intensity on JMA scale	Observed intensity	Forecast time	Warning time	Remarks
9	2008/11/22 00:44	5.2	5L	4	3.6	10.7	
10	2009/08/11 05:07	6.5	5U	6L	3.8	3.8	
11	2009/08/25 06:37	4.1	5L	0	15.3	21.0	False alarm
12	2009/10/30 16:03	6.8	5L	4	4.2	26.8	
13	2010/02/27 05:31	7.2	6L	5L	3.2	4.1	
14	2010/03/14 17:08	6.7	5L	5L	3.2	3.6	
15	2010/09/29 16:59	5.7	5L	4	3.3	7.4	
16	2010/10/03 09:26	4.7	5L	5L	5.8	5.8	
17	2010/12/02 06:44	4.6	5L	3	3.3	8.5	

Columns mean from left to right, the number of the event (the same as Figure 21.2(a)), origin time (Japan Standard Time), magnitude, predicted seismic intensity on JMA scale, observed intensity, forecast time measure form the first trigger, warning time measure form the first trigger, and remarks, respectively.
¹Measure from the first trigger.

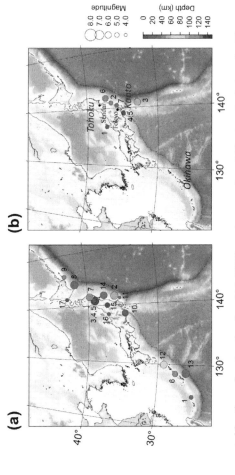

FIGURE 19.2 (a) Hypocentral locations of earthquakes for which EEW warnings were issued (that is, seismic intensity 5 lower or greater was predicted), and (b) those of the events for which warning was not issued but intensity 5 lower or greater was observed, for the period October 2007 to February 2011. Numbers of the events in Figure 21.2(a) and (b) correspond to entries of Table 21.1(a) and (b), respectively.

TABLE 19.1(b) Events for which EEW Warnings were Not Issued (that is, Predicted Intensity was Smaller Than 5 Lower), but Intensity 5 Lower or Greater was Actually Observed (October 2007 to February 2011)

No.	Origin Time (JST)	Mag.	Seismic Intensity Predicted	Observed	Forecast[1] (s)
1	2008/01/26 04:33	4.8	4	5L	5.4
2	2008/07/05 16:49	5.2	4	5L	4.2
3	2009/08/13 07:48	6.6	4	5L	20.1
4	2009/12/17 23:45	5.0	4	5L	4.5
5	2009/12/18 08:45	5.1	4	5L	4.4
6	2010/06/13 12:32	6.2	4	5L	5.2
7	2010/07/23 06:06	5.0	4	5L	3.3

Columns mean from left to right, the number of the event (the same as Figure 19.2(b)), origin time (Japan Standard Time), magnitude, predicted seismic intensity on JMA scale, observed intensity, and forecast time measure from the first trigger, respectively.
[1]Measure from the first trigger.

observed intensity was 5 lower. The prediction error was one degree of intensity, which should be an acceptable precision for intensity predictions.

In general, the JMA EEW system functioned as expected for the 41 months before the $M_w9.0$ Tohoku Earthquake.

19.3.2. During the $M_w9.0$ Tohoku Earthquake

The performance of JMA EEW for the Tohoku Earthquake is summarized briefly here; detailed reports were published by Hoshiba et al. (2011) and Hoshiba and Ozaki (2012, 2013).

During the $M_w9.0$ mainshock of the Tohoku Earthquake, a total of 15 EEW messages were issued within 2 min (Table 19.2). The first EEW forecast announcement was issued 5.4 s after the system was triggered (Figure 19.3(a)). The magnitude was initially estimated to be 4.3 because the waveform started with very small amplitude (Hoshiba and Iwakiri, 2011). By the fourth forecast, 8.6 s after the first trigger, seismic intensity was predicted to be 5 lower for central Miyagi Prefecture (around Sendai City); thus, the fourth forecast was a warning announcement to the general public in the Tohoku district (Figure 19.3(c)). NHK, a nonprofit broadcasting company, broadcast the warning nationwide, and other TV and radio companies did so locally. Cellular phone companies sent the warning to all users in the warning regions using broadcasting technology. The warning was sent earlier than the S wave arrival

TABLE 19.2 Issuance of JMA EEW Announcement during the 2011 Tohoku Earthquake (M_w9.0)

No.	Hour:Min:Sec HH:MM:SS.S[1] March 11, 2011	Lapse Time From Trigger (s)	Lapse Time From OT (s)	Mag.	Label in Figure 19.3
1	14:46:45.6	5.4	27.5	4.3	a
2	14:46:46.7	6.5	28.6	5.9	
3	14:46:47.7	7.5	29.6	6.8	b
4	14:46:48.8	8.6	30.7	7.2	c
5	14:46:49.8	9.6	31.7	6.3	
6	14:46:50.9	10.7	32.8	6.6	
7	14:46:51.2	11.0	33.1	6.6	
8	14:46:56.1	15.9	38.0	7.2	
9	14:47:02.4	22.2	44.3	7.6	d
10	14:47:10.2	30.0	52.1	7.7	f
11	14:47:25.2	45.0	67.1	7.7	
12	14:47:45.3	65.1	87.2	7.9	g
13	14:48:05.2	85.0	107.1	8.0	
14	14:48:25.2	105.0	127.1	8.1	
15	14:48:37.0	116.8	138.9	8.1	h

For this event, 15 forecasts were issued. Forecast number 4 was a warning. The last column indicates the timings corresponding approximately to the panels in Figure 19.3.
[1]Japan Standard Time.
Source: Modified from Hoshiba and Ozaki (2012, 2013).

(Figure 19.3(c)) and 15 s before strong ground motion (intensity 5 lower) was recorded at the station closest to the epicenter (Figure 19.3(e)). Thus, the JMA EEW system functioned well in the Tohoku district as designed.

The JMA EEW system predicted an intensity of 4 in Tokyo in the Kanto district in the fifteenth (final) issue (Figure 19.3(h)). The timing of the final issue, 139 s from the origin time, was before the start of strong ground motion in the Kanto district. Actual observations, however, eventually reached 5 upper in Tokyo, and 6 upper and 6 lower at many observation sites in the Kanto district (Figure 19.3(i)), which exceeds the criterion for a JMA EEW warning; thus, it was underpredicted. The underprediction can be attributed to the greater extent of the eventual fault rupture of the M_w9.0 event than that

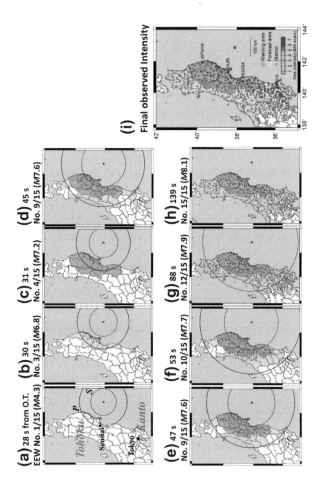

FIGURE 19.3   **Regions of JMA EEW warning and forecast, and distribution of seismic intensities at several timings from the trigger for the 2011 Tohoku Earthquake (M$_w$9.0).** The pink area indicates regions where warnings were issued, and yellow areas are those specified in forecasts. Wave fronts of P and S waves are shown by circles. Seismic intensities were measured using waveforms from K-NET, KiK-net (NIED), and JMA seismic stations (triangles). Panels (a)–(h) correspond approximately to the timing indicated in Table 19.2. Panel (i) shows the finally observed seismic intensity; note the late shaking in the Kanto district. *Modified from Hoshiba and Ozaki (2012, 2013).*

expected based on the system's magnitude determination of $M8.1$. Inversion analysis of the source process revealed that the rupture of the $M_w9.0$ event extended to off Kanto (Kurahashi and Irikura, 2011; Yoshida et al., 2011). The greater-than-expected extent of the fault rupture causes stronger ground motion at the Kanto district.

For the northern part of Ibaraki Prefecture in the Kanto district, where the intensity predicted in the first warning (i.e., the fourth forecast) was less than 4, the predicted intensity rose to 5 lower by the fourteenth forecast; however, it was too late to update the warning, because it was issued 105 s after the trigger, which is later than the 60 s criterion at which upgrades are stopped. This cutoff time of 60 s was introduced into JMA EEW after an experience with a too-late warning during the $M7.0$ event of May 8, 2008, as described in a previous subsection.

19.3.3. After the $M_w9.0$ Tohoku Earthquake

After the mainshock of $M_w9.0$ Tohoku Earthquake, aftershock activity was quite high in a wide region off eastern Japan, which was a challenge for the EEW system. Table 19.3 summarizes the monthly count of forecasts and warnings after the $M_w9.0$ Tohoku Earthquake, which greatly exceeded those issued in the 41 months before the Tohoku Earthquake. Some of them were false alarms: a warning was issued but the shaking was too small. Table 19.4

TABLE 19.3 Monthly Number of JMA EEW after the Tohoku Earthquake ($M_w9.0$) Compared with that before the Earthquake

	Forecast	Warning
Oct 2007−Feb 2011 (41 months)	1,880	17
Mar 2011	1,196	45
Apr 2011	770	26
May 2011	425	5
Jun 2011	304	5
Jul 2011	248	5
Aug 2011	239	3
Sep 2011	188	4
Oct 2011	163	1
Nov 2011	135	2
Dec 2011	136	1

TABLE 19.4 Examples of JMA EEW Warning after the M_W9.0 Tohoku Earthquake

No.	Origin Time (JST)	Mag.	Seismic Intensity		Forecast[1] (s)	Warning[1] (s)
			Predicted	Observed		
1	2011/03/14 10:02	6.2	5L	5L	6.2 —	10.2
2	2011/03/14 15:52	5.2	6L	4	3.2	13.7
3	2011/03/14 16:25	5.0	6L	3	16.9	16.9
4	2011/03/15 01:36	2.5	5L	2	20.7	20.7
5	2011/03/15 05:34	1.3	5U	0	5.6	7.5
6	2011/03/15 07:29	4.3	6U	3	14.5	14.5
7	2011/03/15 22:31	6.4	5L	6L	3.5	3.5
8	2011/03/16 02:40	4.0	5U	2	4.4	9.9
9	2011/03/16 12:23	4.7	5L	2	8.4	8.8
10	2011/03/16 12:52	6.1	6L	5L	3.6	13.9

Columns mean from left to right the number of the event, origin time (Japan Standard Time), magnitude, predicted seismic intensity on JMA scale, observed intensity, forecast time measure form the first trigger, and warning time, respectively.
[1]Measure from the first trigger.

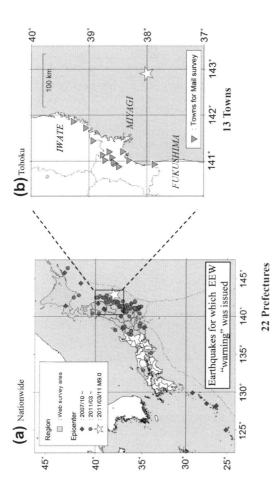

FIGURE 19.4 Regions where the questionnaire survey was conducted. (a) 22 prefectures for the nationwide survey, and (b) 13 towns in Tohoku. Epicenters for which warnings of the JMA EEW were issued last 5 years are plotted in panel (a). Star shows the epicenter of the M_w9.0 Tohoku Earthquake. The blue diamonds and red circles indicate the epicenters of earthquakes before and after the Tohoku Earthquake, respectively, for which JMA EEW warnings were issued.

lists the performance of JMA EEW warnings between March 14 and 16, as an example of this period. Seismic intensity was appropriately predicted for some cases (a warning was issued and strong shaking was observed), but over-predicted for some (a warning was issued but strong shaking was not observed). When multiple aftershocks occurred simultaneously, the system became confused, and did not always determine the location and magnitude correctly. For example, at the time of the $M1.3$ earthquake of March 15 at 5:34 (No. 5 in Table 19.4), two other small events occurred simultaneously else-where (totally three events). The system interpreted them as one big event, which led to an overestimation of M and then an overprediction of seismic intensity. In the 49 days from the mainshock to April 28, 2011, JMA appro-priately issued EEW warnings for 26 of the 46 events, for which seismic in-tensity 5 lower or greater was actually observed. In contrast, during the same time, 70 EEW warnings were issued, but actual observed intensities did not exceed 2 at any observation stations in 17 of the 70 events (JMA, 2011). This ratio is much higher than that during 41 months before the $M_w9.0$ Tohoku earthquake: 1 of 17 events did not exceed intensity 2 at any stations (due to a software error in the amplitude parameters (JMA, 2009), as described in Section 21.3.1).

The number of JMA EEW warnings decreased with time: 45 for March, 26 for April, 5 for May, 5 for June, 5 for July, and 3 for August (Table 19.3). During this period, JMA reduced the distance within which observations are attributed to a single event when multiple stations detect seismic waves. JMA also modified its procedure to stop analysis as soon as possible when the event is much smaller than the criteria for EEW forecast described in Section 21.2. The main reason for the decreased number of warnings, however, is the decrease in aftershock activity itself.

19.4. FEEDBACK ABOUT EEW FROM THE GENERAL PUBLIC

In February 2012, JMA conducted questionnaire surveys in the Tohoku district and nationwide (JMA, 2012), asking residents how they obtained the EEW service, how they acted after receiving EEW announcements, how useful they considered EEW as a safety measure against strong ground shaking, and related questions. Respondents numbered 817 in the Tohoku district survey and 2000 in the nationwide survey. This section presents a brief summary of these surveys.

The questionnaire survey was conducted by a survey company using two different modalities: mail and the Web. The mail survey targeted 5500 resident families in 13 towns in Iwate, Miyagi, and Fukushima prefectures in the Tohoku district (Figure 19.4) that experienced strong ground motion during and after the $M_w9.0$ Tohoku Earthquake. The questionnaires were sent from February 3–8, 2012, using "town mail service", in which the town post office delivered mail to all resident families, even those who had temporarily relocated uphill outside the town because of the damage or collapse of their resident house due to the tsunami or strong ground shaking. The targeted families returned 817 responses. The Web survey was conducted nationwide, but limited to the 22 prefectures (Figure 19.4) where JMA EEW warnings had been issued during the previous 5 years. The questionnaire was issued to 10,000 families who had preregistered for such surveys, and 2,000 responses were received. However, 400 of these 2,000 families were advanced users who received EEW forecasts as well as public warnings; therefore, they were excluded, leaving 1,600 families, because it was not appropriate to compare responses from advanced users to the 817 families in the Tohoku district who did not receive EEW forecasts.

Figure 19.5 shows the seismic intensities of the most impressive earthquake the respondents had experienced since March 2011 (when the $M_w 9.0$ Tohoku Earthquake occurred). More than 55 percent of respondents in Tohoku, but 20 percent nationwide, had experienced seismic intensity 5 lower or larger. People in Tohoku had experienced strong ground shaking and many EEW messages because of active aftershocks. This difference between Tohoku and nationwide respondents is an important factor in accounting for the difference in their questionnaire answers.

The methods used to obtain EEW messages are shown in Figure 19.6. Most respondents had received EEW messages from TV or cell phone broadcast mail service. The electrical blackout in the Tohoku district just after the $M_w 9.0$ Tohoku Earthquake, which curtailed TV service, might account for the

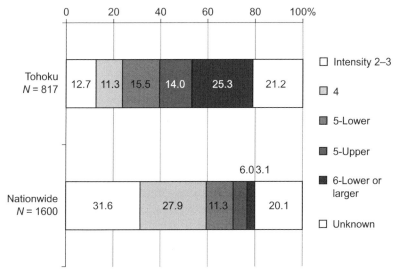

FIGURE 19.5 Seismic intensity (JMA scale) of the most impressive earthquake that respondents had experienced after March 2011. *N* represents the numbers of respondents.

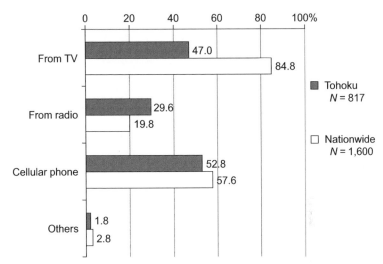

FIGURE 19.6   Method by which survey respondents obtained EEW messages (multiple answers).

difference for "from TV" between Tohoku and nationwide. The cell phone is a powerful carrier for EEW messages with which people can receive it even when TV is switched off. Indeed, the cell phone can alert sleeping people with its audible signal.

Figure 19.7 shows the actions that respondents took after receiving EEW messages. Most respondents took some actions in the Tohoku district (74 percent) and nationwide (54 percent); 16 percent and 17 percent, respectively, tried to take action but could not; and 10 percent and 29 percent, respectively, did nothing (Figure 19.7(a)). Respondents in the Tohoku district could take some actions when receiving EEW more than nationwide. Figure 19.7(b) lists the actual actions taken. There was a large difference between the Tohoku and nationwide in the percentage of respondents who avoided dangerous situations outside the house. The very long duration of strong shaking during the $M_w9.0$ Tohoku earthquake (Hoshiba et al., 2011) may be a factor in this difference, and damage to houses by strong shaking may have motivated this behavior during the very active aftershock period. Figure 19.7(c) indicates the responses to a question whether the action had been determined beforehand. Tohoku respondents were twice as likely as nationwide respondents to have determined beforehand. This may reflect the fact that respondents in the Tohoku district received many warnings because of the aftershocks. Figure 19.7(d) indicates whether the predetermined actions could be taken.

Figure 19.8 summarizes the respondents' evaluation of the usefulness of the EEW service. Figure 19.8(a) shows the response to a question whether they thought EEW is useful. More than 90 percent and 80 percent of respondents in the Tohoku district and nationwide, respectively, said that EEW was useful to

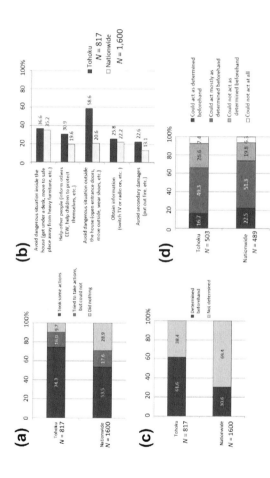

FIGURE 19.7 **Actions taken by survey respondents after receiving EEW.** Questions posed were about (a) whether the respondents actually took actions or not (single answer), (b) what specific actions they took (multiple answer), (c) whether the respondents have determined the action beforehand or not (single answer), and (d) to what degree they could take action as had been determined beforehand (single answer).

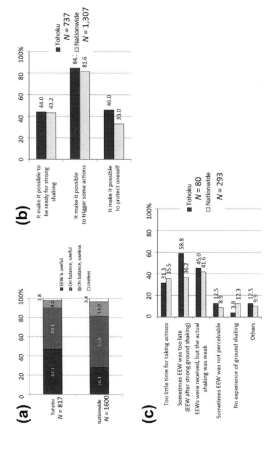

FIGURE 19.8   **Evaluation of usefulness of the EEW services.** Questions posted were about (a) whether the respondents thought EEW was useful or not (single answer), (b) the reasons why they thought EEW was useful, and (c) the reasons why they thought EEW was useless (multiple answer).

some extent (useful + on balance, useful). Tohoku residents thought so more than nationwide. Among the reasons why EEW was considered useful (Figure 19.8(b)), primary was that EEW can be served as a trigger for taking safety measures against strong shaking. Many people stated that EEW helped them prepare for strong shaking, even if they did not actually take specific actions. Psychological relief is a major factor in the perception of usefulness in addition to actual actions. Among reasons cited for considering EEW useless, the main ones were "Too little time for taking action", "Sometimes EEW was too late", and "EEWs were received, but the actual shaking was weak" (Figure 19.8(c)). The latter two reasons reflect the number of false alarms after the M_w9.0 Tohoku Earthquake.

Ohara et al. (2012) conducted their own questionnaire survey in Tokyo Metropolitan area in August 2011, after the Tohoku Earthquake, and reported that respondents who evaluated EEW to be useful amounted to be about 90 percent. Another survey was conducted by Dentsu Research asked respondents which services they regard as useful in cell phones. Its results indicated that 24.8 percent of 1200 respondents in the Kanto district regard the EEW service as a useful function of cell phone (article in Asahi Shinbun (newspaper), June 4, 2011). This percentage is the second largest response; the first is an e-mail function (58.4 percent), the third is a cell-phone TV (22.5 percent), and the fourth is twitter and blogs (17.0 percent). The above two independent questionnaire surveys also indicates the respondents' positive attitude toward EEW.

19.5. SUMMARY AND REMARKS

The JMA EEW system generally functioned as expected before the M_w9.0 Tohoku Earthquake. During the M_w9.0 Tohoku Earthquake, EEW warnings from JMA were issued as rapidly as designed, but there were problems: underprediction of seismic intensity for the Kanto district during the mainshock, and some false alarms during aftershock activity. We should learn lessons from this disaster and endeavor to improve the warning system.

JMA issued a warning to the Tohoku region, even though the initial part of the waveforms was very small because JMA adopted an updating method. This experience indicates that an updating procedure is important using ongoing waveforms for EEW purposes. On the other hand, the JMA EEW system fell short in two respects: it underpredicted the seismic intensity of the mainshock in the Kanto region because of the large extent of the fault rupture, and it issued some false alarms because of confusion when multiple aftershocks occurred simultaneously. The false alarms were a more serious problem than the underprediction for actual operation. In the Kanto district, the shaking grew more and more severe after people first sensed it during the mainshock (Figure 19.9), and it took approximately 100 s to reach the peak ground acceleration (Hoshiba and Ozaki, 2013). The long-lasting onset allowed people

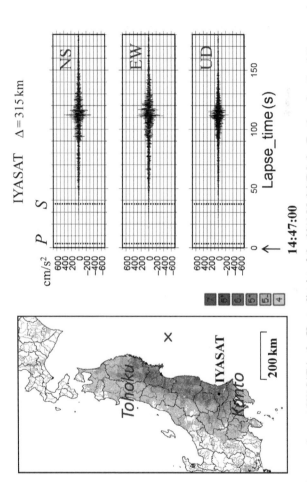

FIGURE 19.9 **Acceleration waveform at IYASAT in the Kanto district of the 2011 M_w9.0 Tohoku Earthquake.** *P* and *S* wave arrival times are shown by dotted lines. Left panel shows the seismic intensity distribution (JMA scale). *Modified from Hoshiba and Ozaki (2012, 2013).*

to be aware of the earthquake much earlier than the strong shaking, which gave them enough time to take appropriate measures (e.g., duck under desk) before strong shaking. That is, people could serve as their own forecaster through the "on-site" method of EEW technique. As for the false alarms, they were issued repeatedly to the general public for several days after the mainshock, whereas the underprediction of the mainshock was a one-time occurrence. The repeated experience of the false alarms might be a major factor which decreases the reliability of the system. Commentators in the mass media mentioned the false alarms more often than the underprediction problems.

These results suggest that in addition to determining the hypocenter and M, real-time observation of current ground motions is important for improving the predictions of ground motion. JMA is enhancing its seismic intensity observation by introducing real-time transmission, as the first stage in addressing the above problems of large extent of fault rupture and multiple simultaneous earthquakes (Nakamura et al., 2013), where ground motion is predicted from the real-time observation of neighbor stations around the target site. The theoretical background is discussed in Hoshiba (2013), in which a new method was proposed to predict ground motion using real-time monitoring of current ground motions without hypocenter and M. From the current situation (initial condition), the future is predicted time-evolutionally based on the physics of wave propagation theory.

The questionnaire survey indicates that most people have found EEW to be useful to some degree. This opinion was more prevalent among Tohoku residents than nationwide. Likewise, the percentage of people who were able to take useful actions was larger in Tohoku than nationwide. The regional difference may be the results of repeated experience, as is the more positive evaluation of EEW in Tohoku. The perceived value of the EEW system is its usefulness both as a trigger of actual actions and as an aid to mental preparedness before strong shaking begins. Most people considered the EEW system useful despite some false alarms. Although it is necessary to improve the EEW system to reduce false alarms and make the predictions more precise, the results of this survey should be encouraging to the community of EEW researchers.

ACKNOWLEDGMENTS

The author thanks reviewer and Prof. M. Wyss (editor) for their useful comments. He thanks N. Hayashimoto, S. Aoki, K. Hirano, Y. Yamada, T. Shimoyama, M. Nakamura, M. Matsui, and A. Wakayama for their help in completing the manuscript, and other colleagues at JMA for their efforts in EEW operation. The JMA EEW system uses a combination of several techniques developed by joint research with the Japan Railway Technical Research Institute, and also by NIED. It also uses real-time data from Hi-net of NIED in addition to the JMA own network for hypocenter determination. Waveform data of K-NET and KiK-net of NIED as well as those of JMA were used in this study. He thanks all these organizations for their effort in maintaining these observations and providing the data. The questionnaire survey was conducted through an independent survey company under contract with JMA. He thanks all respondents to the survey. Some figures were made using Generic Mapping Tools (Wessel and Smith, 1995). This research is partially supported by JSPS KAKENHI grand number 25282114 (Real time prediction of earthquake ground motion: method utilizing the current observation).

REFERENCES

Allen, R., Gasparini, P., Kamigaichi, O., Böse, M., 2009. The status of earthquake early warning around the world: an introductory overview. Seismol. Res. Lett. 80, 682−693.

Doi, K., 2011. The operation and performance of earthquake early warnings by the Japan Meteorological Agency. Soil Dyn. Earthquake Eng. 31, 119−126.

Hoshiba, M., 2013. Real-time prediction of ground motion by Kirchhoff-Fresnel boundary integral equation method: extended front detection method for earthquake early warning. J. Geophys. Res. 118. http://dx.doi.org/10.1002/jgrb.50119.

Hoshiba, M., Iwakiri, K., 2011. Initial 30 seconds of the 2011 off the Pacific coast Tohoku earthquake (M_w9.0) − amplitude and τ_c for magnitude estimation for earthquake early warning. Earth Planets Space 63, 553−557.

Hoshiba, M., Iwakiri, K., Hayashimoto, N., Shimoyama, T., Hirano, K., Yamada, Y., Ishigaki, Y., Kikuta, H., 2011. Outline of the 2011 off the Pacific coast of Tohoku earthquake (M_w 9.0). Earth Planets Space 63, 547−551.

Hoshiba, M., Kamigaichi, O., Saito, M., Tsukada, S., Hamada, N., 2008. Earthquake early warning starts nationwide in Japan, EOS Trans. Am. Geophys. Union 89, 73−74.

Hoshiba, M., Ohtake, K., Iwakiri, K., Aketagawa, T., Nakamura, H., Yamamoto, S., 2010. How precisely can we anticipate seismic intensities? A study of uncertainty of anticipated seismic intensities for the Earthquake Early Warning method in Japan. Earth Planets Space 62, 611−620.

Hoshiba, M., Ozaki, T., 2012. Earthquake early warning and tsunami warning of JMA of the 2011 off the Pacific coast of Tohoku earthquake. Zisin 2 (64), 155−168 (in Japanese with English abstract).

Hoshiba, M., Ozaki, T., 2013. Earthquake Early Warning and Tsunami Warning of the Japan Meteorological Agency, and their Performance in the 2011 Off the Pacific coast of Tohoku Earthquake (M_w9.0), 'Advanced Technologies in Earth Sciences'. Springer, pp. 1−28.

Japan Meteorological Agency, 1996. Seismic Intensity. Gyosei, Tokyo, p. 238 (in Japanese).

Japan Meteorological Agency, 2009. On the False Alarm of JMA EEW (2nd Issue of Press Release). http://www.jma.go.jp/jma/press/0908/25b/200908251700.html (in Japanese, last accessed 18.04.13.).

Japan Meteorological Agency, 2011. Report of the Earthquake Early Warning after the 2011 off the Pacific Coast of Tohoku Earthquake (Press Release). http://www.jma.go.jp/jma/press/1103/29a/eew_hyouka.html (in Japanese, last accessed 18.04.13.).

Japan Meteorological Agency, 2012. Survey on Utilization of Earthquake Early Warning. http://www.jma.go.jp/jma/press/1203/22c/manzokudo201203.htm (in Japanese, last accessed 18.04.13.).

Kamigaichi, O., Saito, M., Doi, K., Matsumori, T., Tsukada, S., Takeda, K., Shimoyama, T., Nakamura, K., Kiyomoto, M., Watanabe, Y., 2009. Earthquake early warning in Japan − warning the general public and future prospects. Seismol. Res. Lett. 80, 717−726.

Kurahashi, S., Irikura, K., 2011. Source model for generating strong ground motions during the 2011 off the Pacific coast of Tohoku earthquake. Earth Planets Space 63, 571–576.

Nakamura, M., Hirano, K., Yamada, Y., Kikuta, H., Hoshiba, M., 2013. The earthquake early warning of Japan Meteorological Agency. Abstr. Meet. Am. S42A-01.

Nakamura, Y., Saita, J., 2007. UrEDAS, the earthquake early warning system: today and tomorrow. In: Gasparini, P., Manfredi, F., Zschau, J. (Eds.), Earthquake Early Warning Systems. Springer, pp. 249–281.

Noda, S., Yamamoto, S., Sato, S., Iwata, N., Korenaga, M., Ashiya, K., 2012. Improvement of back-azimuth estimation in real-time by using a single station record. Earth Planets Space 64, 305–308.

Odaka, T., Ashiya, K., Tsukada, S., Sato, S., Ohtake, K., Nozaka, D., 2003. A new method of quickly estimating epicentral distance and magnitude from a single seismic record. Bull. Seismol. Soc. Am. 93, 526–532.

Ohara, M., Meguro, K., Tanaka, A., 2012. A study on people's awareness of earthquake early warning before and after the 2011 off the Pacific coast of the Tohoku earthquake, Japan. In: Proceeding the 15th World Conference on Earthquake Engineering, 4092.

Ozaki, T., 2012. JMA's tsunami warning for the 2011 great Tohoku earthquake and tsunami warning improvement plan. J. Disaster Res. 7, 439–445.

Takamatsu, K., 2009. Application of the earthquake early warning system for the OKI semiconductor factory. In: Proceedings of the 2nd International Workshop on Earthquake Early Warning, Uji, Kyoto, 17–20.

Tokyo Gas, 2013. Quick Stop of Gas Supply and Prevention of Secondary Disaster. http://www.tokyo-gas.co.jp/safety/safety_common/pdf/bousaitaisaku02.pdf (in Japanese, last accessed 18.04.13.).

Wessel, P., Smith, W.H.F., 1995. New version of the generic mapping tool released. EOS Trans. AGU 76, 329.

Yamamoto, S., Sato, S., Iwata, N., Korenaga, M., Ito, Y., Noda, S., 2011. Improvement of seismic parameter estimation for the earthquake early warning system. Quart. Rep. RTRI 52, 206–209.

Yoshida, Y., Ueno, H., Muto, D., Aoki, S., 2011. Source process of the 2011 off the Pacific coast of Tohoku earthquake with the combination of teleseismic and strong motion data. Earth Planets Space 63, 565–569.

APPENDIX

JMA Seismic Intensity

The JMA intensity scale is widely used in Japan to measure seismic intensity. Since 1996, this scale has been based on instrumental measurements in which not only the amplitude but also the frequency and duration of the shaking are considered. In this appendix, the JMA intensity is briefly explained. More detailed explanation is given by JMA (1996), Hoshiba et al. (2010), and Hoshiba and Ozaki (2013).

Acceleration records (measured in cm/s^2 are applied by the band pass filter with central frequency of 0.5 Hz to characterize the ground motions leading to damage of wooden frame houses and felt shaking. The three components of the filtered acceleration are summed to obtain the time series of vector amplitude, $a(t) = \sqrt{\left(a_{ns}(t)^2 + a_{ew}(t)^2 + a_{ud}(t)^2\right)}$. Then a_c is measured, defined as the value satisfying the condition that the total duration of $a(t) > a_c$ is 0.3 s, where a_c corresponds to the 30th largest value of $a(t)$ for 100 samples/s data, or the fifteenth largest for 50 samples/s data. JMA instrumental intensity, I, is defined as

$$I = 2 \log_{10}\left(a_c\right) + 0.94. \qquad (19.A1)$$

The 10° JMA intensity scale rounds off the instrumental intensity value to the integer (Figure 19.1). For example, instrumental intensities in the range $2.5 \leq I < 3.5$ and $3.5 \leq I < 4.5$ correspond to 3 and 4 on the 10 degree scale, respectively. Intensities of 5 and 6 are divided into two degrees, namely 5 lower $(4.5 \leq I < 5.0)$, 5 upper $(5.0 \leq I < 5.5)$, and 6 lower and 6 upper, respectively. Intensity 1 corresponds to ground motion that people can barely detect, and 7 is the upper limit. At present, seismic intensity is measured at approximately 4300 places throughout Japan by JMA, municipalities, local governments, and NIED.

Chapter 20

To What Extent Can Engineering Reduce Seismic Risk?

Stavros V. Tolis

Geoseismic G.P., Athens, Greece

ABSTRACT

Extensive areas worldwide, many of them heavily urbanized, are susceptible to seismic risk, that is, the probability that people will suffer loss and that their built environment will be damaged. Due to our deficient knowledge of nature and to financial reasons, seismic risk cannot be eliminated. Societies accept a certain level of seismic risk, which depends mainly on culture and information available. The difference between actual and acceptable seismic risk can and should be managed.

Engineers can offer several technological remedies for most seismic hazards and can minimize vulnerability by applying advanced design and construction reducing susceptibility to strong ground motion of any structure. They can also contribute to seismic risk reduction by informing, educating, and organizing communities.

In their effort to reduce seismic risk, engineers face major challenges. Although the trend in the number of lives lost each year to catastrophic seismic events is decreasing in some countries, but not in others, the corresponding costs are continuing to increase rapidly. Today, the globalization of trade and commerce enhanced by wide-ranging transportation systems and modern communications had as result the impact of a damaging earthquake to reach far beyond the epicentral region.

20.1 INTRODUCTION

All human structures are subject to seismic hazard, and global seismic risk is high. Earthquakes around the world continuously remind us how vulnerable we human beings are: enormous losses of life, extensive collapses, heavy damage of human structures, and huge economic damage can result from them. Most recently, the M_w 9.1 Tohoku, Japan, earthquake of March 11, 2011, caused nearly 20,000 deaths, >300 billion US dollars in damage, and it is impossible to estimate the long-term environmental disaster; it ranks among

Earthquake Hazard, Risk, and Disasters. http://dx.doi.org/10.1016/B978-0-12-394848-9.00020-1

the worst natural disasters ever recorded. Considering that it took place in a well-organized and well-prepared country that made great efforts reducing seismic risk, it was a sobering experience.

Engineers are involved in the design, construction, operation, and maintenance of structures. They try to estimate in close cooperation with earth sciences communities, the structure's behavior for a given seismic hazard, the consequences in the case of failure, the available options for minimizing the consequences, the related economic cost, and alternative ways to accomplish their objectives. Nevertheless, seismic risk is not possible to be eliminated entirely. This is due to the significant uncertainties embedded in assessing seismic hazard and the high cost for minimizing seismic vulnerability; however, it is possible to reduce seismic risk to an acceptable level.

After World War II, due to the rapid growth and development of urban settlements around the world, especially in earthquake-prone areas, the level of seismic risk is increasing globally. This trend will remain in the near future as the number of people and assets exposed to the hazardous influence of earthquakes will be growing fast (it is estimated that in the next 20 years, the urban population of developing countries will increase by one billion). It will reverse in the far future only when changes in land use will stabilize and the new engineering technology advances to come will be able to spread worldwide.

20.2 BASIC DEFINITIONS

In the course of daily life, as well as engineering practice, we routinely encounter situations that involve some event that might occur and that, if it did, would bring with it some adverse consequence. The combination of uncertain events and adverse consequences is the determinant of risk (Baecher and Christian, 2003).

In common usage, the word *risk* has a broad variety of meanings as (exposure to) the possibility of loss, injury, or other adverse or unwelcome circumstance; a chance or situation involving such a possibility. According to http://www.merriam-webster.com/dictionary, *risk* is defined as (1) possibility of loss or injury (peril); (2) someone or something that creates or suggests a hazard; (3a) the chance of loss or the perils to the subject matter of an in-surance contract (also, the degree of probability of such loss); (3b) a person or thing that is a specified hazard to an insurer; (3c) an insurance hazard from a specified cause or source (war risk); (4) the chance that an investment (as a stock or commodity) will lose value.

In engineering applications, risk is defined as the product of failure probability and consequences (Vick, 2002). Expressed another way, risk is taken as the expectation of adverse outcome (Baecher and Christian, 2003).

Seismic risk denotes the probability that some humans will incur loss or that their built environment will be damaged (McGuire, 2004). Specifically, it describes the probability that social or economic consequences of earthquakes will equal or exceed specified values at a site, at several sites, or in an area, during a specified exposure time (Algermissen, 2007), that is, it represents the probability of occurrence of a specific level of seismic hazard or loss over a certain time. Generally, in seismology and earthquake engineering, *seismic risk* is defined as the convolution of seismic hazard and vulnerability, expressed as

$$\text{seismic risk} = \text{seismic hazard} \times \text{vulnerability}$$

Quantification of seismic risk is complicated, involving probability, the level of seismic hazard or loss, exposure time, and how vulnerability and seismic hazard interact in time and space.

Seismic hazard describes the natural phenomena caused by an earthquake that has the potential to cause damage and loss (e.g., ground motion, ground-motion amplification, surface rupture, liquefaction, induced landslide, or tsunami); it can be evaluated from instrumental, historical, and geological observations and is usually quantified by three parameters: a level of severity, a spatial measurement, and its recurrence interval or frequency (Wang, 2009).

Vulnerability is generally understood as the inability to withstand the effects of a hazard. *Seismic vulnerability* is defined as the susceptibility of a structure to damage by a certain level of ground shaking.

Commonly, seismic risk has been used interchangeably with seismic hazard and uncertainty; however, they are fundamentally different concepts.

20.3 WHY CANNOT SEISMIC RISK BE ELIMINATED ENTIRELY?

Engineers have worked on developing methods to reduce seismic risk for decades. Some believe that the end of vulnerability will arrive during the next few decades because of fundamental advances in structural design and building technology (Lomnitz, 1999).

Several companies develop and produce viscous dampers, shock absorbers, and base isolation systems that significantly decrease earthquake-induced motions. Viscous dampers and shock absorbers can be installed in structures both as diagonal braces and in the foundation; they are built on technology based on the compression of highly viscous elastomeric fluids. Base isolation systems consist of lead plugs, rubber, and steel placed in layers. All the above-mentioned technologies are still considered expensive and are difficult to be implemented in the design.

Nevertheless, no matter how much engineering advances, seismic risk cannot be eliminated entirely. The two reasons for this are our incomplete knowledge of the physical world and the limitations in financing abilities. There will always be the uncertainty in construction quality.

20.3.1 The Deficient Knowledge of Nature

Seismology is a very young science with its origins in the late nineteenth century. Although great effort has been made in understanding the composition and dynamics of the Earth, we have yet much to learn about tectonic processes and earthquakes. Our seismological records span over an extremely short period compared to the time scales of Earth processes. Still we do not know the most basic question about earthquake recurrence: whether it is time dependent or time independent (Stein, 2006).

Seismic hazard assessments still include significant uncertainties, because they are based on earthquake histories much shorter than the return periods applied in the calculations and on a limited number of available instrumental strong ground-motion records, especially in low seismicity areas. The most recent examples are as follows: scientific communities were surprised by the December 26, 2004, Sumatra earthquake; the segment of the trench between Sumatra and the Andaman islands was not particularly active seismically, and was not considered particularly dangerous or indicated as high-risk on seismic hazard maps (Stein, 2006), and the maximum earthquake size in the Tohoku area was dramatically underestimated before March 11, 2011 (Kagan and Jackson, 2013). Additionally, earth scientists when assessing seismicity generally underestimate the uncertainty in their evaluations, because they seem to be tied to particular interpretations, so that it is difficult for them to make unbiased evaluations based on other alternatives (McGuire, 1992).

Limitations exist as well in our knowledge about the materials with, and on which we build. Even for "industrialized" materials such as concrete and steel uncertainties occur about their characteristics and behavior. The case becomes more complicated, and the uncertainties increase significantly for soil (e.g., Parvez and Rosset, 2014). We rarely are able to know in detail the soil profile and the characteristics of the soil on which we construct our projects. We have still much to learn about soil behavior under both static and seismic loads.

Even though the calculation procedures and the computational tools we use to design structures significantly improve with time, we are far from being able to model realistically the highly random nature of earthquake ground motions and the behavior of materials under dynamic loads. Moreover, many and major assumptions are necessary for modeling in detail the geometry and the interconnectivity of the different elements that complex structures consist of.

20.3.2 Financial Inability

Currently, financial criteria (e.g., amount of money needed, cost of construction, or future profits) dominate decisions for future investments and policies. Therefore, seismic safety decisions are value laden. Once this is recognized, they cannot be made scientifically (Cornell, 1988).

New materials, novel inventions, and improved engineering tools can be applied, overconservative approaches in design can be adopted, and seismic hazards can be avoided if construction developments and new investments are prevented in areas with a known high seismicity. Technology can offer several solutions to minimize seismic vulnerability of structures and some remedies for many seismic hazards today, but societies do not judge seismic risk reduction as high priority; instead, they direct most investments and wealth-producing sources to economic development. Governments, owners, investors, or insurers (e.g., Michel, 2014) are not ready to allocate the necessary resources and spend the corresponding time, effort, and money. They are willing to tolerate significant seismic risk as long as the extra cost for further reduction exceeds what they feel as improvement gained.

Then, no matter how strongly new design codes and legislations may enforce improvements of new structures, it is the existing building stock and lifelines that need immense effort and funds for retrofitting.

20.4 ACCEPTABLE SEISMIC RISK

Societies are always prepared to tolerate a certain level of risk depending on the loss its members are willing to suffer for the corresponding benefits. Generally, the acceptability of risk appears to be exponentially proportional to the corresponding benefits. Awareness of the benefits is determined by advertising, usefulness, and the number of people involved. Moreover, the public is willing to accept "involuntary" risks (collapse of a structure during a natural hazard) roughly 1000 times less than "voluntary" risks (like driving a car) (Starr, 1969).

This is true for seismic risk as well. Most people recognize that it exists in their lives, but awareness of it is biased, varies among individuals, and rarely conforms to common sense. This is due to several interacting factors, including ability of a person to estimate risk, experience with seismic events, level of hazard solutions already in place, social differentials (e.g., education and mentality), and information. The most important factor is the actual experience of an earthquake that enhances the perception of seismic risk, because people believe that future seismic events will be comparable to those experienced; however, the effect of experience diminishes as time passes. It is common for people to overestimate the probability of small seismic events and underestimate the probability of large or great earthquakes (Mileti, 1982).

Therefore, how safe is safe enough in the case of an earthquake or what should be considered as acceptable seismic risk? Most important of all, to whom is it acceptable?

A nonnegligible seismic risk is considered acceptable if by taking further action to reduce it, the result is not guaranteed or the resulting cost is grossly disproportionate compared to the improvement gained. The public intuitively acknowledge a seismic risk as acceptable, when it feels that every reasonable step to reduce it has been taken and that the cost for further reduction is not justified by the possible future benefits expected.

Although several groups of people such as investors and insurers have a greater and usually contradicting interest on acceptable seismic risk level, everybody should be aware of it, because earthquakes cause major disasters that affect, directly or indirectly, everybody in the felt areas and beyond. Today, given the globalization, the heavily specialized and mass producing industries, as well as the highly interdependent economies, when a stop or even disruption of production locally occurs, a reduction of supply may result, and thus prices may increase worldwide (high-order economic losses (Rose, 2002)).

20.5 HOW CAN ENGINEERING REDUCE SEISMIC RISK?

A great deal of effort is still necessary for reducing seismic risk down to acceptable levels. In many areas around the world, the actual seismic risk is significantly greater than the acceptable one. Even countries that are highly developed technologically realize after a strong earthquake that they are more vulnerable than estimated.

In general, seismic risk can be reduced by mitigating seismic hazard and reducing vulnerability. Engineers contribute mainly to vulnerability reduction. Engineering can offer several technological remedies for most seismic hazards as well; although plate tectonic movements cannot be stopped, liquefaction or landslide potential can be mitigated by engineering measures.

Vulnerability can be minimized by applying advanced design and construction against earthquakes to every structure. Cutting-edge design and construction against earthquakes can be achieved by

- Investing, in close cooperation with earth scientists, on basic research to understand the natural processes that produce seismic hazards and how earthquakes impact the built environment.
- Learning from past experiences. All lessons from previous seismic events are valuable. Engineers should visit the affected areas immediately after an earthquake, collect as much as possible data (sometimes made available by remote sensing techniques operated by specialists (Huyck et al., 2014)), investigate the real conditions, and interpret the actual performance of structures and compare it with the estimated one.

- Developing better models, tools, techniques, and methods, for example, the Earth-Defense shake table in Miki City near Kobe, Japan, currently the biggest in the world, measures approximately 20 by 15 m and can support building experiments weighing up to 1100 tn; it gives researchers the ability to test full-size structures.
- Improving existing and creating new materials in construction practice. These materials should have not only improved performance under seismic loading but should be more durable, environmental friendly, and affordable as well.
- Enhancing monitoring procedures during operation, instrumentation schemes and testing methods during construction, and investigation techniques during design.

Appropriate design and construction against earthquakes can be applied to every structure by

- Incorporating research findings into building codes, regulations and recommendations. The engineering community should always be alert by being dynamically involved in regulation enforcement and supervision improvement in both design and construction. For example, columns typically fail due to the absence of adequate transverse reinforcement (shear reinforcing steel) and confinement of the concrete causing buckling of the vertical reinforcement; all modern codes address the issue in detail.
- Rehabilitating, upgrading, and retrofitting existing structures and lifeline systems.
- Promoting better education and training on earthquake-related subjects and supporting knowledge transfer to all people involved in design and construction procedure (making educational and informative materials conveniently accessible, disseminating publications of concise, practical seismic-design guidance for practicing engineers and builders (e.g., Dixit et al., 2014), and organizing lectures and conferences). At the same time, engineers should study the traditional construction methods and try to learn from the centuries old know-how still existing in cultures frequently exposed to earthquakes.

Additionally, engineers can contribute to seismic risk reduction by informing, educating, and organizing communities.

Perception of proper decisions toward seismic risk mitigation is strongly influenced by the available information and the way the problem is presented. Decisions guided by intuition generally are not logical when comparing alternatives of low versus very low risk, or high versus very high confidence, since most people can understand probabilities only poorly, especially near the extremes (McGuire, 1992). The engineering community should offer vigorous information and education to the public to help people perceive earthquake hazards accurately, increase awareness, and develop riskwise behavior. Information must be consistent, specific, and able to be believed above and beyond any wrong information that may get in the way (Mileti, 1982; Yanev and Thompson, 2009). The targeted audiences should include people, organizations and businesses, policy makers and planners, managers, and state officials. A characteristic example is the action of the Nepalese Society for Earthquake Technology; it provides free every Friday evening consultation to everybody, house owners, masons, contractors, technicians, and others who want to learn about the earthquake-resistant technology.

Engineers should actively contribute to the enhancement of preparedness at all levels of government and in the private sector, and improvement of skills and technical capacity of relevant emergency response agencies.

20.6 WHO SHOULD APPLY SEISMIC RISK MITIGATION MEASURES?

In most cases, the public recognizes that a seismic risk generally exists; however, few people personalize that risk or do much about it (Mileti, 1982). Moreover, determination of what constitutes acceptable risk for a community is, by its very nature, a political exercise. Therefore, it is the authorities that should apply all necessary measures to mitigate seismic risk and specify minimum levels of seismic performance for all structures.

This can be achieved by legislating and introducing policies that establish all necessary design codes, construction methods, and operating procedures. They should act in an educated way, based on experts' opinion, compromising social benefits and costs, and trying to gain the consensus of the great majority of the citizens; they should avoid following public's beliefs and wishes, and not allow dynamic minorities to command their will.

20.7 EARTHQUAKE PREDICTION

Much discussion and controversy exists about earthquake prediction (Schorlemmer and Gerstenberger, 2014; Kossobokov, 2014; Sobolev and Chebrov, 2014). Today, all agree that we are still far away from what is considered in scientific terms as an accurate prediction. For several scientists and amateurs, it is the Holy Grail; for many people, it is awaited as a panacea, while some believe it will never be achieved.

Seismologists would like to be able to predict earthquakes; their criticisms are driven by their recognition of the need for a rigorous method (Hough, 2002). On the other hand, for the engineering field, prediction is an irrelevant issue; engineers work and will keep on working with long-term predictions of earthquake rates (i.e., seismic hazard). Of course, they will welcome earthquake prediction whenever it arrives, but they remain completely indifferent, since their efforts in reducing seismic risk can never cease.

Although earthquake prediction, if put into good practice, can significantly reduce human life losses, it cannot alter the vulnerability of structures, neither can it substantially mitigate the consequences of seismic hazards; nor can it relieve after-effects to the economy and everyday life. Soil will liquefy no matter if it is known in advance, a nonadequately designed dam or bridge will fail and should be reconstructed, and damaged buildings should be retrofitted to become habitable again.

20.8 CONCLUSIONS

No matter how hard engineers together with earth scientists work, we are still far away from minimizing seismic risk down to acceptable levels. On the contrary, demographic changes, development pressures, environmental constraints, as well as technical challenges increase seismic risk globally. Notably in developing countries, fast increasing urbanization and population growth in areas of moderate and high seismicity, as well as the aging built environment and mainly infrastructure result in the increase in seismic risk at an alarming rate.

Additionally, the engineering community faces several major challenges in seismic risk reduction:

- Understanding the natural processes that produce seismic hazards and cause vulnerability; limiting fragmented research and promoting interdisciplinary effort; managing the diminishing research funds and, if possible, finding new research sponsors.
- Understanding and reliably estimating the consequences of a seismic event, not only to the built environment but also to the society and the economy as well; recognizing the importance of interdependent critical structures to the resilience of communities after a damaging earthquake.
- Inventing and developing efficient, effective, and affordable technologies that both protect human lives and avoid excessive economic losses; these technologies should be applicable with social equity, with priority given to the most sensitive (schools and hospitals) or important sectors (fire department, rescue teams) during an earthquake.
- Disseminating information, increasing awareness, and enhancing education of communities; motivating people to act implementing vulnerability reduction technologies before damaging seismic events occur and discouraging them from just buying insurance in rich developed countries and praying in poor developing countries.
- Assisting authorities and policy makers by proposing priorities, supporting strategies, and promoting policies for vulnerability reduction and preparedness; convincing authorities and policy makers that seismic risk

reduction should be an organized and continuous effort before a disastrous earthquake and that every earthquake-prone area and any kind of structure, either simple or complex like lifelines, new or existing should be included.

• Increasing cooperation among neighboring countries and adopting a global approach in seismic risk-reduction efforts.

The above challenges reflect the difficult task engineers have in the multidisciplinary effort of seismic risk reduction.

REFERENCES

Algermissen, S.T., 2007. Seismic risk. In: AccessScience. McGraw-Hill. http://www.accessscience.com.

Baecher, G.B., Christian, J.T., 2003. Reliability and Statistics in Geotechnical Engineering. Wiley.

Cornell, C.A., 1988. On the seismology-engineering interface. Bull. Seismol. Soc. Am. 78 (2), 1020−1026.

Dixit, A.M., Acharya, S.P., Shrestha, S.N., Dhungel, R., 2014. How to render schools safe in developing countries. In: Shroder, J., Wyss, M. (Eds.), Earthquake Hazard, Risk, and Disasters. Elsevier, London, pp. 183−202.

Hough, S.E., 2002. Earthshaking Science. Princeton University Press.

Huyck, C., Verrucci, E., Bevington, J., 2014. Remote sensing for disaster response: a rapid, image-based perspective. In: Shroder, J., Wyss, M. (Eds.), Earthquake Hazard, Risk, and Disasters. Elsevier, London, pp. 1−24.

Kagan, Y.Y., Jackson, D.D., 2013. Tohoku earthquake: a surprise? Bull. Seismol. Soc. Am. 103 (2B), 1181−1194.

Kossobokov, V.G., 2014. Times of increased probabilities for the occurrence of catastrophic earthquakes: 25 years of the hypothesis testing in real time. In: Shroder, J., Wyss, M. (Eds.), Earthquake Hazard, Risk, and Disasters. Elsevier, London, pp. 477−504.

Lomnitz, C., 1999. The end of earthquake hazard. Seismol. Res. Lett. 70 (4), 387−388.

McGuire, R.K., 1992. Perceptions of earthquake risk. Bull. Seismol. Soc. Am. 82 (4), 1977−1982.

McGuire, R.K., 2004. Seismic Hazard and Risk Analysis. Earthquake Engineering Research Institute. Monograph No. 10.

Michel, G., 2014. Decision making under uncertainty: insuring and reinsuring earthquake risk. In: Shroder, J., Wyss, M. (Eds.), Earthquake Hazard, Risk, and Disasters. Elsevier, London pp. 543−568.

Mileti, D.S., 1982. Public perceptions of seismic hazards and critical facilities. Bull. Seismol. Soc. Am. 72 (6), S13−S18.

Parvez, I.A., Rosset, P., 2014. The role of microzonation in estimating earthquake risk. In: Shroder, J., Wyss, M. (Eds.), Earthquake Hazard, Risk, and Disasters. Elsevier, London pp. 273−308.

Rose, A., 2002. Model Validation in Estimating High-Order Economic Losses from Natural Hazards. Technical Council on Lifeline Earthquake Engineering. Monograph No. 21, pp. 105−131.

Schorlemmer, D., Gerstenberger, M.C., 2014. Quantifying improvements in earthquake rupture forecasts through testable models. In: Shroder, J., Wyss, M. (Eds.), Earthquake Hazard, Risk, and Disasters. Elsevier, London, pp. 405−429.

Sobolev, G., Chebrov, V., 2014. The experience of real time earthquake predictions in Kamchatka. In: Shroder, J., Wyss, M. (Eds.), Earthquake Hazard, Risk, and Disasters. Elsevier, London pp. 449−475.

Starr, C., 1969. Social benefit versus technological risk. Science 165, 1232−1238.

Stein, S., 2006. Limitations of a young science. Seismol. Res. Lett. 77 (3), 351−353.

Vick, S.G., 2002. Degrees of Belief. A.S.C.E. Press.

Wang, Z., 2009. Seismic hazard vs. seismic risk. Seismol. Res. Lett. 80 (5), 673−674.

Yanev, P., Thompson, A.C.T., 2009. Peace of Mind in Earthquake Country: How to Save Your Home, Business, and Life, third ed. Chronicle Books.

Chapter 21

Decision Making under Uncertainty: Insuring and Reinsuring Earthquake Risk

Gero W. Michel

CRO & Head of Risk Analytics, Montpelier Re, Hamilton, HM HX Bermuda

ABSTRACT

Earthquake insurance has helped pay losses and increased resilience after recent events in Chile and Japan. Although those events were large, the world insurance market has not been tested by a catastrophic earthquake that depleted surplus and required capital influx into the market since Northridge in 1994. Earthquake risk is characteristically seen as "remote". Rare earthquake events result in low risk awareness and low risk reward. Insurance fees are competitive and appear insignificant compared to the exorbitantly large losses that could occur. Earthquake insurance penetration varies markedly around the globe, ranging at below 1 percent for most developing countries and between zero and almost complete market coverage for some Western societies. These differences are not necessarily due to the risk itself but rather result from insurance history, wealth, tradition, government involvement, and politics. Insurers and reinsurers assess risk using risk models. With little available loss history, earthquake risk models are subject to a large degree of uncertainty. As for the capital market, only the systemic part of the risk (i.e., the part that cannot be diversified away) is paid for. Uncertainty is referred to as "unsystematic" and is hence not part of the premium. Uncertainty is retained by insurers and reinsurers. Insurers and reinsurers have created strategies to help manage risk from the bottom-up (i.e., deal by deal) as well as top-down (i.e., optimizing their overall portfolio). The latter includes summing up exposed values as well as calculating probabilistic portfolio losses. Over the last 20 years, risk model development has been dominated by a few vendors who have created comparatively similar tools concentrating on detailed exposure input. Insurers have avoided considering variability and ranges of possible results and the market has made decisions assuming that risk numbers are highly precise and accurate. Over the years, vendor models have increasingly been used as a "currency" for trading and hedging risk. A recent model update doubled risk results for hurricane risk in the United States. This was unprecedented, and decision processes have started to change.

Earthquake Hazard, Risk, and Disasters. http://dx.doi.org/10.1016/B978-0-12-394848-9.00021-3
543

21.1 INTRODUCTION: ARE EARTHQUAKES INSURABLE?

Are earthquakes insurable? With >70 percent of Californians living within 30 miles of a fault where high ground shaking could occur in the next 50 years (http://www.consrv.ca.gov/index/Earthquakes/Pages/qh_earthquakes.aspx), a major earthquake in California is likely to be a tragedy for many. This is not only due to a potentially large number of casualties (>3,000) but it is also due to the fact that many will lose their homes, their largest single asset, without the chance of recovering the loss. Homeowners' earthquake insurance penetration (i.e., the amount of insured through insurable assets) is just 13 percent in that state. Although this figure is higher than the 1 percent homeowners' insurance penetration in Italy, it contrasts dramatically with the >30—50 percent insurance penetration seen in some Latin American countries or almost 100 percent earthquake insurance penetration for the United Kingdom and New Zealand. Although commercial earthquake insurance penetration reaches almost 50 percent in Japan, insured commercial loss limits are as low as US $50 billion vis-à-vis potential commercial economic losses of more than US $1 trillion for an event similar to the 1923 Tokyo earthquake. What drives these differences and how does catastrophic risk insurance work?

This chapter focuses on earthquake insurance, how insurers and reinsurers assess, hedge, and trade earthquake risk and what processes help insurers to manage large potential catastrophic losses.

21.1.1 Risk Keeps the Insurance Market in Business

Resilience—the ability to recover the current status quo after a future loss—is the product insurers offer their clients. The product insurers trade with is however "risk" rather than resilience (see for a wider discussion Taleb, 2012). Dealing with risk and risk management is far different from maintaining status quo. Insurance companies, with the exception of some cooperatives or mutual companies, are concerned with return on capital and growth not unlike others in the financial market. Risk modeling is part of the business because the price and payout of the product is uncertain at the beginning of each contract term.

Insurance involves pooling funds from many insured entities to pay for the losses that some may incur. The insurance product is successful if (1) the number of similar exposure units is large compared to the actual losses which occur; (2) losses are well defined and finite; (3) insurance fees are affordable and risk adjusted; and (4) loss events are accidental.

21.1.2 Insurance Companies Rise and Fall with the Capital and Its Cost

Insurance has helped many homeowners around the world. For instance, a conflagration (fire) loss could endanger our social as well as financial well-being since our homes tend to be our largest individual assets. Insurance makes such events bearable. Conflagration affects in general only a few assets and is (at least theoretically) randomly distributed in space. Natural or man-made events—or perils as the insurance world calls them—such as storms, floods, earthquakes, or terror attacks can affect many. In addition, catastrophic events tend to cluster in certain areas such as hurricanes along the mid-to-lower latitudes and along the coast, earthquakes along major fault lines, floods along rivers, and terror attacks in areas that attract a large number of people. For these widespread events, insurance premiums might not suffice to pay for the loss. Risk managers and regulators have hence demanded insurance companies hold capital to pay for a large loss. The amount of capital needed is defined by the probability of a severe loss and/or a percent of insured assets or a multiple to the annual premium. Capital has become the largest asset for insurers since companies rise and fall with the amount of available capital, its cost, and the return on capital.

Most regional insurance companies are unable to hold and deploy sufficient capital to insure major catastrophes. Hedging their low-frequency/high-severity "tail" risk and minimizing capital cost is hence one of two major concerns for primary insurance companies. The other concern is "performance" meaning the quarterly reported profit and growth of the company. Reinsurance companies help insurance companies minimize capital costs because insurers transfer predominantly tail risk shares of insurance portfolios to the reinsurer. Why is it cheaper for reinsurers to take regional tail risk than it is for insurers? Government regulation allows insurance and reinsurance companies to diversify risk within their portfolios. Diversification means that the same capital can be used to cover more than one large loss as long as losses are not, or are only loosely, correlated. Diversification makes sense, as long as losses are rare and unlikely to clash within any 1 year. Without reinsurance, catastrophe insurance would be significantly more expensive and possibly unaffordable.

21.1.3 Earthquake Risk Awareness and Insurance Penetration

The more affordable the insurance product and the higher the risk awareness within a territory, the higher is the insurance penetration and hence the cheaper the insurance policy. Insurance penetration and availability varies around the globe between below 1 percent in developing countries and some countries in the southern European Union for perils such as earthquake, up to almost 100 percent for homeowners' physical damage/property losses in the United Kingdom. Insurance penetration is as much a function of politics, culture, and history as it is based on the price of the product. The higher the insurance penetration in a country, the better diversified the risk for the insurers, and the larger the resilience of a country and its inhabitants. Obligatory insurance guaranteeing full insurance penetration could result in a working solution. Indeed, before 1990, mostly socialist countries such as the former East

Germany, China, and Poland established state insurers with almost complete government guaranties. Governments can afford to offer catastrophe insurance by raising taxes after an event. Government guarantees still exist in various territories for "outsized" loss events that are likely to be insupportable by the private market, alone. These events include large terror attacks and/or nuclear incidents and/or very unlikely "Black Swan" tail events that are outside the scope of otherwise unaffordable insurance fees (Taleb, 2007/2010). Several countries have agreed to "hybrid solutions" with private companies managing state insurance pools (see e.g., Turkey Catastrophe Insurance Pool) and at least some part of the catastrophe insurance being private. The United Kingdom has achieved high insurance penetration across all relevant perils without a state pool for natural perils (Pool Re has been set up by the insurance industry in cooperation with the UK government so that insurers can continue to cover losses resulting from damage caused by acts of terrorism to commercial property in Great Britain). This was helped by the "Gentlemen's Agreement", a contract between local and regional governments and the insurance industry, which dates back to 1961 and ended in 2013. It states that the government takes care of defenses and countermeasures including local or regional flood levees such that all risks in the country can be reasonably well insured against natural catastrophes (see also http://www.jrf.org.uk/sites/files/jrf/vulnerable-households-flood-insurance-summary.pdf). Homeowners' insurance policies are "all-risk policies" in the United Kingdom. Policies are affordable and risk rates are reasonable. Similarly, all risk policies are common for homeowners' building insurance in Baden-Württemberg (DE), Australia, as well as a few other territories in the world. The all-risk homeowners' buildings policies in Baden-Württemberg started from a former state insurer (Sparkassenversicherung) and were triggered by the Hohenzollern Earthquake in 1978. Similar policies from the former DDR went to Allianz after the political change in Germany in 1990. This is opposed to other areas in Europe, the United States, or Asia where windstorm risk tends to be covered along with fire while flood or earthquake risks are often only endorsed, sublimited, or not at all covered. This leads to (1) high costs for these perils; (2) low insurance penetration; and (3) eventually "adverse selection" since only those in high-risk zones are likely to afford insurance. Earthquake insurance penetration is as low as 12−13 percent among homeowners in California. This has led to a deductible of tens of thousands of dollars and a few thousand dollars in premium each year for earthquake and flood insurance for an average home. High deductibles and large insurance fees are likely to have reduced take-up rates further.

High insurance fees might also mean that only the higher net worth individuals can afford to buy cover. This apparent inequality in insurance has led to government wind pools in the United States that concentrate on high-risk areas such as the hurricane-prone coast. The population in Italy relies on its government to bail out homeowners after a large loss. Why should we pay if the government does this for us? This has resulted in a next to nothing penetration of homeowners' earthquake insurance in Italy. In countries such as Italy insurance has to be funded by tax. This arguably works as long as the economy is healthy and the population is prepared to pay losses with higher taxes after the fact. It will be interesting to see how this will work over the next years in a sluggish economy. In developing countries, the World Bank provides low-interest loans after large events. The European Commission, along with the United Nations and other organizations, has also helped developing countries after major events.

Although commercial insurance penetration in Japan reaches 50 percent, earthquake insurance is capped at an overall limit of approximately US $50 billion. Insurance companies therefore pay only for a small fraction of potentially much larger economic earthquake losses of up to several US $ trillion. Indeed, Japanese earthquakes are deemed capable of creating among the highest potential economic losses for any possible catastrophic event topped only by nuclear incidences, meteorite impact, war, pandemic or solar events   (http://science1.nasa.gov/science-news/science-at-nasa/2008/06may_carringtonflare/). Japanese insurers deem several USD trillion losses "uninsurable" considering the available premium to pay for them. Global reinsurers are unlikely to sustain losses of over US $300–$400 billion which is the global amount of insurance capital available for property and casualty events. The risk that companies cannot recover their losses from the companies that insure their risk is referred to as credit risk (Bluhm et al., 2002). The Japanese society and government have made sure that homeowners are "protected" with minimal "credit risk" in the case of a large earthquake. In order to do this, Japan has set up a pool of government earthquake homeowners' insurance that is administered by insurance companies and a specific insurance vehicle. This cover is independent of international insurers or reinsurers outside Japanese regulation.

21.1.4 Allowing Failure Means Keeping Insurance Affordable

Homeowners' insurance coverage in general (exceptions do exist) means full replacement guarantee. Full replacement for all homeowners' risks in an "unlimited" hypothetical event is unfeasible, unaffordable, and in general unlikely, as catastrophes tend to be finite. In order to keep premiums affordable, insurers need to be allowed to fail in very rare loss events. Unlimited cover leads to unaffordable fees and/or impossible amounts of capital and capital costs. Failing is hence an intrinsic part of insurance. Making insurance failure as unlikely and as harmless for the society as possible is the task of government regulation. The likelihood of allowed/regulated failure of a company varies across the world and within markets. It depends on regulation, company rating, and company strategy. Rating agencies (CRA, Credit Rating Agencies, e.g., S&P, A.M. Best, Fitch, and Moody's) that evaluate countries or other financial institutions also rate insurance companies. Rating agencies care about the likelihood of default and credit security herewith making sure investors have a reasonable understanding of the security of the counterparty they are trading with. Regulators care more about citizens and the ability of an insurance company to pay losses and hence maintain stability in a society. Regulators have created country-specific solutions. In order to coordinate regulation further, the EU (Solvency II, (http://ec.europa.eu/internal_market/insurance/solvency/index_en.htm)) as well as Japan, Bermuda, and others have initiated harmonized insurance risk solutions. Unifying jurisdictions across otherwise sovereign countries and practices has however proven difficult and Solvency II has been postponed several times. Those that criticize these initiatives mention the systemic risk that comes with applying the same formulas and principals across states or countries. They argue that inefficiencies are likely to multiply rather than vanish.

21.1.5 Diversification, More is Better?

Insurance is based on spreading risk. Diversification is therefore an intrinsic part of insurance and reinsurance. Diversification can however lead to systemic risk. In other words, the more leveraged one dollar of capital is and the more often a dollar can be reused for deemed independent risks, the less it is available if risks are more interdependent than originally assumed. This was evident in the September 11th terror attacks where property, aviation, as well as casualty losses all happened in the same event. Catastrophe risk premium (both primary and reinsurance) around the world reaches less than US $100 billion, in contrast to around US $40 trillion traded every year in the asset market. The insurance market is strongly regulated and small compared to the financial market and is unlikely to cause systemic risk for the society close to what the financial market has been able to produce.

Systemic risk and diversification are controversially discussed in our market. Some market participants believe that the strong push from rating agencies to further diversify companies is counterproductive and diminishes the ability of companies to payout after events especially if diversification becomes an incentive to enter into low premium/risk ratio lines of business. Climate change has been viewed as reducing the value of diversification with upward trends in event frequencies and/or severity. It is also expected to lead to an increase in events clashing in any 1 year.

The recent Thailand floods (2011) or the Tohoku (2011) earthquake have highlighted the potential for indirect and supply chain losses. Major earthquakes or floods in countries with large foreign direct investments such as Thailand can cause a hiatus in the global supply of semiconductor car manufacturing, food, and other commodities or specialized products if redundancies in the supply are limited. Interdependencies of western societies with emerging markets are growing. Challenges seem to be greater for countries where growth and foreign direct investment are large, and loss history and regulation might not suffice to manage future losses. These countries might also not be on our radar screen for large international losses: who would have expected Thailand could produce the largest ever flood loss across the globe? In our competitive world companies that minimize inventory are far better off (Lee, 2004; Slone et al., 2010). The need for "just in time" product supply has been increasing our dependency on supply chain efficiency and failure herewith increasing the likelihood of indirect losses. The insurance market has been trying to keep indirect losses such as business interruption manageable by tying payouts to physical losses. This means that policies only pay business interruption if the risk has suffered physical damage. "Contingent Business Interruption" policies are however paying supply chain-related business downtime. These products seem to be on the rise.

Other possible global events include a repeat of the 1859 geomagnetic superstorm, the Carrington event. A solar flare or coronal mass ejection hit Earth's magnetosphere and induced the largest known solar storm. A repeat of this storm could potentially affect large parts of North America, the EU, as well as South Africa resulting in several months of business interruption. A solar event can create both physical losses such as fires and business interruption in a single event. Solar event losses are not excluded from most insurance policies, and event scenarios suggest potential insured losses of several hundred billion USDs. Large pandemic events or a further decrease in the state of the economy might also create global losses. A distressed economy leads (1) to decreasing insurance penetration with less available money for most of us to buy cover; (2) reduction in government or individual spending for loss mitigation; or (3) increased likelihood of riots or moral hazard such as an increased number of fraudulent claims. A distressed economy is however likely to reduce the surge of demand versus supply of construction workers after an event as more workforce is available for reconstruction (http://www. willisresearchnetwork.com/assets/templates/wrn/files/WRN%20Demand% 20Surge%20Handout.pdf, Anna Olsen and Keith Porter).

21.2 INSURANCE RISK MANAGEMENT

Recent insurance failure has been dominated by mistakes in underwriting and pricing (P. Kovacs, 2012, http://tools.cia-ica.ca/meetings/annual/2012/ Presentations/Session_7_Kovacs.pdf). Failure due to large catastrophic losses is usually the result of mistakes in risk management as opposed to mistakes in pricing. Risk management errors include (1) underestimation of the frequency of events and/or (2) miscalculation of the correlation of risks and/or (3) underestimating the possible size of a loss footprint, and/or (4) too high a credit given to diversification and mitigation or hedging benefits. The Thailand flood losses in 2011 or the September 11th World Trade Center losses are examples of an unexpected clash of losses. Clash events might combine business lines such as casualty, personal accident, aviation, and property otherwise deemed independent.

Overexposure to underpriced high-frequency losses is the reason for most of the recent failures of insurance companies around the globe. High-frequency catastrophic business is concerned with smaller events affecting the annual performance rather than the tail risk of an insurance company. Capital is often not allocated to these high-frequency losses and pricing is competitive. Changes in the frequency of these small losses can undermine performance. Senior management is less concerned with these risks as they are not viewed as affecting the capital base. Underwriters price risk with too small or misleading recent history. Risk managers tend to assume Poissonian distributions (http://en.wikipedia.org/wiki/Poisson_distribution) as best describing frequency. This results in negligible clash potentials (Silver, 2012). Processes including spatial or temporal clustering can lead to skewed frequency distributions and resulting frequencies are vastly different from a Poissonian distribution. Examples include (1) triggered earthquakes and/or aftershocks, (2) favorable conditions for windstorms in a territory or year that can lead to clustering (http://www.willisresearchnetwork.com/assets/templates/wrn/files/WRN%20European%20Windstorm%20Clustering%20-%20Briefing%20Paper.pdf, David Stephenson, 2012), and/or (3) clash potential that might increase with severity as losses become increasingly dependent on global infrastructure or supply chains. Examples include the series of devastating earthquakes in New Zealand in 2010 and 2011, the cluster of the 2004 hurricanes, the 2011 Thailand loss, the unprecedented clash of US tornadoes in 2011 (http://www.noaanews.noaa.gov/2011_tornado_information.html) and the recent mortgage crises.

21.2.1 Pricing and Exposure

21.2.1.1 Pricing Risk

Catastrophe risk pricing is derived using risk zone rates agreed within a company and/or based on rates suggested by specific consulting and rating organizations. Traditionally, high-resolution modeling did not have a major effect on pricing schemes. Insurance pricing tends to be "smoothed" across a portfolio resulting in policy owners in lower risk zones paying for those in higher risk areas. As a result of more recent risk-based pricing, larger rating differentials have however entered the market. Risk-based pricing makes sense especially for those perils that produce large individual claims. Examples include rating schemes for assets (1) close to the coast, or on (2) flood plains, or (3) property close to well-known active faults. Differential pricing schemes are increasingly used to manage exposure accumulation in the highest risk zones. Risks located on a flood plain or in close proximity to a well-known active fault can lead to the insurer asking for exorbitantly high rates leading clients to reject the cover. How much can be rejected might be regulated depending on the country and/or peril. The smaller the insurance penetration and the higher the differences in risk per zone, the larger the chance of "adverse" selection. Adverse selection potentially results in too little available premium for high-risk territories.

The promise to pay and a long-term commitment are intrinsic parts of the risk-sharing concept. Insurance decision should therefore be based on stable solutions. Large changes in pricing schemes or model results clash with this fundamental principle of insurance. Many insurance companies own tools that translate catastrophe model results into pricing schemes. These schemes can be based on loss costs (Average Annual Losses) or indices derived using hazard maps. Examples of hazard pricing include the use of flood maps for the United Kingdom or Germany (https://msc.fema.gov/webapp/wcs/stores/servlet/FemaWelcomeView?storeId=10001&catalogId=10001&langId=-1). Some insurers base catastrophe pricing on tail risk rather than on loss cost or on a combination of various methods. Tail risk measures rather than loss costs can be useful for pricing if risk curves are steep and loss costs hence are unlikely to represent risk adequately. Earthquake insurance rates are either uniformly distributed or differentiate risk within limited steps. Differential earthquake risk rating is common in high-risk zones such as California or Japan. More complex differentiation schemes however exist for such areas as flood insurance in the United Kingdom or for commercial property. Although mitigation features are increasingly included in models, they are rarely used to price insurance risk.

21.2.1.2 Exposure and Underwriting

Replacement values or the monetary amounts needed to replace an asset are referred to as "exposure" in our market. Exposure values are not "absolute" but rather based on indices such as square footage or the number of rooms in a house. Other measures inflate historical benchmark data over time using construction indices and growth. These measures might not be unique within a territory and can change. Replacement costs of a building are subject to market prices and a function of current construction costs. Vendor catastrophe models link exposure to construction cost changes triggered by large loss events. This is explained by changes in demand for construction work after a large event. Inflated replacement costs due to higher demand after a catastrophic event are referred to as demand surge. Some recent ideas however suggest that demand surge is neither uniquely distributed in an area nor is it a direct function of the size of an industry loss. Demand surge might rather be influenced by local and/ or regional politics, claims handling practices, as well as the state of the economy.

Exposure is not closely related to the value of a house in the open market because house prices are a function of the neighborhood, the value of the ground and more, rather than the sheer cost of rebuilding it. Content exposure is often based on customer estimates and/or is a function of the value of a building and/or the apparent wealth of an owner. Examples of recent changes in exposure values include >10 percent inflated exposure values from 1 year to the next in otherwise stagnant markets in Asia and Oceania. Changes were explained to be caused by inefficiencies in existing indices after losses. Exposure in developing as well as emerging markets can vary markedly from year to year due to growth and/or changes in reporting. Exposure measures might also reveal little about the quality of the exposure and/or how a property owner handles her asset or how claims are handled. Larger commercial or industrial risks are often augmented in more detail with engineers and surveyors employed by, or working as, consultants for insurance companies. Exposure values are derived at different resolution levels from high-resolution location measures (latitude and longitude) or street address levels over several digit postcodes to "Cresta Zones" (Catastrophe Risk Evaluation and Standardizing Target Accumulations) or even down to country resolution.

Underwriting is bound to cumulative exposure limits that are calculated based on risk tolerance in defined territories. Internal rules define underwriting capacity for how much risk a company can take of large buildings, building blocks, streets, districts, countries, and perils. Reinsurers have increasingly demanded high-resolution exposure information from insurers. This is mainly due to the fact that vendor catastrophe model results depend on exposure distribution and resolution. Insurers as well as intermediaries augment and process data further before passing it onto reinsurers. One main reason for this is to erase errors in data and double counting. Exposure processing is often done using catastrophe model results. Some brokers in two European markets have revealed that exposure indices are augmented based on inflated actual versus modeled historical losses. This means that exposure values can be changed if models show results higher than what the inflated historical losses are. The argument is that this is an efficient way to take out model biases. Even if this would be correct, this would only be true if only one model is used in the market, which is of course incorrect. In other cases, low-resolution exposure is disaggregated by insurers, intermediaries, or modeling companies. It is not always easy to restore the original data and/or determine how much of the high-resolution information is genuine or constructed. Within the framework of Solvency II, regulators have required a reduction of "fraud" in data generation and exposure processes are likely to be further regulated going forward. Risk managers have started considering exposure "stochastic" rather than fixed. Reference tools serving as benchmarks for company exposure data have been entering the market recently. Exposure reference tools range from aggregate measures to detailed exposure attributes, construction types, and mitigation features, among others. These reference values help risk managers to manage and adjust exposure values. New open source and open development initiatives such as "OpenStreetMap" (http://www.openstreetmap.org/) will help this process further. Current risk models are driven solely by exposure. Human behavior and how a property is used and maintained along with claims management procedures are not considered. Recent losses suggest that these "soft" factors can change losses by single digit multiples depending on peril and area (http://www.vce.at/SYNER-G/pdf/reference_reports/RR5-LB-NA-25882-EN-N.pdf).

Exposure measures without underwriting judgment and scrutiny are deemed an inefficient proxy for risk. For some lines of business, premium and loss history are the sole measure used to assess risk. This is also common for the high-frequency low severity loss range that is often dominated by company idiosyncrasies or much too small numbers of events in common modeling software.

21.2.2 Risk Accumulation, Limits, and Probable Maximum Losses

21.2.2.1 Deterministic Accumulation of Risk Limits

Most companies define their target risk accumulation in catastrophic risk zones ahead of underwriting. This is done for each quarter, annually, or for the next few years. Targets are based on capital availability, current pricing, as well as company risk tolerance. Maximum limits are defined deterministically based on premium multiples or exposure multiples and/or based on stochastic model results. Multiples have been regulated in many markets and have been determined using either historical losses and/or based on market practice. Insurance companies in Chile had managed their earthquake risk to 10−14 percent of the overall sum of the Santiago de Chile exposure, their largest single Cresta (https://www.cresta.org/) zone accumulation. The idea was that a large single event is likely to destroy up to about 10−12 percent of the larger Santiago de Chile area. These measures had limited insurance company growth. Risk concentration was hence capped earlier than what some risk models would have allowed. With these measures, company failure was largely prevented after the 2010 8.8 magnitude earthquake in Chile despite the fact that economic losses reached between 8 and 10 percent of the country gross domestic product. Such loss levels would have left many companies bankrupt in other territories of the world. Reinsurance pricing was however based on less conservative stochastic models. Chile exposure is a minor risk for most reinsurers keeping calculated capital cost for Chile at minimal levels. Reinsurance pricing was hence competitive. Coinsurance or the sharing of losses between insurers and reinsurers is limited in Chile. Asia, Oceania, and Eastern Europe are other areas where insurers can afford to hedge their catastrophe risk down to minimal levels with reinsurers.

Early on worldwide Cresta zones were considered merely independent by some insurance companies. Cresta zones hence allowed monitoring accumulation of limits across the world. Ten to 15 years ago, insurers and reinsurers were measuring their aggregation based on several hundred Cresta zone limits around the globe. Cresta zones are independent for small events, only. "Realistic Disaster Scenarios" (RDS) have been used by the syndicated Lloyd's Market (http://www.lloyds.com/the-market). Lloyd's maintains a set of mandatory RDSs to stress test both individual syndicates and the market as a whole. The event scenarios are regularly reviewed to ensure that they represent material catastrophe risks (http://www.lloyds.com/~/media/Files/ The%20Market/Tools%20and%20resources/Exposure%20management/RDS_ Scenario_Specification_January_2013.pdf). These scenarios capture exposure limits with respect to large "designer" or actual historical events. They relate exposure to losses. Losses are derived using combined loss costs for individual subterritories or lines of business often using specifically constructed events provided by model vendors. The downside is that these rely on a limited set of events. Large losses for areas or risk combinations outside the RDS can cause surprises. More recently, a company (http://www.artemis.bm/blog/2013/05/ 28/karen-clark-co-launches-windfieldbuilder-enhances-visibility-of-hurricane-threats/) has created tools that seamlessly distribute "characteristic" designer events in order to determine a distribution of limits as a function of "realistic" event footprints. Rating agencies and regulators require companies to report exposure zone information for defined zones and territories.

21.2.2.2 Probable Maximum Losses

"PMLs" originally stem from the notion of "probable maximum loss" or "possible maximum loss" (http://www.casact.org/pubs/proceed/proceed69/ 69031.pdf. Is PML a Useful Concept?). These measures considered calculation of certain tail losses that are unlikely to be exceeded. Recognizing the subjectivity of these measures, PMLs were then related to certain nonexceedance probabilities. These probabilities are referred to as return periods (RPs) such that $RP = 1/(1 - \text{nonexceedance probability})$. Benchmark RPs were commonly set to 1 in 100 "Value at Risk" (VAR) events (http://en.wikipedia.org/wiki/Value_at_risk). During recent years, the term evolved further and PMLs now include various RPs as well as Tail Value at Risk (TVAR) measures (Bargès et al., 2009; Jorion, 2007). These measures are commonly derived using stochastic in-house tools or vendor models and can be both relative (loss divided by replacement cost) and absolute. Measures range from companywide PMLs to PMLs broken down to regions, peril, Line of Business, among others. Insurance companies manage their risk to regional PMLs and limits. Global reinsurers' earthquake PMLs often includes California N (larger San Francisco) and S (larger Los Angeles), the larger New Madrid zone as well as individual country risk around the world. The resolution used by a company to manage their PMLs depends on senior management (CEOs and CROs), risk appetite, concentration, and available capital along with rating agency demand and regulation.

Regulators and rating agencies also demand certain "stress tests" for which theoretical losses of different regions and perils are combined within a year. Their aim is to ensure that the company can sustain extreme loss years with event clash. The upside of these measures is that they allow comparison of company risk. The downside is that such measures might not suffice to describe idiosyncrasies of companies well. The year 2011 was one of the largest ever catastrophe loss years for the insurance market. It included various catastrophic losses such as the Thailand and Australian floods, the Tohoku JP Tsunami, the New Zealand earthquake series, and the largest ever US tornado losses. Many of these losses were not specified in most of these tests. PMLs are compared to premium and capital. Today, PMLs depend largely on the model used, model completeness, and model settings and loading. Loading is specific to model users and include inflating exposure and loss costs due to nonmodeled perils seasonal forecasting and/or uncertainty. Loading depends largely on senior management's risk appetite and has been used to manage underwriting appetite as well. Until recently, most companies did not or only very crudely considered model error. Even now, only very few companies consider differentiating epistemic (knowledge) and aleatoric (random) uncertainty in their underwriting (Vose, 2001; Hubbard, 2009; Kahneman et al., 1982). Understanding what is random and/or learnable might however be one of the largest differentiators among companies in terms of risk management. PMLs are considered without error bars. This might sound odd for many outside this market given the fact that model error and uncertainty is likely to range in the same order of magnitude as the mean values. One reason for the "nutshell numbers' concept" is the fact that PMLs are used as quarterly performance measures. Companies plan to reduce or increase the risk in certain territories based on the current status of the market, availability of risk, as well as capital strategy. Freezing model parameters and erasing variability allows for the monitoring of even small changes. Single values also help intercompany communication and trading. Rating agencies and regulators have demanded certain model settings and also seem to require companies manage to "point estimates" (which is not necessarily their intent). The author assumes that this will change in the near future with increasing awareness of the uncertainty inherent in these measures.

21.3 INSURANCE AND REINSURANCE, TWO SIDES OF THE SAME COIN?

We have treated the insurance and reinsurance market as merely one and the same in the above sections. Reinsurers are more diversified than insurers and often global (Kiln and Kiln, 2001; Vaughan and Vaughan, 2003; De Weert, 2011). Reinsurers can leverage catastrophe risk at higher levels than local or regional insurers. The smaller an insurer, the steeper the risk curve and hence the smaller the average annual loss (loss cost) as opposed to a more "heavy" tail. Fifteen to 20 years ago, the fortunes of insurers and reinsurers were largely aligned. This was due to the dominance of pro-rata deals in the earlier reinsurance market. Pro-rata reinsurance programs were constructed such that reinsurers would take a share of the insurance risk across all levels for an acquisition fee. Both insurers and reinsurers hence shared profit and losses

(Riley, 2001). Unlimited tail cover for various deals around the globe was however becoming dangerous for even large reinsurers. A 1-in-1,000 year loss might not be a concern for a single territory, but it might become hazardous if several of these deemed independent territories are combined in a reinsurance portfolio. An outsized loss at a probability of 0.1 percent might be bearable for a large reinsurer, but 10 equal and independent zones would shift outsized losses into the 1 percent probability range. Reinsurers therefore demanded "capping" the high-severity, low-frequency "tail" risk. Pro-rata deals were also expensive for insurers as their main need was to reduce their tail risk rather than the higher frequency risk range that could be paid by the annual fees. Pro-rata reinsurance programs were hence increasingly replaced by an excess of loss deals. Excess of loss deals cover only certain "risk layers" capped at attachment and exhaustion points. Intermediaries further boosted this trend by linking insurance and reinsurance companies. Intermediaries or so-called "brokers" who specialized in catastrophe risk assessment and trading added value by helping small insurers to make up for a deficit in risk knowledge compared to globally exposed and experienced reinsurers. This trend toward excess reinsurance decoupled insurance from reinsurance company profits.

A pronounced example of this bifurcation is what we might refer to as the "Florida bubble" (see for comparison: Wiedemer et al., 2011). High catastrophe losses coupled with regulation and rating agency demand in high surplus levels have kept primary "insurance" profits in Florida at low margins for most of the last 20 years. This trend was boosted further by high vendor model risk results. With reinsurance demand larger than supply, Florida reinsurance prices skyrocketed after the 2004 and 2005 losses. Reinsurers have since seen strong profitability in Florida with the absence of hurricane losses after 2005. This offset in reinsurance versus insurance profit was further boosted by "climate change discussions" and the so-called "near-term increases" in modeled risk. Trends toward higher losses were explained by climate change and/or multidecadal variability (Mingfang et al., 2009). Nature, at least so far, did however not agree with these concepts and losses have been significantly below long-term average over the last 8 years. The "Florida bubble" began to burst in 2013 as traditional reinsurers faced losing their share as new financial investors and collateral insurance entered and added greater than 30 percent of new capital into the market.

Greed and fear are part of the insurance and reinsurance market, not unlike other financial sectors. Fear means having too much risk in the book and/or losing precious business to competitors. Greed arises after a loss when companies that had not participated in the loss try to take advantage of the distracted market. Fear of potentially large losses increases the demand and price for insurance and reinsurance products, whereas greed diminishes it. Fear and greed can lead to nonrational, short-term behavior. Price reductions to nonprofitable levels are one of the effects of this.

21.4 MANAGING THE UNKNOWN, INSURANCE RISK MODELING

Risk models in the insurance market date back to the 1970s when a company called Wiggins Co. created the first earthquake risk software. Companies that dominate risk modeling such as AIR (http://www.air-worldwide.com/Home/AIR-Worldwide/), EQE (http://www.eqecat.com/), and RMS (http://www.rms.com/) were founded in the late 1980s. Reinsurers such as Munich Re and Swiss Re as well as brokers (EW Blanch, Greig Fester) and a few larger insurance companies started to create their own risk models in the early 1990s. Earthquake risk models entered the market following the 1994 Northridge earthquake. Catastrophe model use was rare until 1992 when Hurricane Andrew caused unprecedented losses close to US $20 billion. Actuarial models commonly used in insurance companies had taken a decade of hurricane history as basis for risk assessment (Figure 21.1). The decade ending in the early 1990s had enjoyed a "low hurricane activity regime (Figure 21.1)". Changes due to inflation and high growth rates had however masked higher activity rates that had happened decades before. AIR had used longer term storm frequency and severity as a proxy for forecasting large future losses.

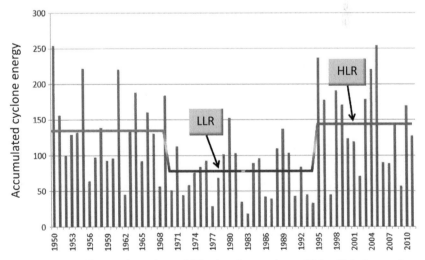

FIGURE 21.1 Changes in regimes: LLR = low loss regimes, HLR = high loss regime. V_{max} = maximum 6-hourly wind speed. The Accumulated Cyclone energy was derived using Hurdat information (http://www.aoml.noaa.gov/hrd/hurdat/). Large changes in activity in the mid-1990s coincided with a several-fold change in insured Hurricane losses. Reasons for the changes in activity are controversially discussed and range from Climate Change to oscillations in the Atlantic Sea surface temperature (AMO, Schlesinger and Ramankutty, 1994) and variability.

21.4.1 Risk Regimes

With Hurricane Andrew, a new "activity regime" was about to start that caught insurance companies by surprise (Figure 21.1). "Regimes" here are considered being levels of higher or lower activity stable over a certain amount of time interrupted by short transition periods (see also Scheffer and Carpenter, 2003). Different reasons are considered for the apparent change in activity in the mid-to-late 1990s. Changes in regimes leading to potentially unconsidered large losses are not uncommon in recent history although the pronounced change in hurricane activity of the 1990s remains unique in its impact within the insurance world. Less pronounced examples of apparent changes include increased windstorm and flood loss activity in Australia, apparently changing flood risk in Central-Eastern Europe and clusters of earthquakes such as those in Christchurch, New Zealand in 2010/2011. Climate change has been blamed for changes in hurricane activity. This has led to various well-known debates about increasing risk in a changing climate. Modeling companies as well as intermediaries have been alerted to various apparent regime changes including decreasing and increasing frequencies. Apparent recent downward trends in European windstorm and Japanese typhoon are examples of these changes. Many of the so-called regime changes might however not have been sufficiently backed up with data, and the topic has been discussed controversially. Unprecedented 1992 hurricane Andrew losses of $40—$50 billion (adjusted for 2013) have since been topped by $65 billion hurricane Katrina losses in 2005 and cumulative losses of US $120 billion in 2005 by >120 billion cumulative losses in 2011. Climate variability and change (Knutson et al., 2010) as well as worldwide penetration and exposure growth might be the reason for these changes.

21.4.2 Catastrophe Models

The change in hurricane activity caused an increase in demand for reinsurance as well as an increase in primary windstorm risk premium. The year 1993 saw the advent of analytical (Re)insurance Companies (Davenport and Harris, 2007), which were founded on catastrophe models. Companies started to hire scientists for natural catastrophic risk analyses with the main purpose of determining the maximum possible downside for all relevant catastrophic risks for a country. Early risk model development migrated from reinsurance companies to dedicated risk modeling vendors which employed well-educated model developers and engineers. Until today, insurers relied largely on recent historical losses as benchmarks. Catastrophe models were expected to follow historical loss experience closely. Insurers were also interested in simplified "nutshell" single risk factors and demanded, at least early on, "black box solutions" that could not easily be manipulated by internal staff or external business partners. Publicly available or government models had little access to our industry. Many of the government models concentrated efforts on hazard (HERP, (http://www.jishin.go.jp/main/index-e.html) USGS (http://earthquake. usgs.gov/hazards/)) rather than risk (http://www.fema.gov/hazus) or were simply too complex or time consuming to set up and run for reinsurance purposes. With little available historical loss data, models had to be (1) calibrated to recent loss experience; (2) easy to use; (3) as globally comparable as possible; (4) as fast as possible; and (5) run on personal computers. Speed remains an issue in our market as companies aim to run millions of exposure data points for tens of programs every day. Most of these five points were not major concerns for government science.

21.4.2.1 The More the Better

Following early ideas that hazard and vulnerability are independent and scalable, vendors disaggregated models and model results further. The easiest way to do this included modeling risk at progressively smaller and more detailed levels of exposure. High-resolution loss data are scarce even in the insurance industry, and it has been difficult to allocate high-resolution losses to high-resolution exposure. Data to calibrate loss models are in general restricted. Dividing scarce data for model calibration and validation has been deemed inefficient. Model validation was hence based on making sure that the model reflected historical loss occurrences as close as possible. This resulted in almost perfect coincidence between recent history and model results given the growing skill in parameterization and calibration. Deviations of 15–20 percent of modeled loss to historical outcome were hence deemed as indicating model bias or as a result of not sufficiently detailed exposure data. Exposure quality and resolution have since become a major focus of the insurance industry. As of today, street-level exposure resolution has become popular for the United States, Canada, Europe, and Australia, among others.

The only way to run exposure through risk models at the speed demanded by reinsurers was to strip out most of the simulation process. Risk models were hence constructed as giant lookup tables that allowed feeding a fixed and deemed complete stochastic event set with company exposure. Vulnerability modifiers or loss modifiers were used to account for further detailed exposure attributes such as occupancy types and mitigation features, among others. Higher demand for high-speed, high-resolution platforms created a challenge for model vendors. High-performance computing technology was not yet available in the world of Microsoft and insurance companies were not prepared to switch from Microsoft to Linux or Unix. High-resolution models required detailed modeling of the variability inherent in the catastrophic losses. The best way to achieve this would have been to increase the number of modeled events. The larger the number of events included in the model, the slower the speed of the "lookup process". Vendors were therefore forced into compromises. They sacrificed variability and kept the number of events to manageable levels. This meant that insurance companies were able to feed high-resolution exposure into risk models. This allowed modifying vulnerabilities and losses according to specific exposure attributes. Proxies were used to help compute primary policy conditions. This worked reasonably well for territories and perils that created large loss footprints at small average claim sizes. However, it created challenges if the number of modeled assets and/or if event footprints were small. Tail risk measures were difficult to prove "inaccurate" given the ambiguity in losses and/or RPs. Examples of years and perils that created actual events far outside modeled events included the 2004 clash of hurricanes, the 2005 Katrina year, the 2010 and 2011 tornado events, and the New Zealand and Tohoku earthquakes.

21.4.2.2 All Inclusive Hedging and Trading Platforms

The three major vendors (AIR, RMS, and EQECAT) started to educate companies further about hazard and natural catastrophe risk and companies increasingly outsourced their risk assessment to the same vendor model. Rating agencies and regulators started to demand "harmonized" model use. Companies required financial tools and complex deal structures to be covered with these tools. By now the models administer all relevant levels of risk from (1) simple homeowners' policies; to (2) complex structured commercial deals; to (3) composite and structured reinsurance deals with various levels of inuring business; to (4) holistic portfolio management; and (5) incremental capital pricing. The major purpose of these risk models changed from deal-by-deal, "bottom-up" risk assessment, to "top-down" portfolio management, and finally to hedging ad trading. These models had become far more important than original pricing models had anticipated, as they started to provide a currency, a "virtual trading platform" for hedging and trading complex risk products the market was unable to trade before. With these models the three major vendors had created a need for a new "model product" that had not been considered relevant for insurance business in the past. With only three relevant modeling companies and high entrance hurdles, companies were able to demand exorbitant fees. Models did not replace any earlier products hence creating a "blue ocean product" (Kim and Mauborgne, 2005). Higher model penetration created a need for a new breed of insurance analysts called "modelers". Modelers are trained to run vendor models, structure and manipulate gigantic exposure data bases, and help underwriters and risk managers calculate loss curves. Hiring an increasing number of modelers increased costs and anchored modeling deep within company risk management. This process was however not without criticism. Critics stated that (1) models need to be forecasting rather than "hint-casting" tools to create value; (2) models are overfitted meaning the number of variables is too high for the data available to calibrate models; (3) model accuracy is only loosely defined; (4) the number of events is far too small to even model county resolution for many perils; (5) systemic risk is increasing in the market as most companies base their assessment on the same models; and (6) most companies seem to have "outsourced" their "brains" and hence part of their competitive advantage to products they might not sufficiently understand.

21.4.2.3 A Model Bubble?

The "model bubble" at least partly burst in 2011 when the largest model vendor released a new model version that changed risk far beyond what users had estimated to be possible. How could we believe in risk differentials at address levels if the vendor can change overall country risk results by up to >100 percent? The new model version also violated an "unwritten" model development notion stating that "a risk model is only successful if model risk results for the average insurance or reinsurance deal do not exceed pricing levels in the market". The new model version created tail risk losses and loss costs larger than what the "hard" (high relative prices) market was prepared to afford. Since then companies have started to question model accuracy further and various new initiatives (see e.g., http://www.oasislmf.org/the-oasis-community/e-marketplace/) concerned with model skill and uncertainty have entered the market. "Parsimonious" or "Occam's razor" (Baker, 2004, retrieved July 25, 2012) are newly introduced terms in our industry. Insurance companies have increased investment in internal model development. Tools that simulate events at various resolutions have started to replace "lookup" tables. The market is opening up for variability and decision making under uncertainty as opposed to managing risk using "all-inclusive nutshell values". The Public Private Academic (PPA) insurance and government partnership GEM (http://www.globalquakemodel.org/), the Global Earthquake Model has started implementing many of these ideas into their model development. Other PPA initiatives such as the GVM, the Global Volcano Model include insurance, private, and government science and highlight this trend further (http://www.globalvolcanomodel.org/). Recent changes in regulation such as Solvency II seem to foster rather than hinder the trend of taking risk assessment back into the insurance companies.

21.5 EARTHQUAKE INSURANCE, HAS IT BEEN SUCCESSFUL?

Covering high severity, low-frequency earthquake risk has been a challenge for many years. How can you explain to a policyholder that she should pay for losses perhaps even her grandparents might not have experienced? Why should I pay a fee for something that is unlikely to happen in my lifetime? Even if policyholders pay the fee, are fees ever likely to be sufficient given the fact that covers are expected to amortize over decades to centuries and require long-term investments? Switzerland, one of the most developed countries with the highest living standards and security, is almost completely uninsured against earthquake risk despite the fact that the 1356 Basel earthquake (Mayer-Rosa and Cadiot, 1979) was among the most devastating earthquakes in the central European history. Insurance subject to very large loss events seems to only make sense if risk can be largely spread among capital providers. Global reinsurers and the capital market have to play a major role in taking earthquake risk. The more the earthquake risk is diversified within an insurance portfolio, the more the loss is bearable at a small premium for earthquake covers. Consider 100 independent risk zones with each zone creating a large and devastating earthquake loss of US $1 billion at an average of every 100 (1 percent probability) or more years. Assume that the premium is roughly two times the expected loss and 3 percent of the limit for each zone. Let us assume further that higher frequent earthquakes are unlikely to cause significant losses to the portfolio and the loss is capped at US $1 billion for each zone. A global risk taker who takes 10 percent of all these 100 zones would have at average one $100 million loss in a year. This loss could be paid, with the collected $300 million in premium (100 million × 100 zones × 3 percent). This would leave the risk taker a healthy profit recognizing common expenses of between 25 and 35 percent. The challenge is that (1) there are no 100 equal earthquake risk zones available in our market; and (2) even if there were, earthquakes might cluster in time and more than only one earthquake is possible in each year. This part of the risk related to uncertainty rather than the "mean" is considered "random and unsystematic". Companies are remunerated for systemic risk—the risk that is deemed impossible to diversify away. Unsystematic risk is, in general, not compensated (see e.g., Porras, 2011). The assumption of random distributions in time and space might not hold true if earthquakes cluster spatially or temporarily. The less likely a risk and the further away from experience, the more a risk taker has to rely on models. Such reliance did not always work.

21.5.1 2011 Losses in New Zealand and Japan

New Zealand's Earthquake Commission (EQC) is a Government-owned fund, which provides natural disaster insurance to the owners of residential properties in New Zealand. EQC was established in 1945. Coverage includes earthquake as well as landslides, volcanic eruptions, hydrothermal activity, and tsunamis. For residential properties, land, storm and flood damage is covered. Prior to the 2010 quake, the EQC had collected a fund of NZ$5.9 billion, with NZ$4.4 billion left prior to the 2011 quake, after taking off the NZ$1.5 billion paid for the 2010 earthquake. As of April 2012, the EQC has paid out NZ$3 billion in claims for both earthquakes, and is expected to pay a further $9 billion, which will deplete their funds and may require a further NZ$1 billion from the government. Insurance penetration is high for NZ and indirect losses are partly covered (e.g., land covers). The insured economic loss is deemed >60 percent for NZ. The earthquake cover was originally built for a repeat of the 1855 Wellington Weirarapa Magnitude 8.1−8.3 earthquake rather than a 6.3 or 7.3 magnitude quake in Christchurch.

Although we might argue that the pool concept has worked for many policyholders in Christchurch, the lack of available cover for Wellington and other areas needed a new solution. Two companies, AMI, a large Christchurch-based mutual company, as well as Western Pacific, a small Queenstown-based private company were in trouble after the event due to large number of claims (http://www.treasury.govt.nz/publications/informationreleases/canterbury earthquakes/insurance/t2011-662.pdf). The public was unhappy with the recovery speed and performance of the insurance industry (Miles, 2012). Dunmore Publishing, NZ.

The magnitude 9 earthquake and tsunami in Tohoku, Japan in 2011 seemed to have occurred as expected for a large earthquake for the industry. Tohoku losses reached <30 percent of the overall insurance protection available in the market. The covered insurance loss amounted to roughly US $35 billion. Most of this loss (~80 percent) was due to ground acceleration rather than flooding. The insured loss of US $35 billion compares to an economic loss reaching almost US $300 billion making it the costliest natural disaster in world history (http://www.cedim.de/english/2232.php). The Bank of Japan offered US $183 billion to the banking system to help normalize market conditions, which were already in a difficult situation due to the ongoing global mortgage crisis. Mutual companies (cooperatives), including those that specialize in agriculture or fishery products, paid a large part of the loss. Being Japanese, these companies had large solvency margin ratios meaning they had relatively high amounts of available capital compared to the losses that occurred. This is due to the fact that Japanese clients compare solvency margin ratios from insurance companies, giving those with higher ratios a competitive advantage. This leads to low returns on capital in the Japanese market. The homeowners' government pool (http://www.sonpo.or.jp/protection/disaster/earthquake/0015. html) seemed to have worked as anticipated and was not exhausted by the event. Despite significant government aid, large parts of the loss were covered by private commercial and industrial companies. A repeat of the Tokyo 1923 event is expected to deplete all funds leaving large parts of the homeowners and commercial companies in Japan at a loss.

21.5.2 Pools Around the World

Earthquake insurance is compulsory in Turkey. This has led to an insurance penetration of roughly 50 percent. The homeowners' insurance cover includes replacement of the building. Older buildings that do not follow the current construction code might require extra funding in the case of a loss, but insurers do not tend to offer to adjust a building to new and better codes and standards due to the notion of "moral hazard". Are policyholders tempted to allow for unnecessary losses if they believe their asset will gain value after a loss? Extra costs for adhering to the code after a loss might lead to uncertainty and delays and/or to a more sluggish enforcement of the code (the notion is that this might have happened in several recent earthquakes especially in developing and/or emerging markets). Returning to "normal" quickly has often been the priority over what we might deem to be "right". Not taking the opportunity to build at a higher standard after an event means missing the chance to increase sustainability in a country. Model results are vastly different for Turkey bringing up questions whether or not the cover limits will hold for earthquake insurance in Turkey in a large Istanbul event. Other high earthquake risk areas covered by pools include Taiwan, Romania, and California. Australia, the United Kingdom, and Canada show high earthquake insurance penetration. However, most areas of the world including many areas in Europe and the United States are largely uncovered or not sufficiently covered against earthquake losses.

21.6 GOVERNMENT EARTHQUAKE POOLS

Governments are the insurers of the last resort for almost all countries of the world. As mentioned above, this is due to the fact that insurers need to be allowed to fail as fees would otherwise be unaffordable. Specific agreements between governments and the insurance industry exist in many countries. Examples of this are the Gentlemen's Agreement in the United Kingdom, various terror and windstorm pools in the United States, or terror and earthquake, natural perils, or aviation pools in several countries in Europe and Australia. Terror pools are generally covered to a certain limit by the insurance industry, above which the Government provides a guarantee for exceeding losses. The reason for this is that losses from events such as nuclear or biological terror attacks or wars might become too large to be financed by the industry with its current resources. Joint ventures between government and private companies have proven successful in keeping insurance covers alive and people protected. The NFIP, the US national flood insurance program bears a promise similar to the UK Gentlemen's agreement. The program is based on an agreement between local communities and the federal government, which states that if a community adopts and enforces a flood plain management ordinance to reduce future flood risks to new construction in Special Flood Hazard Areas, the federal government will make flood insurance available (http://www.fas.org/sgp/crs/misc/R42850.pdf). Managing the NFIP has however not been without challenges (see e.g., The Evaluation of the National Flood Insurance Program, 2006). A number of earthquake pools exist around the world but only New Zealand's has been tested so far. The California Earthquake Authority (CEA) and the Turkish Catastrophe Insurance Pool (TCIP), the Turkish earthquake pool, are the oldest earthquake pools existing. The CEA was established after the Northridge event. The TCIP was established in 2000 after the 1999 Izmit and Ducze earthquakes as a compulsory homeowners' pool. In 2013, the pool tapped into new capital sources with a cat bond placed with both, the global insurance and capital

markets. Other pools include PAID (http://www.ccrif.org/partnerships/WFCP/
Sessions/Day2/Romania_PAID_WFCP_Meeting_Oct_2011.pdf), the Romanian
earthquake Pool as well as TREIF, The Taiwanese earthquake pool. The CEA,
the California earthquake pool was established as a result of the 1994
Northridge earthquake. Before then, California companies offering homeowners' insurance were also required to offer earthquake insurance. After
Northridge, they refused to write either. Eventually, the state Legislature
created a "mini policy", intended to protect a policyholder's dwelling while
excluding nonessential structures that could be sold by any insurer to comply
with the mandatory offer law. The Legislature also created a quasipublic
(privately funded, publicly managed) agency called the CEA. Membership in
the CEA by insurers is voluntary and member companies satisfy the mandatory offer law by selling the CEA mini policy. The state of California specifically states that it will not be held liable for CEA's earthquake insurance in
the event that claims from a major earthquake drain all CEA funds, nor will it
cover claims from non-CEA insurers if they become insolvent due to earthquake losses (http://www.earthquakeauthority.com/index.aspx?id=7&pid=1).
All these pools with the exception of NFIP are covered at least in part by the
reinsurance market and hence depend upon global reinsurance capacity for
pricing. In terms of penetration, the CEA is one of the least successful pools at
just 13 percent. It is difficult to foresee what will happen in California if a large
earthquake strikes the area. It is also difficult to imagine governments bailing
out homeowners and private commercial companies in the current challenging
financial market in many other countries in the world. Portugal, Italy, and
many countries in Asia do not have earthquake pools, nor do they enjoy large
insurance penetration. Latin America has increasingly built on the private
insurance market and/or created a cover for the government (Fonden,
Mexico Cat bond (http://www.gfdrr.org/sites/gfdrr.org/files/documents/DRFI_
MexicoMultiCat_Jan11.pdf)) that uses various instruments to support local
states and entities responding to natural disasters (see also Akula, 2011). These
instruments include reserve funds and risk transfer solutions.

Emerging and developing country needs for earthquake insurance are
deemed large. Global insurers and reinsurers generally have little access to
these markets given their restricted insurance infrastructure. New PPA initiatives are needed to increase resilience in these territories. No lack of capital
availability exists in our market if solutions make sense, risk triggers are
sound, and payment channels are efficient.

21.7 CONCLUSION

Insurance and reinsurance markets have been effective in covering large
catastrophic events such as the 2005 hurricane losses. Prudent risk management, regulation and ratings have led to comparatively minor, recent company
failure despite unprecedented catastrophic losses in 2005 and 2011.

Government insurance pools, with either hybrid (private−government), or solely government solutions, have proven successful in creating significant earthquake insurance penetration in some territories. How well they perform after a large event is yet to be seen. Insurance and reinsurance risk management and pricing are based on various measures including probabilistic loss or damage ratios as well as monitoring and regulating cumulative limits in exposed zones. Earthquake insurance is treated differentially across the world. This is due to differences in insurance infrastructure, politics, history of earthquakes, as well as the size and capabilities of countries and risk awareness of the population. Earthquake insurance is, in general, sublimited and not part of average homeowners' policies. Earthquake risk curves are steep, and uncertainties in loss estimates are significant due to losses being relatively rare and average claims sizes being large. Some countries (such as Japan) believe that earthquake losses can be too large and hence premiums unaffordable to provide full cover. Other countries (1) have to deal with low earthquake insurance penetration; (2) have created earthquake pools; and/or (3) do rely on the government to bail out homeowners as well as the private industry after a large loss. This means increasing taxes after an event, which is likely to become increasingly difficult in the current challenging economic environment.

Earthquake risk modeling in the insurance world dates back to the early 1990s. Stochastic, risk model penetration increased after the September 11th terror attacks in 2001 and the hurricane events in 2004 and 2005. Risk models allow for the creation of new "derivative" products based on limited information and serve as a currency or platform for trading and hedging risk among insurance and reinsurance companies. More recently, this has allowed trading directly with the capital market. Probabilistic loss measures were largely outsourced to three vendors that had provided risk platforms. Over the last 2 years (triggered by model changes and backed up by changes in regulation), companies have started to bring risk management back into their core in-house activities. With the growing need for understanding and managing model risk comes the need for modeling uncertainty and an increased demand for government science in the insurance. The current drive for change seems to be large compared to what it was over the last 10 years, and we can only speculate how earthquake risk management, insurance penetration, and government involvement will evolve over the next decade. Overcapacity has increased the demand for new insurance products outside the current market. PPA initiatives are needed to channel this capacity toward those areas that have the highest needs and herewith increase earthquake penetration around the globe. This includes the need for governments, politicians, scientists, insurers, reinsurers, broker, and the capital market to work closely together in order to create new and tradable products that can balance risk and reward for a longer term.

REFERENCES

Akula, V., 2011. A Fistful of Rice, My Unexpected Quest to End Poverty through Profitability. Harvard Business Review Press, 191 pp.

Baker, A., 2004 Revised 2010. "Simplicity", Stanford Encyclopedia of Philosophy. Stanford University, California. ISSN 1095-5054.

Bargès, M., Cossette, H., Marceau, E., 2009. TVaR-based capital allocation with copulas. Insur. Math. Econ. 45, 348–361. Retrieved July 20, 2012.

Bluhm, C., Overbeck, L., Wagner, C., 2002. An Introduction to Credit Risk Modeling. Chapman & Hall/CRC, ISBN 978-1-58488-326-5.

Davenport, T.H., Harris, J.G., March 2007. Competing on Analytics: The New Science of Winning. Harvard Business School Press.

De Weert, F., 2011. Bank and Insurance Capital Management Wiley Finance, 246 pp.

Hubbard, D., 2009. The Future of Risk Management, Why it's Broken and How to Fix It. John Wiley and Sons, Inc., 281 pp.

Jorion, P., 2007. Value at Risk, The New Benchmark for Managing Financial Risk, third ed. McGraw-Hill. 602 pp.

Kahneman, D., Slovic, P., Tversky, A., 1982. Judgment under Uncertainty: Heuristics and Biases. Cambridge University Press, 555 pp.

Kiln, R., Kiln, S., 2001. Reinsurance in Practice, fourth ed. Witherby Publisher. ISBN 1 85609 179 1 471 p.

Kim, W.C., Mauborgne, R., 2005. Blue Ocean Strategy. Harvard Business School Press.

Knutson, T., et al., 2010. Nat. Geosci. 3, 157–163. http://dx.doi.org/10.1038/ngeo779. Published online: February 21, 2010.

Lee, H.L., 2004. The Triple-A supply chain. Harv. Bus. Rev. 82 (10), 102–112, 157.

Mayer-Rosa, D., Cadiot, B., 1979. A review of the 1356 Basel earthquake: basic data. Tectonophysics 53, 325–333.

Miles, S., 2012. The Christchurch Fiasco: The Insurance Aftershock and Its Implications for New Zealand and beyond.

Mingfang, T., Yochanan, K., Richard, S., Cuihua, L., 2009. Forced and internal twentieth-century SST trends in the North Atlantic. J. Clim. 22 (6), 1469–1481. http://dx.doi.org/10.1175/2008JCLI2561.1. Bibcode: 2009JCli...22.1469T.

Porras, E.R., 2011. The Cost of Capital. Polgrave Macmillan.

Riley, K., 2001. Reinsurance The Nuts and Bolts, second ed. Witherby & Co Ltd. 247 pp.

Scheffer, M., Carpenter, S.R., 2003. Catastrophic regimes shifts in ecosystems: linking theory to observation. Trends Ecol. Evol. 18 (12). http://eaton.math.rpi.edu/csums/papers/Ecostability/scheffercatastrophe.pdf.

Schlesinger, M.E., Ramankutty, N., 1994. An oscillation in the global climate system of period 65–70 years. Nature 367 (6465), 723–726. http://dx.doi.org/10.1038/367,723a0.

Silver, N., 2012. The Signal and the Noise, Why So Many Predictions Fail – but Some Don't. The Penguin Press.

Slone, R.E., Dittmann, J.P., Mentzer, J.T., 2010. The New Supply Chain Agenda. Harvard Business Press, 216 pp.

Taleb, N., 2007/2010. The Black Swan: The Impact of the Highly Improbable. Random House and Penguin, New York. ISBN 978-1-4000-6351.

Taleb, N., 2012. Antifragile: Things That Gain from Disorder. Random House, New York, ISBN 978-1-4000-6782-4.

The Evaluation of the National Flood Insurance Program, 2006. Final Report, NFIP Evaluation Final Report Working Group, American Institute for Research.

Vaughan, E.J., Vaughan, T., 2003. Fundamental of Risk and Insurance, ninth ed. John Wiley & Sons. 686 pp.

Vose, D., 2001. Risk Analysis, a Quantitative Guide. Wiley and Sons, 418 pp.

Wiedemer, D., Wiedemer, R.A., Spitzer, C., 2011. Aftershock, Protect Yourself and Profit in the Next Global Financial Meltdown. Revised and Updated, second ed. John Wiley & Sons. 304 pp.

Index

Printed and bound by CPI Group (UK) Ltd, Croydon, CR0 4YY

13/05/2025

01869552-0001